Springer Proceedings in Mathematics & Statistics

Volume 170

Springer Proceedings in Mathematics & Statistics

This book series features volumes composed of selected contributions from workshops and conferences in all areas of current research in mathematics and statistics, including operation research and optimization. In addition to an overall evaluation of the interest, scientific quality, and timeliness of each proposal at the hands of the publisher, individual contributions are all refereed to the high quality standards of leading journals in the field. Thus, this series provides the research community with well-edited, authoritative reports on developments in the most exciting areas of mathematical and statistical research today.

More information about this series at http://www.springer.com/series/10533

Scott Chapman · Marco Fontana
Alfred Geroldinger · Bruce Olberding
Editors

Multiplicative Ideal Theory and Factorization Theory

Commutative and Non-commutative Perspectives

Springer

Editors
Scott Chapman
Department of Mathematics and Statistics
Sam Houston State University
Huntsville, TX
USA

Alfred Geroldinger
Institut für Mathematik
und Wissenschaftliches Rechnen
Universität Graz
Graz
Austria

Marco Fontana
Dipartimento di Matematica e Fisica
Università degli Studi Roma Tre
Rome
Italy

Bruce Olberding
Department of Mathematical Sciences
New Mexico State University
Las Cruces, NM
USA

ISSN 2194-1009 ISSN 2194-1017 (electronic)
Springer Proceedings in Mathematics & Statistics
ISBN 978-3-319-38853-3 ISBN 978-3-319-38855-7 (eBook)
DOI 10.1007/978-3-319-38855-7

Library of Congress Control Number: 2016939917

Mathematics Subject Classification (2010): 20M14, 13H10, 13A30, 14H20, 13G05, 13F05, 13C10, 11R11

Printed on acid-free paper

This Springer imprint is published by Springer Nature
The registered company is Springer International Publishing AG Switzerland

Contents

Introduction

This volume is a collection of articles concerning topics in multiplicative ideal theory and factorization theory. While it is not strictly a conference proceedings, most of its papers were solicited from the speakers at a fall 2014 Algebra conference held in Graz, Austria (more details on the conference can be found below). All of these articles were invited and all have been refereed.

While some of these manuscripts contain new results, most are expository in nature. These papers survey the state of the art of their various topics, reveal new relationships between areas which were so far seemingly unconnected, and pose problems for further research. Topics in these papers include topological aspects in ring theory (such as spectral spaces and spectral representations), Prüfer domains of integer-valued polynomials and their monadic submonoids, semigroup algebras (both commutative and non-commutative), the arithmetic and ideal theory of Mori domains and monoids, the arithmetic of Krull monoids and its role in additive combinatorics, and the computational aspects of factorization theory. A special feature of this volume are surveys on both multiplicative ideal theory and factorization theory in non-commutative rings (with a focus on non-commutative Krull rings and monoids).

This compendium is also intended to be a tribute to our friend and colleague, Franz Halter-Koch, whose work and ideas have inspired not only the editors of this volume, but countless more colleagues in this area. His influence is omnipresent in many articles of this tome. We thank the authors for their contributions, the referees for their work, and the Editorial Staff at Springer-Verlag for their guidance and patience in directing this volume to its publication.

The Conference in Graz

About eighty mathematicians from 20 different countries gathered at the University of Graz during the period September 22–26, 2014, for the meeting "Arithmetic and Ideal Theory of Rings and Semigroups." The meeting featured 5 days of lectures on topics including the theory of commutative rings and their modules with a focus on Prüfer, Krull and Mori domains, topological aspects of ring theory, ring-inspired graph theory, rings of integer-valued polynomials, module theory with a focus on direct-sum decompositions, and factorization and divisibility theory in rings and semigroups. The conference was part of a long series of conferences focusing on ideal and factorization theory which took place during the last few decades, mainly in Austria, France, Italy, Korea, Morocco, and the USA (this is documented by a series of conference proceedings starting in the late 1990s; see [1, 5, 7, 14–16, 19, 22, 24, 26–28]).

The conference was organized by Alfred Geroldinger, Florian Kainrath, Andreas Reinhart, and Daniel Smertnig. Support for the conference came from the University of Graz, the Institute for Mathematics and Scientific Computing, NAWI Graz, the City of Graz, and the Austrian Science Fund FWF (Project Number P26036-N26 and W1230 Doctoral Program Discrete Mathematics). We thank all our sponsors. Without their assistance and support the conference would not have been possible.

The September 24th session of this meeting was dedicated to Franz Halter-Koch in honor of his 70th birthday. In the following section, we briefly review his career and influence in the general areas of number theory and algebra, especially with regard to the development of multiplicative ideal theory and factorization theory. While many mathematicians consider a long vita and publication list a sign of a successful career, such an analysis of Franz Halter-Koch merely scratches the surface of his impact in the various areas of algebra and number theory in which he works. This was demonstrated at the September 24th session with three welcoming addresses by Horst Brunotte (the first doctoral student of Halter-Koch), Ulrich Krause (University of Bremen), and Scott Chapman. These addresses were followed by detailed lectures on Halter-Koch's work by Marco Fontana (who reviewed his work in multiplicative ideal theory) and Alfred Geroldinger (who reviewed his work in factorization theory).

A Tribute to Franz Halter-Koch

Franz Halter-Koch began his study of mathematics and physics at the University of Graz in 1962. Some years later, he moved to the University of Hamburg where he did his master thesis under the supervision of Helmut Hasse. He returned to the University of Graz, wrote his doctoral thesis under the guidance of the number theorist Alexander Aigner, and received his Ph.D. in 1968. After positions at the Graz University of Technology and at the University of Cologne, he became a full professor at the University of Essen in 1973. He moved back to Graz in 1981, where he served as full professor at the University of Graz until his retirement in 2008. During this period, he was the head of the Algebra and Number Theory Group as well as the head of the mathematics department's group in charge of the education of high school teachers. In addition, Halter-Koch served as Chair of the mathematics department from 1994 to 2004.

While the week-long program at Graz touched on many different areas of algebra, it is safe to say that a large number of the talks would not have been possible without the prior work of Franz Halter-Koch. Over an almost fifty-year research career, he has published more than 150 papers in refereed journals and conference proceedings. The areas of his research comprise classical algebraic number theory, commutative algebra in rings and monoids, factorization theory, classical elementary number theory, and functional equations.

Apart from his original papers, Halter-Koch has written the following three monographs:

I. *Ideal Systems. An Introduction to Multiplicative Ideal Theory*, Marcel Dekker, 421 p., 1998, [50];
II. *Non-Unique Factorizations. Algebraic, Combinatorial and Analytic Theory* (with A. Geroldinger), Chapman & Hall/CRC, 700 p., 2006, [34];
III. *Quadratic Irrationals. An Introduction to Classical Number Theory*, Chapman & Hall/CRC, 431 p., 2013, [58].

Instead of a systematic evaluation of Halter-Koch's scientific oeuvre, we offer the reader a short and friendly tour through his work and use his monographs as the link between sights.

Multiplicative ideal theory has its origin during the first half of the twentieth century in the work of Krull, Lorenzen, Noether, and Prüfer. The first systematic study of its topics in an abstract context was given by the monograph of Jaffard ([65], published 1960). However, its extreme style greatly limited its diffusion and influence. In the 1970s, the monographs of Larsen and McCarthy ([69], published 1971) and of Gilmer ([36], published 1972 on the basis of his 1968 Queen's Notes) provided a first systematic treatment of valuation, Dedekind, Prüfer, and Krull domains. Based on these books, multiplicative ideal theory has flourished. In monograph #I, Halter-Koch offers a treatise of multiplicative ideal theory based on the concept of ideal systems (generalizing the concept of star operations and Krull's *Strich* operations), which is valid both for rings and monoids. Based on this work, Halter-Koch then introduced the concept of module systems, which are a common generalization of ideal systems and of semistar operations, and built a purely multiplicative theory on the basis of this concept. In this abstract framework, he developed an axiomatic theory of Kronecker function rings and discovered the connection between valuation domains, Kronecker function rings, and Lorenzen monoids in an utmost general context (see [51, 52, 54, 60] and the survey [56]).

Factorization theory has its roots in algebraic number theory. W. Narkiewicz began the study of this area in a systematic way in the 1960s (documented in his monograph [70]; the first edition published in 1974), and he initiated the study of factorizations in algebraic number rings in the 1970s and 1980s (see early papers by Kaczorowski, Krause, Rush, Salce and Zanardo [67, 68, 73, 74]). Also in the 1980s, Halter-Koch began to consider the behavior of non-unique factorizations, first in the setting of algebraic number rings [38, 39], and then, with his doctoral student Alfred Geroldinger, in the more general case of commutative (Krull) monoids [31, 42]. In the early 1990s, several groups of ring theorists became interested in factorization properties of various classes of integral domains (see early papers by Anderson et al. and by Chapman and Smith [2, 3, 18]). Since that time, factorization theory has flourished, and in 2006 Geroldinger and Halter-Koch published the monograph #II which quickly became the area's standard reference.

We would be remiss not to mention a key contribution of Halter-Koch to factorization theory. In Ref. [49], he introduced the concept of transfer homomorphisms, which turned out to be fundamental for the further development of the area. Transfer homomorphisms were further studied in Ref. [33] and generalized only recently to the non-commutative setting by Baeth and Smertnig [9]. This in turn led to the concept of transfer Krull monoids [30]. Factorization theory has widened its scope to areas such as additive combinatorics [35], module theory [8, 10, 20], and rings with zero-divisors [6, 29], but yet has remained anchored in ideal theory [25, 64, 71, 72] and in analytic theory [66]. The influence of Halter-Koch in any of these directions is omnipresent (to mention a couple of papers covering all these directions we cite [4, 17, 21, 32, 44–46, 59, 63]).

Binary quadratic forms, continued fractions, and power residues are among the classical concepts of number theory tracing back to the famous works of Gauß, Dirichlet, and Dedekind. Starting with his doctoral thesis [37], Halter-Koch has remained interested in these topics of classical elementary number theory throughout his life (see for example [40, 41, 43, 47, 48, 53, 55, 57, 61, 62]). In his recent monograph #III, he presents the classical theory of continued fractions, quadratic orders, and binary quadratic forms in a unified way, based on the concept of quadratic irrationals and their equivalence. On this basis, he also presents some more recent developments such as rational reciprocity laws, Z.-H. Sun's biquadratic class group characters, cyclic 2-class groups of order divisible by 8 and 16, applications of quadratic orders to binary Diophantine equations, and others.

Halter-Koch's work has inspired not only his own doctoral students (Horst Brunotte 1977, Alfred Geroldinger 1987, Otto Wurnig 1994, Florian Kainrath 1997, Wolfgang Hassler 2001, Maximilian Pacher 2001, Andreas Reinhart 2010, and Andreas Philipp 2011), but many colleagues in the community also. In 1995, Scott Chapman (supported by a Fulbright Fellowship) spent part of the Spring Semester in Graz to study factorization theory. His stay later heavily influenced the work of his many American students in a decade of Research Experience for Undergraduates (REU) programs funded by the National Science Foundation. In 2003, a group of these students spent some weeks in Graz (under supervision of Chapman and Vadim Ponomarenko) for a summer school in factorization theory. Several publications of this group of students on factorization theory show the success of this program (see, for example, [11–13, 23]). Halter-Koch's influence on these and many other REU connected publications from Chapman's group cannot be overestimated.

Despite his retirement 8 years ago, Halter-Koch has remained active. He has published 16 papers and one monograph during this period and has continued to give lectures on advanced topics in algebra and number theory within the Graz doctoral program.

Huntsville, TX, USA — Scott Chapman
Rome, Italy — Marco Fontana
Graz, Austria — Alfred Geroldinger
Las Cruces, NM, USA — Bruce Olberding

References

1. D.D. Anderson (ed.), Factorization in integral domains. Lecture notes in pure and applied mathematics, vol. 189 (Marcel Dekker, 1997)
2. D.D. Anderson, D.F. Anderson, M. Zafrullah, Factorization in integral domains. J. Pure Appl. Algebra **69**, 1–19 (1990)
3. D.D. Anderson, D.F. Anderson, M. Zafrullah, Factorization in integral domains II. J. Algebra **152**, 78–93 (1992)

4. D.D. Anderson, S.T. Chapman, F. Halter-Koch, M. Zafrullah, Criteria for unique factorization in integral domains. J. Pure Appl. Algebra **127**, 205–218 (1998)
5. D.D. Anderson, I.J. Papick (eds.), Ideal theoretic methods in commutative algebra. Lecture notes in pure and applied mathematics, vol. 220 (Marcel Dekker, 2001)
6. D.D. Anderson, S. Valdes-Leon, Factorization in commutative rings with zero divisors. Rocky Mt. J. Math. **26**, 439–480 (1996)
7. Badawi (ed.), in *Focus on Commutative Rings Research* (Nova Science Publishers, 2006)
8. N.R. Baeth, A. Geroldinger, Monoids of modules and arithmetic of direct-sum decompositions. Pac. J. Math. **271**, 257–319 (2014)
9. N.R. Baeth, D. Smertnig, Factorization theory: From commutative to noncommutative settings. J. Algebra **441**, 475–551 (2015)
10. N.R. Baeth, R. Wiegand, Factorization theory and decomposition of modules. Am. Math. Monthly **120**, 3–34 (2013)
11. M. Banister, J. Chaika, S.T. Chapman, W. Meyerson, On a result of James and Niven concerning unique factorization in congruence semigroups. Elem. Math. **62**, 68–72 (2007)
12. M. Banister, J. Chaika, S.T. Chapman, W. Meyerson, On the arithmetic of arithmetical congruence monoids. Colloq. Math. **108**, 105–118 (2007)
13. M. Banister, J. Chaika, S.T. Chapman, W. Meyerson, A theorem on accepted elasticity in certain local arithmetical congruence monoids. Abh. Math. Semin. Univ. Hamb. **79**, 79–86 (2009)
14. J.W. Brewer, S. Glaz, W. Heinzer, B. Olberding (eds.), in *Multiplicative Ideal Theory in Commutative Algebra* (Springer, Berlin, 2006)
15. S.T. Chapman (ed.), Arithmetical properties of commutative rings and monoids. Lecture notes in pure and applied mathematics, vol. 241 (Chapman & Hall/CRC, 2005)
16. S.T. Chapman and S. Glaz (eds.), in *Non-Noetherian Commutative Ring Theory*, Kluwer (Academic Publishers, 2000)
17. S.T. Chapman, F. Halter-Koch, U. Krause, Inside factorial monoids and integral domains. J. Algebra **252**, 350–375 (2002)
18. S.T. Chapman, W.W. Smith, Factorization in Dedekind domains with finite class group. Isr. J. Math. **71**, 65–95 (1990)
19. D.E. Dobbs, M. Fontana, S. Kabbaj (eds.), Advances in commutative ring theory. Lecture notes in pure and applied mathematics, vol. 205 (Marcel Dekker, 1999)
20. A. Facchini, K. Fuller, C.M. Ringel, C. Santa-Clara (eds.), *Krull Monoids and Their Application in Module Theory, Algebras, Rings and their Representations* (World Scientific, 2006), pp. 53–71
21. A. Facchini, F. Halter-Koch, Projective modules and divisor homomorphisms. J. Algebra Appl. **2**, 435–449 (2003)
22. A. Facchini, E. Houston, L. Salce (eds.), Rings, Modules, Algebras, and Abelian Groups. Lecture Notes in Pure and Applied Mathematics, vol. 236 (Marcel Dekker, 2004)
23. B.W. Finklea, T. Moore, V. Ponomarenko, Z.J. Turner, Invariant polynomials and minimal zero sequences. Involve **1**, 159–165 (2008)
24. M. Fontana, S. Frisch, S. Glaz (eds.), *Commutative Algebra: Recent Advances in Commutative Rings, Integer-Valued Polynomials, and Polynomial Functions* (Springer, 2014)
25. M. Fontana, E. Houston, T. Lucas, *Factoring Ideals in Integral Domains*. Lecture Notes of the Unione Matematica Italiana, vol. 14 (Springer, 2013)
26. M. Fontana, S. Kabbaj, S. Wiegand (eds.), *Commutative Ring Theory and Applications*. Lecture Notes in Pure and Applied Mathematics, vol. 231 (Marcel Dekker, 2003)
27. M. Fontana, S.-E. Kabbaj, B. Olberding, I. Swanson (eds.), *Commutative Algebra and its Applications*. de Gruyter Proceedings in Mathematics (de Gruyter, 2009)
28. M. Fontana, S.-E. Kabbaj, B. Olberding, I. Swanson (eds.), *Commutative Algebra: Noetherian and Non-Noetherian Perspectives* (Springer, 2011)
29. C. Frei, S. Frisch, Non-unique factorization of polynomials over residue class rings of the integers. Commun. Algebra **39**, 1482–1490 (2011)

30. A. Geroldinger, *Sets of lengths*. American Math. Monthly, 2016, to appear.
31. A. Geroldinger, *Über nicht-eindeutige Zerlegungen in irreduzible Elemente*, Math. Z. **197** (1988), 505–529.
32. A. Geroldinger, F. Halter-Koch, Congruence monoids. Acta Arith. **112**, 263–296 (2004)
33. A. Geroldinger, F. Halter-Koch, *Transfer principles in the theory of non-unique factorizations*, Arithmetical Properties of Commutative Rings and Monoids, Lecture Notes in Pure and Applied Mathematics, vol. 241 (Chapman & Hall/CRC, 2005), pp. 114–142
34. A. Geroldinger, F. Halter-Koch, *Non-unique Factorizations. Algebraic, Combinatorial and Analytic Theory*. Pure and Applied Mathematics, vol. 278 (Chapman & Hall/CRC, 2006)
35. A. Geroldinger, I. Ruzsa, *Combinatorial Number Theory and Additive Group Theory*. Advanced Courses in Mathematics (CRM Barcelona, Birkhäuser, 2009)
36. R. Gilmer, *Multiplicative Ideal Theory*. Pure and Applied Mathematics, vol. 12 (Marcel Dekker, 1972)
37. F. Halter-Koch, Kriterien zum 8. Potenzcharakter der Reste 3, 5 und 7. Math. Nachr. **44**, 129–144 (1970)
38. F. Halter-Koch, Factorization of algebraic integers. Grazer Math. Berichte **191** (1983)
39. F. Halter-Koch, On the factorization of algebraic integers into irreducibles. Coll. Math. Soc. János Bolyai **34**, 699–707 (1984)
40. F. Halter-Koch, Konstruktion von Klassenkörpern und Potenzrestkriterien für quadratische Einheiten. Manuscr. Math. **54**, 453–492 (1986)
41. F. Halter-Koch, Binäre quadratische Formen und Diederkörper. Acta Arith. **51**, 141–172 (1988)
42. F. Halter-Koch, Halbgruppen mit Divisorentheorie. Expo. Math. **8**, 27–66 (1990)
43. F. Halter-Koch, On a class of insoluble binary quadratic Diophantine equations. Nagoya Math. J. **123**, 141–151 (1991)
44. F. Halter-Koch, Chebotarev formations and quantitative aspects of non-unique factorizations. Acta Arith. **62**, 173–206 (1992)
45. F. Halter-Koch, A generalization of Davenport's constant and its arithmetical applications. Colloq. Math. **63**, 203–210 (1992)
46. F. Halter-Koch, A characterization of Krull rings with zero divisors. Arch. Math. Brno **29**, 119–122 (1993)
47. F. Halter-Koch, Quadratische Ordnungen mit groβer Klassenzahl, II. J. Number Theory **44**, 166–171 (1993)
48. F. Halter-Koch, A Theorem of Ramanujan concerning binary quadratic forms. J. Number Theory **44**, 209–213 (1993)
49. F. Halter-Koch, Finitely generated monoids, finitely primary monoids and factorization properties of integral domains, *Factorization in Integral Domains*. Lecture Notes in Pure and Applied Mathematics, vol. 189 (Marcel Dekker, 1997), pp. 73–112
50. F. Halter-Koch, *Ideal Systems. An Introduction to Multiplicative Ideal Theory* (Marcel Dekker, 1998)
51. F. Halter-Koch, Localizing systems, module systems and semistar operations. J. Algebra **238**, 723–761 (2001)
52. F. Halter-Koch, Kronecker function rings and generalized integral closures. Commun. Algebra **31**, 45–59 (2003)
53. F. Halter-Koch, Representation of prime powers in arithmetical progressions by binary quadratic forms. J. Théor. Nombres Bordx. **15**, 141–149 (2003)
54. F. Halter-Koch, Weak module systems and applications: a multiplicative theory of integral elements and the Marot property, *Commutative Ring Theory and Applications*. Lecture Notes in Pure and Applied Mathematics, vol. 231 (Marcel Dekker, 2003), pp. 213–231
55. F. Halter-Koch, Diophantine equations of Pellian type. J. Number Theory **131**, 1597–1615 (2011)

56. F. Halter-Koch, Multiplicative ideal theory in the context of commutative monoids, in *Commutative Algebra: Noetherian and Non-Noetherian Perspectives*, eds. by M. Fontana, S.-E. Kabbaj, B. Olberding, I. Swanson (Springer, 2011), pp. 203–231
57. F. Halter-Koch, Lucas sequences and quadratic orders. Arch. Math. **100**, 417–430 (2013)
58. F. Halter-Koch, *Quadratic Irrationals*. Pure and Applied Mathematics, vol. 306 (Chapman & Hall/CRC, 2013)
59. F. Halter-Koch, Arithmetical interpretation of Davenports constants with weights. Arch. Math. **103**, 125–131 (2014)
60. F. Halter-Koch, Lorenzen Monoids: a multiplicative approach to Kronecker function rings. Commun. Algebra **43**, 3–22 (2015)
61. F. Halter-Koch, P. Kaplan, K.S. Williams, An Artin character and representation of primes by binary quadratic forms II. Manuscr. Math. **37**, 357–381 (1982)
62. F. Halter-Koch, G. Lettl, A. Pethő, R. Tichy, Thue equations associated with Ankeny-Brauer-Chowla number fields. J. Lond. Math. Soc. **60**, 1–20 (1999)
63. F. Halter-Koch, W. Müller, Quantitative aspects of non-unique factorization: a general theory with applications to algebraic function fields. J. Reine Angew. Math. **421**, 159–188 (1991)
64. O.A. Heubo-Kwegna, B. Olberding, A. Reinhart, Topological invariants and ideal factorization for one-dimensional Prüfer domains, J. Pure. Appl. Algebra. to appear http://arxiv.org/pdf/1512.03312.pdf
65. P. Jaffard, *Les Systémes d'Idéaux* (Dunod, 1960)
66. J. Kaczorowski, Analytic monoids and factorization problems, Semigroup Forum, to appear.
67. J. Kaczorowski, A pure arithmetical characterization for certain fields with a given class group. Colloq. Math. **45**, 327–330 (1981)
68. U. Krause, A characterization of algebraic number fields with cyclic class group of prime power order. Math. Z. **186**, 143–148 (1984)
69. M.D. Larsen, P.J. McCarthy, *Multiplicative Theory of Ideals* (Academic Press, 1971)
70. W. Narkiewicz, *Elementary and Analytic Theory of Algebraic Numbers* (PWN—Polish Scientific Publishers, 1974)
71. Olberding, Factorization into prime and invertible ideals. J. Lond. Math. Soc. **62**, 336–344 (2000)
72. Olberding, Factorization into prime and invertible ideals II. J. Lond. Math. Soc. **80**, 155–170 (2009)
73. D.E. Rush, An arithmetic characterization of algebraic number fields with a given class group. Math. Proc. Camb. Philos. Soc. **94**, 23–28 (1983)
74. L. Salce, P. Zanardo, Arithmetical characterization of rings of algebraic integers with cyclic ideal class group. Boll. Unione. Mat. Ital., VI. Ser., D, Algebra Geom. **1**, 117–122 (1982)

Multiplicative Ideal Theory in Noncommutative Rings

Evrim Akalan and Hidetoshi Marubayashi

Abstract The aim of this paper is to survey noncommutative rings from the viewpoint of multiplicative ideal theory. The main classes of rings considered are maximal orders, Krull orders (rings), unique factorization rings, generalized Dedekind prime rings, and hereditary Noetherian prime rings. We report on the description of reflexive ideals in Ore extensions and Rees rings. Further we give necessary and sufficient conditions (or sufficient conditions) for well-known classes of rings to be maximal orders, and we propose a polynomial-type generalization of hereditary Noetherian prime rings.

Keywords Maximal order · Reflexive ideal · Krull ring (order) · Unique factorization ring · Generalized Dedekind · Generalized Noetherian prime ring

1 Introduction

Multiplicative (arithmetic) ideal theory in algebraic number fields originated by Dedekind was developed by M. Sono, W. Krull, E. Noether, H. Prüfer, E. Artin during the period 1910–1930. In particular, E. Noether gave an axiomatic foundation on Dedekind's theory.

In the noncommutative setting, Dedekind–Noether's ideal theory was first extended to algebras by A. Speiser, H. Brandt, E. Artin, and H. Hasse (e.g., [9, 45, 69] and see also [31]), and then, in [10] K. Asano extended Noether's axiomatic foundation to noncommutative rings: Let R be a bounded order in its

E. Akalan (✉)
Department of Mathematics, Hacettepe University, Beytepe Campus,
06800 Ankara, Turkey
e-mail: eakalan@hacettepe.edu.tr

H. Marubayashi
Faculty of Sciences and Engineering, Tokushima Bunri University,
Sanuki, Kagawa 769-2193, Japan
e-mail: marubaya@naruto-u.ac.jp

S. Chapman et al. (eds.), *Multiplicative Ideal Theory and Factorization Theory*,
Springer Proceedings in Mathematics & Statistics 170,
DOI 10.1007/978-3-319-38855-7_1

quotient ring Q. Then the set of all fractional R-ideals is a group if and only if the following three conditions hold:

(a1) R is a bounded maximal order in Q.
(a2) R satisfies the ascending chain conditions on integral ideals.
(a3) Any nonzero prime ideal is maximal,

which is the same axiomatic foundation as one of Noether in case of commutative domains. Furthermore, he extended many important ideal theories to orders satisfying (a1), (a2) and (a3) which is nowadays called bounded Asano rings (orders) [11, 12].

The aim of this article is to survey noncommutative rings from the viewpoint of multiplicative ideal theory.

In Sect. 2, we give the definitions of maximal orders, Asano, Dedekind, and hereditary which are the main topics in multiplicative ideal theory and give a classical ideal theory of maximal orders. Furthermore we discuss the maximal order properties of well-known noncommutative rings such as group rings, polynomial rings, universal enveloping algebras, and so on.

In Sect. 3, we define the concept of Krull orders in the sense of Chamarie and study the algebraic structure of Krull orders as well as the ideal theories of polynomial rings and Ore–Rees rings over Krull orders.

We give in Sect. 4 an overview of noncommutative unique factorization rings (UFRs for short) which is one of the important classes of maximal orders.

Section 5 contains a generalization of Dedekind and Asano orders which is called G-Dedekind (or G-Asano) and we give several characterizations of G-Dedekind. We also consider polynomial rings and Rees rings over G-Dedekind.

Hereditary prime rings are one of the most successful subjects in noncommutative rings during the years 1960–1970. In Sect. 6, we only discuss the ideal theory in HNP rings and propose a polynomial-type generalization of HNP rings.

We refer the readers to the books [63, 68] for terminologies not defined in this article.

Because of the page limit, we do not give the proofs of Propositions and Theorems and we quote the original papers or books for reader's convenience.

In the case of commutative rings and monoids we refer the reader to the books [34, 39] for commutative rings and [41] for monoids.

2 Maximal Orders

Throughout this paper, R is a prime Goldie ring unless otherwise stated with its quotient ring Q, which is a simple Artinian ring (in other words, R is an *order* in Q).

In this section, we define the concept of maximal orders in Q and give a classical ideal theory in maximal orders. Furthermore we give necessary and sufficient conditions (or sufficient conditions) for some well-known noncommutative rings to be maximal orders.

Definition 2.1 (1) Orders R and S in Q are *equivalent* if $aRb \subseteq S$ and $cSd \subseteq R$ for some units a, b, c, d in Q.
(2) An order is maximal if it is maximal in the set of all equivalent orders.

For a fractional right R-ideal I, $O_r(I) = \{q \in Q \mid Iq \subseteq I\}$, which is called a *right order* of I. It is easy to see that $O_r(I) \supseteq R$ and is equivalent to R. Similarly for a fractional left R-ideal J, $O_l(J) = \{q \in Q \mid qJ \subseteq J\}$, the *left order* of J, which contains R and is equivalent to R. Thus we have the following ideal theoretic characterizations of maximal orders:

Proposition 2.2 ([63, 68]) *Let R be an order in Q. The following conditions are equivalent:*

(1) R is a maximal order in Q.
(2) $O_l(J) = R$ for all fractional left R-ideals J and $O_r(I) = R$ for all fractional right R-ideals I.
(3) $O_l(A) = R = O_r(A)$ for all fractional R-ideals A.
(4) $O_l(A) = R = O_r(A)$ for all nonzero ideals A of R.

For a fractional right R-ideal I, let $I^* = \{q \in Q \mid qI \subseteq R\}$ and for a fractional left R-ideal J, let $J^+ = \{q \in Q \mid Jq \subseteq R\}$. If R is a maximal order, then for a fractional R-ideal A in Q, $A^* = A^{-1} = A^+$ by Proposition 2.2, here $A^{-1} = \{q \in Q \mid AqA \subseteq A\}$. Thus $A^{*+} = A^{+*}$, which contains A. If $A = A^{**}$, then A is called a *reflexive* fractional R-ideal in Q (some say a *divisorial* fractional R-ideal in Q).

Let $D(R) = \{A \mid A$ is a reflexive fractional R-ideal$\}$. For any A, B in $D(R)$, we define the multiplication "$\circ$" by $A \circ B = (AB)^{**}$. Then we have the following theorem which extends Asano's result:

Theorem 2.3 ([63, 68]) *Suppose R is a maximal order in Q.*

(1) R is a group with the multiplication "$\circ$".
(2) If R satisfies the ascending chain condition on reflexive ideals of R, then
(i) $D(R)$ is an Abelian group generated by maximal reflexive ideals.
(ii) Any maximal reflexive ideal is a minimal prime ideal (height-1 prime).
(3) The center of R is a completely integrally closed domain.

Theorem 2.3 (3) shows maximal orders are nothing but completely integrally closed in case of commutative domains.

A fractional R-ideal A is said to be *invertible* if $A^*A = R = AA^+$. An order in Q is said to be *Asano* if each nonzero ideal is invertible, and is said to be *Dedekind* if it is Asano and hereditary (see [68] for more detailed results on Asano and Dedekind orders).

In case of commutative domains, invertible ideal is equivalent to projective. Hence Dedekind, Asano, and hereditary are all same. However, in the noncommutative setting, invertible ideal is projective and the converse does not necessarily hold. Thus Dedekind orders imply Asano and hereditary. The converse implications do not necessarily hold and there are no implications between Asano and hereditary (see [63, 68] for such examples). However, if we assume that R is *bounded*, that is any essential one-sided ideal contains a nonzero ideal (this concept is defined by Asano), then we have

Proposition 2.4 ([56]) *Bounded Noetherian Asano orders are Dedekind.*

In the rest of this section, we give necessary and sufficient conditions for some well-known noncommutative rings to be maximal orders (or a sufficient condition for well-known noncommutative rings to be maximal orders).

Proposition 2.5 (Algebra case) *Let Q be a simple Artinian ring with its center F and R as a subring of Q with its center D. R is called a D-order in Q if the following two conditions are satisfied:*

(i) F is the quotient field of D and $Q = FR$, that is, R is an order in Q.
(ii) Every element of R is integral over D.
(1) There always exists a maximal D-order by Zorn's lemma.
(2) If D is a Dedekind domain, then every maximal D-order is a bounded noncommutative Dedekind order [73].
(3) If D is a Krull domain, then every maximal D-order is a bounded noncommutative Krull order ([35, 63], see Sect. 3 for the definition of Krull orders).

Let σ be an automorphism of R and δ be a left σ-derivation on R. The noncommutative polynomial ring $R[x; \sigma, \delta] = \{f(x) = a_n x^n + \cdots + a_0 \mid a_i \in R\}$ in an indeterminate x with multiplication : $xa = \sigma(a)x + \delta(a)$ for any $a \in R$ is called an *Ore extension*.

In [72], Ore defined noncommutative polynomial rings in case R is a skew field and studied the structure of them. It is easy to see that σ and δ are extended to an automorphism σ of Q and a left σ-derivation δ on Q. Since $Q[x; \sigma, \delta]$ is a principal ideal ring, that is, any one-sided ideal is principal [22], it has a quotient ring which is a simple Artinian ring and so $R[x; \sigma, \delta]$ has a quotient ring which is the same quotient ring of $Q[x; \sigma, \delta]$.

Let I be an invertible ideal of R with $\sigma(I) = I$. A subring $R[Ix; \sigma, \delta] = \sum_{n=0}^{\infty} \bigoplus I^n x^n$ of $R[x; \sigma, \delta]$ is called an *Ore–Rees* ring associated to I, where $I^0 x^0 = R$.

Proposition 2.6 (Ore extensions and Ore–Rees ring) *If R is a maximal order in Q, then so is $R[x; \sigma, \delta]$, and if R is a Noetherian maximal order then so is $R[Ix; \sigma, \delta]$ [23, 47].*

Proposition 2.7 (Strongly graded rings) *(1) Let $S = \sum_{n \in \mathbb{Z}} \bigoplus R_n$ be a strongly $\mathbb{Z}$-graded ring, where $\mathbb{Z}$ is the ring of integers. If R_0, the part of degree zero, is a maximal order, then so is S [65].*

(2) Let R be a semiprime $\mathbb{Z}$-graded ring. R is an Asano order if and only if
(i) Every gr-R-ideal is invertible, and
(ii) Every essential gr-maximal ideal is maximal [53].

A commutative Noetherian local ring D is *regular* if and only if $gl.dim(D) < \infty$. If D is regular, then it is a UFD and so it is a maximal order. In noncommutative setting, we have

Proposition 2.8 (Rings of finite global dimensions) *(1) Any local Noetherian ring of finite global dimension which is integral over its center is a maximal order [40].*
(2) Any Noetherian, prime, AR-ring of finite global dimension with enough invertible ideals is a maximal order [19].
(3) Let F be a field and R be a Noetherian F-algebra.
(i) If R is Auslander-regular, Cohen–Macaulay and stably free, then R is a maximal order in its quotient division ring [77].
(ii) If R is a graded ring of finite global dimension such that R is integral over its center, then R is a maximal order in its quotient division ring [77].

Let $T = \begin{pmatrix} R & V \\ W & S \end{pmatrix}$ be a ring of Morita contexts which is a prime Goldie ring, where R and S are prime Goldie rings with the quotient rings $Q(R)$ and $Q(S)$, respectively, V, W are an (R, S)-bimodule, an (S, R)-bimodule, respectively. It follows that $Q(R)V = VQ(S)$ and $Q(S)W = WQ(R)$, which are denoted by $Q(V)$ and $Q(W)$, respectively. Then the quotient ring of T is $\begin{pmatrix} Q(R) & Q(V) \\ Q(W) & Q(S) \end{pmatrix}$, denoted by $Q(T)$. Similar to maximal orders, we can define an (R, S)-*maximal module* in $Q(V)$ and an (S, R)-*maximal module* in $Q(W)$ (see [7] for the definition of maximal modules). Put $V^* = \{\tilde{w} \in Q(W) \mid \tilde{w}V \subseteq S\}$ and $V^+ = \{\tilde{w} \in Q(W) \mid V\tilde{w} \subseteq R\}$. Similarly we define W^* and W^+.

Proposition 2.9 (Rings of Morita contexts) *The following conditions are equivalent:*

(1) T is a maximal order in $Q(T)$.
(2) (i) R and S are maximal orders in $Q(R)$ and $Q(S)$, respectively, and
(ii) $V^* = W = V^+$ and $W^* = V = W^+$.
(3) (i) V is an (R, S)-maximal module in $Q(V)$ and W is an (S, R)-maximal module in $Q(W)$, and
(ii) $V^* = W = V^+$ and $W^* = V = W^+$ [7, 66].

As a generalization of universal enveloping algebras, in [15], Bell and Goodearl defined a PBW extension as follows: An over-ring S of R is called a *Poincare–Birkhoff–Witt extension* of R (*PBW extension* for short) if there exist elements $x_1, x_2, \ldots, x_n \in S$ such that

(i) the ordered monomials $x_1^{v_1} \ldots x_n^{v_n}$, where v_i are non-negative integers, form a basis for S as a free left R-module,
(ii) $x_i r - r x_i \in R$ for each $i = 1, \ldots, n$ and any $r \in R$, and
(iii) $x_i x_j - x_j x_i \in R + Rx_1 + \cdots + Rx_n$ for all $i, j = 1, \ldots, n$.

Proposition 2.10 (Enveloping algebras) *Let D be a Noetherian integrally closed domain and $\mathfrak{g}$ be a Lie D-algebra which is a finite free D-module. Then the enveloping algebra $U(\mathfrak{g})$ is a maximal order [23].*
(2) If R is a maximal order in $Q(R)$, then the PBW extension $R < x_1, x_2, \ldots, x_n >$ is a maximal order [64].

Proposition 2.11 (Semigroup algebras) *Let F be a field and S a submonoid of a torsion free finitely generated abelian-by-finite group. The semigroup algebra $F[S]$ is a Noetherian maximal order if and only if the following conditions are satisfied.*
(1) S satisfies the ascending chain condition on left and right ideals.
(2) For every minimal prime P in S the semigroup S_P is a maximal order with only one minimal prime ideal.
(3) $\cap S_P = S$, where P runs over all minimal prime ideals of S [49].

3 Krull Orders

Several noncommutative ring theorists defined Krull orders (Krull rings) and studied the ideal theory and polynomial extensions during the period 1970–1980 [23, 24, 51, 52, 54, 55, 58–61]. However, in case of orders having polynomial identities, these Krull orders coincide.

In this section, we only give a definition of Krull orders due to Chamarie and study ideal theory, polynomial extensions, and Ore–Rees rings over Krull orders.

Let $\mathcal{F}$ be a right Gabriel topology on R and $R_{\mathcal{F}} = \{q \in Q \mid qF \subseteq R$ for some $F \in \mathcal{F}\}$, which is called the *right quotients of R with respect to $\mathcal{F}$*. If I is a right ideal of R, then $I_{\mathcal{F}} = \{q \in Q \mid qF \subseteq I$ for some $F \in \mathcal{F}\}$ is a right ideal of $R_{\mathcal{F}}$, and I is said to be *$\mathcal{F}$-closed* if $I_{\mathcal{F}} \cap R = I$. Similarly for a left Gabriel topology $\mathcal{F}'$ on R we denote the left quotients of R with respect to $\mathcal{F}'$ by ${}_{\mathcal{F}'}R$ (see [79] for Gabriel topologies and quotients).

We now introduce a special Gabriel topology on R as follows.

Put $\mathcal{F}_{\mathcal{R}} = \{F \mid F$ is a right ideal such that $(r^{-1} \cdot F)^* = R$ for any $r \in R\}$ which is a right Gabriel topology on R, where $r^{-1} \cdot F = \{a \in R \mid ra \in F\}$. Similarly $\mathcal{F}'_{\mathcal{R}} = \{F' \mid F'$ is a left ideal such that $(F' \cdot r^{-1})^+ = R$ for any $r \in R\}$ is a left Gabriel topology on R.

A right (left) ideal $I(J)$ of R is called *τ-closed* if $I = I_{\mathcal{F}_{\mathcal{R}}} \cap R$ $(J = {}_{\mathcal{F}'_{\mathcal{R}}}J \cap R)$. An order in Q is said to be *τ-Noetherian* if it satisfies the ascending chain conditions on τ-closed left ideals as well as τ-closed right ideals.

Definition 3.1 An order in a simple Artinian ring is called a Krull order (ring) in the sense of Chamarie [23, 24] if it is a maximal order and τ-Noetherian.

Note that Noetherian maximal orders are Krull orders. We start with ideal theory between a Krull order and its over-ring.

Proposition 3.2 ([23, 63]) *Let R be a Krull order in Q and R' be an over-ring of R such that $R_{\mathcal{F}} = R' = {}_{\mathcal{F}'}R$ for some right (left) Gabriel topology $\mathcal{F}(\mathcal{F}')$ on R. Then*
(1) R' is a Krull order in Q.
(2) For any fractional right R-ideal I, ${}_{\mathcal{F}'}(I^{-1}) = (IR')^{-1} = (I_{\mathcal{F}})^{-1}$, where $I^{-1} = \{q \in Q \mid IqI \subseteq I\}$.
(3) The map: $I \longrightarrow I_{\mathcal{F}}$ is a bijection between the set of reflexive $\mathcal{F}$-closed right ideals I of R and the set of reflexive right ideals of R' (I is called reflexive if $I = I^{*+}$).

Theorem 3.3 (Structure theorem for Krull orders, [63]) *Let R be a Krull order in Q. Then*

(1) The center of R is a Krull domain.
(2) Any reflexive prime ideal P is localizable and R_P, the localization of R at P is a local principal ideal ring.
(3) $R = \cap R_P \cap S(R)$, where P ranges over all maximal reflexive ideals and $S(R)$ = $\bigcup\{A^{-1} \mid A$ is nonzero ideal of $R\}$ is a reflexively simple Krull order in Q.
(4) R has a finite character property, that is, any regular element $c \in R$ is a non-unit in only finitely many of R_P.
(5) For any essential right ideal I, $I^{*+} = \cap I R_P \cap (I S(R))^{*+}$.

In the remainder of this section, R is an order in Q with an automorphism σ and a left σ-derivation δ, and put $T = Q[x; \sigma, \delta]$.

We denote the prime spectrum of R by Spec(R) and $\mathrm{Spec}_0^*(R[x; \sigma, \delta])= \{P$: reflexive prime ideals $\mid P \cap R = (0)\}$. It is shown in [63] that R is τ-Noetherian if and only if so is $R[x; \sigma, \delta]$ (in [23], Chamarie proved that R is τ-Noetherian, then so is $R[x; \sigma, \delta]$).

Proposition 3.4 ([63]) *Suppose R is τ-Noetherian and put $S = R[x; \sigma, \delta]$.*
(1) There is a one-to-one correspondence between $\mathrm{Spec}_0^*(S)$ *and* Spec(T) *which is given by: $P' \longrightarrow P = P' \cap S$, where $P' \in$* Spec(T).
(2) If $P \in \mathrm{Spec}_0^*(S)$, then P is localizable and $S_P = T_{P'}$ which is a local principal ideal ring, where $P' = PT$.

A fractional R-ideal $\mathfrak{a}$ is called *σ-stable* if $\sigma(\mathfrak{a}) \subseteq \mathfrak{a}$ and it is *σ-invariant* if $\sigma(\mathfrak{a}) = \mathfrak{a}$. An order R is called a *σ-maximal order* if $O_l(\mathfrak{a}) = R = O_l(\mathfrak{a})$ for any σ-invariant ideal $\mathfrak{a}$ of R, and R is a *σ-Krull order* if it is a σ-maximal order and τ-Noetherian. Similarly, a fractional R-ideal $\mathfrak{a}$ is called *δ-stable* if $\delta(\mathfrak{a}) \subseteq \mathfrak{a}$ and R is called a *δ-maximal order* in Q if $O_l(\mathfrak{a}) = R = O_r(\mathfrak{a})$ for any δ-stable ideal $\mathfrak{a}$ of R. An order is said to be a *δ-Krull order* if it is a δ-maximal order and τ-Noetherian.

In case $\delta = 0$ or $\sigma = 1$, we denote $R[x; \sigma, \delta]$ by $R[x; \sigma]$ or $R[x; \delta]$, respectively.

Theorem 3.5 ([23, 63]) *Let R be an order in Q.*

(1) If R is a Krull order, then so is $R[x; \sigma, \delta]$ (there are examples of orders R not Krull such that $R[x; \sigma, \delta]$ is a Krull order) [1, 62, 67].
(2) R is a σ-Krull order if and only if $R[x, \sigma]$ is a Krull order.
(3) R is δ-Krull order if and only if $R[x; \delta]$ is a Krull order.

Let $S = R[x; \sigma]$ or $S = R[x; \delta]$. We describe all the reflexive fractional S-ideals in case S is a Krull order.

Proposition 3.6 ([63]) *(1) Let $S = R[x; \sigma]$ and suppose R is a σ-Krull order in Q. Let P be an ideal of S with $P \cap R \neq 0$. Then P is a reflexive prime ideal if and only if $P = \mathfrak{p}[x; \sigma]$ for some σ-invariant reflexive ideal $\mathfrak{p}$ of R which is σ-prime* ($\mathfrak{p}$ *is σ-prime if, for σ-stable ideals $\mathfrak{a}$, $\mathfrak{b}$, $\mathfrak{ab} \subseteq \mathfrak{p}$ implies either $\mathfrak{a} \subseteq \mathfrak{p}$ or $\mathfrak{b} \subseteq \mathfrak{p}$*).

(2) Let $S = R[x; \delta]$ and suppose R is a δ-Krull order in Q. Let P be an ideal of S with $P \cap R \neq 0$. Then P is a reflexive prime ideal if and only if $P = \mathfrak{p}[x; \delta]$ for some δ-stable reflexive ideal $\mathfrak{p}$ of R which is δ-prime ($\mathfrak{p}$ is δ-prime if, for δ-stable ideals $\mathfrak{a}$, $\mathfrak{b}$, $\mathfrak{ab} \subseteq \mathfrak{p}$ implies either $\mathfrak{a} \subseteq \mathfrak{p}$ or $\mathfrak{b} \subseteq \mathfrak{p}$).

By Propositions 3.4 and 3.6, any maximal reflexive ideal P of S is either $P \in \mathrm{Spec}_0^*(S)$ or $P = \mathfrak{p}[x; \sigma]$ for a reflexive σ-prime ideal $\mathfrak{p}$ of R (in case $S = R[x; \delta]$, $P \in \mathrm{Spec}_0^*(S)$ or $P = \mathfrak{p}[x; \delta]$ for a reflexive δ-prime ideal $\mathfrak{p}$ of R).

We denote the set of σ-invariant reflexive fractional R-ideals by $D_\sigma(R)$ and by $D_\delta(R) = \{\mathfrak{a} \mid \mathfrak{a}$ is a δ-stable reflexive fractional R-ideal$\}$. Then $D_\sigma(R)$ is an abelian group generated by maximal σ-invariant reflexive ideals of R. Similarly, $D_\delta(R)$ is an abelian group generated by maximal δ-stable reflexive ideals of R.

Thus we have the following which describe all reflexive fractional S-idealsIdeal!sideal@s-ideal.

Theorem 3.7 ([63]) *(1) Suppose R is a σ-Krull order in Q and $S = R[x; \sigma]$, $T = Q[x; \sigma]$. Then*

$$D(S) \cong D_\sigma(R) \oplus D(T).$$

(2) Suppose R is a δ-Krull order in Q and let $S = R[x; \delta]$, $T = Q[x; \delta]$. Then

$$D(S) \cong D_\delta(R) \oplus D(T).$$

Let R be a Krull order in Q. The set of principal fractional R-ideals forms a subgroup $P(R)$ of $D(R)$, where a fractional R-ideal $\mathfrak{a}$ is principal if $\mathfrak{a} = aR = Ra$ for some $a \in \mathfrak{a}$. The factor group $D(R)/P(R)$ is called the *divisor class group* of R, which is denoted by $Cl(R)$. In case R is a σ-Krull order (δ-Krull order), we can similarly define $Cl_\sigma(R) = D_\sigma(R)/P_\sigma(R)$ ($Cl_\delta(R) = D_\delta(R)/P_\delta(R)$) which is called the *$\sigma$-divisor class group* of R (*δ-divisor class group* of R), respectively, where $P_\sigma(R)$ is the subgroup of σ-invariant principal fractional R-ideals ($P_\delta(R)$ is the subgroup of δ-stable principal fractional R-ideals).

Proposition 3.8 ([63]) *(1) Suppose R is a σ-Krull order in Q and let $S = R[x; \sigma]$. Then the map $\phi : D_\sigma(R) \longrightarrow D(S)$ defined by $\phi(\mathfrak{a}) = \mathfrak{a}[x; \sigma]$, where $\mathfrak{a} \in D_\sigma(R)$ induces an isomorphism: $Cl_\sigma(R) \cong Cl(S)$.*
(2) Suppose R is a δ-Krull order in Q and let $S = R[x; \delta]$. Then the map ψ : $D_\delta(R) \longrightarrow D(S)$ defined by $\psi(\mathfrak{a}) = \mathfrak{a}[x; \delta]$, where $\mathfrak{a} \in D_\delta(R)$ induces a surjective map: $Cl_\delta(R) \longrightarrow Cl(S)$. If R is a domain, then $Cl_\delta(R) \cong Cl(S)$.

Let $S = R[Ix; \sigma, \delta]$ be an Ore–Rees ring, where R is a Noetherian prime ring as in Sect. 2. A fractional R-ideal $\mathfrak{a}$ is called *$(\sigma; I)$-invariant* if $I\sigma(\mathfrak{a}) = \mathfrak{a}I$.

An order R is a *$(\sigma; I)$-maximal order* if $O_l(\mathfrak{a}) = R = O_r(\mathfrak{a})$ for any $(\sigma; I)$-invariant ideal $\mathfrak{a}$ of R. If R is a $(\sigma; I)$-maximal order, then $D_{\sigma;I}(R)$, the set of all $(\sigma; I)$-invariant reflexive fractional R-ideals, is an Abelian group generated by maximal $(\sigma; I)$-invariant reflexive ideals of R (this is proved by standard way).

A fractional R-ideal $\mathfrak{a}$ is said to be *$(\delta; I)$-stable* if $I\delta(\mathfrak{a}) \subseteq \mathfrak{a}$ and $I\mathfrak{a} = \mathfrak{a}I$. We can define a $(\delta; I)$-maximal order in an obvious way and denote the set of all $(\delta; I)$-stable reflexive fractional R-ideals by $D_{\delta;I}(R)$. If R is a $(\delta; I)$-maximal order, then $D_{\delta;I}(R)$ is an Abelian group generated by maximal $(\delta; I)$-stable reflexive ideals of R.

In case $\delta = 0$ or $\sigma = 1$, we write $R[Ix; \sigma]$ for $R[Ix; \sigma, 0]$ and $R[Ix; \delta]$ for $R[Ix; 1, \delta]$, respectively. If S is a maximal order (in case $\delta = 0$ or $\sigma = 1$), then we completely describe the structure of reflexive fractional S-ideals as follows:

Theorem 3.9 ([47]) *Let R be a Noetherian prime ring and $S = R[Ix; \sigma]$ or $S = R[Ix; \delta]$. Then*
(1) In case $\delta = 0$.
(i) S is a maximal order if and only if R is a $(\sigma; I)$-maximal order.
(ii) If R is a $(\sigma; I)$-maximal order, then any reflexive fractional S-idealIdeal!sideal@s-ideal is of the form:

$$x^n w\mathfrak{a}[Ix; \sigma]$$

for some $\mathfrak{a} \in D_{\sigma;I}(R)$, $w \in \mathbb{Z}(Q(T))$, the center of $Q(T)$, and n is an integer.
(2) In case $\sigma = 1$.
(i) S is a maximal order if and only if R is a $(\delta; I)$-maximal order.
(ii) If R is a $(\delta; I)$-maximal order, then any reflexive fractional S-ideal is of the form:

$$w\mathfrak{a}[Ix; \delta]$$

for some $\mathfrak{a} \in D_{\delta;I}(R)$, $w \in \mathbb{Z}(Q(T))$.

Let G be a polycyclic-by-finite group and $R[G]$ be the group ring. A subset of G is called *orbital* if it has only finite many distinct G-conjugates. G is called *dihedral free* if it contains no orbital subgroup isomorphic to the infinite dihedral group $< a, b \in G \mid aba = b^{-1}, a^2 = 1 >$.

Proposition 3.10 (Group rings) *Let G be a polycyclic-by-finite group. The group ring $R[G]$ is a prime Krull order if and only if*

(i) R is a prime Krull order,
(ii) G has no nontrivial finite normal subgroup, and
(iii) G is dihedral free [17, 18, 20].

Remark (1) In the first paragraph of Sect. 3, we did not give the definitions of Krull rings different from Krull rings due to Chamarie. See [54, 55] for the definition of Ω-Krull rings, and see [61] for the definition in the sense of Marubayashi.

It is natural, in a sense, from the viewpoint of multiplicative ideal theory that an order is a Krull order (ring) if it is a maximal order and satisfies the ascending chain condition on integral reflexive ideals ([38, 49, 70] for monoids).

In case of rings having polynomial identities, those Krull rings all coincide, which is proved by using Posner's theorem [68, 13.6.5].

Krull orders in the sense of Chamarie are Krull orders in the sense of [70] ([63, Lemma 2.2.3]). However, it is still open whether each reflexive prime ideal of Krull orders in the sense of [70] is localizable or not, which is important to study the structure of orders. It is a remarkable result that an order R is Krull in the sense

of [70] if and only if the monoid of regular elements of R is a Krull monoid [38, Proposition 5.1].
(2) See [38, 75] for multiplicative ideal theory in noncommutative monoids.

4 Unique Factorization Rings

Noncommutative unique factorization rings were defined by various ring theorists with two different approaches. In 1963, P.M. Cohn generalized the notion of commutative unique factorization domains (UFD) to noncommutative rings with an element-wise approach, [29]. In 1984, A.W. Chatters introduced unique factorization for elements in the context of Noetherian rings which are not necessarily commutative with both element-wise approach and ideal theoretic approach, [25], and published a series of papers on the subject with his co-authors (D.A. Jordan, M.P. Gilchrist, and D. Wilson). In [1], authors gave a more general definition to noncommutative unique factorization rings and introduced connections to Krull orders. In this section, we give a summary of all approaches mentioned above.

A commutative unique factorization domain (UFD) is an integral domain satisfying the following three conditions (e.g. [81]):

1. Every element of R which is neither zero nor unit is a product of primes.
2. Any two prime factorizations of a given element have same number of factors.
3. The primes occurring in any factorization of a are completely determined by a, except for their order and for multiplication by units.

In [29], Cohn generalizes the notion of UFD to noncommutative rings by taking 1–3 as starting point. By an integral domain we understand a ring (not necessarily commutative) in which $1 \neq 0$, and without zero-divisors. Thus in an integral domain R, the nonzero elements form a semigroup under multiplication which will be denoted by R^*. Two elements a, b of a ring R are said to be associated, if $b = uav$, where u, v are units in R. An irreducible element in R is a non-unit which is not a product of two non-units. Clearly, if a is irreducible, or unit, or zero, then so is any element associated to a. Two elements a, b of R are said to be right similar, if $R/aR \cong R/bR$, as right R-modules [48].

Lemma 4.1 ([33]) *Two elements in an integral domain are right similar if and only if they are left similar.*

Let $a, b \in R$ and consider any factorizations of a and b:

$$a = u_1 u_2 \dots u_r,$$
$$b = v_1 v_2 \dots v_s.$$

These factorizations are said to be isomorphic, if $r = s$ and there is a permutation π of (1,...,r) such that u_i is similar to $v_{i\pi}$.

Proposition 4.2 ([29, Proposition 2.2]) *Let a, b be nonzero elements of an integral domain R which are similar. Then any factorization of a gives rise to an isomorphic factorization of b.*

A factorization of a may be regarded as a chain of cyclic submodules from R to aR, and by the isomorphism $R/aR \cong R/bR$ this gives a chain from R to bR, in which corresponding factors are isomorphic.

Definition 4.3 ([29]) A unique factorization domain (UFD for short) is an integral domain R such that every non-unit of R^* has a factorization into irreducibles and any two factorizations of a given element are isomorphic.

Since in a commutative integral domain R, a and b are associated if and only if $R/aR \cong R/bR$ holds, we have the following theorem:

Theorem 4.4 ([29, Theorem 2.3]) *A commutative integral domain is a UFD if and only if it satisfies 1–3 above.*

Noncommutative principal ideal domains [48] are given as an example of a noncommutative UFD. This includes in particular the skew polynomial rings studied by Ore [72] and the ring of integral quaternions. Moreover, any free associative algebra is a UFD [29, Theorem 6.3].

In 1984, A.W. Chatters defined unique factorization domains in the context of (not necessarily commutative) Noetherian rings which also has an equivalent elementwise definition.

Let R be a prime ring. A height-1 prime ideal of R is a prime ideal P of R such that P is minimal among nonzero prime ideals of R. An element p of R is *completely prime* if $pR = Rp$ is a height-1 prime of R and R/pR is a domain. If I is an ideal of R then $\mathcal{C}(I)$ is the set of elements of R which are regular (i.e. not zero-divisors) modulo I. Set $\mathcal{C} = \bigcap \mathcal{C}(P)$, where P ranges over the height-1 primes of R.

Proposition 4.5 ([25, Proposition 2.1]) *Let R be a prime Noetherian ring with at least one height-1 prime ideal, then the following conditions on R are equivalent:*
1. Every height-1 prime of R is of the form pR for some completely prime element p of R.
2. R is a domain and every nonzero element of R is of the form $cp_1p_2 \dots p_n$ for some $c \in \mathcal{C}$ (as defined above) and for some finite sequence $p_1, \dots, p_n$ of completely prime elements of R.

Definition 4.6 ([25]) A Noetherian unique factorization domain (Noetherian UFD for short) is a Noetherian integral domain which has at least one height-1 prime ideal and which satisfies the equivalent conditions of Proposition 4.5.

Examples of Noetherian UFDs include Noetherian UFDs of commutative algebra and also the universal enveloping algebras of solvable Lie algebras.

We can deduce from Sect. 2 that a commutative Noetherian domain is a maximal order if and only if it is integrally closed. In the case of Noetherian UFDs we have the following theorem:

Theorem 4.7 ([25, Theorem 2.10]) *Let R be a Noetherian UFD such that every nonzero prime ideal of R contains a height-1 prime; then R is a maximal order.*

The Noetherian UFDs defined as in [25] has one respect which is not analogous to the commutative case, and that is the existence of Noetherian UFDs R such that the polynomial ring $R[x]$ is not a UFD. Because of this reason, Chatters and Jordan gave a more general definition of Noetherian unique factorization rings.

Definition 4.8 ([27]) A ring R will be called a Noetherian unique factorisation ring (Noetherian UFR, for short) if R is a prime Noetherian ring such that every nonzero prime ideal of R contains a nonzero principal prime ideal.

The class of Noetherian UFRs includes all Noetherian UFDs as defined in [25]. If D is the division algebra of rational quaternions and $R = D[x]$ then R is a Noetherian UFR and $(x^2 + 1)R$ is a height-1 prime of R, but R is not a Noetherian UFD because $R/(x^2 + 1)R$ is not a domain.

Following are some of the important results obtained by Chatters and Jordan.

Theorem 4.9 ([27]) *If R is a Noetherian UFR then R is a maximal order.*

Theorem 4.10 ([27]) *If R is a Noetherian UFR then $R[x]$ is a Noetherian UFR.*

Let $R[x; \sigma]$ and $R[x; \delta]$ be defined as in Sect. 2. Then;

Theorem 4.11 ([27]) *Let R be a Noetherian UFR with an automorphism of finite order. Then $R[x; \sigma]$ is a Noetherian UFR.*

Theorem 4.12 ([27]) *Let R be a Noetherian UFR and let δ be a derivation of R such that every nonzero δ-prime ideal contains a nonzero principal δ-ideal. Then $R[x; \delta]$ is a Noetherian UFR.*

However, if R is a Noetherian UFR in the sense of [27], then $R[x; \sigma]$ and $R[x; \delta]$ are not necessarily Noetherian UFRs in the sense of [27].

Let G be a polycyclic-by-finite group. A *plinth* in G is a torsion-free abelian orbital subgroup H of G such that $H \otimes_{\mathbb{Z}} \mathbb{Q}$ is an irreducible $\mathbb{Q}T$-module for every subgroup T of a finite index in $N_G(H)$, where $\mathbb{Q}$ is the field of rationals. The plinth H is *centric* if its centralizer $C_G(H)$ has a finite index in G. We denote by $\Delta(G)$ the FC-subgroup, that is $\Delta(G) = \{g \in G : |G : C_G(g)| < \infty\}$, where $C_G(g)$ is the centralizer of g in G.

Proposition 4.13 ([17, 26]) *Let R be a commutative ring and G be a polycyclic-by-finite group. Then $R[G]$ is a Noetherian UFR in the sense of [27] if and only if*

(1) R is a UFD,
(2) G has no nontrivial finite normal subgroup,
(3) G is dihedral free, and
(4) Every plinth of G is centric.

Proposition 4.14 ([17, 26]) *Let R be a commutative ring and G be a polycyclic-by-finite group. Then $R[G]$ is a UFD in the sense of [25] if and only if*

(1) R is a UFD,
(2) G is torsion free,
(3) All plinths of G are central, and
(4) $G/\Delta(G)$ is torsion free.

Let S be a monoid with a polycyclic-by-finite group of quotients G. S is called *normalizing* if every element in S is normal, that is, $cS = Sc$ for all $c \in S$ and S is called *UF-monoid* if every prime ideal of S contains a principal prime ideal P, that is, $P = Sr = rS$ for some $r \in S$ as one in [27].

Proposition 4.15 ([50]) *Let S be a normalizing monoid with a torsion-free polycyclic-by-finite group of quotients G and let K be a field. Assume that S satisfies the ascending chain condition on left and right ideals. Then the semigroup algebra $K[S]$ is a Noetherian UFR in the sense of [27] if and only if $K[G]$ is a Noetherian UFR in the sense of [27] and S is a UF-monoid.*

In [28] Chatters, Gilchrist and Wilson developed a theory of noncommutative UFRs without the Noetherian condition.

Let R be an associative ring with identity element. An element x of R is *normal* if $xR = Rx$. A principal ideal of R is an ideal of the form xR for some normal element x of R. Let R be a prime ring, a prime element of R is a nonzero normal element p such that pR is a prime ideal.

Definition 4.16 ([28]) A ring R is called a unique factorization ring (UFR for short) if every nonzero prime ideal of R contains a prime element.

If R is a UFR as in [28] then the set of principal ideals of R is closed under finite intersections and satisfies the ascending chain condition, and the polynomial ring over R in an arbitrary number of central indeterminates is also a UFR. Restricting to the case of UFRs which satisfy a polynomial identity (PI) gives several genuinely noncommutative examples such as trace rings of generic matrix rings [21], the ring of n by n matrices over a commutative Dedekind domain of finite class number n; and the group ring $R[G]$ where R is any UFR which satisfies a PI and G is a torsion-free abelian group which satisfies the ascending chain condition for cyclic subgroups [28].

Another definition of unique factorization rings and its connections to Krull orders are given in Abbasi et al. [1]. Noetherian UFRs in the sense of [27] are Krull orders in the sense of Marubayashi [61] by [1, Proposition 1.9], and Krull orders in the sense of [61] are Krull orders in the sense of Chamarie [24]. Existence of examples of Krull orders which are not Krull orders in the sense of [61] and being natural that UFRs are closed under the polynomial extensions were the motivation of the authors of [1] to give a new definition of UFRs.

Definition 4.17 ([1]) Let R be a τ-Noetherian order with an automorphism σ in a simple Artinian ring Q. Then R is called a σ-unique factorization ring (a σ-UFR for short) if any σ-prime ideal P of R such that $P = P^{*+}$ or $P={}^{+*}P$ is principal.

In case σ is the identity mapping on R, R is said to be a UFR.

It turns out [63] that R is a UFR in the sense of [1] if and only if
(i) R is a maximal order and
(ii) Any reflexive ideal is principal.

Noetherian UFRs in the sense of [27] are UFRs in the sense of [1], however the converse is not true in general (see [1] for examples). Furthermore, we have the following:

Proposition 4.18 ([1]) *Let R be a UFR in the sense of [1]. Then R is a Noetherian UFR in the sense of [27] if and only if R_N is a simple ring, where N is Ore set consisting of all normal elements in R.*

Let δ be a derivation on R. Replacing σ-prime ideals by δ-prime ideals, we can naturally define δ-UFRs. Of course, if R is a UFR in the sense of [1], then R is a σ-UFR and δ-UFR, and we have the following characterizations:

Proposition 4.19 ([1]) *(1) R is a σ-UFR if and only if $R[x;\sigma]$ is a UFR in the sense of [1].*
(2) If R is a δ-UFR, then $R[x;\delta]$ is a UFR in the sense of [1]. In case R is a domain, the converse is also true.

In [1], they obtained the following characterizations of group ring $R[G]$ (independent on [26]).

Proposition 4.20 ([1]) *Let R be a UFR in the sense of [1] and G be a polycyclic-by-finite group. Then $R[G]$ is a UFR in the sense of [1] if and only if*
(1) G has no nontrivial finite normal subgroup, and
(2) G is dihedral free.

Proposition 4.21 ([1]) *Let R be a Noetherian UFR in the sense of [27] and G be a polycyclic-by-finite group. Then $R[G]$ is a Noetherian UFR in the sense of [27] if and only if*
(1) G has no nontrivial finite normal subgroup,
(2) G is dihedral free, and
(3) every plinth of G is centric.

We refer the readers to [76] for more examples and detailed survey of unique factorization rings.

5 G-Dedekind Prime Rings

The class of rings in which every reflexive (fractional) R-ideal (right or left) is invertible was first defined by Cozzens and Sandomierski in [30] with the name *RI-orders*. In [2], following the commutative version of the theory, Akalan characterized the class of rings in which $(AB)^* = B^*A^*$ is satisfied for all R-ideals A, B and gave the name *Generalized Dedekind prime rings* (G-Dedekind prime, for short) to this class of rings. It turns out that in a G-Dedekind prime ring every reflexive R-ideal is invertible and therefore is an RI-order.

The class of G-Dedekind prime rings is a broad class including both the class of Dedekind prime rings and the class of Noetherian UFRs [27]. Moreover, Noetherian maximal orders with gld ≤ 2 are examples of G-Dedekind prime rings. This assertion follows from Bass' characterization of Noetherian rings with gld ≤ 2 as rings over which duals of finitely generated modules are projective (see [30] and [14, Proposition 5.2]).

Definition 5.1 ([2]) A prime Noetherian maximal order satisfying $(AB)^* = B^*A^*$ for all R-ideals A and B, is called a generalized Dedekind prime ring (G-Dedekind prime ring).

As we have mentioned in Sect. 2 (Theorem 2.3), the set $D(R)$ of all reflexive R-ideals becomes an Abelian group with multiplication "$\circ$". We denote the divisor class group of R by $Cl(R) = D(R)/P(R)$ where $P(R)$ is the subgroup of $D(R)$ which consists of principal R-ideals, and the Picard group of R by $Pic(R) = Inv(R)/P(R)$ where $Inv(R)$ is the group of invertible R-ideals.

Theorem 5.2 ([2, Theorem 3.1]) *For an order R, the following conditions are equivalent:*
(1) $A^{**}A^* = R$ *and* $A^+A^{++} = R$ *for each R-ideal A.*
(2) R is a maximal order and $(AB)^* = B^*A^*$ *for all R-ideals A and B of R.*
(3) R is a maximal order and the product of reflexive R-ideals is reflexive.
(4) R is a maximal order and $D(R)$ *is a group with the usual product.*
(5) R is a maximal order and every reflexive R-ideal is invertible.
(6) R is a maximal order and $Cl(R) = Pic(R)$.

In [8, 80], many examples of commutative maximal orders with reflexive ideals which are not invertible are given. Following is a noncommutative example of a prime Noetherian maximal order with a reflexive ideal which is not invertible.

Example 5.3 By [16, Example 35] there exists a prime Noetherian smooth PI ring R which is also a maximal order with a unique height one prime ideal P which is not a projective R-module on either side. This height one prime ideal P is maximal reflexive by [3, Theorem 3.1]. However since P is not projective, it is not invertible.

The class of G-Dedekind prime rings is closed under the formation of $n \times n$ full matrix rings and moreover if R is a G-Dedekind prime ring then so is the ring eRe where e is an idempotent such that $ReR = R$. Thus, being a G-Dedekind prime ring is a Morita invariant.

Theorem 5.4 ([2, Theorem 5.4]) *If R is a PI G-Dedekind prime ring then so is the polynomial ring* $R[x]$.

In [4], Akalan showed that the PI condition can be waived from Theorem 5.4.

Theorem 5.5 ([2, Theorem 6.2]) *If R is a PI G-Dedekind prime ring then so is the Rees ring* $R[Ix]$ *where I is an invertible ideal of R.*

In Marubayashi et al. [67], authors use the terminology "generalized Asano prime rings" for "generalized Dedekind prime rings". Let σ be an automorphism of R, they call R a *σ-generalized Asano* prime ring (a σ-G-Asano prime ring for short) if it is a σ-Krull prime ring whose σ-invariant reflexive R-ideals are invertible. In case σ is identity, R is said to be a *G-Asano* prime ring.

Theorem 5.6 ([67, Theorem 2.8]) *Let R be an order in Q. R is a σ-G-Asano prime ring if and only if $R[x;\sigma]$ is a G-Asano prime ring.*

Definition 5.7 ([46]) A ring R is called δ-generalized Asano prime ring if R is a δ-Krull prime ring whose δ-stable reflexive R-ideals are invertible.

Theorem 5.8 ([46, Theorem 2.6]) *Let R be an order in Q. Then R is a δ-generalized Asano prime ring if and only if $S = R[x;\delta]$ is a generalized Asano prime ring.*

A generalized Asano prime ring is a Krull prime ring, but the converse of this does not necessarily hold [67] and [36, Example 1.10].

6 Hereditary Noetherian Prime Rings (HNP Rings) and a Generalization of HNP Rings

Hereditary Noetherian prime rings (HNP for short) are a very interesting class of rings and a lot of research has been done on them, especially for 1960–1990. In 1960, Auslander and Goldman found an example of HNP rings which is not Dedekind in crossed product algebras [13]. Since then, in case of algebras, Harada had studied the structure of HNP rings including ideal theory [42–44]. In 1970, Eisenbud and Robson studied the ideal theory of HNP rings which are not necessarily algebras. In this section, we mainly discuss the ideal theory in HNP rings and propose a polynomial-type generalization of HNP rings.

One of the important results on HNP rings is that the invertible ideals in an HNP ring generate an Abelian group as in Dedekind orders, which is obtained under the condition: every ideal is projective (left and right projective). The followings are the key propositions to prove this result.

Proposition 6.1 ([32, Proposition 2.1]) *Let R be an order in a simple Artinian ring such that each ideal of R is projective. Then every invertible ideal of R is a product of maximal invertible ideals (ideals maximal amongst the invertible ideals).*

Proposition 6.2 ([32, Proposition 2.2]) *Let R be an order in a simple Artinian ring such that each ideal of R is projective. Then each maximal ideal of R is either idempotent or invertible.*

A finite set of distinct idempotent maximal ideals $M_1, \ldots, M_n$ such that $O_r(M_1) = O_l(M_2), \ldots, O_r(M_n) = O_l(M_1)$ is called a *cycle*. An invertible maximal ideal is considered to be a trivial case of a cycle.

Theorem 6.3 ([32, Theorem 2.6]) *Let R be an order in a simple Artinian ring such that each ideal of R is projective. Then each maximal invertible ideal of R is the intersection of a cycle.*

Theorem 6.4 ([32, Theorem 2.9]) *Let R be an order in a simple Artinian ring such that each ideal of R is projective. Then the invertible ideals of R generate an Abelian group.*

An ideal A is called *eventually idempotent* if A^k is idempotent for some $k \geq 1$.

Proposition 6.5 ([32, Proposition 4.5]) *Let R be an HNP ring and A be an ideal of R which is not contained in any invertible ideal. Then A is eventually idempotent. More precisely, there are only a finite number of idempotent ideals* $M_1, \ldots, M_k$ *containing A and* $A^k = (M_1 \cap \cdots \cap M_k)^k$ *is idempotent. (see [37] for more detail results on eventually idempotent.)*

Theorem 6.6 ([32, Theorem 4.2]) *Let R be an HNP ring and I an ideal of R. Then* $I = XA$, *where X is an invertible ideal and A is an eventually idempotent ideal.*

Let A be a right ideal of R. The subring $\mathbf{I}(A) = \{r \in R | rA \subseteq A\}$ of R is called the *idealizer* of A in R. A is said to be *generative* if $RA = R$. The idealizer is one of the powerful tools to study HNP rings.

Theorem 6.7 ([74, Theorem 5.3]) *Let R be an HNP ring and A be an essential right ideal which is generative. Then* $\mathbf{I}(A)$ *is an HNP ring if and only if A is semimaximal, that is, A is a finite intersection of maximal right ideals.*

Theorem 6.8 ([74, Theorem 6.3]) *The following conditions on an HNP ring R are equivalent.*

(1) R is contained in and is equivalent to a Dedekind prime ring.
(2) R has finitely many idempotent ideals.
(3) R has finitely many idempotent maximal ideals.
(4) R is obtained as an iterated idealizer from a Dedekind prime ring.

It was an interesting question that any HNP ring has only finitely many idempotent ideals or not. In [78], they obtained examples of HNP rings in which there are infinite many idempotent maximal ideals.

A right ideal A is called *isomaximal* if R/A is a finite direct sum of isomorphic simple modules. In case A is isomaximal and generative, we have the following correspondence between Spec(R) and Spec(S), where $S = \mathbf{I}(A)$.

Theorem 6.9 ([68, Theorem 5.6.11]) *Let R be an HNP ring, A be a generative isomaximal right ideal and* $S = \mathbf{I}(A)$. *Then there is a set embedding* ϕ: $\{P \in Spec(R) | P \nsubseteq A\} \rightarrow \mathrm{Spec}(S)$ *given by* $P \rightarrow P \cap S$. *This preserves idempotence and invertibility. Further:*
(1) If there is no nonzero prime ideal P of R with $P \subseteq A$, *then there is only one nonzero prime of S not in the image of* ϕ, *that is, A, which is idempotent.*
(2) If there is a (necessarily unique) nonzero prime ideal $P \subseteq A$, *then there are exactly two nonzero primes of S not in the image of* ϕ, *A and, say* A'. *Both are idempotent and* A' *is an isomaximal generative left ideal of R containing P.*

We refer the readers to [57, 68] for more information about ideal theory in HNP rings.

Finally we discuss the ideal theory of polynomial rings over an HNP ring and propose a generalization of HNP rings. Let R be an HNP ring and $S = R[x]$ be the polynomial ring. Then S is not necessarily an HNP ring. In fact S is an HNP ring if and only if $R = Q$.

Note: any one sided reflexive ideal of S is projective since $\mathrm{gl.dim}(S) \leq 2$.

Let A be a nonzero ideal of S such that $A = A^{*+}$ or $A = A^{+*}$, equivalently, A is right projective or A is left projective. Then we have the following [5]:
(a) If $\mathfrak{a} = A \cap R \neq (0)$, then $A = \mathfrak{a}[x]$.
(b) If $A \cap R = (0)$, then $A = B\mathfrak{a}[x]$ for an invertible ideal B of S and an ideal $\mathfrak{a}$ of R.

In both cases, A is left and right projective.

These properties suggest us to define the following which are, in some sense, a polynomial-type generalization of HNP rings.

Definition 6.10 ([6]) (1) A τ-Noetherian prime Goldie ring R is called a generalized HNP ring (a G-HNP ring for short) if each ideal A with $A = A^{*+}$ or $A = A^{+*}$ is left and right projective.
(2) A G-HNP ring is said to be a strongly G-HNP ring if each essential right (left) ideal $I\,(J)$ with $I = I^{*+} (J = J^{+*})$ is right (left) projective, respectively.

If R is an HNP ring, then $R[x]$ is a strongly G-HNP ring. The following is a structure theorem for G-HNP rings (compare with Theorem 3.3).

Theorem 6.11 (Structure theorem for G-HNP rings, [6]) *Let R be a G-HNP ring. Then*

(1) any maximal invertible ideal P is localizable and R_P is a semi-local HNP ring.
(2) $R = \cap R_P \cap S(R)$, where P ranges over all maximal invertible ideals of R and $S(R)$ is a G-HNP ring with no invertible ideals.
(3) R has a finite character property.

We end the paper with the following questions.

Let σ be an automorphism of R and δ be a left σ-derivation on R.

Question 6.12 *(1) What are necessary and sufficient conditions for $R[x; \sigma, \delta]$ to be a G-HNP ring and describe all projective ideals of $R[x; \sigma, \delta]$.*
(2) Let I be an invertible ideal of R. What are necessary and sufficient conditions for $R[Ix; \sigma, \delta]$ to be a G-HNP ring and describe all projective ideals of $R[Ix; \sigma, \delta]$.

Let H be a monoid with quotient group Q. By adopting dual basis lemma for projective modules [68, (3.5.2)], we can define the concept of right hereditary monoids as follows: H is *right hereditary* if $II^* = O_l(I)$ for any right ideal I of H, where $I^* = \{q \in Q | qI \subseteq H\}$. Similarly we can define left hereditary monoids.

Question 6.13 *Is it possible to obtain ideal theories (as ones in HNP rings) in left and right hereditary monoids?*

Acknowledgments This work has been supported by TUBITAK (project no: 113F032) and by JSPS (project no: 24540058). We would like to thank TUBITAK and JSPS for their support. We would also like to thank Professor Alfred Geroldinger and his students for their warm hospitality and efforts to organize the conference. Our thanks go to the referee who carefully checked the manuscript and gave us so many valuable comments.

References

1. G.Q. Abbasi, S. Kobayashi, H. Marubayashi, A. Ueda, Non-commutative unique factorization rings. Commun. Algebra **19**, 167–198 (1991)
2. E. Akalan, On generalized Dedekind prime rings. J. Algebra **320**, 2907–2916 (2008)
3. E. Akalan, Rings with enough invertible ideals and their divisor class groups. Commun. Algebra **37**(12), 4374–4390 (2009)
4. E. Akalan, On rings whose reflexive ideals are principal. Commun. Algebra **38**(9), 3174–3180 (2010)
5. E. Akalan, P. Aydoğdu, H. Marubayashi, B. Saraç, A. Ueda, Projective ideals of polynomial rings over HNP rings. Commun. Algebra (to appear)
6. E. Akalan, P Aydoğdu, H. Marubayashi, B. Saraç, A. Ueda, Generalized HNP rings (preprint)
7. E. Akalan, P. Aydoğdu, H. Marubayashi, B. Saraç, Rings of Morita contexts which are maximal orders. J. Algebra Appl. **15**(6) (2016)
8. D.D. Anderson, B.G. Kang, Pseudo Dedekind domains and divisorial ideals in $R[X]_T$. J. Algebra **122**, 323–336 (1989)
9. E. Artin, Zur Arithmetik hyperkomplexer Zahlen, Abh. Math. Semin. Hamburg Univ. **5**, 261–289 (1928)
10. K. Asano, Arithmetische Idealtheorie in nichtkommutativen Ringen. Jpn. J. Math. **16**, 1–36 (1939)
11. K. Asano, Zur Arithmetik in Schiefringen I. Osaka Math. J. **1**(2), 98–134 (1949)
12. K. Asano, Zur Arithmetik in Schiefringen II. J. Inst. Polytech. Osaka City Univ. **1**, 1–27 (1950)
13. M. Auslander, O. Goldman, Maximal orders. Trans. Am. Math. Soc. **97**, 1–24 (1960)
14. H. Bass, Finitistic dimension and a homological generalization of semiprimary rings. Trans. Am. Math. Soc. **95**, 466–488 (1960)
15. A.D. Bell, K.R. Goodearl, Uniform rank over differential operator rings and Poincare-Birkhoff-Witt extensions. Pac. J. Math. **131**(1), 13–37 (1988)
16. A. Braun, C.R. Hajarnavis, Smooth polynomial identity algebras with almost factorial centers. J. Algebra **299**(1), 124–150 (2006)
17. K.A. Brown, Height one primes of polycyclic group rings. J. Lond. Math. Soc. **32**(2), 426–438 (1985)
18. K.A. Brown, Corrigendum and addendum to 'Height one primes of polycyclic group rings'. J. Lond. Math. Soc. **38**(2), 421–422 (1988)
19. K.A. Brown, C.R. Hajarnavis, A.B. MacEacharn, Noetherian rings of finite global dimension. Proc. Lond. Math. Soc. **44**, 349–371 (1982)
20. K.A. Brown, H. Marubayashi, P.F. Smith, Group rings which are v-HC orders and Krull orders. Proc. Edinb. Math. Soc. **34**, 217–228 (1991)
21. L. le Bruyn, Trace rings of generic matrices are unique factorization domains. Glasg. Math. J. **28**, 11–13 (1986)
22. G. Cauchon, Les T-anneaux et les anneaux *à* identit*é*s polyn*ô*miales Noeth*é*riens, Th*è*se de doctorat, Universit*é* Paris XI, 1977
23. M. Chamarie, Anneaux de Krull non commutatifs, Th*è*se, Uni. de Lyon, 1981
24. M. Chamarie, Anneaux de Krull non commutatifs. J. Algebra **72**, 210–222 (1981)
25. A.W. Chatters, Non-commutative unique factorization domains. Math. Proc. Camb. Philos. Soc. **95**, 49–54 (1984)

26. A.W. Chatters, J. Clark, Group rings which are unique factorization rings. Commun. Algebra **19**(2), 585–598 (1991)
27. A.W. Chatters, D.A. Jordan, Non-commutative unique factorization rings. J. Lond. Math. Soc. **33**(2), 22–32 (1986)
28. A.W. Chatters, M.P. Gilchrist, D. Wilson, Unique factorization rings. Proc. Edinb. Math. Soc. **35**, 255–269 (1992)
29. P.M. Cohn, Non-commutative unique factorization domains. Trans. Am. Math. Soc. **109**, 313–331 (1963)
30. J.H. Cozzens, F.L. Sandomierski, Maximal orders and localization. I. J. Algebra **44**, 319–338 (1977)
31. M. Deuring, *Algebren* (Springer, Berlin, 1935) (Revised, 1968)
32. D. Eisenbud, J.C. Robson, Hereditary Noetherian prime rings. J. Algebra **16**, 86–104 (1970)
33. H. Fitting, Über den Zusammenhang zwischen dem Begriff der Gleichartigkeit zweier Ideale und dem Äquivalenzbegriff der Elementarteilertheorie. Math. Ann. **112**, 572–582 (1936)
34. M. Fontana, J.A. Huckaba, I.J. Papick, *Prüfer Domains, Monographs and Textbooks in Pure and Applied Mathematics*, vol. 203 (Marcel Dekker, New York, 1997)
35. R.M. Fossum, Maximal orders over Krull domains. J. Algebra **10**, 321–332 (1968)
36. R.M. Fossum, *The Divisor Class Group of a Krull Domain* (Springer, Berlin, 1973)
37. H. Fujita, K. Nishida, Ideals of hereditary noetherian prime rings. Hokkaido Math. J. **11**, 286–294 (1982)
38. A. Geroldinger, Non-commutative Krull monoids: a divisor theoretic approach and their arithmetic. Osaka J. Math. **50**(2), 503–539 (2013)
39. R. Gilmer, *Multiplicative Ideal Theory, Queen's Papers in Pure and Applied Mathematics*, vol. 90 (Queen's University, Kingston ON, 1992) (Corrected reprint of the 1972 edition)
40. A.J. Gray, A class of maximal orders integral over their centres. Glasg. Math. J. **24**, 177–180 (1983)
41. F. Halter-Koch, An introduction to multiplicative ideal theory, *Ideal systems, Monographs and Textbooks in Pure and Applied Mathematics*, vol. 211 (Marcel Dekker, New York, 1998)
42. M. Harada, Hereditary orders. Trans. Am. Math. Soc. **107**, 273–290 (1963)
43. M. Harada, Structure of hereditary orders over local rings. J. Math. Osaka City Univ. **14**, 1–22 (1963)
44. M. Harada, Multiplicative ideal theory in hereditary orders. J. Math. Osaka City Univ. **14**, 83–106 (1963)
45. H. Hasse, Über p-adische Schiekörper und ihre Bedeutung für die Arithmetik hyperkomplexer Zahlsysteme. Math. Ann. **104**, 495–534 (1931)
46. M.R. Helmi, H. Marubayashi, A. Ueda, Differential polynomial rings which are generalized Asano prime rings. Indian J. Pure Appl. Math. **44**(5), 673–681 (2013)
47. M.R. Helmi, H. Marubayashi, A. Ueda, Ore-Rees rings which are maximal orders. J. Math. Soc. Jpn. **68**(1), 405–423 (2016)
48. N. Jacobson, *Theory of Rings* (American Mathematical Society, Providence, R.I, 1943)
49. E. Jespers, J. Okninski, *Noetherian Semigroup Algebras, Algebra and Applications*, vol. 7 (Springer, Heidelberg, 2007)
50. E. Jespers, Q. Wang, Noetherian unique factorization semigroup algebras. Commun. Algebra **29**(12), 5701–5715 (2001)
51. E. Jespers, P. Wauters, On central Ω-Krull rings and their class group. Can. J. Math. **36**(2), 206–239 (1984)
52. E. Jespers, P. Wauters, Marubayashi-Krull orders and strongly graded rings. J. Algebra **86**, 511–521 (1984)
53. E. Jespers, P. Wauters, Asano-orders and graded rings. Commun. Algebra **13**, 811–833 (1985)
54. E. Jespers, L. Le Bruyn, P. Wauters, Ω-Krull rings I. Commun. Algebra **10**, 1801–1818 (1982)
55. E. Jespers, L. Le Bruyn, P. Wauters, A characterization of central Ω-Krull rings. J. Algebra **81**, 165–179 (1983)
56. T.H. Lenagan, Bounded Asano orders are hereditary. Bull. Lond. Math. Soc. **3**, 67–69 (1971)

57. L.S. Levy, J.C. Robson, *Hereditary Noetherian prime rings and idealizers, Mathematical Surveys and Monographs*, vol. 174 (American Mathematical Society, Providence, RI, 2011)
58. H. Marubayashi, Non commutative Krull rings. Osaka J. Math. **12**, 703–714 (1975)
59. H. Marubayashi, On bounded Krull prime rings. Osaka J. Math. **13**, 491–501 (1976)
60. H. Marubayashi, A characterization of bounded Krull prime rings. Osaka J. Math. **15**, 13–20 (1978)
61. H. Marubayashi, Polynomial rings over Krull orders in simple Artinian rings. Hokkaido Math. J. **9**, 63–78 (1980)
62. H. Marubayashi and A. Ueda, Examples of Ore extensions which are maximal orders but the based rings are not maximal orders (preprint)
63. H. Marubayashi, F. Van Oystaeyen, *Prime Divisors and Noncommutative Valuation Theory*, vol. 2059, Lecture Notes in Mathematics (Springer, Heidelberg, 2012)
64. H. Marubayashi, Y. Zhang, Maximality of PBW extensions. Commun. Algebra **24**(4), 1377–1388 (1996)
65. H. Marubayashi, E. Nauwelaerts, F. Van Oystaeyen, Graded rings over arithmetical orders. Commun. Algebra **12**(6), 745–775 (1984)
66. H. Marubayashi, Y. Zhang, P. Yang, On the rings of Morita contexts which are some well-known orders. Commun. Algebra **26**(5), 1429–1444 (1998)
67. H. Marubayashi, I. Muchtadi-Alamsyah, A. Ueda, Skew polynomial rings which are generalized Asano prime rings. J. Algebra Appl **7**(12), 1–8 (2013)
68. J.C. McConnell, J.C. Robson, *Noncommutative Noetherian Rings*, Pure and Applied Mathematics (A Wiley-Interscience Publication, New York, 1987)
69. E. Noether, Abstrakter Aufbau der Idealtheorie in algebraischen Zahl-und Funktionenkörpern, Math. Ann. **96**, (1926)
70. J. Okninski, Noetherian semigroup algebras and beyond, in *Multiplicative Ideal Theory and Factorization Theory*, ed. by S.T. Chapman, M. Fontana, A. Geroldinger, B. Olberding (Springer, Heidelberg, 2016)
71. O. Ore, Linear equations in noncommutative fields. Ann. Math. **32**, 463–477 (1931)
72. O. Ore, Theory of noncommutative polynomials. Ann. Math. **34**, 480–508 (1933)
73. I. Reiner, *Maximal Orders*, vol. 5 (Academic, Cambridge, 1975)
74. J.C. Robson, Idealizers and hereditary Noetherian prime rings. J. Algebra **22**, 45–81 (1972)
75. D. Smertnig, Sets of lengths in maximal orders in central simple algebras. J. Algebra **390**, 1–43 (2013)
76. D. Smertnig, Factorizations of elements in noncommutative rings: a survey, in *Multiplicative Ideal Theory and Factorization Theory*, ed. by S.T. Chapman, M. Fontana, A. Geroldinger, B. Olberding (Springer, Heidelberg, 2016)
77. J.T. Stafford, Auslander-regular algebras and maximal orders. J. Lond. Math. Soc. **50**(2), 276–292 (1994)
78. J.T. Stafford, R.B. Warfield Jr., Constructions of hereditary Noetherian rings and simple rings, Proc. Lond. Math. Soc. **51**, 1–20 (1985)
79. B. Stenström, *Rings of Quotients*, vol. 217 (Springer, Heidelberg, 1975)
80. M. Zafrullah, On generalized Dedekind domains. Mathematika **33**, 285–295 (1986)
81. O. Zariski, P. Samuel, *Commutative Algebra*, vol. I (Van Nostrand, Princeton, N.J, 1958)

About Number Fields with Pólya Group of Order ≤2

David Adam and Jean-Luc Chabert

Abstract Carlitz characterized the number fields K with class number ≤ 2 by the equality of the lengths of all the factorizations of every integer of K into irreducible elements. Analogously, we study the links between the order of the Pólya group $\mathcal{P}o(K)$ of a number field K and the factorizations into irreducible elements of some rational numbers. Our main results concern quadratic fields where we prove some equivalences between, on the one hand, $|\mathcal{P}o(K)| = 1$ and uniqueness of factorizations, on the other hand, $|\mathcal{P}o(K)| = 2$ and uniqueness of lengths of factorizations. We also show how analogous results may be formulated in the case of function fields.

1 Introduction

Let K be a number field. Denote its ring of integers by $\mathcal{O}_K$ and its class group by $\mathcal{C}l(K)$. If the group $\mathcal{C}l(K)$ is trivial it means that $\mathcal{O}_K$ is a principal ideal domain. As $\mathcal{O}_K$ is a Dedekind domain, to be a principal ideal domain is equivalent to be a unique factorization domain. From this point of view, Carlitz [4] proved in a very short paper the following result which says that, to weaken the hypothesis by allowing $\mathcal{C}l(K)$ to have not one but two elements is equivalent to weaken the factorization property in $\mathcal{O}_K$ in the following way:

Theorem 1 (Carlitz) *The class number of a number field K is ≤ 2 if and only if, for every integer x of K, all the factorizations of x into irreducible elements of $\mathcal{O}_K$ have the same length.*

We are interested here in a subgroup of $\mathcal{C}l(K)$ called the Pólya group of K. Let us recall its definition.

D. Adam
GAATI, Université de la Polynésie Française, BP 6570, 98702 Faa'a, Tahiti, French Polynesia
e-mail: david.adam@upf.pf

J.-L. Chabert (✉)
LAMFA CNRS-UMR 7352, Université de Picardie, 80039 Amiens, France
e-mail: jean-luc.chabert@u-picardie.fr

S. Chapman et al. (eds.), *Multiplicative Ideal Theory and Factorization Theory*, Springer Proceedings in Mathematics & Statistics 170,
DOI 10.1007/978-3-319-38855-7_2

Notation. If an integer q is the norm of at least one maximal ideal of $\mathcal{O}_K$, we denote by $\Pi_q(K)$ the ideal product of all maximal ideals of $\mathcal{O}_K$ with norm q

$$\Pi_q(K) = \prod_{\substack{\mathfrak{m}\in\mathrm{Max}(\mathcal{O}_K)\\ N_{K/\mathbb{Q}}(\mathfrak{m})=q}} \mathfrak{m}. \tag{1}$$

Definition 1 [3, Sect. II.3] The *Pólya group* of K is the subgroup $\mathcal{P}o(K)$ of the class group $\mathcal{C}l(K)$ of K generated by the classes of all the ideals $\Pi_q(K)$ defined by Formula (1).

The Pólya group could also be defined as the subgroup of the class group generated by the classes of Bhargava's factorial ideals (which are defined in [2]).

The idea for this article comes from a remark by Jesse Elliott: the hypothesis $\mathrm{Card}(\mathcal{C}l(K)) \leq 2$ corresponds to an interesting property, it could also be the case for the similar hypothesis $\mathrm{Card}(\mathcal{P}o(K)) \leq 2$. Noticing that the Pólya group of K is trivial, if and only if, for every $n \in \mathbb{N}$, the $\mathcal{O}_K$-module

$$\mathrm{Int}_n(\mathcal{O}_K) = \{f \in \mathrm{Int}(\mathcal{O}_K) \mid \deg(f) \leq n\}$$

is free [15, 16], Elliott [7] suggests the following conjecture:

Conjecture. For every number field K, if $\mathrm{Card}(\mathcal{P}o(K)) \leq 2$, then

$$\overline{\lim}_{N\to+\infty} \frac{1}{N}\mathrm{Card}\{n \leq N \mid \mathrm{Int}_n(\mathcal{O}_K) \text{ is free}\} \geq \frac{1}{2}.$$

For our part, always with the assumption $\mathrm{Card}(\mathcal{P}o(K)) \leq 2$, returning to the spirit of the result of Carlitz, we consider the factorizations of rational integers into irreducible elements of $\mathcal{O}_K$, because there are natural links between the rational integers and the ideals $\Pi_q(K)$ whose classes generate $\mathcal{P}o(K)$. We will see that we have to exclude the prime numbers which are decomposed in the extension $K/\mathbb{Q}$.

Recall that a prime number p is said to be *decomposed* in the number field K if there are at least two prime ideals of the ring of integers $\mathcal{O}_K$ lying over p. Consequently, the prime p is *undecomposed* in K if and only if $p\mathcal{O}_K$ is a primary ideal of $\mathcal{O}_K$, that is, is a power of a maximal ideal of $\mathcal{O}_K$.

In Sect. 2, we prove that $|\mathcal{P}o(K)| = 1$ (resp., $|\mathcal{P}o(K)| \leq 2$) implies the uniqueness of the factorization (resp., of the length of the factorizations) into irreducible elements of $\mathcal{O}_K$ of all products of undecomposed primes numbers (Theorem 2). In Sect. 3, we study the obstructions for the converses of the previous assertions. In Sect. 4, we obtain characterizations in the particular case of Galois number fields of odd prime degree. In Sect. 5, we obtain equivalences for quadratic number fields (Theorems 3 and 4). Finally, in the last section, we end with some analogous results in the function fields case.

2 The Hypothesis Card($\mathscr{P}o(K)$) ≤ 2

In this section, we describe consequences of the hypothesis Card($\mathscr{P}o(K)$) ≤ 2. We consider rational integers m which are product of primes which are themselves undecomposed in the extension $K/\mathbb{Q}$ and the factorizations of these rational integers m into irreducible elements of $\mathcal{O}_K$.

Theorem 2 *Let K be a number field. We denote its ring of integers by $\mathcal{O}_K$ and its Pólya group by $\mathscr{P}o(K)$. Let m be any rational integer which is a product of undecomposed primes.*

1. *If $|\mathscr{P}o(K)| = 1$, then the factorization of m into irreducible elements of $\mathcal{O}_K$ is unique.*
2. *If $|\mathscr{P}o(K)| \leq 2$, then all the factorizations of m into irreducible elements of $\mathcal{O}_K$ have the same length.*

Let us be precise: in 'a product of primes' the primes are not necessarily distinct, and 'the uniqueness of a factorization' in $\mathcal{O}_K$ is always up to units of $\mathcal{O}_K$ and up to the order of the elements in the product.

Proof Note first that, if the prime p is undecomposed in the extension $K/\mathbb{Q}$ and if $\mathfrak{p}$ denotes the unique prime ideal of $\mathcal{O}_K$ lying over p, then

$$p\mathcal{O}_K = \mathfrak{p}^e \ , \ [\mathcal{O}_K/\mathfrak{p} : \mathbb{F}_p] = f \ \text{ where } \ ef = [K : \mathbb{Q}] \, , \ \text{and } \mathfrak{p} = \Pi_{p^f}(K).$$

Now, let

$$m = p_1^{h_1} \dots p_k^{h_k}$$

where, for $i = 1, \dots, k$, the prime p_i lies under a unique maximal ideal $\mathfrak{p}_i$ of $\mathcal{O}_K$. Let

$$p_i\mathcal{O}_K = \mathfrak{p}_i^{e_i} \quad \text{with} \quad e_i \geq 1.$$

Then,

$$m\mathcal{O}_K = \mathfrak{p}_1^{h_1e_1} \cdots \mathfrak{p}_k^{h_ke_k}. \tag{2}$$

By hypothesis on m, the ideals $\mathfrak{p}_i$ are the ideals $\Pi_{p_i^{f_i}}(K)$.

1- Assume that $|\mathscr{P}o(K)| = 1$ (in this case, K is called a *Pólya field* [20]). Then, the ideals $\mathfrak{p}_i = \Pi_{p_i^{f_i}}(K)$ are principal and $\mathfrak{p}_i = \pi_i\mathcal{O}_K$ where π_i is an irreducible element of $\mathcal{O}_K$. Consequently, $p_i\mathcal{O}_K = \pi_i^{e_i}\mathcal{O}_K$, that is, $p_i = u_i\pi_i^{e_i}$ where u_i is a unit in $\mathcal{O}_K$. Finally,

$$m = u\pi_1^{h_1e_1} \cdots \pi_k^{h_ke_k} \text{ where } u \in \mathcal{O}_K^\times.$$

If π is an irreducible element of $\mathcal{O}_K$ which divides m, then

$$\pi\mathcal{O}_K = \prod_{j\in J} \mathfrak{p}_j^{\gamma_j} = \prod_{j\in J} \pi_j^{\gamma_j}\mathcal{O}_K \quad \text{where } J \subseteq \{1, \dots, k\} \text{ and } 1 \leq \gamma_j \leq h_je_j.$$

The irreducibility of π implies the existence of some index $j \in \{1, \dots, k\}$ such that $\pi\mathcal{O}_K = \mathfrak{p}_j = \pi_j\mathcal{O}_K$, that is, such that π and π_j are associated. One may easily conclude by iteration that the factorization of m is unique.

2- Assume that $|\mathscr{P}o(K)| \leq 2$. Then, for each i, either the ideal $\mathfrak{p}_i$ is principal, or the ideal $\mathfrak{p}_i^2$ is principal. Let π be an irreducible of $\mathcal{O}_K$ which divides m and consider the factorization of the ideal $\pi\mathcal{O}_K$ in a product of maximal ideals of $\mathcal{O}_K$. If in this factorization there is an ideal $\mathfrak{p}_i$ which is principal, then necessarily $\pi\mathcal{O}_K = \mathfrak{p}_i$. Otherwise, there are at least two maximal ideals (not necessarily distinct) $\mathfrak{p}_i$ and $\mathfrak{p}_j$ which are not principal and the hypothesis on $\mathscr{P}o(K)$ implies that $\mathfrak{p}_i\mathfrak{p}_j$ is principal, and hence, necessarily $\pi\mathcal{O}_K = \mathfrak{p}_i\mathfrak{p}_j$.

Finally, the number of irreducible elements which appear in the factorization of m may be computed in the following way: if ν denotes the number of principal ideals $\mathfrak{p}_i$ which appear in the right hand side of Eq. (2) taking into account their multiplicity and if μ denotes the number of nonprincipal ideals $\mathfrak{p}_i$ still taking into account their multiplicity, then the number of irreducible elements in a factorization of m is necessarily $\nu + \frac{1}{2}\mu$, which is a fixed integer for a given m.

The following examples show that we cannot admit decomposed primes in Theorem 2, neither when $|\mathscr{P}o(K)| = 1$, nor when $|\mathscr{P}o(K)| \leq 2$.

Example 1 Let $K = \mathbb{Q}(\sqrt{-31})$. We know that $|\mathscr{P}o(K))| = 1$ (see for instance [3, Corollary II.4.5]). On the other hand, $5\mathcal{O}_K = \mathfrak{p}\mathfrak{q}$ where $\mathfrak{p}$ and $\mathfrak{q}$ are not principal (there are no integers of $\mathcal{O}_K$ with norm 5). Consequently, 5 is irreducible in $\mathcal{O}_K$ and the order of the classes of $\mathfrak{p}$ and $\mathfrak{q}$ is 3 (the class number of K is 3). In other words, $\mathfrak{p}^3 = \pi\mathcal{O}_K$ and $\mathfrak{q}^3 = \pi'\mathcal{O}_K$ where π and π' are irreducible. Finally, we have

$$5^3\mathcal{O}_K = \pi\pi'\mathcal{O}_K$$

with 3 irreducible elements on the left side and 2 on the right side.

Example 2 Even in the cyclotomic case, one has to exclude the decomposed primes. For instance, let $K = \mathbb{Q}(\zeta_{39})$ where $\zeta_{39} = e^{2i\pi/39}$. Then, $\mathscr{P}o(K)$ is trivial as for every cyclotomic number field [20, Proposition 2.6]. Let us consider the factorization of 13 in $\mathcal{O}_K : e_{K/\mathbb{Q}}(13) = 12$ and $f_{K/\mathbb{Q}}(13) = 1$ since $13 \equiv 1 \pmod 3$, and hence,

$$13\mathcal{O}_K = (\mathfrak{q}\mathfrak{q}')^{12}.$$

We show now that the ideals $\mathfrak{q}$ and $\mathfrak{q}'$ are not principal by considering the containments $\mathbb{Q} \subset \mathbb{Q}(\sqrt{-39}) \subset K$. For instance, if $\mathfrak{q}$ were a principal ideal, then the ideal

$$N_{K/\mathbb{Q}(\sqrt{-39})}(\mathfrak{q}) = (\mathfrak{q} \cap \mathcal{O}_{\mathbb{Q}(\sqrt{-39})})^{f_{K/\mathbb{Q}(\sqrt{-39})}(\mathfrak{q})} = \mathfrak{q} \cap \mathcal{O}_{\mathbb{Q}(\sqrt{-39})},$$

which is the prime ideal of $\mathcal{O}_{\mathbb{Q}(\sqrt{-39})}$ lying over 13, would be principal, but it is not. On the other hand, $h_K = 2$, and hence, $\mathfrak{q}^2 = \pi\mathcal{O}_K$, $\mathfrak{q}'^2 = \pi'\mathcal{O}_K$, $\mathfrak{q}\mathfrak{q}' = \sigma\mathcal{O}_K$, and π, π', σ are irreducible elements of $\mathcal{O}_K$. The equality $(\mathfrak{q}\mathfrak{q}')^2 = \mathfrak{q}^2\mathfrak{q}'^2$ leads to two distinct factorizations $\sigma^2\mathcal{O}_K = \pi\pi'\mathcal{O}_K$.

3 Toward Reciprocal Assertions

Note that the uniqueness of the factorization (resp., the uniqueness of the length of the factorizations) of the products of undecomposed primes is equivalent to the uniqueness of the factorization (resp., the length of the factorizations) of the products of undecomposed primes which are (at least partially) ramified.

Indeed, an undecomposed prime p which is not ramified is totally inert, and hence, $p\mathcal{O}_K$ is a prime ideal, which means that p is not only an irreducible element of $\mathcal{O}_K$, but it is a prime element of $\mathcal{O}_K$. Consequently, if such an element p appears in some factorization of an integer m, necessarily it appears in all the factorizations of m.

3.1 Counterexamples

The converse of both implications in Theorem 2 are false as shown by the following examples of quadratic fields.

Example 3 The field $K = \mathbb{Q}(\sqrt{-5})$ is an example of a non-Pólya field whereas the factorizations are unique. The ramified primes are 2 and 5. Let $2\mathcal{O}_K = \mathfrak{p}^2$ and $5\mathcal{O}_K = \mathfrak{q}^2$. Then, $\mathfrak{q} = \sqrt{-5}\mathcal{O}_K$ while $\mathfrak{p}$ is not principal. Consequently, on the one hand 2 is irreducible in $\mathcal{O}_K$, on the other hand $\mathcal{P}o(K)$ is not trivial. Let us prove the uniqueness of the factorization of every product $m = p_1 \dots p_k$ of undecomposed primes. As previously said, we may assume for our proof that all the p_i's are ramified, that is, that the product is of the form $m = 2^a 5^b$. Clearly, m admits the unique factorization $2^a(\sqrt{-5})^{2b}$.

Example 4 The field $K = \mathbb{Q}(\sqrt{-21})$ is an example where $|\mathcal{P}o(K)| = 4$ while the factorizations have the same length. The ramified primes are 2, 3, and 7. Let

$$2\mathcal{O}_K = \mathfrak{p}^2\,,\ 3\mathcal{O}_K = \mathfrak{q}^2 \text{ and } 7\mathcal{O}_K = \mathfrak{r}^2. \tag{3}$$

The ideals $\mathfrak{p}$, $\mathfrak{q}$, and $\mathfrak{r}$ are not principal. Consequently, 2, 3, and 7 are irreducible elements of $\mathcal{O}_K$. Since the field K is not real, we know with Hilbert [12] that the relations between the classes of $\mathfrak{p}$, $\mathfrak{q}$, and $\mathfrak{r}$ are all given by relations (3) and by

$$\mathfrak{q}\,\mathfrak{r} = \sqrt{-21}\,\mathcal{O}_K. \tag{4}$$

The Pólya group of K which is generated by the classes of $\mathfrak{p}$, $\mathfrak{q}$, and $\mathfrak{r}$ is then of order 4. Let us prove that all the factorizations of every product $m = p_1 \dots p_k$ of undecomposed primes have the same length. We still assume that all the p_i's are ramified, and hence, that $m = 2^a 3^b 7^c$. Then, one has

$$m\mathcal{O}_K = 2^a 3^b 7^c \mathcal{O}_K = \mathfrak{p}^{2a}\mathfrak{q}^{2b}\mathfrak{r}^{2c}.$$

The only irreducibles which can divide m are $2, 3, 7$ and $\sqrt{-21}$, and hence, the factorizations of m into irreducible elements are of the form $m = u\, 2^{\alpha}\, 3^{\beta}\, 7^{\gamma}\, \sqrt{-21}^{\delta}$ where $u \in \mathcal{O}_K^{\times}$, $\alpha = a$, $2\beta + \delta = 2b$ and $2\gamma + \delta = 2c$. Consequently, $\alpha + \beta + \gamma + \delta = a + b + c$ and the lengths of all the factorizations of m are equal.

These examples show that the hypotheses $|\mathcal{P}o(K)| = 1$ and $|\mathcal{P}o(K)| \leq 2$ are too strong.

3.2 Nontrivial Relations in $\mathcal{P}O(K)$

The following notation will be used in the sequel.

Notation. Denote by $p_1, \ldots, p_t$ the prime numbers which are undecomposed and ramified in the extension $K/\mathbb{Q}$, and by $\mathfrak{p}_1, \ldots, \mathfrak{p}_t$ the corresponding prime ideals of $\mathcal{O}_K$ lying over these p_j. For $1 \leq j \leq t$, we have $p_j\mathcal{O}_K = \mathfrak{p}_j^{e_j}$ where $e_j = e_{K/\mathbb{Q}}(p_j)$ and $|\mathcal{O}_K/\mathfrak{p}_j| = p_j^{f_j}$ where $f_j = f_{K/\mathbb{Q}}(p_j)$. Clearly, $e_j \times f_j = [K : \mathbb{Q}]$.

Since $\mathfrak{p}_j = \Pi_{p_j^{f_j}}(K)$, we are interested in the relations between the classes $\overline{\mathfrak{p}}_j$ of the $\mathfrak{p}_j$'s in $\mathcal{P}o(K)$. Finally, denote by ε_j the order of $\overline{\mathfrak{p}}_j$. Clearly, ε_j divides e_j and $\mathfrak{p}_j^{\varepsilon_j} = \pi_j\mathcal{O}_K$ where π_j is an irreducible element of $\mathcal{O}_K$. The relation $\overline{\mathfrak{p}}_j^{\varepsilon_j} = 1$ in $\mathcal{P}o(K)$ will be said to be *trivial* and we introduce the following definition:

Definition 2 We say that there is a *non-trivial relation* in $\mathcal{P}o(K)$ between the classes $\overline{\mathfrak{p}}_j$ if there exists a sequence $\alpha_1, \ldots, \alpha_t$ of integers such that

$$\mathfrak{p}_1^{\alpha_1} \ldots \mathfrak{p}_t^{\alpha_t} = y\mathcal{O}_K \tag{5}$$

for some $y \in \mathcal{O}_K$, where $0 \leq \alpha_j < \varepsilon_j$, and where at least two α_j are nonzero. Moreover, we say that such a nontrivial relation is *minimal* if there is no other nontrivial relation with exponents β_j such that $0 \leq \beta_j \leq \alpha_j$ with $\beta_{j_0} < \alpha_{j_0}$ for at least one j_0.

Proposition 1 *The factorization into irreducible elements of every product of undecomposed primes is unique if and only if there is no nontrivial relation between the classes* $\overline{\mathfrak{p}}_j$.

Proof Assume that there exists a nontrivial relation of the form (5). Clearly, $\alpha_j \neq 0$ implies that the ideal $\mathfrak{p}_j$ is not principal, that is, $\varepsilon_j \neq 1$. Let us prove that $m = \prod_{j=1}^{t} p_j^{\frac{n}{e_j}\alpha_j}$ where $n = [K : \mathbb{Q}]$ admits two distinct factorizations. First,

$$m\mathcal{O}_K = \prod_{j=1}^{t} p_j^{\frac{n}{e_j}\alpha_j}\mathcal{O}_K = \prod_{j=1}^{t} \mathfrak{p}_j^{n\times\alpha_j} = y^n\mathcal{O}_K.$$

Using a factorization of y, we will obtain a factorization for m in product of irreducible elements whose exponents are nonzero multiples of n.

On the other hand, we have the equality

$$m\mathcal{O}_K = \prod_{j=1}^{t} (\mathfrak{p}_j^{\varepsilon_j})^{\frac{n}{\varepsilon_j}\alpha_j} = \left(\prod_{j=1}^{t} \pi_j^{\frac{n}{\varepsilon_j}\alpha_j}\right) \mathcal{O}_K.$$

Assume, for instance, that $\alpha_1 \neq 0$, and hence, that $1 \leq \alpha_1 < \varepsilon_1$. Then, we have another factorization of m where the exponent of π_1 is $< n$.

Conversely, assume that there is no nontrivial relation. Then, the only irreducible elements which can divide $m = p_1^{h_1} \ldots p_t^{h_t}$ are the π_j's. Thus, we have the uniqueness of the factorization of m.

Proposition 2 *The lengths of the factorizations into irreducible elements of every product of undecomposed primes are equal if and only if, for every minimal nontrivial relation of the form (5), we have*

$$\sum_{j=1}^{t} \frac{\alpha_j}{\varepsilon_j} = 1. \qquad (6)$$

Proof Assume that there exists a nontrivial relation of the form (5) and consider such a minimal relation. Then, $\mathfrak{p}_1^{\alpha_1}\mathfrak{p}_2^{\alpha_2} \ldots \mathfrak{p}_t^{\alpha_t} = y\mathcal{O}_K$ and the minimality of the relation implies the irreducibility of y. With the notation of the previous proof, we have

$$m\mathcal{O}_K = y^n \mathcal{O}_K = \left(\prod_{j=1}^{t} \pi_j^{\frac{n}{\varepsilon_j}\alpha_j}\right) \mathcal{O}_K.$$

The uniqueness of the length of the factorizations implies equality (6).

Conversely, assume that every minimal nontrivial relation of the form (5) satisfies equality (6). Let us consider these relations

$$\mathfrak{p}_1^{\alpha_{1,k}}\mathfrak{p}_2^{\alpha_{2,k}} \ldots \mathfrak{p}_t^{\alpha_{t,k}} = \sigma_k \mathcal{O}_K \quad (1 \leq k \leq s)$$

where the elements σ_k are irreducible in $\mathcal{O}_K$. Let $m = p_1^{h_1} \ldots p_t^{h_t}$. The only irreducible elements which can divide m are the π_j's $(1 \leq j \leq t)$ and the σ_k's $(1 \leq k \leq s)$. From

$$m\mathcal{O}_K = \prod_{j=1}^{t} \mathfrak{p}_j^{\beta_j} = \prod_{j=1}^{t} \pi_j^{\gamma_j} \times \prod_{k=1}^{s} \sigma_k^{\delta_k}\ \mathcal{O}_K,$$

we deduce:

$$\beta_j = h_j e_j = \varepsilon_j \gamma_j + \sum_{k=1}^{s} \alpha_{j,k}\delta_k \quad (1 \leq j \leq t).$$

Thus,

$$\sum_{j=1}^{t}\frac{\beta_j}{\varepsilon_j}=\sum_{j=1}^{t}h_j\frac{e_j}{\varepsilon_j}=\sum_{j=1}^{t}\gamma_j+\sum_{k=1}^{s}\left(\sum_{j=1}^{t}\frac{\alpha_{j,k}}{\varepsilon_j}\right)\delta_k=\sum_{j=1}^{t}\gamma_j+\sum_{k=1}^{s}\delta_k$$

which shows that the number of irreducible elements in the factorization, that is, $\sum_j \gamma_j + \sum_k \delta_k$ is a constant equal to $\sum_j h_j \frac{e_j}{\varepsilon_j}$ which depends only on m.

3.3 Factorizations in Monoids

While our aim was to emphasize on the group $\mathcal{P}o(K)$ and, in the spirit of Carlitz' theorem, to find links with factorizations of rational integers, the previous propositions show that we have the uniqueness of factorizations or of the lengths of the factorizations only by considering relations between the classes of the ramified primes which are not decomposed. As the classes of ramified primes which are decomposed may take part to the group $\mathcal{P}o(K)$, we understand that the sufficient conditions $|\mathcal{P}o(K)| = 1$ or $|\mathcal{P}o(K)| \leq 2$ may be not necessary for the uniqueness.

Let us consider for a while the question of the uniqueness from the point of view of the factorization theory in commutative monoids (see [8]). We said that the Pólya group is generated by the classes of the ideals $\Pi_q(K)$ (given by formula (1)). Let us consider the ideals $\Pi_q(K)$ themselves, they generate a free submonoid of the monoid of nonzero ideals of $\mathcal{O}_K$, and the undecomposed ramified primes $\mathfrak{p}_1, \ldots, \mathfrak{p}_t$ (which are some particular ideals $\Pi_q(K)$) generate a smaller free submonoid F:

$$F = \{\mathfrak{p}_1^{\beta_1} \cdots \mathfrak{p}_t^{\beta_t} \mid \beta_1, \ldots, \beta_t \in \mathbb{N}\}.$$

Now we introduce the following submonoid of the monoid $\mathcal{O}_K^* = \mathcal{O}_K \setminus \{0\}$:

$$H = \{\alpha \in \mathcal{O}_K^* \mid \alpha\mathcal{O}_K \in F\}$$

As H is divisor-closed $[\forall \alpha \in H\ \forall \beta \in \mathcal{O}_K\ (\beta|\alpha \Rightarrow \beta \in H)]$, the factorization of an element $\alpha \in H$ into irreducible elements of $\mathcal{O}_K$ is the same as the factorization into irreducible elements of H. Recall that the monoid H is said to be factorial if the factorization of every element of H into irreducible elements of H is unique up to the units. Then, we may formulate a stronger version of Proposition 1

Proposition 3 *The monoid H is factorial if and only if there is no nontrivial relation between the classes $\overline{\mathfrak{p}}_j$.*

The fact that the condition is necessary follows from Proposition 1, while the proof of the fact that the condition is sufficient is similar to those given in the proof of Proposition 1. We can made analogous remarks with respect to Proposition 2.

Recall that the monoid H is said to be half-factorial if the factorizations of every element of H into irreducible elements have the same length.

Proposition 4 *The monoid H is half-factorial if and only if relation (6) is satisfied by every minimal nontrivial relation of the form (5).*

Proof Let $G_0 = \{\overline{\mathfrak{p}}_1, \ldots, \overline{\mathfrak{p}}_t\} \subseteq \mathscr{C}l(K)$ and let $\mathscr{B}(G_0)$ be the block monoid of G_0, that is, the free abelian monoid formed by the sums $\beta_1\overline{\mathfrak{q}_1} + \cdots + \beta_r\overline{\mathfrak{q}_r}$ (where $\overline{\mathfrak{q}}_1, \ldots, \overline{\mathfrak{q}}_r$ denote the distinct elements of G_0) such that $\overline{\mathfrak{q}_1}^{\beta_1} \cdots \overline{\mathfrak{q}_r}^{\beta_r} = 1$. Clearly, the canonical homomorphism of monoids $H \to \mathscr{B}(G_0)$ is surjective; in fact it is a transfer homomorphism. Thus, H is half-factorial if and only if $\mathscr{B}(G_0)$ is half-factorial and, by Zacks-Skula theorem, $\mathscr{B}(G_0)$ is half-factorial if and only if every irreducible block in $\mathscr{B}(G_0)$ has cross-number 1 (see [8, Proposition 6.7.3]), this is just relation (6).

Putting together Propositions 1 and 3 on the one hand, and Propositions 2 and 4 on the other hand, we have:

Corollary 1 *Let K be a number field. The following assertions are equivalent:*

(i) For every rational integer m which is not a multiple a prime number decomposed in $\mathscr{O}_K$, the factorization (resp., the lengths of the factorizations) of m into irreducible elements of $\mathscr{O}_K$ is unique (resp., are equal).

(ii) For every algebraic integer α of $\mathscr{O}_K$ not contained in a prime ideal of K lying over a decomposed prime number, the factorization (resp., the lengths of the factorizations) of α into irreducible elements of $\mathscr{O}_K$ is unique (resp., are equal).

Proof In fact, this corollary is obvious. Let H_0 denote the submonoid of H formed by the rational integers which are product of undecomposed primes. The corollary says that H is factorial (resp., half-factorial) if and only if H_0 is factorial (resp., half-factorial). This is a clear consequence of the fact that $H_0 \subset H$ and, for each $\alpha \in H$, $\alpha^{[K:\mathbb{Q}]}$ is in H_0.

3.4 Tame Ramification

Back to classical algebraic number theory, we consider now a case where there does exist a nontrivial relation. Noticing that in both examples of Sect. 3.1, the prime 2 is ramified with ramification index 2, we may try to exclude this case by assuming that ramifications are tame, that is, no ramified prime divides one of its ramification indices. With such an hypothesis and assuming moreover that the extension $K/\mathbb{Q}$ is Galois, we know that the different δ_K of K is equal to

$$\delta_K = \prod_{\mathfrak{p}\in\mathrm{Max}(\mathscr{O}_K)} \mathfrak{p}^{e_{K/\mathbb{Q}}(\mathfrak{p})-1} = \prod_{j=1}^{w} \Pi_j^{e_j-1} = \prod_{j=1}^{w} p_j \times \prod_{j=1}^{w} \Pi_j^{-1} \tag{7}$$

where $p_1, \ldots, p_w$ denotes the ramified primes in the extension $K/\mathbb{Q}$ and $\Pi_1, \ldots, \Pi_w$ the corresponding ideals $\Pi_q(K)$, that is the products of the maximal ideals of $\mathcal{O}_K$ lying over p_j. As a consequence, we have

Proposition 5 *Let K be a Galois number field with tame ramifications. The ideal* $\prod_{j=1}^{w} \Pi_j$ *is principal if and only if the different* δ_K *is principal. This is the case, in particular, either if the* $\mathbb{Z}$*-algebra* $\mathcal{O}_K$ *is monogenic, or if the exponent of* $\mathcal{P}o(K)$ *is* ≤ 2.

Proof The fact that $\prod_{j=1}^{w} \Pi_j$ is principal if and only if δ_K is principal is an obvious consequence of (7). Assume first that the $\mathbb{Z}$-algebra $\mathcal{O}_K$ is monogenic, that is, that $\mathcal{O}_K$ is of the form $\mathbb{Z}[\alpha]$ for some $\alpha \in \mathcal{O}_K$. Then, the ideal δ_K is principal since $\delta_K = f'(\alpha)\,\mathcal{O}_K$ where f denotes the minimal polynomial of α over $\mathbb{Q}$.

Assume now that the exponent of $\mathcal{P}o(K)$ is ≤ 2. We know that the class of δ_K in the class group $\mathcal{C}l(K)$ is a square (see [19, Chap. XIII, Theorem 13]). As, by (7), the class of δ_k belongs to $\mathcal{P}o(K)$, we may conclude.

In order to be able to obtain links between the equivalences given by Propositions 1 and 2 and conditions on the Pólya group, we have to avoid the ramified primes which are decomposed. Thus, we restrict our study to Galois number fields K of prime degree.

4 Galois Number Fields of Prime Degree

From now on, we assume that K is a Galois number field of prime degree l. Then, every prime p is either totally ramified, or totally inert, or totally decomposed. Consequently, if p is ramified, $p\mathcal{O}_K = \mathfrak{p}^l$ and $\Pi_p(K) = \mathfrak{p}$ is maximal; if p is decomposed, $p\mathcal{O}_K = \mathfrak{p}_1 \ldots \mathfrak{p}_l = \Pi_p(K)$ and $\Pi_p(K)$ is principal; and if p is inert, $p\mathcal{O}_K = \mathfrak{p}$, and $\Pi_{p^l}(K)$ is both maximal and principal. As we do not want to consider decomposed primes p, we only have to consider ideals $\Pi_q(K)$ which are maximal. Moreover, if p is inert, p is a prime element of $\mathcal{O}_K$, thus it cannot lead to distinct factorizations of products of undecomposed primes. Thus, we have

Lemma 1 *If K is a Galois number field of prime degree l, the following assertions are equivalent:*

(i) For every rational integer which is a product of undecomposed primes, the factorization (resp., the length of the factorizations) into irreducible elements of $\mathcal{O}_K$ *is unique.*
(ii) For every rational integer whose radical divides the discriminant d_K *of K, the factorization (resp., the length of the factorizations) into irreducible elements of* $\mathcal{O}_K$ *is unique.*

About the Pólya group, we have the following:

Proposition 6 *Let K be a Galois number field of prime degree l. Then,*

$$|\mathscr{P}o(K)| = \begin{cases} 2^{t-2} \text{ if } l = 2\,, K \subset \mathbb{R}\,, N_{K/\mathbb{Q}}(\mathcal{O}_K^\times) = \{+1\} \\ l^{t-1} \text{ otherwise} \end{cases} \tag{8}$$

where t denotes the number of ramified primes.

Proof Recall that, in a cyclic extension $K/\mathbb{Q}$ of degree n where there are t ramified primes $p_1, \ldots, p_t$ with ramification indices $e_1, \ldots, e_t$, the order of the Pólya group satisfies

$$|\mathscr{P}o(K)| = \frac{\prod_{i=1}^t e_i}{n} \text{ or } \frac{\prod_{i=1}^t e_i}{2n} \quad \text{(cf. [5, Corollary 3.11])}$$

and the second equality occurs exactly when K is real and $N_{K/\mathbb{Q}}(\mathcal{O}_K^\times) = \{+1\}$. Here, we may conclude since the ramification indices are necessarily equal to l.

We denote by $p_1, \ldots, p_t$ the primes which are ramified in the extension $K/\mathbb{Q}$ and by $\mathfrak{p}_1, \ldots, \mathfrak{p}_t$ the corresponding prime ideals of $\mathcal{O}_K$. Clearly, $p_j\mathcal{O}_K = \mathfrak{p}_j^l$ for $1 \le j \le t$. The following morphism is well defined and surjective:

$$\varphi : (\overline{k_1}, \ldots, \overline{k_t}) \in (\mathbb{Z}/l\mathbb{Z})^t \mapsto \overline{\mathfrak{p}}_1^{k_1} \cdots \overline{\mathfrak{p}}_t^{k_t} \in \mathscr{P}o(K). \tag{9}$$

If $l \neq 2$, it follows from Proposition 6 that $\mathrm{Ker}(\varphi)$ is of order l. Consequently,

Corollary 2 *If K is a Galois number field of odd prime degree l, then one and only one of the following assertions holds: either the kernel of the morphism φ defined in (9) is generated by one class $\overline{\mathfrak{p}}_j$, that is, $\mathfrak{p}_j$ is principal (and this is the only ramified prime ideal which is principal), or Ker(φ) is generated by a nontrivial relation.*

Proposition 7 *Let K be a Galois number field of odd prime degree l. The following assertions are equivalent:*

(i) Every rational integer which is a product of undecomposed primes admits a unique factorization into irreducible elements of $\mathcal{O}_K$.
(ii) There is a ramified prime ideal of $\mathcal{O}_K$ which is principal.

Proof By Proposition 1, assertion (i) is equivalent to the nonexistence of nontrivial relation between the $\overline{\mathfrak{p}}_j$ and, by Corollary 2, this nonexistence is equivalent to the existence of a principal ramified prime ideal.

Corollary 3 *Let K be a Galois number field of odd prime degree l. Assume that the prime l is not ramified in K and that the different δ_K is a principal ideal. Then, the following assertions are equivalent:*

(i) Every product of undecomposed primes admits a unique factorization into irreducible elements of $\mathcal{O}_K$.
(ii) $|\mathscr{P}o(K)| = 1$.

Here the fact that $|\mathcal{P}o(K)| = 1$ is equivalent to the fact that there is only one ramified prime.

Proof (*i*) $\Rightarrow$ (*ii*): The ramifications are tame since we assume that l is not ramified, then, by Proposition 5, the ideal $\mathfrak{p}_1 \dots \mathfrak{p}_t$ is principal. By Proposition 1, if (*i*) holds, this relation is trivial, that is, all the ramified prime ideals $\mathfrak{p}_j$ are principal. (In fact, by Corollary 2, there is exactly one ramified prime.)

(*ii*) $\Rightarrow$ (*i*) follows from Theorem 2.

Example 5 Following [6, Theorem 6.4.6], the field $K = \mathbb{Q}(\theta)$ where θ is a root of the equation

$$X^3 - 57X + 19 = 0$$

is a cyclic cubic field where the ramified primes are 3 and 19. Clearly, θ is a generator of the prime ideal $\mathfrak{p}$ lying over 19. This is an example where we have the uniqueness of the factorizations (cf. Proposition 7) while K is not a Pólya field (cf. Proposition 6).

Proposition 8 *Let K be a Galois number field of odd prime degree l. The following assertions are equivalent:*

(*i*) *All the factorizations into irreducible elements of $\mathcal{O}_K$ of a rational integer which is a product of undecomposed primes have the same lengths.*
(*ii*) *Either there is a ramified prime ideal which is principal, or there is a nontrivial relation between the classes of ramified prime ideals of the form $\overline{\mathfrak{p}}_1^{\alpha_1} \dots \overline{\mathfrak{p}}_t^{\alpha_t} = 1$ with $\alpha_j \geq 0$ where $\sum_{j=1}^t \alpha_j = l$.*

Proof By Corollary 2, either there is a ramified prime ideal which is principal, or there is a nontrivial relation between the classes of ramified prime ideals. In this latter case, by Proposition 2, if (*i*) holds, such a minimal nontrivial relation satisfies $\sum_j \frac{\alpha_j}{\varepsilon_j} = 1$, which means here $\sum_j \alpha_j = l$.

Conversely, assume that (*ii*) holds. Taking into account Proposition 7, we may assume that the ideals $\mathfrak{p}_i$ are not principal, and hence, that there is a relation $\mathfrak{p}_1^{\alpha_1} \dots \mathfrak{p}_t^{\alpha_t} = w\mathcal{O}_K$ with $\alpha_j \geq 0$ where $\sum_{j=1}^t \alpha_j = l$. Clearly, this nontrivial relation is minimal and, by Proposition 2, (*i*) holds.

Corollary 4 *Let K be a Galois number field of odd prime degree l. Assume that l is not ramified and that the different δ_K is principal. The following assertions are equivalent:*

(*i*) *All the factorizations into irreducible elements of $\mathcal{O}_K$ of a rational integer which is a product of undecomposed primes have the same lengths.*
(*ii*) *$|\mathcal{P}o(K)| = 1$ or l^{l-1} or, equivalently, there are 1 or l ramified primes in K.*

Proof This is an obvious consequence of Propositions 5 and 8.

Unfortunately, following [11], there are very few number fields K of prime degree l such that $\mathcal{O}_K$ is monogenic. In particular, the only cyclic number fields of prime degree $l \geq 5$ are real subfields of cyclotomic fields. More precisely, if l is a Sophie

Germain's prime, that is, if l and $2l+1$ are primes, the real subfield $\mathbb{Q}(\cos\frac{2\pi}{2l+1})$ of the cyclotomic field $\mathbb{Q}(e^{\frac{2i\pi}{2l+1}})$ is of degree l and its ring of integers $\mathbb{Z}[\cos\frac{2\pi}{2l+1}]$ is monogenic. We know that in this case $|\mathscr{P}o(\mathbb{Q}(\cos\frac{2\pi}{2l+1}))| = 1$ [20, Proposition 2.6].

On the other hand, there exist infinite families of cyclic cubic number fields whose ring of integers is monogenic (see [10]) and, of course, the ring of integers of every quadratic number field is monogenic.

5 Quadratic Number Fields

Let $K = \mathbb{Q}(\sqrt{d})$ be a quadratic field where d is a square-free integer. What about the converses of the implications in Theorem 2?

Let $p_1, \ldots, p_s$ be the prime numbers which divide d. The ramified primes are $p_1, \ldots, p_s$, and 2 in the case where $d \equiv 3 \pmod 4$. From $d = \pm p_1 \ldots p_s$, we have $\sqrt{d}\mathscr{O}_K = \mathfrak{p}_1 \ldots \mathfrak{p}_s$, which is a nontrivial relation between the $\overline{\mathfrak{p}}_j$'s if and only if there are nonprincipal prime ideals dividing $d\mathscr{O}_K$. This leads us to introduce the following notation:

Notation. In this section, $\mathscr{P}o^*(K)$ denotes the subgroup of $\mathscr{P}o(K)$ generated by the classes of the $\mathfrak{p}_j$'s which divide d.

Theorem 3 *Let $K = \mathbb{Q}(\sqrt{d})$ be a quadratic field where d is a square-free integer. The following assertions are equivalent:*

(i) Every product of undecomposed primes admits a unique factorization into irreducible elements of $\mathscr{O}_K$.
(ii) $\mathscr{O}_K$ has at most one ramified prime ideal which is not principal.
(iii) $|\mathscr{P}o^*(K)| = 1$.

Proof Assume that (*i*) holds. Then, by Proposition 1, there is no nontrivial relation. Consequently, the relation $\sqrt{d}\mathscr{O}_K = \mathfrak{p}_1 \ldots \mathfrak{p}_s$ implies that all the prime ideals $\mathfrak{p}_1, \ldots, \mathfrak{p}_s$ are principal, that is, $|\mathscr{P}o^*(K)| = 1$. Thus, (*i*) implies (*iii*).

Clearly, (*iii*) implies (*ii*) since all the ramified primes divide d except when $d \equiv 3 \pmod 4$: 2 is ramified and the corresponding prime ideal may be nonprincipal.

Finally, assume that (*ii*) holds. Then, all the prime ideals $\mathfrak{p}_j$ dividing d are principal: $\mathfrak{p}_j = \pi_j\mathscr{O}_K$ where π_j is a prime element in $\mathscr{O}_K$. If $d \equiv 3 \pmod 4$, 2 is ramified and the corresponding prime ideal may be nonprincipal, in this case 2 is an irreducible element of $\mathscr{O}_K$. Thus, if m denotes an integer whose radical divides the discriminant d_K of K ($d_K = d$ or $4d$), then all the irreducible elements of $\mathscr{O}_K$ dividing m are prime elements except in the case where 2 is irreducible. Consequently, (*i*) holds.

Note that the field $\mathbb{Q}(\sqrt{-5})$ studied in Counterexample 3 corresponds to this case where 2 is irreducible in $\mathscr{O}_K$.

Theorem 4 *Let $K = \mathbb{Q}(\sqrt{d})$ be a quadratic field where d is a square-free integer. The following assertions are equivalent:*

(i) *All the factorizations into irreducible elements of $\mathcal{O}_K$ of any product of undecomposed primes have the same lengths.*
(ii) *Either* $|\mathcal{P}o^*(K)| \leq 2$, *or* $|\mathcal{P}o^*(K)| = 4$ *and there is a product of two ramified prime ideals which is a principal ideal.*

Proof We first recall Formula (8) in the case of quadratic number fields:

$$|\mathcal{P}o(K)| = \begin{cases} 2^{t-2} \text{ if } K \subset \mathbb{R} \text{ and } N_{K/\mathbb{Q}}(\mathcal{O}_K^\times) = \{+1\} \\ 2^{t-1} \text{ otherwise} \end{cases}$$

where t denotes the number of ramified primes (see [12, Sect. 73] or [3, Sect. II.4]). Recall also that we have

$$d = \pm p_1 \dots p_s \text{ with } p_i \mathcal{O}_K = \mathfrak{p}_i^2.$$

First case: $|\mathcal{P}o(K)| = 2^{t-1}$
The relations between the classes of the $\mathfrak{p}_i$'s are all deduced from (see also [12, Sect. 73]):

$$\mathfrak{p}_i^2 = p_i \mathcal{O}_K \ (1 \leq i \leq t) \text{ and } \mathfrak{p}_1 \dots \mathfrak{p}_s = \sqrt{d}\mathcal{O}_K.$$

By Proposition 2, assertion (*i*) means that either there is no nontrivial relation between the $\mathfrak{p}_j$'s, that is $s \leq 1$, or every minimal nontrivial relation satisfies (6), that is here, $s = 2$ (since $\alpha_j = 1$ and $\varepsilon_j = 2$). Finally, (*i*) $\Leftrightarrow s \leq 2 \Leftrightarrow |\mathcal{P}o^*(K)| \leq 2$.

Second case: $|\mathcal{P}o(K)| = 2^{t-2}$
There is another fundamental relation between the classes of the $\mathfrak{p}_j$'s $(1 \leq j \leq t)$.

The first subcase. The prime 2 does not divide d, but is ramified and the prime ideal lying over 2 is principal. Then, the relations between the $\mathfrak{p}_j$ $(1 \leq j \leq s)$ are as in the first case and, analogously, we may conclude (*i*) $\Leftrightarrow s \leq 2 \Leftrightarrow |\mathcal{P}o^*(K)| \leq 2$.

The other subcase. The other relation is then between the prime ideals which divide d. Thus, by renumbering the p_i's, it may be written (see [12, Sect. 73]):

$$\alpha\mathcal{O}_K = \mathfrak{p}_1 \dots \mathfrak{p}_r \text{ with } 1 \leq r \leq \frac{s}{2}.$$

Then, the relations between the classes of the $\mathfrak{p}_i$'s are all deduced from

$$\mathfrak{p}_i^2 = p_i \mathcal{O}_K \ (1 \leq i \leq t)\ ,\ \alpha\mathcal{O}_K = \mathfrak{p}_1 \dots \mathfrak{p}_r \text{ and } \beta\mathcal{O}_K = \mathfrak{p}_{r+1} \dots \mathfrak{p}_s.$$

By Proposition 2, assertion (*i*) means that either there is no nontrivial relation or each minimal nontrivial relation is a product of two prime ideals, equivalently, either $s \leq 3$, or $s = 4$ and $r = 2$. These latter assertions mean that either $|\mathcal{P}o^*(K)| = 2$, or $|\mathcal{P}o^*(K)| = 4$ and there is a product of two prime ideals which is principal.

Finally, we have proved that (*i*) implies (*ii*). To be sure that (*ii*) implies (*i*), it remains to see that the assertion '$|\mathcal{P}o^*(K)| = 4$ and there is a product of two primes which is principal' may only occur in the second subcase. Indeed, if we are not in the second subcase, $|\mathcal{P}o^*(K)| \leq 4$ implies $s \leq 3$. If $s = 3$, the fact that there is a product of two prime ideals which is principal implies that the third prime ideal dividing d is principal, which itself implies that $|\mathcal{P}o^*(K)| \leq 2$. Finally, $s \leq 2$ and $|\mathcal{P}o^*(K)| \leq 2$.

Note that, for the field $\mathbb{Q}(\sqrt{-21})$ studied in Example 4, we have $|\mathcal{P}o^*(K)| = 2$ while $|\mathcal{P}o(K)| = 4$. The following example shows that we may have $|\mathcal{P}o^*(K)| = 4$ with a product of two prime ideals which is principal, while $|\mathcal{P}o(K)| = 8$.

Example 6 Let $K = \mathbb{Q}(\sqrt{3 \times 7 \times 17 \times 79})$. Since $28203 \equiv 3 \pmod 4$, one has $\mathcal{O}_K = \mathbb{Z}[\sqrt{28203}]$. The group $\mathcal{P}o^*(K)$ is generated by the classes of ideals $\mathfrak{P}_3$, $\mathfrak{P}_7$, $\mathfrak{P}_{17}$ and $\mathfrak{P}_{79}$ where $\mathfrak{P}_p$ denotes the prime ideal of $\mathcal{O}_K$ above the prime p. As ± 3 and ± 79 are not quadratic residues modulo 17, the ideals $\mathfrak{P}_3$ and $\mathfrak{P}_{79}$ are not principal. The equality $168^2 - 28203 \times 1^2 = 21$ implies that $\mathfrak{P}_3\mathfrak{P}_7 = (168 + \sqrt{28203})\mathcal{O}_K$ is principal. From the equality $\sqrt{28203}\mathcal{O}_K = \mathfrak{P}_3\mathfrak{P}_7\mathfrak{P}_{17}\mathfrak{P}_{79}$, one deduces that $\mathfrak{P}_{17}\mathfrak{P}_{79}$ is principal. Finally, $\mathfrak{P}_3\mathfrak{P}_{17}$ is not principal because the equality $x^2 - 28203y^2 = 51$ is impossible (modulo 4), while the equality $x^2 - 28203y^2 = -51$ is impossible (modulo 7). Then we may conclude that

$$\mathcal{P}o^*(K) = \{\overline{\mathcal{O}_K}, \overline{\mathfrak{P}_3}, \overline{\mathfrak{P}_{17}}, \overline{\mathfrak{P}_3\mathfrak{P}_{17}}\}$$

is of order 4. Moreover, since -1 is not a square modulo 3, the norm of the fundamental unit of K is 1 and, as 2 is ramified, Formula (8) gives $|\mathcal{P}o(K)| = 8$.

6 A Few Words About the Function Fields Case

Let q be a power of a prime p and $K/\mathbb{F}_q(T)$ be a finite extension of function fields. Denote the integral closure of $\mathbb{F}_q[T]$ in K by by $\mathcal{O}_K$. Analogously to Definition 1, one defines the Pólya group of $\mathcal{O}_K$

Definition 3 The *Pólya group* of $\mathcal{O}_K$ is the subgroup $\mathcal{P}o(\mathcal{O}_K)$ of the class group $\mathcal{C}l(\mathcal{O}_K)$ of $\mathcal{O}_K$ generated by the classes of the ideals $\Pi_{q^r}(\mathcal{O}_K)$ defined by

$$\Pi_{q^r}(\mathcal{O}_K) = \prod_{\substack{\mathfrak{m}\in \mathrm{Max}(\mathcal{O}_K)\\ N(\mathfrak{m})=q^r}} \mathfrak{m}.$$

The following proposition shows that the naive function field analog of Theorem 2 does not hold.

Proposition 9 *Assume that q is odd and let $\beta \in \mathbb{F}_q \setminus \mathbb{F}_q^2$.*

(1) *Let $K := \mathbb{F}_q(T)[y]$ where $y^2 = \beta T(T+1)$. Then $|\mathcal{P}o(\mathcal{O}_K)| = 1$, while $T(T+1)$ admits two distinct factorizations into irreducible elements of $\mathcal{O}_K$.*
(2) *Let $K := \mathbb{F}_q(T)[y]$ where $y^2 = \beta T(T+1)Q(T)$ and $Q(T) \in \mathbb{F}_q[T]$ is a monic irreducible polynomial of degree 2. Then $|\mathcal{P}o(\mathcal{O}_K)| = 2$, while $T(T+1)Q(T)$ admits two factorizations into irreducible elements of $\mathcal{O}_K$ with different lengths.*

Proof In both cases, the extension $K/\mathbb{F}_q(T)$ is an imaginary extension. As a consequence $\mathcal{O}_K^\times = \mathbb{F}_q^*$ (see [17]).
(1) The fact that $|\mathcal{P}o(K)| = 1$ is a consequence of [1, Theorem 12]. It follows from [18, Proposition VI.3.1] that the ramified prime ideals of $\mathcal{O}_K$ are the ideals $\mathfrak{p}_T$ and $\mathfrak{p}_{T+1}$ lying over T and $T+1$ respectively. Thus, $y\mathcal{O}_K = \mathfrak{p}_T\mathfrak{p}_{T+1}$. The ideal $\mathfrak{p}_T$ is not principal. Indeed, assume that $\mathfrak{p}_T = \alpha\mathcal{O}_K$ with $\alpha = A + yB$ $(A, B \in \mathbb{F}_q[T])$. This implies that $A^2 - \beta T(T+1)B^2 = vT$ where $v \in \mathbb{F}_q^*$, that is $A^2 = T(v + \beta(T+1)B^2)$. Obviously, $B = 0$ is impossible. The comparison of the leading coefficients of both sides leads to a contradiction since $\beta \notin \mathbb{F}_q^2$. In the same way, one could show that $\mathfrak{p}_{T+1}$ is not principal. Consequently y, T and $T+1$ are irreducible elements of $\mathcal{O}_K$, and $y^2 = \beta T(T+1)$ are two different factorizations into irreducible elements of $\mathcal{O}_K$.
(2) Analogously, the ramified prime ideals of $\mathcal{O}_K$ are the ideals $\mathfrak{p}_T$, $\mathfrak{p}_{T+1}$, and $\mathfrak{p}_Q$ lying over T, $T+1$, and $Q(T)$ respectively. Clearly, $\mathcal{P}o(\mathcal{O}_K)$ is generated by the classes of $\mathfrak{p}_T\mathfrak{p}_{T+1}$ and of $\mathfrak{p}_Q$. From the equalities

$$T(T+1)\mathcal{O}_K = \mathfrak{p}_T^2\mathfrak{p}_{T+1}^2, \quad y\mathcal{O}_K = (\mathfrak{p}_T\mathfrak{p}_{T+1})\mathfrak{p}_Q, \quad Q\mathcal{O}_K = \mathfrak{p}_Q^2,$$

one deduces that $\mathcal{P}o(\mathcal{O}_K) = \{[\mathcal{O}_K], [\mathfrak{p}_Q]\}$. As in (1), one proves that the six ideals $\mathfrak{p}_T$, $\mathfrak{p}_{T+1}$, $\mathfrak{p}_Q$, $\mathfrak{p}_T\mathfrak{p}_{T+1}$, $\mathfrak{p}_T\mathfrak{p}_Q$, and $\mathfrak{p}_{T+1}\mathfrak{p}_Q$ are not principal. Consequently, $T+1$, T, Q, and y are irreducible elements of $\mathcal{O}_K$. The equality

$$y^2 = \beta T(T+1)Q(T)$$

corresponds to two factorizations with different lengths.

Nevertheless, the introduction of Sect. 3 and the whole Sect. 3.2 are still true when we replace 'prime number' by 'irreducible polynomial' (in $\mathbb{F}_q[T]$). In particular, Propositions 1 and 2 still hold for any extension $K/\mathbb{F}_q(T)$. But, to go further and in order to retrieve in the function fields case other results analogous to those of the zero characteristic, we are led to consider the group of classes of ambiguous ideals instead of the Pólya group.

Definition 4 Let $K/\mathbb{F}_q(T)$ be a Galois extension with Galois group G.

1. An ideal I of $\mathcal{O}_K$ is said to be *ambiguous* if for every $\sigma \in G$, $\sigma(I) = I$.
2. A class $\mathcal{C}$ of $\mathcal{C}l(\mathcal{O}_K)$ is said to be *ambiguous* if, for every $\sigma \in G$, one has $\sigma(\mathcal{C}) = \mathcal{C}$, that is, for every ideal $I \in \mathcal{C}$, one has $\sigma(I) \in \mathcal{C}$.
3. A class $\mathcal{C}$ of $\mathcal{C}l(\mathcal{O}_K)$ is said to be *strongly ambiguous* if $\mathcal{C}$ contains an ambiguous ideal I.

One denotes by $\mathcal{A}m_{str}(K)$ the subgroup of $\mathcal{C}l(\mathcal{O}_K)$ formed by the strongly ambiguous classes.

Remark 1 1. Clearly, a strongly ambiguous class is an ambiguous class, but the converse does not hold: [21, Theorem 2] shows that in the class group of the field $\mathbb{F}_3(T)[y]$ with $y^2 = -(T^2+1)(T^2+2T+2)$ there exists a class that is ambiguous but not strongly ambiguous.

2. When the extension of function fields $K/\mathbb{F}_q(T)$ is Galois, the group $\mathscr{A}m_{str}(K)$ is generated by the classes of the following ideals:

$$\prod_{\substack{\mathfrak{p}\in \mathrm{Max}(\mathcal{O}_K)\\ \mathfrak{p}|P}} \mathfrak{p} \qquad (P \in \mathbb{F}_q[T] \text{ ramified in } K).$$

Then, we have the containments

$$\mathscr{P}o(K) \subseteq \mathscr{A}m_{str}(K) \subseteq \mathscr{C}l(K)$$

which may be strict, while for a Galois number field K we have

$$\mathscr{P}o(K) = \mathscr{A}m_{str}(K) \subseteq \mathscr{C}l(K).$$

Thus, from now on, we assume that the extension of function fields $K/\mathbb{F}_q(T)$ is Galois with Galois group G. Since the proofs follow closely those of the characteristic zero case, we will sketch them only. Here is an analog of Theorem 2.

Theorem 5 *Let $K/\mathbb{F}_q(T)$ be a Galois extension of function fields. Let m be a product of irreducible polynomials of $\mathbb{F}_q[T]$ which are undecomposed in the extension.*

(1) *If $|\mathscr{A}m_{str}(K)| = 1$, the factorization of m into irreducible elements of $\mathcal{O}_K$ is unique.*
(2) *If $|\mathscr{A}m_{str}(K)| \leq 2$, all the factorizations of m into irreducible elements of $\mathcal{O}_K$ have the same length.*

Proof If $P \in \mathbb{F}_q[T]$ is an irreducible polynomial which is undecomposed in the extension, then there exists only one maximal ideal $\mathfrak{p}$ of $\mathcal{O}_K$ lying over P, and hence, for every $\sigma \in \mathrm{Gal}(K/\mathbb{F}_q(T))$, one has $\mathfrak{p}^\sigma = \mathfrak{p}$. The proof ends as in Theorem 2.

Now, we prove the converses of Theorem 5 for quadratic separable extensions $K/\mathbb{F}_q(T)$. In this case, the group of classes of ambiguous ideals is generated by the ramified primes of $\mathcal{O}_K$. Recall that a quadratic extension of function fields $K/\mathbb{F}_q(T)$ is said to be *real* if the infinite place $\left(\frac{1}{T}\right)$ of $\mathbb{F}_q(T)$ is split in K. Recall also

Proposition 10 *Let $K/\mathbb{F}_q(T)$ be a quadratic extension. If t denotes the number of ramified primes in the extension, then one has*

$$|\mathscr{A}m_{str}(K)| = \begin{cases} q \text{ odd} \begin{cases} 2^{t-2} \text{ if } K \text{ real and } N_{K/\mathbb{F}_q(T)}(\mathcal{O}_K^\times) = \mathbb{F}_q^{*2} \\ 2^{t-1} \text{ otherwise} \end{cases} & \text{(see [21])} \\ q \text{ even} \begin{cases} 2^{t-1} \text{ if } K \text{ real} \\ 2^{t} \quad \text{otherwise.} \end{cases} & \text{(see [13])} \end{cases}$$

Theorem 3 about the uniqueness of the factorizations translates into the two following theorems.

Theorem 6 *If q is odd and if $K/\mathbb{F}_q(T)$ is a quadratic extension, then the following assertions are equivalent:*

(i) Every product of irreducible polynomials of $\mathbb{F}_q[T]$ which are undecomposed in the extension admits a unique factorization into irreducible elements of $\mathcal{O}_K$.
(ii) All the ramified prime ideals of $\mathcal{O}_K$ are principal.
(iii) $|\mathscr{A}m_{str}(K)| = 1$.

Proof One can write $K := \mathbb{F}_q(T)[y]$ with $y^2 = D(T)$ where $D(T) \in \mathbb{F}_q[T]$ is squarefree. Assume that D owns $t \geq 2$ primes divisors $P_1, \ldots, P_t$ in $\mathbb{F}_q[T]$. The following equality holds:

$$\sqrt{D}\mathcal{O}_K = \mathfrak{p}_1 \cdots \mathfrak{p}_t, \tag{10}$$

where each $\mathfrak{p}_i \in \mathrm{Max}(\mathcal{O}_K)$ divides P_i. Analog of Proposition 1 proves that (i) implies (ii). Clearly, $(ii) \Leftrightarrow (iii)$, and $(iii) \Rightarrow (i)$ follows from Theorem 5.

Theorem 7 *If q is even and $K/\mathbb{F}_q(T)$ is a quadratic separable extension, then the following assertions are equivalent:*

(i) Every product of irreducible polynomials of $\mathbb{F}_q[T]$ which are undecomposed in the extension admits a unique factorization into irreducible elements of $\mathcal{O}_K$.
(ii) Denoting by t the number of ramified primes, either $|\mathscr{A}m_{str}(K)| = 2^t$, or $|\mathscr{A}m_{str}(K)| = 2^{t-1}$ and $\mathcal{O}_K$ has a principal ramified prime ideal.

Proof By Proposition 10, the equality $|\mathscr{A}m_{str}(K)| = 2^t$ is obviously equivalent to the nonexistence of trivial relations in $\mathcal{O}_K$. On the other hand, the equality $|\mathscr{A}m_{str}(K)| = 2^{t-1}$ holds if and only if there exists a nontrivial relation in $\mathcal{O}_K$ or one ramified prime ideal of $\mathcal{O}_K$ is principal.

Remark 2 Both cases may occur
(1) The field $\mathbb{F}_2(T)[y]$ with $y^2 + y = \frac{T^3+T^2+1}{T(T+1)}$ is an imaginary function field (see [13]). The ramified irreducible polynomials of $\mathbb{F}_2[T]$ are T and $T+1$ (see [9, Chap. III]). Clearly,

$$\mathscr{A}m_{str}(K) = \{1, [\mathfrak{p}_T], [\mathfrak{p}_{T+1}], [\mathfrak{p}_T\mathfrak{p}_{T+1}]\},$$

where $\mathfrak{p}_T$ and $\mathfrak{p}_{T+1}$ are the primes ideals of $\mathcal{O}_K$ above T and $T+1$.
(2) The field $\mathbb{F}_2(T)[y]$ with $y^2 + (T+1)^2y = T(T+1)$ is a real function field. There is a ramified prime ideal in $\mathcal{O}_K$, the ideal lying over $T+1$. Moreover $\mathscr{C}l(\mathcal{O}_K)$ is trivial (see [14]).

The uniqueness of the length of the factorizations is characterized by the following theorems:

Theorem 8 *If q is odd and $K/\mathbb{F}_q(T)$ is a quadratic extension, the following assertions are equivalent:*

(i) All the factorizations into irreducible elements of $\mathcal{O}_K$ of any product of undecomposed primes of $\mathbb{F}_q[T]$ have the same lengths.
(ii) Either $|\mathcal{A}m_{str}(K)| \leq 2$, or $|\mathcal{A}m_{str}(K)| = 4$ and there is a product of two ramified prime ideals which is a principal ideal.

Proof Write $K = \mathbb{F}_q(T)[y]$ with $y^2 = D(T)$ where $D \in \mathbb{F}_q[T]$ is squarefree with prime factorization $D = P_1 \cdots P_t$. Adapting the proof of Theorem 4, Proposition 2, and Equality (10) lead to the result.

Theorem 9 *If q is even and $K/\mathbb{F}_q(T)$ is a quadratic separable extension, the following assertions are equivalent:*

(i) All the factorizations into irreducible elements of $\mathcal{O}_K$ of any product of undecomposed primes of $\mathbb{F}_q[T]$ have the same lengths.
(ii) Denoting by t the number of ramified primes ideals of $\mathcal{O}_K$, either $|\mathcal{A}m_{str}(K)| = 2^t$, or $|\mathcal{A}m_{str}(K)| = 2^{t-1}$ and there is a principal ramified prime ideal of $\mathcal{O}_K$ or a product of two ramified prime ideals of $\mathcal{O}_K$ which is a principal ideal.

Proof By Theorem 7, one can assume that $|\mathcal{A}m_{str}(K)| = 2^{t-1}$ and there is no ramified principal prime ideal of $\mathcal{O}_K$. Since all the orders of the classes in $\mathcal{A}m_{str}(K)$ of the ramified prime ideals are equal to 2, there is a relation between the ramified prime ideals $\mathfrak{p}_i$ $(1 \leq i \leq t)$ of $\mathcal{O}_K$ which can be written as

$$\prod_{i=1}^{t} [\mathfrak{p}_i]^{\alpha_i} = 1 \qquad (\alpha_i \in \{0, 1\}),$$

with at least two nonzero α_i's. By Proposition 4, if we consider such a minimal nontrivial relation, assertion (i) holds if and only if there are exactly two nonzero α_i's.

Remark 3 Here is an example where $|\mathcal{A}m_{str}(K)| = 2^{t-1}$ and there is a product of two ramified prime ideals which is a principal ideal. Let $K := \mathbb{F}_2(T)[y]$ with $y^2 + y = \frac{1}{T(T+1)}$ (K is an elliptic field following [9]). The ramified prime ideals of $\mathcal{O}_K$ are the primes ideals $\mathfrak{p}_T$ and $\mathfrak{p}_{T+1}$ above T and $T+1$, and they are not principal. Indeed, assume (for instance) that $\mathfrak{p}_T$ is principal. Obviously $\sigma(\mathfrak{p}_T) = \mathfrak{p}_{T+1}$, where σ is the automorphism of K defined by $\sigma(y) = y$ and $\sigma(T) = T + 1$. Hence $\mathfrak{p}_{T+1}$ is also principal and $\mathcal{A}m_{str}(K) = \{1\}$. This is a contradiction. Moreover, we have $y^{-1}\mathcal{O}_K = \mathfrak{p}_T\mathfrak{p}_{T+1}$.

Acknowledgments The authors want to thank the anonymous referee who suggested to study the problem in the framework of the theory of factorization in monoids and proposed almost everything that is contained in Sect. 3.3.

References

1. D. Adam, Pólya and Newtonian function fields. Math. Manuscr. **126**(2), 231–246 (2008)
2. M. Bhargava, Generalized factorials and fixed divisors over subsets of a Dedekind domain. J. Number Theory **72**, 67–75 (1998)
3. P.-J. Cahen, J.-L. Chabert, *Integer-Valued Polynomials*, vol. 48, American Mathematical Society Surveys and Monographs (American Mathematical Society, Providence, 1997)
4. L. Carlitz, A characterization of algebraic number fields with class number two. Proc. Am. Math. Soc. **11**, 391–392 (1960)
5. J.-L. Chabert, Factorial groups and Pólya groups in Galoisian extensions of $\mathbb{Q}$, *Commutative Ring Theory and Applications*, vol. 231, Lecture Notes in Pure and Applied Mathematics (Marcel Dekker, New York, 2003), pp. 77–86
6. H. Cohen, *A Course in Computational Algebraic Number Theory* (Springer, New York, 1993)
7. J. Elliott, The probability that $\mathrm{Int}_n(D)$ is free, *Commutative Algebra, Recent Advances in Commutative Rings, Integer-Valued Polynomials, and Polynomial Functions* (Springer, New York, 2014), pp. 133–151
8. A. Geroldinger, F. Halter-Koch, *Non-Unique Factorizations, Algebraic, Combinatorial and Analytic theory* (Chapman & Hall, Boca-Raton, 2006)
9. D. Goldschmidt, *Algebraic Functions and Projective Curves*, vol. 215, Graduate Texts in Mathematics (Springer, New York, 2002)
10. M.-N. Gras, Sur les corps cubiques cycliques dont l'anneau des entiers est monogène. C. R. Acad. Sci. Paris Sér. A **278**, 59–62 (1974)
11. M.-N. Gras, Non monogénéité de l'anneau des entiers des extensions cycliques de $\mathbb{Q}$ de degré premier $l \geq 5$. J. Number Theory **23**, 347–353 (1986)
12. D. Hilbert, Die Theorie der algebraischen Zahlkörper. Jahresbericht der Deutschen Mathematiker-Vereinigung **4**(1894–95), 175–546 (1897)
13. S. Hu, Y. Li, The genus fields of Artin-Schreier extensions. Finite Fields Appl. **16**(4), 255–264 (2010)
14. D. Lebrigand, Real quadratic extensions of the rational function field in characteristic two, arithmetic, geometry and coding theory (AGCT 2003), Séminaires et Congrès, vol. 11 (2005), pp. 143–169
15. A. Ostrowski, Über ganzwertige Polynome in algebraischen Zahlkörpern. J. reine angew. Math. **149**, 117–124 (1919)
16. G. Pólya, Über ganzwertige Polynome in algebraischen Zahlkörpern. J. reine angew. Math. **149**, 97–116 (1919)
17. M. Rosen, *Number Theory in Function Fields*, vol. 210, Graduate Texts in Mathematics (Springer, New York, 2002)
18. H. Stichtenoth, *Algebraic Function Fields and Codes, Universitext* (Springer, New York, 1993)
19. A. Weil, *Basic Number Theory* (Springer, New York, 1967)
20. H. Zantema, Integer valued polynomials over a number field. Manuscr. Math. **40**, 155–203 (1982)
21. X. Zhang, Ambiguous classes and 2-rank of class group of quadratic function fields. J. China Univ. Sci. Technol. **17**(4), 425–431 (1987)

The Interplay of Invariant Theory with Multiplicative Ideal Theory and with Arithmetic Combinatorics

Kálmán Cziszter, Mátyás Domokos and Alfred Geroldinger

Dedicated to Franz Halter-Koch on the occasion of his 70th birthday

Abstract This paper surveys and develops links between polynomial invariants of finite groups, factorization theory of Krull domains, and product-one sequences over finite groups. The goal is to gain a better understanding of the multiplicative ideal theory of invariant rings, and connections between the Noether number and the Davenport constants of finite groups.

Keywords Invariant rings · Krull monoids · Noether number · Davenport constant · Zero-sum sequences · Product-one sequences

1 Introduction

The goal of this paper is to deepen the links between the areas in the title. Invariant theory is concerned with the study of group actions on algebras, and in the present article we entirely concentrate on actions of finite groups on polynomial algebras via linear substitution of the variables.

K. Cziszter · M. Domokos (✉)
MTA Alfréd Rényi Institute of Mathematics, Reáltanoda u. 13–15, Budapest 1053, Hungary
e-mail: domokos.matyas@renyi.mta.hu

K. Cziszter
e-mail: cziszter.kalman@gmail.com

A. Geroldinger
Institute for Mathematics and Scientific Computing, University of Graz, NAWI Graz, Heinrichstraße 36, 8010 Graz, Austria
e-mail: alfred.geroldinger@uni-graz.at

S. Chapman et al. (eds.), *Multiplicative Ideal Theory and Factorization Theory*, Springer Proceedings in Mathematics & Statistics 170,
DOI 10.1007/978-3-319-38855-7_3

To begin with, let us briefly sketch the already existing links between the mentioned areas. For a finite-dimensional vector space V over a field $\mathbb{F}$ and a finite group $G \le \mathrm{GL}(V)$, let $\mathbb{F}[V]^G \subset \mathbb{F}[V]$ denote the ring of invariants. Since E. Noether we know that $\mathbb{F}[V]^G \subset \mathbb{F}[V]$ is an integral ring extension and that $\mathbb{F}[V]^G$ is a finitely generated $\mathbb{F}$-algebra. In particular, $\mathbb{F}[V]^G$ is an integrally closed noetherian domain and hence a Krull domain. Benson [4] and Nakajima [58] determined its class group. Krull domains (their ideal theory and their class groups) are a central topic in multiplicative ideal theory (see the monographs [46, 51] and the recent survey [52]). B. Schmid [73] observed that the Noether number of a finite abelian group G equals the Davenport constant of G (a constant of central importance in zero-sum theory) and this established a first link between invariant theory and arithmetic combinatorics. Moreover, ideal and factorization theory of Krull domains are most closely linked with zero-sum theory via transfer homomorphisms (see [37, 40] and Sect. 3.2).

These links serve as our starting point. It is well known that a domain R is a Krull domain if and only if its monoid $R^\bullet$ of nonzero elements is a Krull monoid if and only if R (resp. $R^\bullet$) has a divisor theory. To start with Krull monoids, a monoid H is Krull if and only if its associated reduced monoid $H/H^\times$ is Krull, and every Krull monoid H is a direct product $H^\times \times H_0$ where H_0 is isomorphic to $H/H^\times$. A reduced Krull monoid is uniquely determined (up to isomorphism) by its characteristic (roughly speaking by its class group $\mathcal{C}(H)$ and the distribution of the prime divisors in its classes; see the end of Sect. 4.2). By definition of the class group, a Krull monoid H is factorial if and only if $\mathcal{C}(H)$ is trivial. Information on the subset $\mathcal{C}(H)^* \subset \mathcal{C}(H)$ of classes containing prime divisors is the crucial ingredient to understand the arithmetic of H, and hence in order to study the arithmetic of Krull monoids the first and most important issue is to determine $\mathcal{C}(H)^*$. By far the best understood setting in factorization theory are Krull monoids with finite class groups where every class contains a prime divisor. Indeed, there has been an abundance of work on them and we refer the reader to the survey by W.A. Schmid in this proceedings [77]. A canonical method to obtain information on $\mathcal{C}(H)^*$ is to identify explicitly a divisor theory for H. A divisor theory of a monoid (or a domain) H is a divisibility preserving homomorphism from H to a free abelian monoid which satisfies a certain minimality property (Sect. 2.1). The concept of a divisor theory stems from algebraic number theory and it has found far-reaching generalizations in multiplicative ideal theory [51]. Indeed, divisor-theoretic tools, together with ideal-theoretic and valuation-theoretic ones, constitute a highly developed machinery for the structural description of monoids and domains.

All the above-mentioned concepts and problems from multiplicative ideal theory are studied for the ring of invariants. Theorem 4.5 (in Sect. 4.2) provides an explicit divisor theory of the ring of invariants $R = \mathbb{F}[V]^G$. The divisibility preserving homomorphism from $R^\bullet$ goes into a free abelian monoid which can be naturally described in the language of invariant theory, and the associated canonical transfer homomorphism $\theta \colon R^\bullet \to \mathcal{B}(\mathcal{C}(R)^*)$ from the multiplicative monoid of the ring R onto the monoid of zero-sum sequences over the class group of R also has a natural invariant theoretic interpretation. In addition to recovering the result of Benson and Nakajima on the class group $\mathcal{C}(\mathbb{F}[V]^G)$ (our treatment is essentially self-contained), we gain further information on the multiplicative structure of R, and we pose the problem to determine its characteristic (Problem 1). In particular, whenever we can show—for a given ring of invariants—that every class contains at least one prime divisor, then all results of factorization theory (obtained for Krull monoids with finite class group and prime divisors in all classes) apply to the ring of invariants.

In Sect. 4.3 we specialize to abelian groups whose order is not divisible by the characteristic of $\mathbb{F}$. The Noether number $\beta(G)$ is the supremum over all finite dimensional G-modules V of the maximal degree of an element in a minimal homogeneous generating system of $\mathbb{F}[V]^G$, and the Davenport constant $\mathsf{D}(G)$ is the maximal length of a minimal zero-sum sequence over G. We start with a result on the structural connection between $\mathbb{F}[V]^G$ and the monoid of zero-sum sequences over G, that lies behind the equality $\beta(G) = \mathsf{D}(G)$. Clearly, the idea here is well known (as far as we know, it was first used by B. Schmid [73], see also [24]). The benefit of the detailed presentation as given in Proposition 4.7 is twofold. First, the past 20 years have seen great progress in zero-sum theory (see Sect. 3.4 for a sample of results) and Proposition 4.7 allows to carry over all results on the structure of (long) minimal zero-sum sequences to the structure of G-invariant monomials. Second, we observe that the submonoid M^G of $R^\bullet$ consisting of the invariant monomials is again a Krull monoid, and restricting the transfer homomorphism $\theta\colon R^\bullet \to \mathcal{B}(\mathcal{C}(R)^*)$ (mentioned in the above paragraph) to M^G we obtain essentially the canonical transfer homomorphism $M^G \to \mathcal{B}(\mathcal{C}(M^G)^*)$. This turns out to be rather close to the transfer homomorphism $\psi\colon M^G \to \mathcal{B}(\widehat{G})$ into the monoid of zero-sum sequences over the character group of G (see Proposition 4.7), which is responsible for the equality $\beta(G) = \mathsf{D}(G)$. The precise statement is given in Proposition 4.9, which explains how the transfer homomorphism ψ (existing only for abelian groups) relates to the more general transfer homomorphism θ from the above paragraph which exists for an arbitrary finite group. In Proposition 4.9 we point out that every class of $\mathcal{C}(\mathbb{F}[V]^G)$ contains a prime divisor which contributes to Problem 1.

Let now G be a finite non-abelian group. Until recently, the precise value of the Noether number $\beta(G)$ was known only for the dihedral groups and very few small groups (such as A_4). In the last couple of years the first two authors have determined the precise value of the Noether number for groups having a cyclic subgroup of index two and for non-abelian groups of order $3p$ [10, 12, 13]. In this work results on zero-sum sequences over finite abelian groups (for example, information on the structure of long minimal zero-sum sequences and on the kth Davenport constants) were successfully applied. Moreover, a decisive step was the introduction of the kth Noether numbers, a concept inspired by the kth Davenport constants of abelian groups. The significance of this concept is that it furnishes some reduction lemmas (listed in Sect. 5.1) by which the ordinary Noether number of a group can be bounded via structural reduction in the group.

The concept of the kth Davenport constants $\mathsf{D}_k(G)$ has been introduced by Halter-Koch [50] for abelian groups in order to study the asymptotic behavior of arithmetical counting functions in rings of integers of algebraic number fields (see [40, Theorem 9.1.8], [67, Theorem 1]). They have been further studied in [15, 30]. In the last years the third author and Grynkiewicz [39, 48] studied the (small and the large) Davenport constant of non-abelian groups, and among others determined their precise values for groups having a cyclic subgroup of index two. It can be observed that for these groups the Noether number is between the small and the large Davenport constant.

This motivated a new and more abstract view at the Davenport constants, namely kth Davenport constants of BF-monoids (Sect. 2.5). The goal is to relate the Noether number with Davenport constants of suitable monoids as a generalization of the equation $\beta(G) = \mathsf{D}(G)$ in the abelian case. Indeed, the kth Davenport constant $\mathsf{D}_k(G)$ of an abelian group G is recovered as our kth Davenport constant of the monoid $\mathscr{B}(G)$ of zero-sum sequences over G.

We apply the new concept of the kth Davenport constants to two classes of BF-monoids. First, to the monoid $\mathscr{B}(G, V)$ associated to a G-module V in Sect. 4.4 (when G is abelian we recover the monoid M^G of G-invariant monomials from Sect. 4.3), whose Davenport constants provide a lower bound for the corresponding Noether numbers (see Proposition 4.12). Second, we study the monoid of product-one sequences over finite groups (Sects. 3.1 and 3.3). We derive a variety of features of the kth Davenport constants of the monoid of product-one sequences over G and observe that they are strikingly similar to the corresponding features of the kth Noether numbers (see Sect. 5.1 for a comparison).

We pose a problem on the relationship between Noether numbers and Davenport constants of non-abelian groups (Problem 2) and we illustrate the efficiency of the above methods by Examples 5.2–5.4 (appearing for the first time), where the explicit value of Noether numbers and Davenport constants of some non-abelian groups are determined.

Throughout this paper, let G be a finite group, $\mathbb{F}$ be a field, and V be a finite dimensional $\mathbb{F}$-vector space endowed with a linear action of G.

2 Multiplicative Ideal Theory: Krull Monoids, C-Monoids, and Class Groups

We denote by $\mathbb{N}$ the set of positive integers, and we put $\mathbb{N}_0 = \mathbb{N} \cup \{0\}$. For every $n \in \mathbb{N}$, we denote by C_n a cyclic group with n elements. For real numbers $a, b \in \mathbb{R}$, we set $[a, b] = \{x \in \mathbb{Z} : a \le x \le b\}$. If A, B are sets, we write $A \subset B$ to mean that A is contained in B but may be equal to B. In Sects. 2.1–2.4 we gather basic material on Krull monoids and C-monoids. In Sect. 2.5 we introduce a new concept, namely Davenport constants of BF-monoids.

2.1 *Monoids and Domains: Ideal Theoretic and Divisor Theoretic Concepts*

Our notation and terminology follows [40, 51] (note that the monoids in [51] do contain a zero-element, whereas the monoids in [40] and in the present manuscript do not contain a zero-element). By a *monoid*, we mean a commutative, cancellative semigroup with unit element. Then the multiplicative semigroup $R^\bullet = R\setminus\{0\}$ of nonzero elements of a domain is a monoid. Following the philosophy of multiplicative ideal theory we describe the arithmetic and the theory of divisorial ideals of domains by means of their multiplicative monoids. Thus we start with monoids.

Let H be a multiplicatively written monoid. An element $u \in H$ is called

- *invertible* if there is an element $v \in H$ with $uv = 1$.
- *irreducible* (or an *atom*) if u is not invertible and, for all $a, b \in H$, $u = ab$ implies a is invertible or b is invertible.
- *prime* if u is not invertible and, for all $a, b \in H$, $u \mid ab$ implies $u \mid a$ or $u \mid b$.

We denote by $\mathcal{A}(H)$ the set of atoms of H, by $H^\times$ the group of invertible elements, and by $H_{\mathrm{red}} = \{aH^\times : a \in H\}$ the associated reduced monoid of H. We say that H is reduced if $|H^\times| = 1$. We denote by $\mathsf{q}(H)$ a quotient group of H with $H \subset \mathsf{q}(H)$, and for a prime element $p \in H$, let $\mathsf{v}_p \colon \mathsf{q}(H) \to \mathbb{Z}$ be the p-adic valuation. Each monoid homomorphism $\varphi \colon H \to D$ induces a group homomorphism $\mathsf{q}(H) \colon \mathsf{q}(H) \to \mathsf{q}(D)$. For a subset $H_0 \subset H$, we denote by $[H_0] \subset H$ the submonoid generated by H_0, and by $\langle H_0 \rangle \le \mathsf{q}(H)$ the subgroup generated by H_0. We denote by $\widetilde{H} = \big\{x \in \mathsf{q}(H) : x^n \in H \text{ for some } n \in \mathbb{N}\big\}$ the *root closure* of H, and by $\widehat{H} = \big\{x \in \mathsf{q}(H) : \text{there exists } c \in H \text{ such that } cx^n \in H \text{ for all } n \in \mathbb{N}\big\}$ the *complete integral closure* of H. Both $\widetilde{H}$ and $\widehat{H}$ are monoids, and we have $H \subset \widetilde{H} \subset \widehat{H} \subset \mathsf{q}(H)$. We say that H is root closed (completely integrally closed resp.) if $H = \widetilde{H}$ ($H = \widehat{H}$ resp.). For a set P, we denote by $\mathcal{F}(P)$ the free abelian monoid with basis P. Then every $a \in \mathcal{F}(P)$ has a unique representation in the form

$$a = \prod_{p \in P} p^{\mathsf{v}_p(a)}, \quad \text{where } \mathsf{v}_p(a) \in \mathbb{N}_0 \text{ and } \mathsf{v}_p(a) = 0 \text{ for almost all } p \in P.$$

The monoid H is said to be

- *atomic* if every $a \in H\setminus H^\times$ is a product of finitely many atoms of H.
- *factorial* if every $a \in H\setminus H^\times$ is a product of finitely many primes of H (equivalently, $H = H^\times \times \mathcal{F}(P)$ where P is a set of representatives of primes of F).
- *finitely generated* if $H = [E]$ for some finite subset $E \subset H$.

If $H = H^\times \times \mathcal{F}(P)$ is factorial and $a \in H$, then $|a| = \sum_{p \in P} \mathsf{v}_p(a) \in \mathbb{N}_0$ is called the length of a. If H is reduced, then it is finitely generated if and only if it is atomic and $\mathcal{A}(H)$ is finite. Since every prime is an atom, every factorial monoid is atomic. For every non-unit $a \in H$,

$$\mathsf{L}_H(a) = \mathsf{L}(a) = \{k \in \mathbb{N} \colon a \text{ may be written as a product of } k \text{ atoms}\} \subset \mathbb{N}$$

denotes the *set of lengths* of a. For convenience, we set $\mathsf{L}(a) = \{0\}$ for $a \in H^\times$. We say that H is a BF-monoid if it is atomic and all sets of lengths are finite. A monoid homomorphism $\varphi \colon H \to D$ is said to be

- a *divisor homomorphism* if $\varphi(a) \mid \varphi(b)$ implies that $a \mid b$ for all $a, b \in H$.
- *cofinal* if for every $\alpha \in D$ there is an $a \in H$ such that $\alpha \mid \varphi(a)$.
- a *divisor theory* (for H) if $D = \mathcal{F}(P)$ for some set P, φ is a divisor homomorphism, and for every $p \in P$, there exists a finite nonempty subset $X \subset H$ satisfying $p = \gcd\big(\varphi(X)\big)$.

Obviously, every divisor theory is cofinal. Let $H \subset D$ be a submonoid. Then $H \subset D$ is called

- *saturated* if the embedding $H \hookrightarrow D$ is a divisor homomorphism.
- *divisor closed* if $a \in H$, $b \in D$ and $b \mid a$ implies $b \in H$.
- *cofinal* if the embedding $H \hookrightarrow D$ is cofinal.

It is easy to verify that $H \hookrightarrow D$ is a divisor homomorphism if and only if $H = \mathsf{q}(H) \cap D$, and if this holds, then $H^\times = D^\times \cap H$. If $H \subset D$ is divisor closed, then $H \subset D$ is saturated.

For subsets $A, B \subset \mathsf{q}(H)$, we denote by $(A \colon B) = \{x \in \mathsf{q}(H) \colon xB \subset A\}$, by $A^{-1} = (H \colon A)$, and by $A_v = (A^{-1})^{-1}$. A subset $\mathfrak{a} \subset H$ is called an s-ideal of H if $\mathfrak{a}H = \mathfrak{a}$. A subset $X \subset \mathsf{q}(H)$ is called a fractional v-ideal (or a *fractional divisorial ideal*) if there is a $c \in H$ such that $cX \subset H$ and $X_v = X$. We denote by $\mathcal{F}_v(H)$ the set of all fractional v-ideals and by $\mathcal{I}_v(H)$ the set of all v-ideals of H. Furthermore, $\mathcal{I}_v^*(H)$ is the monoid of v-invertible v-ideals (with v-multiplication) and $\mathcal{F}_v(H)^\times = \mathsf{q}\big(\mathcal{I}_v^*(H)\big)$ is its quotient group of fractional invertible v-ideals. The monoid H is completely integrally closed if and only if every nonempty v-ideal of H is v-invertible, and H is called v-noetherian if it satisfies the ACC (ascending chain condition) on v-ideals. If H is v-noetherian, then H is a BF-monoid. We denote by $\mathfrak{X}(H)$ the set of all minimal nonempty prime s-ideals of H.

The map $\partial \colon H \to \mathcal{I}_v^*(H)$, defined by $\partial(a) = aH$ for each $a \in H$, is a cofinal divisor homomorphism. Thus, if $\mathcal{H} = \{aH \colon a \in H\}$ is the monoid of principal ideals of H, then $\mathcal{H} \subset \mathcal{I}_v^*(H)$ is saturated and cofinal.

2.2 Class Groups and Class Semigroups

Let $\varphi \colon H \to D$ be a monoid homomorphism. The group $\mathcal{C}(\varphi) = \mathsf{q}(D)/\mathsf{q}(\varphi(H))$ is called the *class group* of φ. For $a \in \mathsf{q}(D)$, we denote by $[a]_\varphi = a\mathsf{q}(\varphi(H)) \in \mathcal{C}(\varphi)$ the class containing a. We use additive notation for $\mathcal{C}(\varphi)$ and so $[1]_\varphi$ is the zero element of $\mathcal{C}(\varphi)$.

Suppose that $H \subset D$ and that $\varphi = (H \hookrightarrow D)$. Then $\mathcal{C}(\varphi) = \mathsf{q}(D)/\mathsf{q}(H)$, and for $a \in D$ we set $[a]_\varphi = [a]_{D/H} = a\mathsf{q}(H)$. Then

$$D/H = \{[a]_{D/H} : a \in D\} \subset \mathcal{C}(\varphi)$$

is a submonoid with quotient group $\mathsf{q}(D/H) = \mathcal{C}(\varphi)$. It is easy to check that D/H is a group if and only if $H \subset D$ is cofinal. In particular, if D/H is finite or if $\mathsf{q}(D)/\mathsf{q}(H)$ is a torsion group, then $D/H = \mathsf{q}(D)/\mathsf{q}(H)$. Let H be a monoid. Then $\mathcal{H} \subset \mathcal{I}_v^*(H)$ is saturated and cofinal, and

$$\mathcal{C}_v(H) = \mathcal{I}_v^*(H)/\mathcal{H} = \mathcal{F}_v(H)^\times/\mathsf{q}(\mathcal{H})$$

is the *v-class group* of H.

We will also need the concept of class semigroups which are a refinement of ordinary class groups in commutative algebra. Let D be a monoid and $H \subset D$ a submonoid. Two elements $y, y' \in D$ are called H-equivalent, if $y^{-1}H \cap D = y'^{-1}H \cap D$. H-equivalence is a congruence relation on D. For $y \in D$, let $[y]_H^D$ denote the congruence class of y, and let

$$\mathcal{C}(H, D) = \{[y]_H^D : y \in D\} \quad \text{and} \quad \mathcal{C}^*(H, D) = \{[y]_H^D : y \in (D \setminus D^\times) \cup \{1\}\}.$$

Then $\mathcal{C}(H, D)$ is a semigroup with unit element $[1]_H^D$ (called the *class semigroup* of H in D) and $\mathcal{C}^*(H, D) \subset \mathcal{C}(H, D)$ is a subsemigroup (called the *reduced class semigroup* of H in D). The map

$$\theta : \mathcal{C}(H, D) \to D/H, \quad \text{defined by} \quad \theta([a]_H^D) = [a]_{D/H} \quad \text{for all } a \in D,$$

is an epimorphism, and it is an isomorphism if and only if $H \subset D$ is saturated.

2.3 Krull Monoids and Krull Domains

Theorem 2.1 *Let H be a monoid. Then, the following statements are equivalent:*

(a) H is v-noetherian and completely integrally closed,
(b) $\partial : H \to \mathcal{I}_v^*(H)$ is a divisor theory.
(c) H has a divisor theory.
(d) There is a divisor homomorphism $\varphi : H \to D$ into a factorial monoid D.
(e) H_{red} is a saturated submonoid of a free abelian monoid.

If H satisfies these conditions, then H is called a *Krull monoid.*

Proof See [40, Theorem 2.4.8] or [51, Chap. 22].

Let H be a Krull monoid. Then $\mathcal{I}_v^*(H)$ is free abelian with basis $\mathfrak{X}(H)$. Let $\mathfrak{p} \in \mathfrak{X}(H)$. Then $\mathsf{v}_\mathfrak{p}$ denotes the $\mathfrak{p}$-adic valuation of $\mathcal{F}_v(H)^\times$. For $x \in \mathsf{q}(H)$, we set $\mathsf{v}_{\mathfrak{p}}(x) = \mathsf{v}_{\mathfrak{p}}(xH)$ and we call $\mathsf{v}_{\mathfrak{p}}$ the $\mathfrak{p}$-adic valuation of H. Then $\mathsf{v}\colon H \to \mathbb{N}_0^{(\mathfrak{X}(H))}$, defined by $\mathsf{v}(a) = \big(\mathsf{v}_{\mathfrak{p}}(a)\big)_{\mathfrak{p}\in\mathfrak{X}(H)}$ is a divisor theory and $H = \{x \in \mathsf{q}(H)\colon \mathsf{v}_{\mathfrak{p}}(x) \ge 0 \text{ for all } \mathfrak{p} \in \mathfrak{X}(H)\}$.

If $\varphi\colon H \to D = \mathcal{F}(P)$ is a divisor theory, then there is an isomorphism $\Phi\colon \mathcal{I}_v^*(H) \to D$ such that $\Phi \circ \partial = \varphi$, and it induces an isomorphism $\overline{\Phi}\colon \mathcal{C}_v(H) \to \mathcal{C}(\varphi)$. Let $D = \mathcal{F}(P)$ be such that $H_{\mathrm{red}} \hookrightarrow D$ is a divisor theory. Then D and P are uniquely determined by H,

$$\mathcal{C}(H) = \mathcal{C}(H_{\mathrm{red}}) = D/H_{\mathrm{red}}$$

is called the *(divisor) class group* of H, and its elements are called the classes of H. By definition, every class $g \in \mathcal{C}(H)$ is a subset of $\mathsf{q}(D)$ and $P \cap g$ is the set of prime divisors lying in g. We denote by $\mathcal{C}(H)^* = \{[p]_{D/H_{\mathrm{red}}} : p \in P\} \subset \mathcal{C}(H)$ the subset of classes containing prime divisors (for more details we refer to the discussion after Definition 2.4.9 in [40]).

Proposition 2.2 *Let H be a Krull monoid, and let $\varphi\colon H \to D = \mathcal{F}(P)$ be a divisor homomorphism.*

1. *There is a submonoid $C_0 \subset \mathcal{C}(\varphi)$ and an epimorphism $C_0 \to \mathcal{C}_v(H)$.*
2. *Suppose that $H \subset D$ is saturated and that $\mathsf{q}(D)/\mathsf{q}(H)$ is a torsion group. We set $D_0 = \{\gcd_D(X)\colon X \subset H \text{ finite}\}$, and for $p \in P$ define $e(p) = \min\{\mathsf{v}_p(h)\colon h \in H \text{ with } \mathsf{v}_p(h) > 0\}$.*

   (a) D_0 is a free abelian monoid with basis $\{p^{e(p)} : p \in P\}$.
   (b) The embedding $H \hookrightarrow D_0$ is a divisor theory for H.

Proof 1. follows from [40, Theorem 2.4.8], and 2. from [74, Lemma 3.2].

Let R be a domain with quotient field K. Then $R^\bullet = R\setminus\{0\}$ is a monoid, and all notions defined for monoids so far will be applied for domains. To mention a couple of explicit examples, we denote by $\mathsf{q}(R)$ the quotient field of R and we have $\mathsf{q}(R) = \mathsf{q}(R^\bullet) \cup \{0\}$, and for the complete integral closure we have $\widehat{R} = \widehat{R^\bullet} \cup \{0\}$ (where $\widehat{R}$ is the integral closure of R in its quotient field). We denote by $\mathfrak{X}(R)$ the set of all minimal nonzero prime ideals of R, by $\mathcal{I}_v(R)$ the set of divisorial ideals of R, by $\mathcal{I}_v^*(R)$ the set of v-invertible divisorial ideals of R, and by $\mathcal{F}_v(R)$ the set of fractional divisorial ideals of R. Equipped with v-multiplication, $\mathcal{F}_v(R)$ is a semigroup, and the map

$$\iota^\bullet\colon \mathcal{F}_v(R) \to \mathcal{F}_v(R^\bullet), \quad \text{defined by} \quad \mathfrak{a} \mapsto \mathfrak{a}\setminus\{0\},$$

is a semigroup isomorphism mapping $\mathcal{I}_v(R)$ onto $\mathcal{I}_v(R^\bullet)$ and fractional principal ideals of R onto fractional principal ideals of $R^\bullet$. Thus R satisfies the ACC on divisorial ideals of R if and only if $R^\bullet$ satisfies the ACC on divisorial ideals of $R^\bullet$. Furthermore, R is completely integrally closed if and only if $R^\bullet$ is completely integrally closed. A domain R is a Krull domain if it is completely integrally closed and satisfies the ACC on divisorial ideals of R, and thus R is a Krull domain if and only if $R^\bullet$ is a Krull monoid. If R is a Krull domain, we set $\mathscr{C}(R) = \mathscr{C}(R^\bullet)$. The group $\mathscr{F}_v(R)^\times$ is the group of v-invertible fractional ideals and the set $\mathscr{I}_v^*(R) = \mathscr{F}_v(R)^\times \cap \mathscr{I}_v(R)$ of all v-invertible v-ideals of R is a monoid with quotient group $\mathscr{F}_v(R)^\times$. The embedding of the nonzero principal ideals $\mathscr{H}(R) \hookrightarrow \mathscr{I}_v^*(R)$ is a cofinal divisor homomorphism, and the factor group

$$\mathscr{C}_v(R) = \mathscr{F}_v(R)^\times / \{aR \colon a \in K^\times\} = \mathscr{I}_v^*(R)/\mathscr{H}(R)$$

is called the *v-class group* of R. The map $\iota^\bullet$ induces isomorphisms $\mathscr{F}_v(R)^\times \xrightarrow{\sim} \mathscr{F}_v(R^\bullet)^\times$, $\mathscr{I}_v^*(R) \xrightarrow{\sim} \mathscr{I}_v^*(R^\bullet)$, and $\mathscr{C}_v(R) \xrightarrow{\sim} \mathscr{C}_v(R^\bullet)$, and in the sequel we shall identify these monoids and groups.

The above correspondence between domains and their monoids of non-zero elements can be extended to commutative rings with zero-divisors and their monoids of regular elements [45, Theorem 3.5], and there is an analogue for prime Goldie rings [38, Proposition 5.1].

Examples 2.3 1. (Domains) As mentioned above, the multiplicative monoid $R^\bullet$ of a domain R is a Krull monoid if and only if R is a Krull domain. Thus Property (a) in Theorem 2.1 implies that a noetherian domain is Krull if and only if it is normal (i.e. integrally closed in its field of fractions). In particular, rings of invariants are Krull, as we shall see in Theorem 4.1.

2. (Submonoids of domains) Regular congruence submonoids of Krull domains are Krull [40, Proposition 2.11.6].

3. (Monoids of modules) Let R be a (possibly noncommutative) ring and let $\mathscr{C}$ be a class of finitely generated (right) R-modules which is closed under finite direct-sums, direct summands, and isomorphisms. Then the set $\mathscr{V}(\mathscr{C})$ of isomorphism classes of modules is a commutative semigroup with operation induced by the direct sum. If the endomorphism ring of each module in $\mathscr{C}$ is semilocal, then $\mathscr{V}(\mathscr{C})$ is a Krull monoid [19, Theorem 3.4]. For more information we refer to [1, 20, 21].

4. (Monoids of product-one sequences) In Theorem 3.2 we will characterize the monoids of product-one sequences which are Krull.

2.4 C-Monoids and C-Domains

A monoid H is called a C-*monoid* if it is a submonoid of a factorial monoid F such that $H \cap F^\times = H^\times$ and the reduced class semigroup $\mathscr{C}^*(H, F)$ is finite. A domain is called a C-*domain* if $R^\bullet$ is a C-monoid.

Proposition 2.4 *Let F be a factorial monoid and $H \subset F$ a submonoid such that $H \cap F^\times = H^\times$.*

1. *If H is a C-monoid, then H is v-noetherian with $(H \colon \widehat{H}) \neq \emptyset$, and the complete integral closure $\widehat{H}$ is a Krull monoid with finite class group $\mathscr{C}(\widehat{H})$.*

2. *Suppose that $F/F^\times$ is finitely generated, say $F = F^\times \times [p_1, \ldots, p_s]$ with pairwise nonassociated prime elements $p_1, \ldots, p_s$. Then, the following statements are equivalent:*

   (a) *H is a* C*-monoid defined in F.*
   (b) *There exist some $\alpha \in \mathbb{N}$ and a subgroup $W \le F^\times$ such that $(F^\times : W) \mid \alpha$, $W(H \setminus H^\times) \subset H$, and for all $j \in [1, s]$ and $a \in p_j^\alpha F$ we have $a \in H$ if and only if $p_j^\alpha a \in H$.*

Proof For 1., see [40, Theorems 2.9.11 and 2.9.13] and for 2. see [40, Theorems 2.9.7].

Examples 2.5 1. (Krull monoids) A Krull monoid is a C-monoid if and only if the class group is finite [40, Theorem 2.9.12].

2. (Domains) Let R be a domain. Necessary conditions for R being a C-domain are given in Proposition 2.4. Thus suppose that R is a Mori domain (i.e., a v-noetherian domain) with nonzero conductor $\mathfrak{f} = (R : \widehat{R})$ and suppose that $\mathscr{C}(\widehat{R})$ is finite. If $R/\mathfrak{f}$ is finite, then R is a C-domain by [40, Theorem 2.11.9]. This result generalizes to rings with zero-divisors [45], and in special cases we know that R is a C-domain if and only if $R/\mathfrak{f}$ is finite [69].

3. (Congruence monoids) Let R be Krull domain with finite class group $\mathscr{C}(R)$ and $H \subset R$ a congruence monoid such that $R/\mathfrak{f}$ is finite where $\mathfrak{f}$ is an ideal of definition for H. If R is noetherian or $\mathfrak{f}$ is divisorial, then H is a C-monoid [40, Theorem 2.11.8]. For a survey on arithmetical congruence monoids see [2].

4. In Sect. 3.1 we shall prove that monoids of product-one sequences are C-monoids (Theorem 3.2), and we will meet C-monoids again in Proposition 4.11 dealing with the monoid $\mathscr{B}(G, V)$.

Finitely generated monoids allow simple characterizations when they are Krull or when they are C-monoids. We summarize these characterizations in the next lemma.

Proposition 2.6 *Let H be a monoid such that H_{red} is finitely generated.*

1. *Then H is v-noetherian with $(H : \widehat{H}) \ne \emptyset$, $\widetilde{H} = \widehat{H}$, $\widetilde{H}/H^\times$ is finitely generated, and $\widehat{H}$ is a Krull monoid. In particular, H is a Krull monoid if and only if $H = \widehat{H}$.*
2. *H is a* C*-monoid if and only if $\mathscr{C}(\widehat{H})$ is finite.*
3. *Suppose that H is a submonoid of a factorial monoid $F = F^\times \times \mathscr{F}(P)$. Then, the following statements are equivalent:*

   a. *H is a* C*-monoid defined in F, $F^\times/H^\times$ is a torsion group, and for every $p \in P$ there is an $a \in H$ such that $\mathsf{v}_p(a) > 0$.*
   b. *For every $a \in F$, there is an $n_a \in \mathbb{N}$ with $a^{n_a} \in H$.*

   If (a) and (b) hold, then P is finite and $\widetilde{H} = \widehat{H} = \mathsf{q}(H) \cap F \subset F$ is saturated and cofinal.

Proof 1. follows from [40, 2.7.9–2.7.13], and 2. follows from [41, Proposition 4.8].

3. (a) $\Rightarrow$ (b) For every $p \in P$, we set $d_p = \gcd\big(\mathsf{v}_p(H)\big)$, and by assumption we have $d_p > 0$. We set $P_0 = \{p^{d_p} : p \in P\}$ and $F_0 = F^\times \times \mathcal{F}(P_0)$. By [40, Theorem 2.9.11], H is a C-monoid defined in F_0 and there is a divisor theory $\partial \colon \widehat{H} \to \mathcal{F}(P_0)$. By construction of F_0, it is sufficient to prove the assertion for all $a \in F_0$. Since $F^\times/H^\times$ is a torsion group, it is sufficient to prove the assertion for all $a \in \mathcal{F}(P_0)$. Let $a \in \mathcal{F}(P_0)$. Since $\mathcal{C}(\widehat{H})$ is finite, there is an $n'_a \in \mathbb{N}$ such that $a^{n'_a} \in \widehat{H}$. Since $\widehat{H} = \widetilde{H}$, there is an $n''_a \in \mathbb{N}$ such that $(a^{n'_a})^{n''_a} \in H$.

(b) $\Rightarrow$ (a) For every $p \in P$ there is an $n_p \in \mathbb{N}$ such that $p^{n_p} \in H$ whence $\mathsf{v}_p(p^{n_p}) = n_p > 0$. Clearly, we have $\widehat{H} \subset \widehat{F} = F$, and hence $\widehat{H} \subset \mathsf{q}(\widehat{H}) \cap F = \mathsf{q}(H) \cap F$. Since for each $a \in F$ there is an $n_a \in \mathbb{N}_0$ with $a^{n_a} \in H$, we infer that $\mathsf{q}(H) \cap F \subset \widetilde{H} = \widehat{H}$ and hence $\widehat{H} = \mathsf{q}(H) \cap F$. Furthermore, $H \subset F$ and $\widehat{H} \subset F$ are cofinal, and $\mathsf{q}(F)/\mathsf{q}(H) = F/H$ is a torsion group. Clearly, $\mathsf{q}(H) \cap F \subset F$ is saturated, and thus $\widehat{H}$ is Krull. Since $\widehat{H}^\times = \widehat{H} \cap F^\times$ and $H^\times = \widehat{H}^\times \cap H$, it follows that $H^\times = H \cap F^\times$ and then we obtain that $F^\times/H^\times$ is a torsion group.

By 1., $\widehat{H}/H^\times$ is finitely generated, say $\widehat{H}/H^\times = \{u_1H^\times, \ldots, u_nH^\times\}$, and set $P_0 = \{p \in P : p \text{ divides } u_1 \cdot \ldots \cdot u_n \text{ in } F\}$. Then P_0 is finite, and we assert that $P_0 = P$. If there would exist some $p \in P \setminus P_0$, then there is an $n_p \in \mathbb{N}$ such that $p^{n_p} \in H$, and hence $p^{n_p}H^\times$ is a product of $u_1H^\times, \ldots, u_nH^\times$, a contradiction. Therefore P is finite, $F/F^\times$ is a finitely generated monoid, $\mathsf{q}(F)/F^\times$ is a finitely generated group, and therefore $\mathsf{q}(F)/\mathsf{q}(H)F^\times$ is a finitely generated torsion group and thus finite. Since $\varphi \colon \widehat{H} \to F \to F/F^\times$ is a divisor homomorphism and $\mathcal{C}(\varphi) = \mathsf{q}(F)/\mathsf{q}(H)F^\times$, Proposition 2.2.1 implies that $\mathcal{C}(\widehat{H})$ is an epimorphic image of a submonoid of $\mathsf{q}(F)/\mathsf{q}(H)F^\times$ and thus $\mathcal{C}(\widehat{H})$ is finite. Thus 2. implies that H is a C-monoid (indeed, Property 2.(b) of Proposition 2.4 holds and hence H is a C-monoid defined in F).

2.5 Davenport Constants of BF-Monoids

Let H be a BF-monoid. For every $k \in \mathbb{N}$, we study the sets

$$\mathcal{M}_k(H) = \{a \in H : \max \mathsf{L}(a) \le k\} \quad \text{and} \quad \overline{\mathcal{M}}_k(H) = \{a \in H : \max \mathsf{L}(a) = k\}.$$

A monoid homomorphism $|\cdot| \colon H \to (\mathbb{N}_0, +)$ will be called a *degree function* on H. In this section, we study abstract monoids having a degree function. The results will be applied in particular to monoids of product-one sequences and to monoids $\mathcal{B}(G, V)$ (see Sects. 3.3 and 4.4). In all our applications the monoid H will be a submonoid of a factorial monoid F and if not stated otherwise the degree function on H will be the restriction of the length function on F.

If $\theta \colon H \to B$ is a homomorphism and H and B have degree functions, then we say that θ is *degree preserving* if $|a|_H = |\theta(a)|_B$ for all $a \in H$. Suppose we are given a degree function on H and $k \in \mathbb{N}$, then

$$\mathsf{D}_k(H) = \sup\{|a| : a \in \mathcal{M}_k(H)\} \in \mathbb{N}_0 \cup \{\infty\}$$

is called the *large kth Davenport constant* of H (with respect to $|\cdot|_H$). Clearly, $\mathscr{M}_1(H) = \mathscr{A}(H) \cup H^\times$. We call $\mathsf{D}(H) = \mathsf{D}_1(H) = \sup\{|a| : a \in \mathscr{A}(H)\} \in \mathbb{N}_0 \cup \{\infty\}$ the *Davenport constant* of H. For every $k \in \mathbb{N}$, we have $\mathscr{M}_k(H) \subset \mathscr{M}_{k+1}(H)$, $\mathsf{D}_k(H) \le \mathsf{D}_{k+1}(H)$, and $\mathsf{D}_k(H) \le k\mathsf{D}(H)$. Furthermore, we have $|u| = 0$ for every unit $u \in H^\times$. Therefore, the degree function on H induces automatically a degree function $|\cdot| : H_{\text{red}} \to (\mathbb{N}_0, +)$, and so the kth Davenport constant of H_{red} is defined. Obviously we have $\mathsf{D}_k(H) = \mathsf{D}_k(H_{\text{red}})$. Let $\mathsf{e}(H)$ denote the smallest $\ell \in \mathbb{N}_0 \cup \{\infty\}$ with the following property:

> There is a $K \in \mathbb{N}_0$ such that every $a \in H$ with $|a| \ge K$ is divisible by an element $b \in H \setminus H^\times$ with $|b| \le \ell$.

Clearly, $\mathsf{e}(H) \le \mathsf{D}(H)$.

Proposition 2.7 *Let H be a* BF*-monoid and $|\cdot| : H \to (\mathbb{N}_0, +)$ be a degree function.*

1. *If H_{red} is finitely generated, then the sets $\mathscr{M}_k(H_{\text{red}})$ are finite and $\mathsf{D}_k(H) < \infty$ for every $k \in \mathbb{N}$.*
2. *If $\mathsf{D}(H) < \infty$, then there exist constants $D_H, K_H \in \mathbb{N}_0$ such that $\mathsf{D}_k(H) = k\mathsf{e}(H) + D_H$ for all $k \ge K_H$.*
3. *If $\mathsf{D}(H) < \infty$, then the map $\mathbb{N} \to \mathbb{Q}$, $k \mapsto \frac{\mathsf{D}_k(H)}{k}$ is nonincreasing.*
4. *Suppose that H has a prime element. Then*

$$\mathsf{D}_k(H) = \max\{|a| : a \in \overline{\mathscr{M}}_k(H)\} \le k\mathsf{D}(H)$$

and

$$k\mathsf{D}(H) = \max\{|a| : a \in H,\ \min \mathsf{L}(a) \le k\} = \max\{|a| : a \in H,\ k \in \mathsf{L}(a)\}.$$

Proof 1. Suppose that H_{red} is finitely generated. Then $\mathscr{A}(H_{\text{red}})$ is finite whence $\mathscr{M}_k(H)$ is finite for every $k \in \mathbb{N}$. It follows that $\mathsf{D}(H) < \infty$ and $\mathsf{D}_k(H) \le k\mathsf{D}(H) < \infty$ for all $k \in \mathbb{N}$.

2. Suppose that $\mathsf{D}(H) < \infty$ and note that $\mathsf{e}(H) \le \mathsf{D}(H)$. Let $\mathsf{f}(H) \in \mathbb{N}_0$ be the smallest $K \in \mathbb{N}_0$ such that every $a \in H$ with $|a| \ge K$ is divisible by an element $b \in H$ with $|b| \le \mathsf{e}(H)$. We define $A = \{a \in \mathscr{A}(H) : |a| = \mathsf{e}(H)\}$. Let $k \in \mathbb{N}$ and continue with the following assertion.

A. There exist $a_1, \ldots, a_k \in A$ such that $a_1 \ldots a_k \in \mathscr{M}_k(H)$. In particular, $\mathsf{D}_k(H) \ge |a_1 \ldots a_k| = k\mathsf{e}(H)$.

Proof of **A**. Assume to the contrary that for all $a_1, \ldots, a_k \in A$ the product $a_1 \ldots a_k$ is divisible by an atom $u \in \mathscr{A}(H)$ with $|u| < \mathsf{e}(H)$. We set $K = \mathsf{f}(H) + (k-1)\mathsf{e}(H)$ and choose $a \in H$ with $|a| \ge K$. Then a can be written in the form $a = a_1 \ldots a_k b$ where $a_1, \ldots, a_k, b \in H$ and $|a_i| \le \mathsf{e}(H)$ for all $i \in [1, k]$. If there is some $i \in [1, k]$ with $|a_i| < \mathsf{e}(H)$, then a_i is a divisor of a with $|a_i| < \mathsf{e}(H)$. Otherwise, $a_1, \ldots, a_k \in A$ and by our assumption the product $a_1 \ldots a_k$ and hence a has a divisor of degree strictly smaller than $\mathsf{e}(H)$. This is a contradiction to the definition of $\mathsf{e}(H)$. □(Proof of **A**)

Now let $k \ge \mathsf{f}(H)/\mathsf{e}(H) - 1$. Then **A** implies that $\mathsf{D}_k(H) + \mathsf{e}(H) \ge (k+1)\mathsf{e}(H) \ge \mathsf{f}(H)$. Let $a \in H$ with $|a| > \mathsf{D}_k(H) + \mathsf{e}(H)$. Then, by definition of $\mathsf{f}(H)$, there are $b, c \in H$ such that $a = bc$ with $|c| \le \mathsf{e}(H)$ and hence $|b| > \mathsf{D}_k(H)$. This implies that $\max \mathsf{L}(b) > k$, whence $\max \mathsf{L}(a) > k+1$ and $a \notin \mathcal{M}_{k+1}(H)$. Therefore, we obtain that $\mathsf{D}_{k+1}(H) \le \mathsf{D}_k(H) + \mathsf{e}(H)$ and thus

$$0 \le \mathsf{D}_{k+1}(H) - (k+1)\mathsf{e}(H) \le \mathsf{D}_k(H) - k\mathsf{e}(H).$$

Since a non-increasing sequence of nonnegative integers stabilizes, the assertion follows.

3. Suppose that $\mathsf{D}(H) < \infty$. Let $k \in \mathbb{N}$, $a \in \mathcal{M}_{k+1}(H)$ with $|a| = \mathsf{D}_{k+1}(H)$, and set $l = \max \mathsf{L}(a)$. Then $l \le k+1$. If $l \le k$, then $a \in \mathcal{M}_k(H)$ and $\mathsf{D}_{k+1}(H) \ge \mathsf{D}_k(H) \ge |a| = \mathsf{D}_{k+1}(H)$ whence $\mathsf{D}_k(H) = \mathsf{D}_{k+1}(H)$. Suppose that $l = k+1$. We set $a = a_1 \ldots a_{k+1}$ with $a_1, \ldots, a_{k+1} \in \mathcal{A}(H)$ and $|a_1| \ge \cdots \ge |a_{k+1}|$ whence $|a_{k+1}| \le (|a_1| + \cdots + |a_k|)/k$. It follows that

$$\frac{\mathsf{D}_{k+1}(H)}{k+1} = \frac{|a_1| + \cdots + |a_{k+1}|}{k+1} \le \frac{|a_1| + \cdots + |a_k|}{k} \le \frac{\mathsf{D}_k(H)}{k},$$

where the last inequality holds because $a_1 \ldots a_k \in \mathcal{M}_k(H)$.

4. Let $p \in H$ be a prime element. We assert that

$$\mathsf{D}_k(H) \le \max\bigl\{|a| \colon a \in H,\ \max \mathsf{L}(a) = k\bigr\}. \qquad (*)$$

Indeed, if $a \in \mathcal{M}_k(H)$ and $\max \mathsf{L}(a) = l \le k$, then $ap^{k-l} \in \mathcal{M}_k(H)$ and

$$|a| \le |ap^{k-l}| \le \max\bigl\{|a| \colon a \in H,\ \max \mathsf{L}(a) = k\bigr\},$$

and hence $(*)$ follows. Next, we assert that

$$\max\bigl\{|a| \colon a \in H,\ \min \mathsf{L}(a) \le k\bigr\} \le k\mathsf{D}(H). \qquad (**)$$

Let $a \in H$ with $\min \mathsf{L}(a) = l \le k$, say $a = u_1 \ldots u_l$, where $u_1, \ldots, u_l \in \mathcal{A}(H)$. Then $|a| = |u_1| + \cdots + |u_l| \le l\mathsf{D}(H) \le k\mathsf{D}(H)$, and thus $(**)$ follows. Using $(*)$ and $(**)$ we infer that

$$\begin{aligned} \mathsf{D}_k(H) \le \max\bigl\{|a| \colon a \in H,\ \max \mathsf{L}(a) = k\bigr\} &\le \max\bigl\{|a| \colon a \in H,\ \max \mathsf{L}(a) \le k\bigr\} \\ = \mathsf{D}_k(H) &\le \max\bigl\{|a| \colon a \in H,\ \min \mathsf{L}(a) \le k\bigr\} \end{aligned}$$

and that

$$k\mathsf{D}(H) = \max\bigl\{|a| \colon a \in H,\ k \in \mathsf{L}(a)\bigr\} \le \max\bigl\{|a| \colon a \in H,\ \min \mathsf{L}(a) \le k\bigr\} \le k\mathsf{D}(H).$$

Let F be a factorial monoid and $H \subset F$ a submonoid such that $H^\times = H \cap F^\times$. Then H is a BF-monoid by [40, Corollary 1.3.3]. For $k \in \mathbb{N}$, let $\mathcal{M}_k^*(H)$ denote the set of all $a \in F$ such that a is not divisible by a product of k non-units of H. The restriction of the usual length function $|\cdot|\colon F \to \mathbb{N}_0$ on F (introduced in Sect. 2.1) gives a degree function on H. We define the *small kth Davenport constant* $\mathsf{d}_k(H)$ as

$$\mathsf{d}_k(H) = \sup\{|a|\colon a \in \mathscr{M}_k^*(H)\} \in \mathbb{N}_0 \cup \{\infty\}. \tag{1}$$

In other words, $1 + \mathsf{d}_k(H)$ is the smallest integer $\ell \in \mathbb{N}$ such that every $a \in F$ of length $|a| \geq \ell$ is divisible by a product of k non-units of H. We call $\mathsf{d}(H) = \mathsf{d}_1(H)$ the *small Davenport constant* of H. Clearly we have $\mathscr{M}_k^*(H) \subset \mathscr{M}_{k+1}^*(H)$ hence $\mathsf{d}_k(H) \leq \mathsf{d}_{k+1}(H)$.

Furthermore, let $\eta(H)$ denote the smallest integer $\ell \in \mathbb{N} \cup \{\infty\}$ such that every $a \in F$ with $|a| \geq \ell$ has a divisor $b \in H \setminus H^\times$ with $|b| \in [1, \mathsf{e}(H)]$. For $p \in \mathscr{A}(F)$ denote by o_p the smallest integer $\ell \in \mathbb{N} \cup \{\infty\}$ such that $p^{o_p} \in H$. Clearly, we have $o_p \leq \eta(H)$ for all $p \in \mathscr{A}(F)$.

Proposition 2.8 *Let $F = F^\times \times \mathscr{F}(P)$ be a factorial monoid and $H \subset F$ a submonoid such that $H^\times = H \cap F^\times$, and let $k \in \mathbb{N}$.*

1. *If for every $a \in F$ there is a prime $p \in F$ such that $ap \in H$, then $1 + \mathsf{d}_k(H) \leq \mathsf{D}_k(H)$.*
2. *Suppose that H_{red} is finitely generated and that for every $a \in F$ there is an $n_a \in H$ such that $a^{n_a} \in H$. Then H is a* C*-monoid and we have*
   (a) *$\mathsf{e}(H) = \max\{o_p\colon p \in P\}$ and $\eta(H) < \infty$.*
   (b) *$\mathsf{d}_k(H) + 1 \geq k\mathsf{e}(H)$ and there exist constants $d_H \in \mathbb{Z}_{\geq -1}$, $k_H \in \mathbb{N}_0$ such that $\mathsf{d}_k(H) = k\mathsf{e}(H) + d_H$ for all $k \geq k_H$.*

Proof 1. Let $a \in \mathscr{M}_k^*(H)$ such that $|a| = \mathsf{d}_k(H)$. We choose a prime $p \in F$ such that $ap \in H$. Take any factorization $ap = u_1 \ldots u_\ell$ where $u_i \in \mathscr{A}(H)$. We may assume that $p \mid u_1$ in F. Then $u_2 \ldots u_\ell \mid a$ in F, and hence, $\ell - 1 < k$. Thus, it follows that $ap \in \mathscr{M}_k(H)$ and $\mathsf{D}_k(H) \geq |ap| = |a| + 1 \geq \mathsf{d}_k(H) + 1$.

2.(a) By Proposition 2.6.3, H is a C-monoid, P is finite and hence $\mathsf{e}(H) < \infty$. If $p \in P$, then $p^{o_p} \in \mathscr{A}(H)$ and by the minimality of o_p, p^{o_p} does not have a divisor $b \in H \setminus H^\times$ such that $|b| < o_p$. Thus, it follows that $\mathsf{e}(H) \geq \max\{o_p\colon p \in P\}$. For the reverse inequality, note that by Proposition 2.4.2 there exists an $\alpha \in \mathbb{N}$ such that for all $p \in P$ and all $a \in p^\alpha F$ we have $a \in H$ if and only if $p^\alpha a \in H$. Since any multiple of α has the same property, we may assume that α is divisible by o_p for all $p \in P$. Let $b \in H$ with $|b| > |P|(2\alpha - 1)$. Then, there exists a $p \in P$ such that $b \in p^{2\alpha} F \cap H$. Hence b is divisible in H by p^α, implying in turn that $p^{o_p} \in \mathscr{A}(H)$ divides b in H. Therefore, we obtain that $\mathsf{e}(H) \leq \max\{o_p\colon p \in P\}$.

If $a \in F$ with $|a| \geq \sum_{p \in P}(o_p - 1)$, then there is a $p \in P$ such that p^{o_p} divides a in F, and thus $\eta(H) \leq 1 + \sum_{p \in P}(o_p - 1)$.

2.(b) Let $p \in P$ with $o(p) = \mathsf{e}(H)$. Then $p^{ko_p - 1} \in \mathscr{M}_k^*(H)$ and $|p^{ko_p - 1}| = k\mathsf{e}(H) - 1$, showing the inequality $\mathsf{d}_k(H) + 1 \geq k\mathsf{e}(H)$ for all $k \in \mathbb{N}$. Now let $k \in \mathbb{N}$ be such that $1 + \mathsf{d}_k(H) + \mathsf{e}(H) \geq \eta(H)$, and let $a \in F$ with $|a| \geq \mathsf{d}_k(H) + \mathsf{e}(H) + 1$. Then, by definition of $\eta(H)$, there are $b \in F$ and $c \in H \setminus H^\times$ such that $a = bc$ with $|c| \leq$

$\mathsf{e}(H)$ and $|b| > \mathsf{d}_k(H)$. This implies that b is divisible by a product of k non-units of H whence a is divisible by a product of $k+1$ non-units of H. Therefore, it follows that $1 + \mathsf{d}_{k+1}(H) \le \mathsf{d}_k(H) + \mathsf{e}(H) + 1$ and hence

$$0 \le \mathsf{d}_{k+1}(H) - k\mathsf{e}(H) \le \mathsf{d}_k(H) - (k-1)\mathsf{e}(H) \quad \text{for all sufficiently large} \quad k.$$

Since a nonincreasing sequence of nonnegative integers stabilizes, the assertion follows.

3 Arithmetic Combinatorics: Zero-Sum Results with a Focus on Davenport Constants

This section is devoted to Zero-Sum Theory, a vivid subfield of Arithmetic Combinatorics (see [32, 37, 49]). In Sect. 3.1 we give an algebraic study of the monoid of product-one sequences over finite but not necessarily abelian groups. In Sect. 3.2 we put together well-known material on transfer homomorphisms used in Sects. 4.2 and 4.3. In Sects. 3.3 and 3.4 we consider the kth Davenport constants of finite groups. In particular, we gather results which will be needed in Sect. 5.2 and results having relevance in invariant theory by Proposition 4.7.

3.1 The Monoid of Product-One Sequences

Let $G_0 \subset G$ be a subset and let $G' = [G, G] = \langle g^{-1}h^{-1}gh \colon g, h \in G \rangle$ denote the commutator subgroup of G. A *sequence* over G_0 means a finite sequence of terms from G_0 which is unordered and repetition of terms is allowed, and it will be considered as an element of the free abelian monoid $\mathcal{F}(G_0)$. In order to distinguish between the group operation in G and the operation in $\mathcal{F}(G_0)$, we use the symbol $\boldsymbol{\cdot}$ for the multiplication in $\mathcal{F}(G_0)$, hence $\mathcal{F}(G_0) = \big(\mathcal{F}(G_0), \boldsymbol{\cdot}\big)$—this coincides with the convention in the monographs [40, 49]—and we denote multiplication in G by juxtaposition of elements. To clarify this, if $S_1, S_2 \in \mathcal{F}(G_0)$ and $g_1, g_2 \in G_0$, then $S_1 \boldsymbol{\cdot} S_2 \in \mathcal{F}(G_0)$ has length $|S_1| + |S_2|$, $S_1 \boldsymbol{\cdot} g_1 \in \mathcal{F}(G_0)$ has length $|S_1| + 1$, $g_1 \boldsymbol{\cdot} g_2 \in \mathcal{F}(G_0)$ is a sequence of length 2, but $g_1 g_2$ is an element of G. Furthermore, in order to avoid confusion between exponentiation in G and exponentiation in $\mathcal{F}(G_0)$, we use brackets for the exponentiation in $\mathcal{F}(G_0)$. So for $g \in G_0$, $S \in \mathcal{F}(G_0)$, and $k \in \mathbb{N}_0$, we have

$$g^{[k]} = \underbrace{g \boldsymbol{\cdot} \ldots \boldsymbol{\cdot} g}_{k} \in \mathcal{F}(G) \quad \text{with} \quad |g^{[k]}| = k, \quad \text{and} \quad S^{[k]} = \underbrace{S \boldsymbol{\cdot} \ldots \boldsymbol{\cdot} S}_{k} \in \mathcal{F}(G).$$

Now let

$$S = g_1 \cdot \ldots \cdot g_\ell = \prod_{g \in G_0} g^{\mathsf{v}_g(S)},$$

be a sequence over G_0 (in this notation, we tacitly assume that $\ell \in \mathbb{N}_0$ and $g_1, \ldots, g_\ell \in G_0$). Then $|S| = \ell = 0$ if and only if $S = 1_{\mathcal{F}(G_0)}$ is the identity element in $\mathcal{F}(G_0)$, and then S will also be called the *trivial sequence*. The elements in $\mathcal{F}(G_0) \setminus \{1_{\mathcal{F}(G_0)}\}$ are called *nontrivial sequences*. We use all notions of divisibility theory in general free abelian monoids. Thus, for an element $g \in G_0$, we refer to $\mathsf{v}_g(S)$ as the *multiplicity* of g in S. A divisor T of S will also be called a subsequence of S. We call $\operatorname{supp}(S) = \{g_1, \ldots, g_\ell\} \subset G_0$ the *support* of S. When G is written multiplicatively (with unit element $1_G \in G$), we use

$$\pi(S) = \{g_{\tau(1)} \ldots g_{\tau(\ell)} \in G \colon \tau \text{ a permutation of } [1, \ell]\} \subset G$$

to denote the *set of products* of S (if $|S| = 0$, we use the convention that $\pi(S) = \{1_G\}$). Clearly, $\pi(S)$ is contained in a G'-coset. When G is written additively with commutative operation, we likewise let

$$\sigma(S) = g_1 + \cdots + g_\ell \in G$$

denote the *sum* of S. Furthermore, we denote by

$$\Sigma(S) = \{\sigma(T) \colon T \mid S \text{ and } 1 \neq T\} \subset G \quad \text{and} \quad \Pi(S) = \bigcup_{\substack{T \mid S \\ 1 \neq T}} \pi(T) \subset G,$$

the *subsequence sums* and *subsequence products* of S. The sequence S is called

- a *product-one sequence* if $1_G \in \pi(S)$,
- *product-one free* if $1_G \notin \Pi(S)$.

Every map of finite groups $\varphi \colon G_1 \to G_2$ extends to a homomorphism $\varphi \colon \mathcal{F}(G_1) \to \mathcal{F}(G_2)$ where $\varphi(S) = \varphi(g_1) \cdot \ldots \cdot \varphi(g_\ell)$. If φ is a group homomorphism, then $\varphi(S)$ is a product-one sequence if and only if $\pi(S) \cap \operatorname{Ker}(\varphi) \neq \emptyset$. We denote by

$$\mathcal{B}(G_0) = \{S \in \mathcal{F}(G_0) \colon 1_G \in \pi(S)\}$$

the set of all product-one sequences over G_0, and clearly $\mathcal{B}(G_0) \subset \mathcal{F}(G_0)$ is a submonoid. We will use all concepts introduced in Sect. 2.5 for the monoid $\mathcal{B}(G_0)$ with the degree function stemming from the length function on the free abelian monoid $\mathcal{F}(G_0)$. For all notations $*(H)$ introduced for a monoid H we write—as usual—$*(G_0)$ instead of $*(\mathcal{B}(G_0))$. In particular, for $k \in \mathbb{N}$, we set $\mathcal{M}_k(G_0) = \mathcal{M}_k(\mathcal{B}(G_0))$, $\mathsf{D}_k(G_0) = \mathsf{D}_k(\mathcal{B}(G_0))$, $\eta(G_0) = \eta(\mathcal{B}(G_0))$, $\mathsf{e}(G_0) = \mathsf{e}(\mathcal{B}(G_0))$, and so on. By Proposition 2.8.2(a), $\mathsf{e}(G_0) = \max\{\operatorname{ord}(g) \colon g \in G_0\}$. Note that $\mathcal{M}_1^*(G_0)$ is the set of all product-one free sequences over G_0. In particular,

$$\mathsf{D}(G_0) = \sup\{|S| \colon S \in \mathcal{A}(G_0)\} \in \mathbb{N} \cup \{\infty\}$$

is the *large Davenport constant* of G_0, and

$$\mathsf{d}(G_0) = \sup\{|S| \colon S \in \mathcal{F}(G_0) \text{ is product-one free}\} \in \mathbb{N}_0 \cup \{\infty\}$$

is the *small Davenport constant* of G_0. Their study will be the focus of the Sects. 3.3 and 3.4.

Lemma 3.1 *Let $G_0 \subset G$ be a subset.*

1. *$\mathcal{B}(G_0) \subset \mathcal{F}(G_0)$ is a reduced finitely generated submonoid, $\mathcal{A}(G_0)$ is finite, and $\mathsf{D}(G_0) \le |G|$. Furthermore, $\mathcal{M}_k(G_0)$ is finite and $\mathsf{D}_k(G_0) < \infty$ for all $k \in \mathbb{N}$.*
2. *Let $S \in \mathcal{F}(G)$ be product-one free.*
   a. *If $g_0 \in \pi(S)$, then $g_0^{-1} \cdot S \in \mathcal{A}(G)$. In particular, $\mathsf{d}(G) + 1 \le \mathsf{D}(G)$.*
   b. *If $|S| = \mathsf{d}(G)$, then $\Pi(S) = G \setminus \{1_G\}$ and hence $\mathsf{d}(G) = \max\{|S| \colon S \in \mathcal{F}(G) \text{ with } \Pi(S) = G \setminus \{1_G\}\}$.*
3. *If G is cyclic, then $\mathsf{d}(G) + 1 = \mathsf{D}(G) = |G|$.*

Proof 1. We assert that for every $U \in \mathcal{A}(G)$ we have $|U| \le |G|$. Then $\mathcal{A}(G_0) \subset \mathcal{A}(G)$ is finite and $\mathsf{D}(G_0) \le \mathsf{D}(G) \le |G|$. As already mentioned, $\mathcal{B}(G_0) \subset \mathcal{F}(G_0)$ is a submonoid, and clearly $\mathcal{B}(G_0)^\times = \{1_{\mathcal{F}(G_0)}\}$. Since $\mathcal{F}(G_0)$ is factorial and $\mathcal{B}(G_0)^\times = \mathcal{B}(G_0) \cap \mathcal{F}(G_0)^\times$, $\mathcal{B}(G_0)$ is atomic by [40, Corollary 1.3.3]. This means that $\mathcal{B}(G_0) = [\mathcal{A}(G_0) \cup \mathcal{B}(G_0)^\times]$, and thus, $\mathcal{B}(G_0)$ is finitely generated. Since $\mathcal{B}(G_0)$ is reduced and finitely generated, the sets $\mathcal{M}_k(G_0)$ are finite by Proposition 2.7.

Now let $U \in \mathcal{B}(G)$, say $U = g_1 \cdot \ldots \cdot g_\ell$ with $g_1 g_2 \ldots g_\ell = 1_G$. We suppose that $\ell > |G|$ and show that $U \notin \mathcal{A}(G)$. Consider the set

$$M = \{g_1 g_2 \ldots g_i \colon i \in [1, \ell]\}.$$

Since $\ell > |G|$, there are $i, j \in [1, \ell]$ with $i < j$ and $g_1 \ldots g_i = g_1 \ldots g_j$. Then $g_{i+1} \ldots g_j = 1_G$ and thus $g_1 \ldots g_i g_{j+1} \ldots g_\ell = 1_G$ which implies that U is the product of two nontrivial product-one subsequences.

2.(a) If $g_0 \in \pi(S)$, then S can be written as $S = g_1 \cdot \ldots \cdot g_\ell$ such that $g_0 = g_1 \ldots g_\ell$, which implies that $g_0^{-1} \cdot g_1 \cdot \ldots \cdot g_\ell \in \mathcal{A}(G)$.

2.(b) If S is product-one free with $|S| = \mathsf{d}(G)$, and if there would be an $h \in G \setminus \{\Pi(S) \cup \{1_G\}\}$, then $T = h^{-1} \cdot S$ would be product-one free of length $|T| = |S| + 1 > \mathsf{d}(G)$, a contradiction. Thus every product-one free sequence S of length $|S| = \mathsf{d}(G)$ satisfies $\Pi(S) = G \setminus \{1_G\}$. If S is a sequence with $\Pi(S) = G \setminus \{1_G\}$, then S is product-one free and hence $|S| \le \mathsf{d}(G)$.

3. Clearly, the assertion holds for $|G| = 1$. Suppose that G is cyclic of order $n \ge 2$, and let $g \in G$ with $\operatorname{ord}(g) = n$. Then $g^{[n-1]}$ is product-one free, and thus 1. and 2. imply that $n \le 1 + \mathsf{d}(G) \le \mathsf{D}(G) \le n$.

The next result gathers the algebraic properties of monoids of product-one sequences and highlights the difference between the abelian and the non-abelian case.

Theorem 3.2 *Let $G_0 \subset G$ be a subset and let G' denote the commutator subgroup of $\langle G_0 \rangle$.*

1. *$\mathscr{B}(G_0) \subset \mathscr{F}(G_0)$ is cofinal and $\mathscr{B}(G_0)$ is a finitely generated C-monoid. $\widetilde{\mathscr{B}(G_0)} = \widehat{\mathscr{B}(G_0)}$ is a finitely generated Krull monoid, the embedding $\widehat{\mathscr{B}(G_0)} \hookrightarrow \mathscr{F}(G_0)$ is a cofinal divisor homomorphism with class group $\mathscr{F}(G_0)/\mathscr{B}(G_0)$, and the map*

$$\begin{aligned} \Phi \colon \mathscr{F}(G_0)/\mathscr{B}(G_0) &\longrightarrow \langle G_0 \rangle / G' \\ [S]_{\mathscr{F}(G_0)/\mathscr{B}(G_0)} &\longmapsto gG' \quad \text{for any } g \in \pi(S) \end{aligned}$$

is a group epimorphism. Suppose that $G_0 = G$. Then Φ is an isomorphism, every class of $\mathscr{C}(\widehat{\mathscr{B}(G)})$ contains a prime divisor, and if $|G| \ne 2$, then $\widehat{\mathscr{B}(G)} \hookrightarrow \mathscr{F}(G)$ is a divisor theory.
2. *The following statements are equivalent:*
   - *(a) $\mathscr{B}(G_0)$ is a Krull monoid.*
   - *(b) $\mathscr{B}(G_0)$ is root closed.*
   - *(c) $\mathscr{B}(G_0) \subset \mathscr{F}(G_0)$ is saturated.*
3. *$\mathscr{B}(G)$ is a Krull monoid if and only if G is abelian.*
4. *$\mathscr{B}(G)$ is factorial if and only if $|G| \le 2$.*

Proof 1. $\mathscr{B}(G_0)$ is finitely generated by Lemma 3.1. If $n = \operatorname{lcm}\{\operatorname{ord}(g) \colon g \in G_0\}$, then $S^{[n]} \in \mathscr{B}(G_0)$ for each $S \in \mathscr{F}(G_0)$. Thus $\mathscr{B}(G_0) \subset \mathscr{F}(G_0)$ and $\widehat{\mathscr{B}(G_0)} \hookrightarrow \mathscr{F}(G_0)$ are cofinal, $\mathscr{F}(G_0)/\mathscr{B}(G_0)$ is a group and

$$\mathscr{F}(G_0)/\mathscr{B}(G_0) = \mathsf{q}\big(\mathscr{F}(G_0)\big)/\mathsf{q}\big(\mathscr{B}(G_0)\big) = \mathsf{q}\big(\mathscr{F}(G_0)\big)/\mathsf{q}\big(\widehat{\mathscr{B}(G_0)}\big)$$

is the class group of the embedding $\widehat{\mathscr{B}(G_0)} \hookrightarrow \mathscr{F}(G_0)$. All statements on the structure of $\mathscr{B}(G_0)$ and $\widehat{\mathscr{B}(G_0)}$ follow from Proposition 2.6.3, and it remains to show the assertions on Φ.

Let $S, S' \in \mathscr{F}(G_0)$, $g \in \pi(S)$, $g' \in \pi(S')$, and $B \in \mathscr{B}(G_0)$. Then $\pi(S) \subset gG'$, $\pi(S') \subset g'G'$, $\pi(B) \subset G'$, and $\pi(S \boldsymbol{\cdot} B) \subset gG'$. We use the abbreviation $[S] = [S]_{\mathscr{F}(G_0)/\mathscr{B}(G_0)}$, and note that $[S] = [S']$ if and only if there are $C, C' \in \mathscr{B}(G_0)$ such that $S \boldsymbol{\cdot} C = S' \boldsymbol{\cdot} C'$.

In order to show that Φ is well-defined, suppose that $[S] = [S']$ and that $S \boldsymbol{\cdot} C = S \boldsymbol{\cdot} C'$ with $C, C' \in \mathscr{B}(G_0)$. Then $\pi(S \boldsymbol{\cdot} C) = \pi(S' \boldsymbol{\cdot} C') \subset gG' \cap g'G'$, and hence $gG' = g'G'$. In order to show that Φ is surjective, let $g \in \langle G_0 \rangle$ be given. Clearly, there is an $S \in \mathscr{F}(G_0)$ such that $g \in \pi(S)$ whence $\Phi([S]) = gG'$.

Suppose that $G_0 = G$. First, we show that Φ is injective. Let $S, S' \in \mathcal{F}(G)$ with $g \in \pi(S), g' \in \pi(S')$ such that $gG' = g'G'$. Then there are $k \in \mathbb{N}, a_1, b_1, \dots, a_k, b_k \in G$ such that

$$gg'^{-1} = \prod_{i=1}^{k} (a_i^{-1} b_i^{-1} a_i b_i).$$

We define $T = \prod_{i=1}^{k} (a_i^{-1} \cdot b_i^{-1} \cdot a_i \cdot b_i)$ and obtain that

$$S \cdot (S' \cdot g^{-1} \cdot T) = S' \cdot (S \cdot g^{-1} \cdot T) \in \mathcal{F}(G).$$

Since $1 \in \pi(T)$ and $gg'^{-1} \in \pi(T)$, it follows that $1 \in \pi(S' \cdot g^{-1} \cdot T)$ and $1 \in \pi(S \cdot g^{-1} \cdot T)$ which implies that $[S] = [S']$.

If $|G| \le 2$, then 4. will show that $\mathcal{B}(G)$ is factorial and clearly the trivial class contains a prime divisor. Suppose that $|G| \ge 3$. In order to show that $\widehat{\mathcal{B}(G)} \hookrightarrow \mathcal{F}(G)$ is a divisor theory, let $g \in G \setminus \{1_G\}$ be given. Then there is an $h \in G \setminus \{g^{-1}, 1_G\}$, $U = g \cdot g^{-1} \in \mathcal{A}(G) \subset \widehat{\mathcal{B}(G)}$, $U' = g \cdot h \cdot (h^{-1} g^{-1}) \in \mathcal{A}(G) \subset \widehat{\mathcal{B}(G)}$, and $g = \gcd_{\mathcal{F}(G)}(U, U')$. Thus $\widehat{\mathcal{B}(G)} \hookrightarrow \mathcal{F}(G)$ is a divisor theory.

Let $S \in \mathcal{F}(G)$ with $g \in \pi(S)$. Then $g \in \mathcal{F}(G)$ is a prime divisor and we show that $[g] = [S]$. Indeed, if $g = 1_G$, then $S \in \mathcal{B}(G), 1_G \in \mathcal{B}(G), S \cdot 1_G = g \cdot S$ whence $[g] = [S]$. If $\operatorname{ord}(g) = n \ge 2$, then $g^{[n]} \in \mathcal{B}(G), S \cdot g^{[n-1]} \in \mathcal{B}(G), S \cdot g^{[n]} = g \cdot S \cdot g^{[n-1]}$ whence $[S] = [g]$.

2. (a) $\Rightarrow$ (b) Every Krull monoid is completely integrally closed and hence root closed.

(b) $\Rightarrow$ (c) Let $S, T \in \mathcal{B}(G_0)$ with $T \mid S$ in $\mathcal{F}(G_0)$, say $S = T \cdot U$ where $U = g_1 \cdot \ldots \cdot g_\ell \in \mathcal{F}(G_0)$. If $n = \operatorname{lcm}\big(\operatorname{ord}(g_1), \dots, \operatorname{ord}(g_\ell)\big)$, then $(T^{[-1]} \cdot S)^{[n]} = U^{[n]} \in \mathcal{B}(G_0)$. Since $\mathcal{B}(G_0)$ is root closed, this implies that $U = T^{[-1]} \cdot S \in \mathcal{B}(G_0)$ and hence $T \mid S$ in $\mathcal{B}(G_0)$.

(c) $\Rightarrow$ (a) Since $\mathcal{F}(G_0)$ is free abelian, $\mathcal{B}(G_0)$ is Krull by Theorem 2.1.

3. If G is a abelian, then it is obvious that $\mathcal{B}(G) \subset \mathcal{F}(G)$ is saturated, and thus $\mathcal{B}(G)$ is a Krull monoid by 2. Suppose that G is not abelian. Then there are $g, h \in G$ with $gh \ne hg$. Then $ghg^{-1} \ne h, S = g \cdot h \cdot g^{-1} \cdot (ghg^{-1})^{-1} \in \mathcal{B}(G), T = g \cdot g^{-1} \in \mathcal{B}(G)$ divides S in $\mathcal{F}(G)$ but $T^{[-1]} \cdot S = h \cdot (ghg^{-1})^{-1}$ does not have product-one. Thus $\mathcal{B}(G) \subset \mathcal{F}(G)$ is not saturated and hence $\mathcal{B}(G)$ is not Krull by 2.

4. If $G = \{0\}$, then $\mathcal{B}(G) = \mathcal{F}(G)$ is factorial. If $G = \{0, g\}$, then $\mathcal{A}(G) = \{0, g^{[2]}\}$, each atom is a prime, and $\mathcal{B}(G)$ is factorial. Conversely, suppose that $\mathcal{B}(G)$ is factorial. Then $\mathcal{B}(G)$ is a Krull monoid by [40, Corollary 2.3.13], and hence G is abelian by 3. Suppose that $|G| \ge 3$. We show that $\mathcal{B}(G)$ is not factorial. If there is an element $g \in G$ with $\operatorname{ord}(g) = n \ge 3$, then $U = g^{[n]}$, $-U = (-g)^{[n]}$, $W = (-g) \cdot g \in \mathcal{A}(G)$, and $U \cdot (-U) = W^{[n]}$. Suppose there is no $g \in G$ with $\operatorname{ord}(g) \ge 3$. Then there are $e_1, e_2 \in G$ with $\operatorname{ord}(e_1) = \operatorname{ord}(e_2) = 2$ and $e_1 + e_2 \ne 0$. Then $U = e_1 \cdot e_2 \cdot (e_1 + e_2)$, $W_1 = e_1^{[2]}$, $W_2 = e_2^{[2]}$, $W_0 = (e_1 + e_2)^{[2]} \in \mathcal{A}(G)$, and $U^{[2]} = W_0 \cdot W_1 \cdot W_2$.

For a subset $G_0 \subset G$, the monoid $\mathcal{B}(G_0)$ may be Krull or just seminormal but it need not be Krull. We provide examples for both situations.

Proposition 3.3 *Let $G_0 \subset G$ be a subset satisfying the following property* **P***:*

P. *For each two elements $g, h \in G_0$, $\langle h \rangle \subset \langle g, h \rangle$ is normal or $\langle g \rangle \subset \langle g, h \rangle$ is normal.*

Then $\mathscr{B}(G_0)$ is a Krull monoid if and only if $\langle G_0 \rangle$ is abelian.

Proof If $\langle G_0 \rangle$ is a abelian, then it is obvious that $\mathscr{B}(G_0) \subset \mathscr{F}(G_0)$ is saturated, and thus $\mathscr{B}(G_0)$ is Krull by Theorem 3.2.2.

Conversely, suppose that $\mathscr{B}(G_0)$ is Krull and that G_0 satisfies Property **P**. In order to show that $\langle G_0 \rangle$ is abelian, it is sufficient to prove that $gh = hg$ for each two elements $g, h \in G_0$. Let $g, h \in G_0$ be given such that $\langle h \rangle \subset \langle g, h \rangle$ is normal, $\mathrm{ord}(g) = m$, $\mathrm{ord}(h) = n$, and assume to the contrary that $ghg^{-1} \neq h$. Since $g\langle h \rangle g^{-1} = \langle h \rangle$, it follows that $ghg^{-1} = h^\nu$ for some $\nu \in [2, n-1]$. Thus $ghg^{m-1}h^{n-\nu} = 1$ and $S = g \cdot h \cdot g^{[m-1]} \cdot h^{[n-\nu]} \in \mathscr{B}(G_0)$. Clearly, $T = g^{[m]} \in \mathscr{B}(G_0)$ but $S \cdot T^{[-1]} = h^{[n-\nu+1]} \notin \mathscr{B}(G_0)$. Thus $\mathscr{B}(G_0) \subset \mathscr{F}(G_0)$ is not saturated, a contradiction.

Proposition 3.4 *Let $G = D_{2n}$ be the dihedral group, say $G = \langle a, b \rangle = \{1, a, \ldots, a^{n-1}, b, ab, \ldots, a^{n-1}b\}$, where $\mathrm{ord}(a) = n \geq 2$, $\mathrm{ord}(b) = 2$, and set $G_0 = \{ab, b\}$. Then, $\mathscr{B}(G_0)$ is a Krull monoid if and only if n is even.*

Proof Clearly, we have $\mathrm{ord}(ab) = \mathrm{ord}(b) = 2$ and $\langle G_0 \rangle = G$. Suppose that n is odd and consider the sequence $S = (ab)^{[n]} \cdot b^{[n]}$. Since $\big((ab)b\big)^n = 1$, it follows that S is a product-one sequence. Obviously, $S_1 = (ab)^{[n-1]} \cdot b^{[n-1]} \in \mathscr{B}(G_0)$ and $S_2 = (ab) \cdot b \notin \mathscr{B}(G_0)$. Since $S = S_1 \cdot S_2$, it follows that $\mathscr{B}(G_0) \subset \mathscr{F}(G_0)$ is not saturated, and hence $\mathscr{B}(G_0)$ is not Krull by Theorem 3.2.2.

Suppose that n is even. Then $\mathscr{A}(G_0) = \{(ab)^{[2]}, b^{[2]}\}$ and $\mathscr{B}(G_0) = \{(ab)^{[\ell]} \cdot b^{[m]} \colon \ell, m \in \mathbb{N}_0 \text{ even}\}$. This description of $\mathscr{B}(G_0)$ implies immediately that $\mathscr{B}(G_0) \subset \mathscr{F}(G_0)$ is saturated, and hence $\mathscr{B}(G_0)$ is Krull by Theorem 3.2.2.

Remark (*Seminormality of $\mathscr{B}(G_0)$*) A monoid H is called seminormal if for all $x \in \mathsf{q}(H)$ with $x^2, x^3 \in H$ it follows that $x \in H$. Thus, by definition, every root closed monoid is seminormal.

1. Let $n \equiv 3 \mod 4$ and $G = D_{2n}$ the dihedral group, say $G = \langle a, b \rangle = \{1, a, \ldots, a^{n-1}, b, ab, \ldots, a^{n-1}b\}$, where $\mathrm{ord}(a) = n$, $\mathrm{ord}(b) = 2$, and

$$a^k b a^l b = a^{k-l} \quad \text{for all } k, l \in \mathbb{Z}.$$

We consider the sequence

$$S = a^{\left[\frac{n-1}{2}\right]} \cdot b^{[2]} \in \mathscr{F}(G).$$

Then

$$S^{[2]} = \big(a^{\left[\frac{n-1}{2}\right]} \cdot b \cdot a^{\left[\frac{n-1}{2}\right]} \cdot b\big) \cdot (b \cdot b) \text{ and } S^{[3]} = a^{[n]} \cdot \big(a^{\left[\frac{n-3}{4}\right]} \cdot b \cdot a^{\left[\frac{n-3}{4}\right]} \cdot b\big) \cdot b^{[4]}$$

are both in $\mathcal{B}(G)$ whence $S \in \mathsf{q}(\mathcal{B}(G))$, but obviously $S \notin \mathcal{B}(G)$. Thus $\mathcal{B}(\{a, b\})$ and $\mathcal{B}(G)$ are not seminormal.

2. Let $G = H_8 = \{E, I, J, K, -E, -I, -J, -K\}$ be the quaternion group with the relations

$$IJ = -JI = K, \ JK = -KJ = I, \quad \text{and} \quad KI = -IK = J,$$

and set $G_0 = \{I, J\}$. By Theorem 3.2, $\mathcal{B}(G)$ is not Krull and by Proposition 3.3, $\mathcal{B}(G_0)$ is not Krull. However, we assert that $\mathcal{B}(G_0)$ is seminormal.

First, we are going to derive an explicit description of $\mathcal{B}(G_0)$. Since $E = (-E)(-E) = (KK)(II) = (IJ)(IJ)(II)$, it follows that $U = I^{[4]} \boldsymbol{\cdot} J^{[2]} \in \mathcal{B}(G_0)$. Assume that $U = U_1 \cdot U_2$ with $U_1, U_2 \in \mathcal{A}(G_0)$ and $|U_1| \le |U_2|$. Then $|U_1| \in \{2, 3\}$, but U does not have a subsequence with product one and length two or three. Thus $U \in \mathcal{A}(G_0)$ and similarly we obtain that $I^{[2]} \boldsymbol{\cdot} J^{[4]} \in \mathcal{A}(G_0)$. Since $\mathsf{D}(G_0) \le \mathsf{D}(G) = 6$, it is easy to check that

$$\mathcal{A}(G_0) = \{I^{[4]}, J^{[4]}, I^{[2]} \boldsymbol{\cdot} J^{[2]}, I^{[4]} \boldsymbol{\cdot} J^{[2]}, I^{[2]} \boldsymbol{\cdot} J^{[4]}\}.$$

This implies that

$$\mathcal{B}(G_0) = \{I^{[k]} \boldsymbol{\cdot} J^{[l]} \colon k = l = 0 \text{ or } k, l \in \mathbb{N}_0 \text{ are both even with } k + l \ge 4\}.$$

In order to show that $\mathcal{B}(G_0)$ is seminormal, let $x \in \mathsf{q}(\mathcal{B}(G_0))$ be given such that $x^{[2]}, x^{[3]} \in \mathcal{B}(G_0)$. We have to show that $x \in \mathcal{B}(G_0)$. Since $x^{[2]}, x^{[3]} \in \mathcal{B}(G_0) \subset \mathcal{F}(G_0)$ and $\mathcal{F}(G_0)$ is seminormal, it follows that $x \in \mathcal{F}(G_0)$. If $x = I^{[k]}$ with $k \in \mathbb{N}_0$, then $I^{[3k]} \in \mathcal{B}(G_0)$ implies that $4 \mid 3k$, hence $4 \mid k$, and thus $x \in \mathcal{B}(G_0)$. Similarly, if $x = J^{[l]} \in \mathcal{B}(G_0)$ with $l \in \mathbb{N}_0$, then $x \in \mathcal{B}(G_0)$. It remains to consider the case $x = I^{[k]} \boldsymbol{\cdot} J^{[l]}$ with $k, l \in \mathbb{N}$. Since $x^{[3]} = I^{[3k]} \boldsymbol{\cdot} J^{[3l]} \in \mathcal{B}(G_0)$, it follows that k, l are both even, and thus $x \in \mathcal{B}(G_0)$. Therefore, $\mathcal{B}(G_0)$ is seminormal.

3.2 *Transfer Homomorphisms*

A well-established strategy for investigating the arithmetic of a given monoid H is to construct a transfer homomorphism $\theta \colon H \to B$, where B is a simpler monoid than H and the transfer homomorphism θ allows to shift arithmetical results from B back to the (original, more complicated) monoid H. We will use transfer homomorphisms in Sect. 4 in order to show that properties of the monoid of G-invariant monomials can be studied in a monoid of zero-sum sequences (see Propositions 4.7 and 4.9).

Definition 3.5 A monoid homomorphism $\theta \colon H \to B$ is called a *transfer homomorphism* if it has the following properties:

(T1) $B = \theta(H)B^\times$ and $\theta^{-1}(B^\times) = H^\times$.

(T2) If $u \in H$, $b, c \in B$ and $\theta(u) = bc$, then there exist $v, w \in H$ such that $u = vw$, $\theta(v)B^\times = bB^\times$ and $\theta(w)B^\times = cB^\times$.

We will use the simple fact that, if $\theta \colon H \to B$ and $\theta' \colon B \to B'$ are transfer homomorphisms, then their composition $\theta' \circ \theta \colon H \to B'$ is a transfer homomorphism too. The next proposition summarizes key properties of transfer homomorphisms.

Proposition 3.6 *Let $\theta \colon H \to B$ be a transfer homomorphism and $a \in H$.*

1. *a is an atom of H if and only if $\theta(a)$ is an atom of B.*
2. $\mathsf{L}_H(a) = \mathsf{L}_B\big(\theta(a)\big)$, *whence* $\theta\big(\mathcal{M}_k(H)\big) = \mathcal{M}_k(B)$ *and* $\theta^{-1}\big(\mathcal{M}_k(B)\big) = \mathcal{M}_k(H)$.
3. *If θ is degree preserving, then $\mathsf{D}_k(H) = \mathsf{D}_k(B)$ for all $k \in \mathbb{N}$.*

Proof 1. and 2. follow from [40, Proposition 3.2.3]. In order to prove 3., note that for all $k \in \mathbb{N}$ we have

$$\begin{aligned} \mathsf{D}_k(H) &= \sup\{|a|_H \colon a \in \mathcal{M}_k(H)\} = \sup\{|\theta(a)|_B \colon \theta(a) \in \mathcal{M}_k(B)\} \\ &= \sup\{|b|_B \colon b \in \mathcal{M}_k(B)\} = \mathsf{D}_k(B). \end{aligned}$$

The first examples of transfer homomorphisms in the literature start from a Krull monoid to its associated monoid of zero-sum sequences which is a Krull monoid having a combinatorial flavor. These ideas were generalized widely, and there are transfer homomorphisms from weakly Krull monoids to (simpler) weakly Krull monoids (having a combinatorial flavor) and the same is true for C-monoids.

Proposition 3.7 *Let H be a Krull monoid, $\varphi \colon H \to \mathcal{F}(P)$ be a cofinal divisor homomorphism with class group $G = \mathcal{C}(\varphi)$, and let $G^* \subset G$ denote the set of classes containing prime divisors. Let $\widetilde{\theta} \colon \mathcal{F}(P) \to \mathcal{F}(G^*)$ denote the unique homomorphism defined by $\widetilde{\theta}(p) = [p]$ for all $p \in P$, and set $\theta = \widetilde{\theta} \circ \varphi \colon H \to \mathcal{B}(G^*)$.*

1. *θ is a transfer homomorphism.*
2. *For $a \in H$, we set $|a| = |\varphi(a)|$ and for $S \in \mathcal{B}(G^*)$ we set $|S| = |S|_{\mathcal{F}(G^*)}$. Then $|a| = |\theta(a)|$ for all $a \in H$, $\theta(\mathcal{M}_k^*(H)) = \mathcal{M}_k^*(G^*)$ and $\theta^{-1}(\mathcal{M}_k^*(G^*)) = \mathcal{M}_k^*(H)$ for all $k \in \mathbb{N}$. Furthermore, $\mathsf{e}(H) = \mathsf{e}(G^*)$, $\eta(H) = \eta(G^*)$, and $\mathsf{D}_k(H) = \mathsf{D}_k(G^*)$ for all $k \in \mathbb{N}$.*

Proof 1. follows from [40, Sect. 3.4]. By definition, we have $|a| = |\theta(a)|$ for all $a \in H$. Thus, the assertions on $\mathsf{D}_k(H)$ follow from Proposition 2.7, and the remaining statements can be derived in a similar way.

The above transfer homomorphism $\theta \colon H \to \mathcal{B}(G^*)$ constitutes the link between the arithmetic of Krull monoids on the one side and zero-sum theory on the other side. In this way, methods from Arithmetic Combinatorics can be used to obtain precise results for arithmetical invariants describing the arithmetic of H. For an overview of this interplay see [37].

There is a variety of transfer homomorphisms from monoids of zero-sum sequences to monoids of zero-sum sequences in order to simplify specific structural features of the involved subsets of groups. Below we present a simple example of such a transfer homomorphism which we will meet again in Proposition 4.9

(for more of this nature we refer to [74] and to [40, Theorem 6.7.11]). Let G be abelian and let $G_0 \subset G$ be a subset. For $g \in G_0$ we define

$$e(G_0, g) = \gcd\big(\{\mathsf{v}_g(B) \colon B \in \mathscr{B}(G_0)\}\big),$$

and it is easy to check that (for details see [43, Lemma 3.4])

$$\begin{aligned} e(G_0, g) &= \gcd\big(\{\mathsf{v}_g(A) \colon A \in \mathscr{A}(G_0)\}\big) \\ &= \min\big(\{\mathsf{v}_g(A) \colon \mathsf{v}_g(A) > 0, A \in \mathscr{A}(G_0)\}\big) \\ &= \min\big(\{\mathsf{v}_g(B) \colon \mathsf{v}_g(B) > 0, B \in \mathscr{B}(G_0)\}\big) \\ &= \min\big(\{k \in \mathbb{N} \colon kg \in \langle G_0 \setminus \{g\}\rangle\}\big) = \gcd\big(\{k \in \mathbb{N} \colon kg \in \langle G_0 \setminus \{g\}\rangle\}\big). \end{aligned}$$

Proposition 3.8 *Let G be abelian and $G_0, G_1, G_2 \subset G$ be subsets such that $G_0 = G_1 \uplus G_2$. For $g \in G_0$ we set $e(g) = e(G_0, g)$ and we define $G_0^* = \{e(g)g \colon g \in G_1\} \cup G_2$. Then, the map*

$$\begin{aligned} \theta \colon \mathscr{B}(G_0) &\longrightarrow \mathscr{B}(G_0^*) \\ B = \prod_{g \in G_0} g^{[\mathsf{v}_g(B)]} &\longmapsto \prod_{g \in G_1} (e(g)g)^{[\mathsf{v}_g(B)/e(g)]} \prod_{g \in G_2} g^{[\mathsf{v}_g(B)]} \end{aligned}$$

is a transfer homomorphism.

Proof Clearly, θ is a surjective homomorphism satisfying $\theta^{-1}(1_{\mathscr{F}(G_0)}) = \{1_{\mathscr{F}(G_0)}\}$. In order to verify property (**T2**) of Definition 3.5, let $B \in \mathscr{B}(G_0)$ and $C_1, C_2 \in \mathscr{B}(G_0^*)$ be such that $\theta(B) = C_1 \cdot C_2$. We have to show that there are $B_1, B_2 \in \mathscr{B}(G_0)$ such that $B = B_1 \cdot B_2$, $\theta(B_1) = C_1$, and $\theta(B_2) = C_2$. This can be checked easily.

3.3 The kth Davenport Constants: The General Case

Let $G_0 \subset G$ be a subset, and $k \in \mathbb{N}$. Recall that $\mathsf{e}(G) = \max\{\mathrm{ord}(g) \colon g \in G\}$. If G is nilpotent, then G is the direct sum of its p-Sylow subgroups and hence $\mathsf{e}(G) = \mathrm{lcm}\{\mathrm{ord}(g) \colon g \in G\} = \exp(G)$. Let

- $\mathsf{E}(G_0)$ be the smallest integer $\ell \in \mathbb{N}$ such that every sequence $S \in \mathscr{F}(G_0)$ of length $|S| \ge \ell$ has a product-one subsequence of length $|G|$.
- $\mathsf{s}(G_0)$ denote the smallest integer $\ell \in \mathbb{N}$ such that every sequence $S \in \mathscr{F}(G_0)$ of length $|S| \ge \ell$ has a product-one subsequence of length $\mathsf{e}(G)$.

The Davenport constants, together with the Erdős–Ginzburg–Ziv constant $\mathsf{s}(G)$, the constants $\eta(G)$ and $\mathsf{E}(G)$, are the most classical zero-sum invariants whose study (in the abelian setting) goes back to the early 1960s. The kth Davenport constants $\mathsf{D}_k(G)$ were introduced by Halter-Koch [50] and further studied in [40, Sect. 6.1] and [30] (all this work is done in the abelian setting). First results in the non-abelian setting were achieved in [15].

If G is abelian, then W. Gao proved that $\mathsf{E}(G) = |G| + \mathsf{d}(G)$. For cyclic groups this is the Theorem of Erdős–Ginzburg–Ziv which dates back to 1961 [40, Proposition 5.7.9]. W. Gao and J. Zhuang conjectured that the above equality holds true for all finite groups [82, Conjecture 2], and their conjecture has been verified in a variety of special cases [3, 33, 34, 53]. For more in the non-abelian setting see [79, 80].

We verify two simple properties occurring in the assumptions of Propositions 2.7 and 2.8.

- If $S \in \mathscr{F}(G)$ and $g_0 \in \pi(S)$, then $h = g_0^{-1} \in G$ is a prime in $\mathscr{F}(G)$ and $h \cdot S \in \mathscr{B}(G)$.
- Clearly, $1_G \in \mathscr{B}(G)$ is a prime element of $\mathscr{B}(G)$.

Therefore, all properties proved in Propositions 2.7 and 2.8 for $\mathsf{D}_k(H)$ and $\mathsf{d}_k(H)$ hold for the constants $\mathsf{D}_k(G)$ and $\mathsf{d}_k(G)$ (the linearity properties as given in Propositions 2.7.2 and 2.8.2.(b) were first proved by Freeze and W.A. Schmid in case of abelian groups G [30]). We continue with properties which are more specific.

Proposition 3.9 *Let $H \le G$ be a subgroup, $N \lhd G$ be a normal subgroup, and $k, \ell \in \mathbb{N}$.*

1. $\mathsf{d}_k(N) + \mathsf{d}_\ell(G/N) \le \mathsf{d}_{k+\ell-1}(G)$.
2. $\mathsf{d}_k(G) \le \mathsf{d}_{\mathsf{d}_k(N)+1}(G/N)$.
3. $\mathsf{d}_k(G) + 1 \le [G{:}H](\mathsf{d}_k(H) + 1)$.
4. $\mathsf{d}_k(G) + 1 \le k(\mathsf{d}(G) + 1)$.
5. $\mathsf{D}_k(G) \le [G{:}H]\mathsf{D}_k(H)$.

Proof 1. Let $\overline{S} = (g_1N) \cdot \ldots \cdot (g_sN) \in \mathscr{M}_\ell^*(G/N)$ with $|\overline{S}| = s = \mathsf{d}_\ell(G/N)$ and let $T = h_1 \cdot \ldots \cdot h_t \in \mathscr{M}_k^*(N)$ with $t = \mathsf{d}_k(N)$. We consider the sequence $W = g_1 \cdot \ldots \cdot g_s \cdot h_1 \cdot \ldots \cdot h_t \in \mathscr{F}(G)$ and suppose that it is divisible by $S_1 \cdot \ldots \cdot S_a \cdot T_1 \cdot \ldots \cdot T_b$ where $S_i, T_j \in \mathscr{B}(G) \setminus \{1_{\mathscr{F}(G)}\}$, $\operatorname{supp}(S_i) \cap \{g_1, \ldots, g_s\} \ne \emptyset$ and $T_1 \cdot \ldots \cdot T_b \mid h_1 \cdot \ldots \cdot h_t$ for all $i \in [1, a]$ and all $j \in [1, b]$. For $i \in [1, a]$, let $\overline{S_i} \in \mathscr{F}(G/N)$ denote the sequence obtained from S_i by replacing each g_ν by $g_\nu N$ and by omitting the elements of S_i which lie in $\{h_1, \ldots, h_t\}$. Then $\overline{S_1}, \ldots, \overline{S_a} \in \mathscr{B}(G/N) \setminus \{1_{\mathscr{F}(G)}\}$ and $\overline{S_1} \cdot \ldots \cdot \overline{S_a} \mid \overline{S}$ whence $a \le \ell - 1$. By construction, we have $b \le k - 1$ whence $a + b < k + \ell - 1$, $W \in \mathscr{M}_{k+\ell-1}^*(G)$, and $|W| = s + t = \mathsf{d}_k(N) + \mathsf{d}_\ell(G/N) \le \mathsf{d}_{k+\ell-1}(G)$.

2. We set $m = \mathsf{d}_{\mathsf{d}_k(N)+1}(G/N) + 1$. By (1), we have to show that every sequence S over G of length $|S| \ge m$ is divisible by a product of k nontrivial product-one sequences. Let $f \colon G \to G/N$ denote the canonical epimorphism and let $S \in \mathscr{F}(G)$ be a sequence of length $|S| \ge m$. By definition of m, there exist sequences $S_1, \ldots, S_{\mathsf{d}_k(N)+1}$ such that $S_1 \cdot \ldots \cdot S_{\mathsf{d}_k(N)+1} \mid S$ and $f(S_1), \ldots, f(S_{\mathsf{d}_k(N)+1})$ are product-one sequences over G/N. Thus, for each $\nu \in [1, \mathsf{d}_k(N) + 1]$, there are elements $h_\nu \in N$ such that $h_\nu \in \pi(S_\nu)$. Then $T = h_1 \cdot \ldots \cdot h_{\mathsf{d}_k(N)+1}$ is a sequence

over N, and it has k nontrivial product-one subsequences $T_1, \dots, T_k$ whose product $T_1 \cdot \ldots \cdot T_k$ divides T. Therefore we obtain k nontrivial product-one sequences whose product divides S.

3. We set $m = [G:H]$ and start with the following assertion.

A. If $S \in \mathcal{F}(G)$ with $|S| \ge m$, then $\Pi(S) \cap H \neq \emptyset$.

Proof of **A**. Let $S = g_1 \cdot \ldots \cdot g_n \in \mathcal{F}(G)$ with $|S| = n \ge m$. We consider the left cosets $g_1H, g_1g_2H, \dots, g_1 \dots g_mH$. If one of these cosets equals H, then we are done. If this is not the case, then there are $k, \ell \in [1, m]$ with $k < \ell$ such that $g_1 \dots g_kH = g_1 \dots g_kg_{k+1} \dots g_\ell H$ which implies that $g_{k+1} \dots g_\ell \in H$. □(Proof of **A**)

Now let $S \in \mathcal{F}(G)$ be a sequence of length $|S| = [G:H](\mathsf{d}_k(H) + 1)$. We have to show that S is divisible by a product of k nontrivial product-one sequences. By **A**, there are $\mathsf{d}_k(H) + 1$ sequences $S_1, \dots, S_{\mathsf{d}_k(H)+1}$ and elements $h_1, \dots, h_{\mathsf{d}_k(H)+1} \in H$ such that $S_1 \cdot \ldots \cdot S_{\mathsf{d}_k(H)+1} \mid S$ and $h_\nu \in \pi(S_\nu)$ for each $\nu \in [1, \mathsf{d}_k(H) + 1]$. By definition, the sequence $h_1 \cdot \ldots \cdot h_{\mathsf{d}_k(H)+1} \in \mathcal{F}(H)$ is divisible by a product of k nontrivial product-one sequences. Therefore S is divisible by a product of k nontrivial product-one sequences.

4. Let $S \in \mathcal{F}(G)$ be a sequence of length $|S| = k(\mathsf{d}(G) + 1)$. Then S may be written as a product $S = S_1 \cdot \ldots \cdot S_k$ where $S_1, \dots, S_k \in \mathcal{F}(G)$ with $|S_\nu| = \mathsf{d}(G) + 1$ for every $\nu \in [1, k]$. Then each S_ν is divisible by a nontrivial product-one sequence T_ν, and hence, S is divisible by $T_1 \cdot \ldots \cdot T_k$. Thus by (1) we infer that $\mathsf{d}_k(G) + 1 \le k(\mathsf{d}(G) + 1)$.

5. Let $A = g_1 \cdot \ldots \cdot g_\ell \in \mathcal{B}(G)$ with $g_1 \dots g_\ell = 1$ and $\ell > [G : H]\mathsf{D}_k(H)$. We show that $\ell > \mathsf{D}_k(G)$. We set $d = \mathsf{D}_k(H)$ and consider the left H-cosets $C_j = g_1 \dots g_jH$ for each $j \in [1, \ell]$. By the pigeonhole principle there exist $1 \le i_1 < \dots < i_{d+1} \le \ell$ such that $C_{i_1} = \dots = C_{i_{d+1}}$. We set $h_s = g_{i_s+1} \dots g_{i_{s+1}}$ for each $s \in [1, d]$ and $h_{d+1} = g_{i_{d+1}+1} \dots g_\ell g_1 \dots g_{i_1-1}$. Clearly $h_1, \dots, h_{d+1} \in H$, and $g_1 \cdots g_\ell = 1$ implies $h_1 \cdots h_{d+1} = 1$ whence $h_1 \cdot \ldots \cdot h_{d+1} \in \mathcal{B}(H)$. The inequality $d + 1 > \mathsf{D}_k(H)$ implies that $h_1 \cdot \ldots \cdot h_{d+1} = S_1 \cdot \ldots \cdot S_{k+1}$, where $1_{\mathcal{F}(H)} \neq S_i \in \mathcal{B}(H)$ for $i \in [1, k + 1]$. Let $T_i \in \mathcal{F}(G)$ denote the sequence obtained from S_i by replacing each occurrence of h_s by $g_{i_s+1} \cdot \ldots \cdot g_{i_{s+1}}$ for $s \in [1, d]$ and h_{d+1} by $g_{i_{d+1}+1} \cdot \ldots \cdot g_\ell \cdot g_1 \cdot \ldots \cdot g_{i_1-1}$. Then $T_1, \dots, T_{k+1} \in \mathcal{B}(G)$ and $A = g_1 \cdot \ldots \cdot g_\ell = T_1 \cdot \ldots \cdot T_{k+1}$, which implies that $\ell > \mathsf{D}_k(G)$.

Much more is known for the classical Davenport constants $\mathsf{D}_1(G) = \mathsf{D}(G)$ and $\mathsf{d}_1(G) = \mathsf{d}(G)$. We start with metacyclic groups of index two. The following result was proved in [39, Theorem 1.1].

Theorem 3.10 *Suppose that G has a cyclic, index 2 subgroup. Then*

$$\mathsf{D}(G) = \mathsf{d}(G) + |G'| \quad \text{and} \quad \mathsf{d}(G) = \begin{cases} |G| - 1 & \text{if } G \text{ is cyclic} \\ \frac{1}{2}|G| & \text{if } G \text{ is noncyclic,} \end{cases}$$

where $G' = [G, G]$ is the commutator subgroup of G.

The next result gathers upper bounds for the large Davenport constant (for $\mathsf{d}(G)$ see [36]).

Theorem 3.11 *Let $G' = [G, G]$ denote the commutator subgroup of G.*

1. $\mathsf{D}(G) \le \mathsf{d}(G) + 2|G'| - 1$, *and equality holds if and only if G is abelian.*
2. *If G is a non-abelian p-group, then* $\mathsf{D}(G) \le \frac{p^2+2p-2}{p^3}|G|$.
3. *If G is non-abelian of order pq, where p, q are primes with $p < q$, then* $\mathsf{D}(G) = 2q$ *and* $\mathsf{d}(G) = q + p - 2$.
4. *If $N \lhd G$ is a normal subgroup with $G/N \cong C_p \oplus C_p$ for some prime p, then*

$$\mathsf{d}(G) \le (\mathsf{d}(N) + 2)p - 2 \le \frac{1}{p}|G| + p - 2.$$

5. *If G is noncyclic and p is the smallest prime dividing $|G|$, then* $\mathsf{D}(G) \le \frac{2}{p}|G|$.
6. *If G is neither cyclic nor isomorphic to a dihedral group of order $2n$ with odd n, then* $\mathsf{D}(G) \le \frac{3}{4}|G|$.

Proof All results can be found in [48]: see Lemma 4.2, Theorems 3.1, 4.1, 5.1, 7.1, 7.2, and Corollary 5.7.

Corollary 3.12 *The following statements are equivalent:*

(a) *G is cyclic or isomorphic to a dihedral group of order $2n$ for some odd $n \ge 3$.*

(b) $\mathsf{D}(G) = |G|$.

Proof If G is not as in (a), then $\mathsf{D}(G) \le \frac{3}{4}|G|$ by Theorem 3.11.6. If G is cyclic, then $\mathsf{D}(G) = |G|$ by Lemma 3.1.3. If G is dihedral of order $2n$ for some odd $n \ge 3$, then the commutator subgroup G' of G has order n and hence $\mathsf{D}(G) = |G|$ by Theorem 3.10.

3.4 The kth Davenport Constants: The Abelian Case

Throughout this subsection, all groups are abelian and will be written additively.

We have $G \cong C_{n_1} \oplus \cdots \oplus C_{n_r}$, with $r \in \mathbb{N}_0$ and $1 < n_1 \mid \ldots \mid n_r$, $\mathsf{r}(G) = r$ is the *rank* of G and $n_r = \exp(G)$ is the *exponent* of G. We define

$$\mathsf{d}^*(G) = \sum_{i=1}^{r}(n_i - 1).$$

If $G = \{0\}$, then $r = 0 = \mathsf{d}^*(G)$. An s-tuple $(e_1, \ldots, e_s)$ of elements of $G \setminus \{0\}$ is said to be a *basis* of G if $G = \langle e_1 \rangle \oplus \cdots \oplus \langle e_s \rangle$. First, we provide a lower bound for the Davenport constants.

Lemma 3.13 *Let G be abelian.*

1. $\mathsf{D}_k(G) = 1 + \mathsf{d}_k(G)$ *for every* $k \in \mathbb{N}$.
2. $\mathsf{d}^*(G) + (k-1)\exp(G) \le \mathsf{d}_k(G)$.

Proof 1. Let $k \in \mathbb{N}$. By Proposition 2.8.1, we have $1 + \mathsf{d}_k(G) \le \mathsf{D}_k(G)$. Obviously, the map

$$\psi \colon \mathcal{M}_k^*(G) \to \mathcal{M}_k(G) \setminus \{1\}, \quad \text{given by} \quad \psi(S) = (-\sigma(S)) \cdot S,$$

is surjective and we have $|\psi(S)| = 1 + |S|$ for every $S \in \mathcal{M}_k^*(G)$. Therefore, we have $1 + \mathsf{d}_k(G) = \mathsf{D}_k(G)$.

2. Suppose that $G \cong C_{n_1} \oplus \cdots \oplus C_{n_r}$, with $r \in \mathbb{N}_0$ and $1 < n_1 \mid \ldots \mid n_r$. If $(e_1, \ldots, e_r)$ is a basis of G with $\operatorname{ord}(e_i) = n_i$ for all $i \in [1, r]$, then

$$S = e_r^{[n_r(k-1)]} \prod_{i=1}^{r} e_i^{[n_i - 1]}$$

is not divisible by a product of k nontrivial zero-sum sequences whence $\mathsf{d}^*(G) + (k-1)\exp(G) = |S| \le \mathsf{d}_k(G)$.

We continue with a result on the kth Davenport constant which refines the more general results in Sect. 2.5. It provides an explicit formula for $\mathsf{d}_k(G)$ in terms of $\mathsf{d}(G)$ (see [40, Theorem 6.1.5]).

Theorem 3.14 *Let G be abelian,* $\exp(G) = n$, *and* $k \in \mathbb{N}$.

1. *Let* $H \le G$ *be a subgroup such that* $G = H \oplus C_n$. *Then*

$$\mathsf{d}(H) + kn - 1 \le \mathsf{d}_k(G) \le (k-1)n + \max\{\mathsf{d}(G), \eta(G) - n - 1\}.$$

   In particular, if $\mathsf{d}(G) = \mathsf{d}(H) + n - 1$ *and* $\eta(G) \le \mathsf{d}(G) + n + 1$, *then* $\mathsf{d}_k(G) = \mathsf{d}(G) + (k-1)n$.
2. *If* $\mathsf{r}(G) \le 2$, *then* $\mathsf{d}_k(G) = \mathsf{d}(G) + (k-1)n$.
3. *If G is a p-group and* $\mathsf{D}(G) \le 2n - 1$, *then* $\mathsf{d}_k(G) = \mathsf{d}(G) + (k-1)n$.

For the rest of this section, we focus on the classical Davenport constant $\mathsf{D}(G)$. By Lemma 3.13.2, there is the crucial inequality

$$\mathsf{d}^*(G) \le \mathsf{d}(G).$$

We continue with a list of groups for which equality holds. The list is incomplete but the remaining groups for which $\mathsf{d}^*(G) = \mathsf{d}(G)$ is known are of a similar special nature as those listed in Theorem 3.15.3 (see [76] for a more detailed discussion). In particular, it is still open whether equality holds for all groups of rank three (see [76, Sect. 4.1]) or for all groups of the form $G = C_n^r$ (see [47]).

Theorem 3.15 *We have* $\mathsf{d}^*(G) = \mathsf{d}(G)$ *in each of the following cases:*

1. *G is a p-group or has rank* $\mathsf{r}(G) \le 2$.
2. $G = K \oplus C_{km}$ *where* $k, m \in \mathbb{N}$, $p \in \mathbb{P}$ *a prime, m a power of p and* $K \le G$ *is a p-subgroup with* $\mathsf{d}(K) \le m - 1$.
3. $G = C_m^2 \oplus C_{mn}$ *where* $m \in \{2, 3, 4, 6\}$ *and* $n \in \mathbb{N}$.

Proof For 1. see [40] (in particular, Theorems 5.5.9 and 5.8.3) for proofs and historical comments. For 2. see [37, Corollary 4.2.13], and 3. can be found in [5] and [76, Theorem 4.1].

There are infinite series of groups G with $\mathsf{d}^*(G) < \mathsf{d}(G)$. However, the true reason for the phenomenon $\mathsf{d}^*(G) < \mathsf{d}(G)$ is not understood. Here is a simple observation. Suppose that $G = C_{n_1} \oplus \cdots \oplus C_{n_r}$ with $1 < n_1 \mid \ldots \mid n_r$, $I \subset [1, r]$, and let $G' = \oplus_{i \in I} C_{n_i}$. If $\mathsf{d}^*(G') < \mathsf{d}(G')$, then $\mathsf{d}^*(G) < \mathsf{d}(G)$. For series of groups G which have rank four and five and satisfy $\mathsf{d}^*(G) < \mathsf{d}(G)$ we refer to [42, 44]. A standing conjecture for an upper bound on $\mathsf{D}(G)$ states that $\mathsf{d}(G) \le \mathsf{d}^*(G) + \mathsf{r}(G)$. However, the available results are much weaker [6], [40, Theorem 5.5.5].

The remainder of this subsection is devoted to inverse problems with respect to the Davenport constant. Thus the objective is to study the structure of zero-sum free sequences S whose lengths $|S|$ are close to the maximal possible value $\mathsf{d}(G)$.

If G is cyclic of order $n \ge 2$, then an easy exercise shows that S is zero-sum free of length $|S| = \mathsf{d}(G)$ if and only if $S = g^{[n-1]}$ for some $g \in G$ with $\mathrm{ord}(g) = n$. After many contributions since the 1980s, S. Savchev and F. Chen could finally prove a (sharp) structural result. In order to formulate it we need some more terminology. If $g \in G$ is a nonzero element of order $\mathrm{ord}(g) = n$ and

$$S = (n_1 g) \bullet \cdots \bullet (n_\ell g), \quad \text{where} \quad \ell \in \mathbb{N}_0 \quad \text{and} \quad n_1, \ldots, n_\ell \in [1, n],$$

we define

$$\|S\|_g = \frac{n_1 + \cdots + n_\ell}{n}.$$

Obviously, S has sum zero if and only if $\|S\|_g \in \mathbb{N}_0$, and the *index of S* is defined as

$$\mathrm{ind}(S) = \min\{\|S\|_g \colon g \in G \text{ with } G = \langle g \rangle\} \in \mathbb{Q}_{\ge 0}.$$

Theorem 3.16 *Let G be cyclic of order* $|G| = n \ge 3$.

1. *If S is a zero-sum free sequence over G of length* $|S| \ge (n+1)/2$, *then there exist* $g \in G$ *with* $\mathrm{ord}(g) = n$ *and integers* $1 = m_1, \ldots, m_{|S|} \in [1, n-1]$ *such that*
   - $S = (m_1 g) \bullet \cdots \bullet (m_{|S|} g)$
   - $m_1 + \cdots + m_{|S|} < n$ *and* $\Sigma(S) = \{\nu g \colon \nu \in [1, m_1 + \cdots + m_{|S|}]\}$.
2. *If* $U \in \mathcal{A}(G)$ *has length* $|U| \ge \lfloor \frac{n}{2} \rfloor + 2$, *then* $\mathrm{ind}(U) = 1$.

Proof 1. See [71] for the original paper. For the history of the problem and a proof in the present terminology see [37, Chap. 5.1] or [49, Chap. 11].

2. This is a simple consequence of the first part (see [37, Theorem 5.1.8]).

The above result was generalized to groups of the form $G = C_2 \oplus C_{2n}$ by S. Savchev and F. Chen [72]. Not much is known about the number of all minimal zero-sum sequences of a given group. However, the above result allows to give a formula for the number of minimal zero-sum sequences of length $\ell \geq \lfloor \frac{n}{2} \rfloor + 2$ (this formula was first proved by Ponomarenko [66] for $\ell > 2n/3$).

Corollary 3.17 *Let G be cyclic of order $|G| = n \geq 3$, and let $\ell \in \left[\lfloor \frac{n}{2} \rfloor + 2, n \right]$. Then the number of minimal zero-sum sequences $U \in \mathscr{A}(G)$ of length ℓ equals $\Phi(n)\mathsf{p}_\ell(n)$, where $\Phi(n) = |(\mathbb{Z}/n\mathbb{Z})^\times|$ is Euler's Phi function and $\mathsf{p}_\ell(n)$ is the number of integer partitions of n into ℓ summands.*

Proof Clearly, every generating element $g \in G$ and every integer partition $n = m_1 + \cdots + m_\ell$ gives rise to a minimal zero-sum sequence $U = (m_1 g) \cdot \ldots \cdot (m_\ell g)$. Conversely, if $U \in \mathscr{A}(G)$ is of length $|U| = \ell$, then Theorem 3.16.2 implies that there is an element $g \in G$ with $\mathrm{ord}(g) = n$ such that

$$U = (m_1 g) \cdot \ldots \cdot (m_\ell g) \quad \text{where} \quad m_1, \ldots, m_\ell \in [1, n-1] \text{ with } n = m_1 + \cdots + m_\ell. \tag{$*$}$$

Since G has precisely $\Phi(n)$ generating elements, it remains to show that for every $U \in \mathscr{A}(G)$ of length $|U| = \ell$ there is precisely one generating element $g \in G$ with $\|U\|_g = 1$. Let U be as in $(*)$, and assume to the contrary that there are $a \in [2, n-1]$ with $\gcd(a, n) = 1$ and $m_1', \ldots, m_\ell' \in [1, n]$ such that $m_1' + \cdots + m_\ell' = n$ and

$$U = \big(m_1'(ag)\big) \cdot \ldots \cdot \big(m_\ell'(ag)\big).$$

Let $a' \in [2, n-1]$ be such that $aa' \equiv 1 \pmod{n}$. Since

$$\begin{aligned} n &= m_1 + \cdots + m_\ell \geq \mathsf{v}_g(U) + a\mathsf{v}_{ag}(U) + 2(\ell - \mathsf{v}_g(U) - \mathsf{v}_{ag}(U)) \\ &= 2\ell - \mathsf{v}_g(U) + (a-2)\mathsf{v}_{ag}(U) \quad \text{and} \\ n &= m_1' + \cdots + m_\ell' \geq a'\mathsf{v}_g(U) + \mathsf{v}_{ag}(U) + 2(\ell - \mathsf{v}_g(U) - \mathsf{v}_{ag}(U)) \\ &= 2\ell + (a'-2)\mathsf{v}_g(U) - \mathsf{v}_{ag}(U), \end{aligned}$$

it follows that

$$\begin{aligned} (a-1)n &= n + (a-2)n \\ &\geq 2\ell - \mathsf{v}_g(U) + (a-2)\mathsf{v}_{ag}(U) + (a-2)(2\ell + (a'-2)\mathsf{v}_g(U) - \mathsf{v}_{ag}(U)) \\ &= (a-1)2\ell + ((a-2)(a'-2) - 1)\mathsf{v}_g(U), \end{aligned}$$

whence $a = 2, a' = \frac{n+1}{2}$ or $a' = 2, a = \frac{n+1}{2}$ because $\ell \geq \lfloor \frac{n}{2} \rfloor + 2$. By symmetry, we may assume that $a = 2$. Then $\mathsf{v}_g(U) \geq 2\ell - n \geq 2\lfloor \frac{n}{2} \rfloor + 4 - n \geq 3$, and thus $n \geq a'\mathsf{v}_g(U) \geq 3\frac{n+1}{2}$, a contradiction.

The structure of all minimal zero-sum sequences of maximal length $\mathsf{D}(G)$ has been completely determined for rank two groups [31, 35, 68, 75], for groups of the form $G = C_2 \oplus C_2 \oplus C_{2n}$ with $n \geq 2$ [76, Theorem 3.13], and for groups of the form $G = C_2^4 \oplus C_{2n}$ with $n \geq 70$ [8, Theorems 5.8 and 5.9].

4 Multiplicative Ideal Theory of Invariant Rings

After gathering basic material from invariant theory in Sect. 4.1 we construct an explicit divisor theory for the algebra of polynomial invariants of a finite group (see Sect. 4.2). In Sect. 4.3 we present a detailed study of the abelian case as outlined in the Introduction. In Sect. 4.4 we associate a BF-monoid to a G-module whose kth Davenport constant is a lower bound for the kth Noether number.

4.1 *Basics of Invariant Theory*

Let $n = \dim_{\mathbb{F}}(V)$ and let $\rho\colon G \to \mathrm{GL}(n, \mathbb{F})$ be a group homomorphism. Consider the action of G on the polynomial ring $\mathbb{F}[x_1, \ldots, x_n]$ via $\mathbb{F}$-algebra automorphisms induced by $g \cdot x_j = \sum_{i=1}^{n} \rho(g^{-1})_{ji} x_i$. Taking a slightly more abstract point of departure, we suppose that V is a G-module (i.e., we suppose that V is endowed with an action of G via linear transformations). Choosing a basis of V, V is identified with $\mathbb{F}^n$, the group $\mathrm{GL}(n, \mathbb{F})$ is identified with the group $\mathrm{GL}(V)$ of invertible linear transformations of V, and $\mathbb{F}[V] = \mathbb{F}[x_1, \ldots, x_n]$ can be thought of as the symmetric algebra of V^*, the dual G-module of V, in which $(x_1, \ldots, x_n)$ is a basis dual to the standard basis in V. The action on V^* is given by $(g \cdot x)(v) = x(\rho(g^{-1})v)$, where $g \in G, x \in V^*, v \in V$. Note that, if $\mathbb{F}$ is infinite, then $\mathbb{F}[V]$ is the algebra of polynomial functions $V \to \mathbb{F}$, and the action of G on $\mathbb{F}[V]$ is the usual action on functions $V \to F$ induced by the action of G on V via ρ. Denote by $\mathbb{F}(V)$ the quotient field of $\mathbb{F}[V]$, and extend the G-action on $\mathbb{F}[V]$ to $\mathbb{F}(V)$ by

$$g \cdot \frac{f_1}{f_2} = \frac{g \cdot f_1}{g \cdot f_2} \quad \text{for } f_1, f_2 \in \mathbb{F}[V] \quad \text{and } g \in G.$$

We define

$$\mathbb{F}(V)^G = \{f \in \mathbb{F}(V)\colon g \cdot f = f \text{ for all } g \in G\} \subset \mathbb{F}(V) \quad \text{and} \quad \mathbb{F}[V]^G = \mathbb{F}(V)^G \cap \mathbb{F}[V].$$

Then $\mathbb{F}(V)^G \subset \mathbb{F}(V)$ is a subfield and $\mathbb{F}[V]^G \subset \mathbb{F}[V]$ is an $\mathbb{F}$-subalgebra of $\mathbb{F}[V]$, called the *ring of polynomial invariants* of G (the group homomorphism $\rho : G \to \mathrm{GL}(V)$ giving the G-action on V is usually suppressed from the notation). Since every element of $\mathbb{F}(V)$ can be written in the form $f_1 f_2^{-1}$ with $f_1 \in \mathbb{F}[V]$ and $f_2 \in \mathbb{F}[V]^G$, it follows that $\mathbb{F}(V)^G$ is the quotient field of $\mathbb{F}[V]^G$. Next, we summarize some well-known ring theoretical properties of $\mathbb{F}[V]^G$ going back to E. Noether [64].

Theorem 4.1 *Let all notations be as above.*

1. $\mathbb{F}[V]^G \subset \mathbb{F}[V]$ *is an integral ring extension and* $\mathbb{F}[V]^G$ *is normal.*
2. $\mathbb{F}[V]$ *is a finitely generated* $\mathbb{F}[V]^G$*-module, and* $\mathbb{F}[V]^G$ *is a finitely generated* $\mathbb{F}$*-algebra (hence in particular a noetherian domain).*
3. $\mathbb{F}[V]^G$ *is a Krull domain with Krull dimension* $\dim_{\mathbb{F}}(V)$.

Proof 1. To show that $\mathbb{F}[V]^G$ is normal, consider an element $f \in \mathbb{F}(V)^G$ which is integral over $\mathbb{F}[V]^G$. Then f is integral over $\mathbb{F}[V]$ as well, and since $\mathbb{F}[V]$ is normal, it follows that $f \in \mathbb{F}[V] \cap \mathbb{F}(V)^G = \mathbb{F}[V]^G$.

To show that $\mathbb{F}[V]^G \subset \mathbb{F}[V]$ is an integral ring extension, consider an element $f \in \mathbb{F}[V]$ and the polynomial

$$\Phi_f = \prod_{g \in G} (X - gf) \in \mathbb{F}[V][X]. \tag{2}$$

The coefficients of Φ_f are the elementary symmetric functions (up to sign) evaluated at $(gf)_{g \in G}$, and hence, they are in $\mathbb{F}[V]^G$. Thus f is a root of a monic polynomial with coefficients in $\mathbb{F}[V]^G$.

2. For $i \in [1, n]$, we consider the polynomials $\Phi_{x_i}(X)$ (cf. (2)), and denote by $A \subset \mathbb{F}[V]^G \subset \mathbb{F}[V]$ the $\mathbb{F}$-algebra generated by the coefficients of $\Phi_{x_1}, \ldots, \Phi_{x_n}$. By definition, A is a finitely generated $\mathbb{F}$-algebra, and hence, a noetherian domain. Since $x_1, \ldots, x_n$ are integral over A, $\mathbb{F}[V] = A[x_1, \ldots, x_n]$ is a finitely generated (and hence noetherian) A-module. Therefore, the A-submodule $\mathbb{F}[V]^G$ is a finitely generated A-module, and hence, a finitely generated $\mathbb{F}$-algebra.

3. By 1. and 2., $\mathbb{F}[V]^G$ is a normal noetherian domain, and hence a Krull domain by Theorem 2.1.2. Since $\mathbb{F}[V]^G \subset \mathbb{F}[V]$ is an integral ring extension, the Theorem of Cohen–Seidenberg implies that their Krull dimensions coincide, and hence $\dim(\mathbb{F}[V]^G) = \dim(\mathbb{F}[V]) = \dim_{\mathbb{F}}(V)$.

The algebra $\mathbb{F}[V]$ is graded in the standard way (namely, $\deg(x_1) = \cdots = \deg(x_n) = 1$), and the subalgebra $\mathbb{F}[V]^G$ is generated by homogeneous elements. For $\mathbb{F}$-subspaces $S, T \subset \mathbb{F}[V]$ we write ST for the $\mathbb{F}$-subspace in $\mathbb{F}[V]$ spanned by all the products st ($s \in S, t \in T$), and write $S^k = S \ldots S$ (with k factors). The factor algebra of $\mathbb{F}[V]$ by the ideal generated by $\mathbb{F}[V]^G_+$ is usually called the *algebra of coinvariants*. It inherits the grading of $\mathbb{F}[V]$ and is finite dimensional.

Definition 4.2 Let $k \in \mathbb{N}$.

1. Let $\beta_k(G, V)$ be the top degree of the factor space $\mathbb{F}[V]^G_+/(\mathbb{F}[V]^G_+)^{k+1}$, where $\mathbb{F}[V]^G_+$ is the maximal ideal of $\mathbb{F}[V]^G$ spanned by the positive degree homogeneous elements. We call

$$\beta_k(G) = \sup\{\beta_k(G, W) : W \text{ is a } G\text{-module over } \mathbb{F}\}$$

the kth *Noether number* of G.

2. Let $b_k(G, V)$ denote the top degree of the factor algebra $\mathbb{F}[V]/(\mathbb{F}[V]^G_+)^k\mathbb{F}[V]$ and set
$$b_k(G) = \sup\{b_k(G, W) : W \text{ is a } G\text{-module over } \mathbb{F}\}.$$

In the special case $k = 1$ we set

$$\beta(G, V) = \beta_1(G, V), \beta(G) = \beta_1(G), \ b(G, V) = b_1(G, V), \text{ and } b(G) = b_1(G),$$

and $\beta(G)$ is the *Noether number* of G. If $\{f_1, \ldots, f_m\}$ and $\{h_1, \ldots, h_l\}$ are two minimal homogeneous generating sets of $\mathbb{F}[V]^G$, then $m = l$ and, after renumbering if necessary, $\deg(f_i) = \deg(h_i)$ for all $i \in [1, m]$ [61, Proposition 6.19]. Therefore by the Graded Nakayama Lemma [61, Proposition 8.31] we have

$$\beta(G, V) = \max\{\deg(f_i) : i \in [1, m]\},$$

where $\{f_1, \ldots, f_m\}$ is a minimal homogeneous generating set of $\mathbb{F}[V]^G$. Again by the Graded Nakayama Lemma, $b(G, V)$ is the maximal degree of a generator in a minimal system of homogeneous generators of $\mathbb{F}[V]$ as an $\mathbb{F}[V]^G$-module. If $\mathrm{char}(\mathbb{F}) \nmid |G|$, then by [11, Corollary 3.2] we have

$$\beta_k(G) = b_k(G) + 1 \quad \text{and} \quad \beta(G, V) \leq b(G, V) + 1, \tag{3}$$

where the second inequality can be strict. If G is abelian, then $\beta_k(G, V)$ and $b_k(G, V)$ will be interpreted as kth Davenport constants (see Proposition 4.7).

The *regular G-module* V_{reg} has a basis $\{e_g : g \in G\}$ labeled by the group elements, and the group action is given by $g \cdot e_h = e_{gh}$ for $g, h \in G$. More conceptually, one can identify V_{reg} with the space of $\mathbb{F}$-valued functions on G, on which G acts linearly via the action induced by the left multiplication action of G on itself. In this interpretation, the basis element e_g is the characteristic function of the set $\{g\} \subset G$. It was proved in [73] that, if $\mathrm{char}(\mathbb{F}) = 0$, then $\beta(G) = \beta(G, V_{\mathrm{reg}})$. If $\mathbb{F}$ is algebraically closed, each irreducible G-module occurs in V_{reg} as a direct summand with multiplicity equal to its dimension.

Theorem 4.3 *1. If* $\mathrm{char}(\mathbb{F}) \nmid |G|$, *then* $\beta(G) \leq |G|$.
2. *If* $\mathrm{char}(\mathbb{F}) \mid |G|$, *then* $\beta(G) = \infty$.

Proof 1. The case $\mathrm{char}(\mathbb{F}) = 0$ was proved by E. Noether [63] in 1916, and her argument works as well when the characteristic of $\mathbb{F}$ is greater than $|G|$. The general case was shown independently by P. Fleischmann [25] and J. Fogarty [28] (see also [62, Theorem 2.3.3] and [56]. For 2. see [70].

Bounding the Noether number has always been an objective of invariant theory (for recent surveys we refer to [60, 81]; degree bounds are discussed in [10, 17, 26,

54, 78]; see [16] for algorithmic aspects). Moreover, the main motivation to introduce the kth Noether numbers $\beta_k(G)$ [11–13] was to bound the ordinary Noether number $\beta(G)$ via structural reduction (see Sect. 5.1).

4.2 The Divisor Theory of Invariant Rings

Let $G \subset \mathrm{GL}(V)$ and $\chi \in \mathrm{Hom}(G, \mathbb{F}^\bullet)$. Then

$$\mathbb{F}[V]^{G,\chi} = \{f \in \mathbb{F}[V] \colon g \cdot f = \chi(g)f \text{ for all } g \in G\}$$

denotes the space of *relative invariants of weight* χ, and we set

$$\mathbb{F}[V]^{G,\mathrm{rel}} = \bigcup_{\chi \in \mathrm{Hom}(G,\mathbb{F}^\bullet)} \mathbb{F}[V]^{G,\chi}.$$

Clearly, we have $\mathbb{F}[V]^G \subset \mathbb{F}[V]^{G,\mathrm{rel}} \subset \mathbb{F}[V]$, and to simplify notation, we set

$$H = (\mathbb{F}[V]^G \setminus \{0\})_{\mathrm{red}}, \quad D = (\mathbb{F}[V]^{G,\mathrm{rel}} \setminus \{0\})_{\mathrm{red}}, \quad \text{and} \quad E = (\mathbb{F}[V] \setminus \{0\})_{\mathrm{red}}.$$

Since $\mathbb{F}[V]$ is a factorial domain with $\mathbb{F}^\bullet$ as its set of units, $E = \mathcal{F}(P)$ is the free abelian monoid generated by $P = \{\mathbb{F}^\bullet f : f \in \mathbb{F}[V] \text{ is irreducible}\}$. The action of G on $\mathbb{F}[V]$ is via $\mathbb{F}$-algebra automorphisms, so it induces a permutation action of G on E and P. Denote by P/G the set of G-orbits in P. We shall identify P/G with a subset of E as follows: assign to the G-orbit $\{f_1, \ldots, f_r\}$ the element $f_1 \ldots f_r \in E$ (here $f_1, \ldots, f_r \in P$ are distinct).

We say that a nonidentity element $g \in G \subset \mathrm{GL}(V)$ is a *pseudoreflection* if a hyperplane in V is fixed pointwise by g, and g is not unipotent (this latter condition holds automatically if $\mathrm{char}(\mathbb{F})$ does not divide $|G|$, since then a nonidentity unipotent transformation cannot have finite order). We denote by $\mathrm{Hom}^0(G, \mathbb{F}^\bullet) \le \mathrm{Hom}(G, \mathbb{F}^\bullet)$ the subgroup of the character group consisting of the characters that contain all pseudoreflections in their kernels. For each $p \in P$, choose a representative $\tilde{p} \in \mathbb{F}[V]$ in the associate class $p = \mathbb{F}^\bullet \tilde{p}$. We have $\mathfrak{X}(\mathbb{F}[V]) = \{\tilde{p}\mathbb{F}[V] \colon p \in P\}$ because $\mathbb{F}[V]$ is factorial. We set $\mathsf{v}_{\tilde{p}} = \mathsf{v}_p \colon \mathsf{q}(\mathbb{F}[V]^\bullet) = \mathbb{F}(V)^\bullet \to \mathbb{Z}$, and for a subset $X \subset \mathbb{F}(V)$ we write $\mathsf{v}_p(X) = \inf\{\mathsf{v}_p(f) \colon f \in X \setminus \{0\}\}$. The *ramification index* of the prime ideal $\tilde{p}\mathbb{F}[V]$ over $\mathbb{F}[V]^G$ is $e(p) = \mathsf{v}_p(\tilde{p}\mathbb{F}[V] \cap \mathbb{F}[V]^G)$. The ramification index $e(p)$ can be expressed in terms of the *inertia subgroup*

$$I_p = \{g \in G \colon g \cdot f - f \in \tilde{p}\mathbb{F}[V] \quad \text{for all} \quad f \in \mathbb{F}[V]\}.$$

Since $V^\star$ is a G-stable subspace in $\mathbb{F}[V]$, the inertia subgroup I_p acts trivially on $V^\star/(V^\star \cap \tilde{p}\mathbb{F}[V])$. On the other hand I_p acts faithfully on $V^\star$. So if I_p is nontrivial, then $V^\star \cap \tilde{p}\mathbb{F}[V] \neq 0$, implying $\tilde{p} \in V^\star$. Clearly I_p must act trivially on the hyperplane $\mathscr{V}(\tilde{p}) = \{v \in V : \tilde{p}(v) = 0\}$, and hence acts via multiplication by a character $\delta_p \in \mathrm{Hom}(I_p, \mathbb{F}^\bullet)$ on the *one*-dimensional factor space $V/\mathscr{V}(\tilde{p})$. So $\ker(\delta_p)$ is a normal subgroup of I_p (necessarily unipotent hence trivial if $\mathrm{char}(\mathbb{F}) \nmid |G|$) and $I_p = \ker(\delta_p)Z$ decomposes as a semi-direct product of $\ker(\delta_p)$ and a cyclic subgroup Z consisting of pseudoreflections fixing pointwise $\mathscr{V}(\tilde{p})$. So $Z \cong I_p/\ker(\delta_p)$ is isomorphic to a finite subgroup of $\mathbb{F}^\bullet$.

The next Lemma 4.4 is extracted from Nakajima's paper [58].

Lemma 4.4

1. *We have the equality* $e(p) = |Z|$.
2. $\mathsf{v}_p(\mathbb{F}[V]^{G,\chi}) < e(p)$ *for all* $\chi \in \mathrm{Hom}(G, \mathbb{F}^\bullet)$.
3. $\mathsf{v}_p(\mathbb{F}[V]^{G,\chi}) = 0$ *for all* $\chi \in \mathrm{Hom}^0(G, \mathbb{F}^\bullet)$.

Proof 1. By [59, 9.6, Proposition (i)], we have that $e(p) = \mathsf{v}_p(\tilde{p}\mathbb{F}[V] \cap \mathbb{F}[V]^{I_p})$, the ramification index of the prime ideal $\tilde{p}\mathbb{F}[V]$ over the subring of I_p-invariants. Thus, if I_p is trivial, then $e(p) = 1$, and of course $|Z| = 1$. If I_p is nontrivial, then as it was explained above, $\tilde{p}$ is a linear form, which is a relative I_p-invariant with weight δ_p^{-1}, hence $\tilde{p}^{|Z|}$ is an I_p-invariant, implying $\mathsf{v}_p(\tilde{p}\mathbb{F}[V] \cap \mathbb{F}[V]^{I_p}) \le |Z|$. On the other hand $\mathbb{F}[V]^{I_p}$ is contained in $\mathbb{F}[V]^Z$, and the algebra of invariants of the cyclic group Z fixing pointwise the hyperplane $\mathscr{V}(\tilde{p})$ is generated by $\tilde{p}^{|Z|}$ and a subspace of V^* complementary to $\mathbb{F}\tilde{p}$. Thus $\mathsf{v}_p(\tilde{p}\mathbb{F}[V] \cap \mathbb{F}[V]^{I_p}) \ge \mathsf{v}_p(\tilde{p}\mathbb{F}[V] \cap \mathbb{F}[V]^Z) = |Z|$, implying in turn that $e(p) = |Z|$.

2. Take an $h \in \mathbb{F}[V]^G$ with $e(p) = \mathsf{v}_p(h)$. Note that $\mathsf{v}_q(h) = \mathsf{v}_p(h)$ and $\mathsf{v}_q(\mathbb{F}[V]^{G,\chi}) = \mathsf{v}_p(\mathbb{F}[V]^{G,\chi})$ holds for all $q \in G \cdot p$, since $\mathbb{F}[V]^{G,\chi}$ is a G-stable subset in $\mathbb{F}[V]$. Set $S = \{\frac{f}{t} : f \in \mathbb{F}[V], \quad t \in \mathbb{F}[V]^G \setminus \tilde{p}\mathbb{F}[V]\}$. This is a G-stable subring in $\mathsf{q}(\mathbb{F}[V])$ containing $\mathbb{F}[V]$. Consider $S^\chi = S \cap \mathsf{q}(\mathbb{F}[V])^\chi$, where $\mathsf{q}(\mathbb{F}[V])^\chi = \{s \in \mathsf{q}(\mathbb{F}[V]) : g \cdot s = \chi(g)s \quad \text{for all } g \in G\}$. Then $\mathsf{v}_q(S^\chi) = \mathsf{v}_q(\mathbb{F}[V]^{G,\chi})$ for all $q \in G \cdot p$, since for any denominator t of an element $\frac{f}{t}$ of S we have $\mathsf{v}_q(t) = 0$. Now suppose that contrary to our statement we have $e(p) \le \mathsf{v}_p(\mathbb{F}[V]^{G,\chi})$, and hence $\mathsf{v}_q(h) \le \mathsf{v}_q(S^\chi)$ for all $q \in G \cdot p$. In particular this means that $\mathbb{F}[V]^{G,\chi} \neq \{0\}$. Then $\mathsf{v}_q(h^{-1}S^\chi) \ge 0$ holds for all $q \in G \cdot p$. Now S is a Krull domain with $\mathfrak{X}(S) = \{\tilde{q}S : q \in G \cdot p\}$, thus $h^{-1}S^\chi \subset S$ (see the discussion after Theorem 2.1), implying that $S^\chi \subset hS$. Clearly $hS \cap S^\chi = hS^\chi$, so we conclude in turn that $S^\chi \subset hS^\chi$. Iterating this we deduce $\{0\} \neq S^\chi \subset \cap_{n=1}^\infty h^n S$, a contradiction.

3. It is well known that $\mathbb{F}[V]^{G,\chi} \neq \{0\}$ (see the proof of **A4**. below). Write $v = \mathsf{v}_p(\mathbb{F}[V]^{G,\chi})$. Take $f \in \mathbb{F}[V]^{G,\chi}$ with $\mathsf{v}_p(f) = v$, say $f = \tilde{p}^v h$, where $h \in \mathbb{F}[V]$. Note that both f and $\tilde{p}$ are relative invariants of I_p, hence so is h. Therefore $g \cdot h \in \mathbb{F}^\bullet h$, and $\tilde{p} \mid_{\mathbb{F}[V]} (g \cdot h - h)$ for all $g \in I_p$, implying that h is an I_p-invariant. Any $\chi \in \mathrm{Hom}^0(G, \mathbb{F}^\bullet)$ contains I_p in its kernel (the unipotent normal subgroup $\ker(\delta_p)$ of I_p has no non-trivial characters at all, and $Z = I_p/\ker(\delta_p)$ consists of pseudoreflections). Thus f is I_p-invariant as well. Therefore $\tilde{p}^v$ is I_p-invariant, so its weight δ_p^v is trivial. Consequently, the order $|Z|$ of δ_p in $\mathrm{Hom}(I_p, \mathbb{F}^\bullet)$ divides v. We have $e(p) = |Z|$ by 1., and on the other hand $v < e(p)$ by 2., forcing $v = 0$.

For a relative invariant f, we denote by $w(f)$ the weight of f. This induces a homomorphism $w\colon D \to \mathrm{Hom}(G, \mathbb{F}^\bullet)$ assigning to $\mathbb{F}^\bullet f \in D$ the weight $w(f)$ of the relative invariant f. Clearly, w extends to a group homomorphism $w\colon \mathsf{q}(D) \to \mathrm{Hom}(G, \mathbb{F}^\bullet)$. The kernel of w consists of elements of the form $(\mathbb{F}^\bullet h)^{-1}\mathbb{F}^\bullet f$, where $f, h \in \mathbb{F}[V]^{G,\chi}$ for some character χ. Now f/h belongs to $\mathbb{F}(V)^G$, which is the field of fractions of $\mathbb{F}[V]^G$, so there exist $f_1, h_1 \in \mathbb{F}[V]^G$ with $f/h = f_1/h_1$, implying $(\mathbb{F}^\bullet h)^{-1}\mathbb{F}^\bullet f = (\mathbb{F}^\bullet h_1)^{-1}\mathbb{F}^\bullet f_1 \in \mathsf{q}(H)$. Thus $\ker(w) = \mathsf{q}(H)$. Therefore, w induces a monomorphism $\overline{w}\colon \mathsf{q}(D)/\mathsf{q}(H) \to \mathrm{Hom}(G, \mathbb{F}^\bullet)$.

Theorem 4.5 *Let $G \subset \mathrm{GL}(V)$, $H = (\mathbb{F}[V]^G \setminus \{0\})_{\mathrm{red}}$, and $D = (\mathbb{F}[V]^{G,\mathrm{rel}} \setminus \{0\})_{\mathrm{red}}$.*

1. *The embeddings $\mathbb{F}[V]^G \setminus \{0\} \overset{\varphi}{\hookrightarrow} \mathbb{F}[V]^{G,\mathrm{rel}} \setminus \{0\} \overset{\psi}{\hookrightarrow} \mathbb{F}[V]^\bullet$ are cofinal divisor homomorphisms.*
2. *D is factorial, $P/G \subset E$ is the set of prime elements in D, and $\mathcal{C}(\varphi)$ is a torsion group.*
3. *The monoid $D_0 = \{\gcd_D(X)\colon X \subset H \text{ finite}\} \subset D$ is free abelian with basis $\{q^{e(q)}\colon q \in P/G\}$, where $e(q) = \min\{\mathsf{v}_q(h)\colon q \mid_D h \in H\}$, and the embedding $H \hookrightarrow D_0$ is a divisor theory.*
4. *We have $D_0 = \{f \in D\colon w(f) \in \mathrm{Hom}^0(G, \mathbb{F}^\bullet)\}$ and $\overline{w}\mid_{\mathsf{q}(D_0)/\mathsf{q}(H)}\colon \mathcal{C}(\mathbb{F}[V]^G) = \mathsf{q}(D_0)/\mathsf{q}(H) \to \mathrm{Hom}^0(G, \mathbb{F}^\bullet)$ is an isomorphism.*

Theorem 4.5 immediately implies the following corollary which can be found in Benson's book [4, Theorem 3.9.2] and which goes back to Nakajima [58] (see also [27] for a discussion of this theorem).

Corollary 4.6 (Benson–Nakajima) *The class group of $\mathbb{F}[V]^G$ is isomorphic to $\mathrm{Hom}^0(G, \mathbb{F}^\bullet)$, the subgroup of the character group consisting of the characters that contain all pseudoreflections in their kernels.*

Proof (of Theorem 4.5) 1. Since $\mathbb{F}[V]^G = \mathbb{F}(V)^G \cap \mathbb{F}[V]$, the embedding $\psi \circ \varphi\colon \mathbb{F}[V]^G \hookrightarrow \mathbb{F}[V]$ is a divisor homomorphism, and hence φ is a divisor homomorphism. Furthermore, if the quotient of two relative invariants lies in $\mathbb{F}[V]$, then it is a relative invariant whence ψ is a divisor homomorphism. In order to show that the embeddings are cofinal, let $0 \neq f \in \mathbb{F}[V]$ be given. Then $f^* = \prod_{g \in G} gf \in \mathbb{F}[V]^G$ and $f \mid f^*$, so the embedding $\psi \circ \varphi$ is cofinal and hence φ and ψ are cofinal.

2. Suppose that $\{f_1, \ldots, f_r\} \subset \mathbb{F}[V]$ represents a G-orbit in P. Then $g \cdot (f_1 \ldots f_r)$ is a non-zero scalar multiple of $f_1 \ldots f_r$, hence $f_1 \ldots f_r \in \mathbb{F}[V]^{G,\mathrm{rel}}$. This shows that $P/G \subset E$ is in fact contained in D. Conversely, take an irreducible element $\mathbb{F}^\bullet f$ in the monoid D (so f is a relative invariant). Take any irreducible divisor f_1 of f in $\mathbb{F}[V]$. Since $g \cdot f \in \mathbb{F}^\bullet f$, the polynomial $g \cdot f_1$ is also the divisor of f. Denoting by $f_1, \ldots, f_r$ polynomials representing the G-orbit of $\mathbb{F}^\bullet f_1$ in P, we conclude that $f_1 \ldots f_r$ divides f in $\mathbb{F}[V]$, hence $\mathbb{F}^\bullet f_1 \ldots f_r$ divides $\mathbb{F}^\bullet f$ in D as well, so $\mathbb{F}^\bullet f_1 \ldots f_r = \mathbb{F}^\bullet f$. This implies that D is the submonoid of $E = \mathcal{F}(P)$ generated by P/G.

In order to show that $\mathcal{C}(\varphi)$ is a torsion group, let $f \in D$ be given. We have to find an $m \in \mathbb{N}$ such that $f^m \in H$. Clearly, this holds with m being the order in $\mathrm{Hom}(G, F^\bullet)$ of the weight of the relative invariant corresponding to f.

3. Since $\mathscr{C}(\varphi)$ is a torsion group, Proposition 2.2 implies that the embedding $H \hookrightarrow D_0$ is a divisor theory, and that D_0 is free abelian with basis $\{q^{e(q)} : q \in P/G\}$, where $e(q) = \min\{\mathsf{v}_q(h) : q \mid_D h \in H\}$ (note that if $q \in P/G$ is the G-orbit of $p \in P$, then $\mathsf{v}_q(h) = \mathsf{v}_p(h)$, where the latter is the exponent of p in $h \in E = \mathscr{F}(P)$).

4. It remains to prove the following three assertions.

A1. $D_0 = \{f \in D : w(f) \in \mathrm{Hom}^0(G, \mathbb{F}^\bullet)\}$.

A2. $w(D_0) = \mathrm{Hom}^0(G, \mathbb{F}^\bullet)$.

A3. $\overline{w}\mid_{\mathsf{q}(D_0)/\mathsf{q}(H)} : \mathsf{q}(D_0)/\mathsf{q}(H) \to w(D_0)$ is an isomorphism.

Proof of **A1**. Set $D^0 = \{f \in D : w(f) \in \mathrm{Hom}^0(G, \mathbb{F}^\bullet)\}$. We show first $D_0 \subset D^0$. Let χ be a character of G, and assume that $\chi(g) \neq 1$ for some pseudoreflection $g \in G$. Let f be a relative invariant with $w(f) = \chi$. Then for any v with $gv = v$ we have $f(v) = f(g^{-1}v) = (gf)(v) = \chi(g)f(v)$, hence $f(v) = 0$. So $l \mid_{\mathbb{F}[V]} f$, where l is a nonzero linear form on V that vanishes on the reflecting hyperplane of g. Denoting by $l = l_1, \ldots, l_r$ representatives of the G-orbit of $\mathbb{F}^\bullet l$, we find that the relative invariant $q = l_1 \ldots l_r$ divides f. Thus $\gcd_D\{f \in D \mid w(f) = \chi\} \neq 1$. Now suppose that for some $\mathbb{F}^\bullet k \in D_0$ we have that $w(k)$ does not belong to $\mathrm{Hom}^0(G, \mathbb{F}^\bullet)$. By definition of D_0 there exist $h_1, \ldots, h_n \in D$ with $\gcd_D(h_1, \ldots, h_n) = 1$ and $kh_1, \ldots, kh_n \in H$. Clearly $w(h_i) = w(k)^{-1} \notin \mathrm{Hom}^0(G, \mathbb{F}^\bullet)$, hence by the above considerations $\gcd_D(h_1, \ldots, h_n) \neq 1$, a contradiction.

Next we show $D^0 \subset D_0$. Let d be an element in the monoid D^0. By Lemma 4.4.3 for any prime divisor $p \in P$ of d there exists an $h_p \in D$ such that $w(h_p) = w(d)^{-1}$ and $p \nmid_E h_p$. Denote by $m > 1$ the order of $w(d)$ in the group of characters. Clearly $d^m \in H$ and $dh_p \in H$. Moreover, $\gcd_E(d^m, dh_p : p \in P, \quad p \mid_E d) = d$.

Proof of **A2**. The statement follows from **A1**, as soon as we show that $\mathbb{F}[V]^{G,\chi} \neq 0$ for all $\chi \in \mathrm{Hom}(G, \mathbb{F}^\bullet)$. For any character $\chi \in \mathrm{Hom}(G, \mathbb{F}^\bullet)$ the group $\bar{G} = G/\ker(\chi)$ is isomorphic to a cyclic subgroup of $\mathbb{F}^\bullet$, hence its order is not divisible by $\mathrm{char}(\mathbb{F})$. Moreover, $\bar{G}$ acts faithfully on the field $T = \mathbb{F}(V)^{\ker(\chi)}$, with $T^{\bar{G}} = \mathbb{F}(V)^G$. By the Normal Basis Theorem, T as a $\bar{G}$-module over $T^{\bar{G}}$ is isomorphic to the regular representation of $\bar{G}$, hence contains the representation χ as a summand with multiplicity 1. This shows in particular that $T^{\bar{G}}$ contains a relative invariant of weight χ. Multiplying this by an appropriate element of $T^{\bar{G}} \cap \mathbb{F}[V] = \mathbb{F}[V]^G$ we get an element of $\mathbb{F}[V]^{G,\chi}$. So all characters of G occur as the weight of a relative invariant in $\mathbb{F}[V]$.

Proof of **A3**. Since $\overline{w} : \mathsf{q}(D)/\mathsf{q}(H) \to \mathrm{Hom}(G, \mathbb{F}^\bullet)$ is a monomorphism, the map $\overline{w}\mid_{\mathsf{q}(D_0)/\mathsf{q}(H)} : \mathsf{q}(D_0)/\mathsf{q}(H) \to w(\mathsf{q}(D_0))$ is an isomorphism. Note finally that $w(\mathsf{q}(D_0)) = \mathsf{q}(w(D_0)) = w(D_0)$.

As already mentioned, not only the class group but also the distribution of prime divisors in the classes is crucial for the arithmetic of the domain. Moreover, the class group together with the distribution of prime divisors in the classes are characteristic (up to units) for the domain. For a precise formulation we need one more definition.

Let H be a Krull monoid, $H_{\mathrm{red}} \hookrightarrow \mathscr{F}(\mathscr{P})$ a divisor theory, and let G be an abelian group and $(m_g)_{g \in G}$ be a family of cardinal numbers. We say that H has *characteristic*

$(G, (m_g)_{g\in G})$ if there is a group isomorphism $\Phi : G \to \mathcal{C}(H)$ such that $m_g = |\mathcal{P} \cap \Phi(g)|$. Two reduced Krull monoids are isomorphic if and only if they have the same characteristic [40, Theorem 2.5.4]. We pose the following problem.

Problem 1 Let G be a finite group, $\mathbb{F}$ be a field, and V be a finite dimensional $\mathbb{F}$-vector space endowed with a linear action of G. Determine the characteristic of $\mathbb{F}[V]^G$.

Let all assumptions be as in Problem 1 and suppose further that G acts trivially on one variable. Then $\mathbb{F}[V]^G$ is a polynomial ring in this variable and hence every class contains a prime divisor by [29, Theorem 14.3].

4.3 The Abelian Case

Throughout this subsection, suppose that G is abelian, $\mathbb{F}$ *is algebraically closed, and* $\mathrm{char}(\mathbb{F}) \nmid |G|$.

The assumption on algebraic closedness is not too restrictive, since for any field $\mathbb{F}$ the set $\mathbb{F}[V]^G$ spans the ring of invariants over the algebraic closure $\overline{\mathbb{F}}$ as a vector space over $\overline{\mathbb{F}}$. The assumption on the characteristic guarantees that every G-module is completely reducible (i.e., is the direct sum of irreducible G-modules). The dual space V^* has a basis $\{x_1, \ldots, x_n\}$ consisting of G-eigenvectors whence $g \cdot x_i = \chi_i(g)x_i$ for all $i \in [1, n]$ where $\chi_1, \ldots, \chi_n \in \mathrm{Hom}(G, \mathbb{F}^\bullet)$. We set $\widehat{G} = \mathrm{Hom}(G, \mathbb{F}^\bullet)$, $\widehat{G}_V = \{\chi_1, \ldots, \chi_n\} \subset \widehat{G}$, and note that $G \cong \widehat{G}$. Recall that a completely reducible H-module W (for a not necessarily abelian group H) is called *multiplicity free* if it is the direct sum of pairwise non-isomorphic irreducible H-modules. In our case V is multiplicity free if and only if the characters $\chi_1, \ldots, \chi_n$ are pairwise distinct.

It was B. Schmid [73, Sect. 2] who first formulated a correspondence between a minimal generating system of $\mathbb{F}[V]^G$ and minimal product-one sequences over the character group (see also [24]). The next proposition describes in detail the structural interplay. In particular, Proposition 4.7.2 shows that all (direct and inverse) results on minimal zero-sum sequences over $\widehat{G}_V$ (see Sects. 3.3 and 3.4) carry over to $\mathcal{A}(M^G)$.

Proposition 4.7 *Let* $M \subset \mathbb{F}[x_1, \ldots, x_n]$ *be the multiplicative monoid of monomials,* $\psi : M \to \mathcal{F}(\widehat{G}_V)$ *be the unique monoid homomorphism defined by* $\psi(x_i) = \chi_i$ *for all* $i \in [1, n]$, *and let* $M^G \subset M$ *denote the submonoid of G-invariant monomials.*

1. $\mathbb{F}[V]^G$ *has* M^G *as an* $\mathbb{F}$*-vector space basis, and* $\mathbb{F}[V]^G$ *is minimally generated as an* $\mathbb{F}$*-algebra by* $\mathcal{A}(M^G)$.
2. *The homomorphism* $\psi : M \to \mathcal{F}(\widehat{G}_V)$ *and its restriction* $\psi \mid_{M^G} : M^G \to \mathcal{B}(\widehat{G}_V)$ *are degree-preserving transfer homomorphisms. Moreover,* M^G *is a reduced finitely generated Krull monoid, and* $\mathcal{A}(M^G) = \psi^{-1}\big(\mathcal{A}(\widehat{G}_V)\big)$.
3. $\psi \mid_{M^G}$ *is an isomorphism if and only if V is a multiplicity free G-module.*
4. $\beta_k(G, V) = \mathsf{D}_k(M^G) = \mathsf{D}_k(\widehat{G}_V)$ *and* $\beta_k(G) = \mathsf{D}_k(G)$ *for all* $k \in \mathbb{N}$.

Proof 1. Each monomial spans a G-stable subspace in $\mathbb{F}[V]$, hence a polynomial is G-invariant if and only if all its monomials are G-invariant, so M^G spans $\mathbb{F}[V]^G$. The elements of M^G are linearly independent, therefore $\mathbb{F}[V]^G$ can be identified with the monoid algebra of M^G over $\mathbb{F}$, which shows the second statement.

2. M and $\mathscr{F}(\widehat{G}_V)$ are free abelian monoids and ψ maps primes onto primes. Thus $\psi : M \to \mathscr{F}(\widehat{G}_V)$ is a surjective degree-preserving monoid homomorphism and it is a transfer homomorphism. Let $\pi : \mathscr{F}(\widehat{G}) \to \widehat{G}$ be the monoid homomorphism defined by $\pi(\chi) = \chi$ for all $\chi \in \widehat{G}$. Then $\ker(\pi) = \mathscr{B}(\widehat{G})$. Taking into account that for a monomial $m \in M$ G acts on the space $\mathbb{F}m$ via the character $\pi(\psi(m))$, we conclude that for a monomial $m \in M$ we have that $m \in M^G$ if and only if $\psi(m) \in \mathscr{B}(\widehat{G}_V)$. This implies that the restriction $\psi \mid_{M^G}$ of the transfer homomorphism ψ is also a transfer homomorphism. Therefore M^G is generated by $\mathscr{A}(M^G) = \psi^{-1}\big(\mathscr{A}(\widehat{G}_V)\big)$. Since $\mathscr{A}(\widehat{G}_V)$ is finite, and ψ has finite fibers, we conclude that the monoid M^G is finitely generated. Since M is factorial and $\mathbb{F}[V]^G \subset \mathbb{F}[V]$ is saturated by Theorem 4.5, it follows that

$$M \cap \mathsf{q}(M^G) \subset M \cap \mathbb{F}[V] \cap \mathsf{q}(\mathbb{F}[V]^G) \subset M \cap \mathbb{F}[V]^G = M^G$$

whence $M^G \subset M$ is saturated and thus M^G is a Krull monoid.

3. V is a multiplicity free G-module if and only if $\chi_1, \ldots, \chi_n$ are pairwise distinct. Since $\psi : M \to \mathscr{F}(\widehat{G}_V)$ maps the primes $x_1, \ldots, x_n$ of M onto the primes $\chi_1, \ldots, \chi_n$ of $\mathscr{F}(\widehat{G}_V)$, ψ is an isomorphism if and only if $\chi_1, \ldots, \chi_n$ are pairwise distinct.

4. Let $k \in \mathbb{N}$ and $M^G_+ = M^G \setminus \{1\}$. Then $M^G \setminus (M^G_+)^{k+1} = \mathscr{M}_k(M^G)$. Since $\psi \mid_{M^G} : M^G \to \mathscr{B}(\widehat{G}_V)$ is degree-preserving transfer homomorphism, Proposition 3.6.3 implies that $\mathsf{D}_k(M^G) = \mathsf{D}_k(\widehat{G}_V)$. Since $\mathbb{F}[V]^G$ is spanned by M^G, $(\mathbb{F}[V]^G_+)^{k+1}$ is spanned by $(M^G_+)^{k+1}$. Therefore, the top degree of a homogeneous G-invariant not contained in $(\mathbb{F}[V]^G_+)^{k+1}$ coincides with the maximal degree of a monomial in $M^G_+ \setminus (M^G_+)^{k+1} = \mathscr{M}_k(M^G)$. Thus $\beta_k(G, V) = \mathsf{D}_k(M^G)$. For the kth Noether number $\beta_k(G)$ we have

$$\begin{aligned}\beta_k(G) &= \sup\{\beta_k(G, W) : W \text{ is a } G\text{-module over } \mathbb{F}\}\\ &= \sup\{\mathsf{D}_k(\widehat{G}_W) : W \text{ is a } G\text{-module over } \mathbb{F}\} = \mathsf{D}_k(\widehat{G})\end{aligned}$$

because for the regular representation V_{reg} we have $\widehat{G}_{V_{\text{reg}}} = \widehat{G}$.

Recalling the notation of Theorem 4.5, we have

$$H = (\mathbb{F}[V]^G \setminus \{0\})_{\text{red}} \quad \text{and} \quad D_0 = \{\gcd_D(X) : X \subset H \text{ finite}\} \subset D = (\mathbb{F}[V]^{G,\text{rel}} \setminus \{0\})_{\text{red}}.$$

Furthermore, $M \subset \mathbb{F}[V] = \mathbb{F}[x_1, \ldots, x_n]$ is the monoid of monomials, $M^G = M \cap \mathbb{F}[V]^G$, and we can view M as a submonoid of H and then $M^G = M \cap H$. Since $M \subset H$ is saturated, $M = \mathsf{q}(M) \cap H$, and

$$\begin{aligned}\mathsf{q}(M)/\mathsf{q}(M^G) &= \mathsf{q}(M)/\mathsf{q}(M\cap H) = \mathsf{q}(M)/(\mathsf{q}(M)\cap \mathsf{q}(H))\\ &\cong \mathsf{q}(M)\mathsf{q}(H)/\mathsf{q}(H) \subset \mathsf{q}(D)/\mathsf{q}(H),\end{aligned}$$

we consider $\mathsf{q}(M)/\mathsf{q}(M^G)$ as a subset of $\mathsf{q}(D)/\mathsf{q}(H)$.

Proposition 4.8 *Let all notation be as above and set $M_0 = M \cap D_0$.*

1. *$M_0 \subset D_0$ is divisor closed whence M_0 is free abelian, and $\mathcal{A}(M_0) = M \cap \mathcal{A}(D_0) = \{x_1^{e(x_1)}, \dots, x_n^{e(x_n)}\}$.*
2. *We have $e(x_i) = \min\{k \in \mathbb{N} \colon \chi_i^k \in \langle \chi_j \mid j \neq i\rangle\}$.*
3. *$\mathrm{Hom}^0(\rho(G), \mathbb{F}^\bullet)$ is generated by $\{\chi_1^{e(x_1)}, \dots, \chi_n^{e(x_n)}\}$ and $\mathbb{F}[x_1^{e(x_1)}, \dots, x_n^{e(x_n)}] = \mathbb{F}[V]^{G_1}$, where G_1 denotes the subgroup of $\rho(G)$ generated by the pseudoreflections in $\rho(G)$.*
4. *The embedding $M^G \hookrightarrow M_0$ is a divisor theory,*

$$\overline{w}\mid_{\mathsf{q}(M_0)/\mathsf{q}(M^G)} \colon \mathcal{C}(M^G) = \mathsf{q}(M_0)/\mathsf{q}(M^G) \to \mathrm{Hom}^0(\rho(G), \mathbb{F}^\bullet)$$

is an isomorphism, and $\overline{w}(\mathcal{C}(M^G)^*) = \{\chi_1^{e(x_1)}, \dots, \chi_n^{e(x_n)}\}$.

Proof 1. If the product of two polynomials in $\mathbb{F}[V]$ has a single non-zero term, then both polynomials must have only one non-zero term. Thus, if $ab \in M$ for some $a, b \in D$, then both a and b belong to M. Hence $M \subset D$ is divisor closed implying that $M_0 \subset D_0$ is divisor-closed. Therefore $\mathcal{A}(M_0) = M \cap \mathcal{A}(D_0)$.

By Theorem 4.5.3, $\mathcal{A}(D_0) = \{q^{e(q)} \colon q \in \mathcal{A}(D)\}$. The divisor closedness of M in D implies that if $q^{e(q)} \in M$, then $q \in M \cap \mathcal{A}(D) = \mathcal{A}(M) = \{x_1, \dots, x_n\}$. Thus $M \cap \mathcal{A}(D_0) = \{x_1^{e(x_1)}, \dots, x_n^{e(x_n)}\}$.

2. For $i \in [1, n]$, we have

$$e(x_i) = \min\{\mathsf{v}_{x_i}(h) \colon x_i \mid_D h, h \in H\} = \min\{\mathsf{v}_{x_i}(m) \colon x_i \mid_D m, m \in M^G\},$$

where the second equality holds because for all $h \in H$ we have $\mathsf{v}_{x_i}(h) = \min\{\mathsf{v}_{x_i}(m) \colon m$ ranges over the monomials of $h\}$. Note that a monomial $m = \prod_{i=1}^n x_i^{a_i}$ lies in M^G if and only if $\prod_{i=1}^n \chi_i^{[a_i]}$ is a product-one sequence over $\widehat{G}$ if and only if $\chi_i^{a_i} = \prod_{j\neq i} \chi_j^{-a_j}$. Thus $\min\{\mathsf{v}_{x_i}(m) \colon x_i \mid_D m, m \in M^G\} = \min\{k \in \mathbb{N} \colon \chi_i^k \in \langle \chi_j \mid j \neq i\rangle\}$.

3. By Theorem 4.5.4, $\mathrm{Hom}^0(\rho(G), \mathbb{F}^\bullet) = w(D_0)$ and hence $\mathrm{Hom}^0(\rho(G), \mathbb{F}^\bullet)$ is generated by $w(\mathcal{A}(D_0))$. Thus by 1., it remains to show that $\langle w(\mathcal{A}(D_0))\rangle = \langle w(\mathcal{A}(M_0))\rangle$. Since $\mathcal{A}(M_0) \subset \mathcal{A}(D_0)$, it follows that $\langle w(\mathcal{A}(D_0))\rangle \supset \langle w(\mathcal{A}(M_0))\rangle$. To show the reverse inclusion, let $a \in \mathcal{A}(D_0)$. For any monomial m occurring in a, we have $w(m) = w(a)$. By Theorem 4.5.4, $D_0 = \{f \in D \colon w(f) \in \mathrm{Hom}^0(\rho(G), \mathbb{F}^\bullet)\}$ whence $m \in M \cap D_0 = M_0$ and clearly $w(m) \in \langle w(\mathcal{A}(M_0))\rangle$.

Recall that each monomial in $\mathbb{F}[V]$ spans a G-invariant subspace. Thus $f \in \mathbb{F}[V]$ is G_1-invariant if and only if all monomials of f are G_1-invariant. Furthermore, a monomial m is G_1-invariant if and only if $w(m)$ contains G_1 in its kernel; equivalently (by the characterization of D_0) $m \in M \cap D_0 = M_0$. Thus $\mathbb{F}[V]^{G_1}$ is generated by $\mathcal{A}(M_0)$ and hence the assertion follows from 1.

4. Since $M \subset D$, $M_0 \subset D_0$ and $M^G \subset H$ are divisor closed and since the embedding $H \subset D_0$ is a divisor theory (Theorem 4.5.4), $M^G \hookrightarrow M_0$ is a divisor homomorphism into a free abelian monoid. Let $m \in M_0$. Then $m \in D_0$ and there is a finite subset $Y \subset H$ such that $m = \gcd_{D_0}(Y)$. Let $X \subset D_0 \cap M = M_0$ be the set of all monomials occurring in some $y \in Y$. Then $m = \gcd_{D_0}(X) = \gcd_{M_0}(X)$, where the last equality holds because $M_0 \subset D_0$ is divisor closed.

Restricting the isomorphism

$$\overline{w}\,|_{\mathsf{q}(D_0)/\mathsf{q}(H)} \colon \mathcal{C}(\mathbb{F}[V]^G) = \mathsf{q}(D_0)/\mathsf{q}(H) \to \mathrm{Hom}^0(\rho(G), \mathbb{F}^\bullet)$$

from Theorem 4.5, we obtain a monomorphism

$$\overline{w}\,|_{\mathsf{q}(M_0)/\mathsf{q}(M^G)} \colon \mathcal{C}(M^G) = \mathsf{q}(M_0)/\mathsf{q}(M^G) \to \mathrm{Hom}^0(\rho(G), \mathbb{F}^\bullet).$$

By 1. and 3., the image contains the generating set $\{\chi_1^{e(x_1)}, \ldots, \chi_n^{e(x_n)}\}$ of the group $\mathrm{Hom}^0(\rho(G), \mathbb{F}^\bullet)$ and hence the above monomorphism is an isomorphism. The last statement follows from 1. by $\overline{w}(\mathcal{C}(M^G)^*) = \overline{w}(\mathcal{A}(M_0))$.

Proposition 4.9 *Let $M \subset \mathbb{F}[x_1, \ldots, x_n]$ be the multiplicative monoid of monomials, and $M^G \subset M$ the submonoid of G-invariant monomials.*

1. *Every class of $\mathcal{C}(\mathbb{F}[V]^G)$ contains a prime divisor.*
2. *We have the following commutative diagram of monoid homomorphisms*

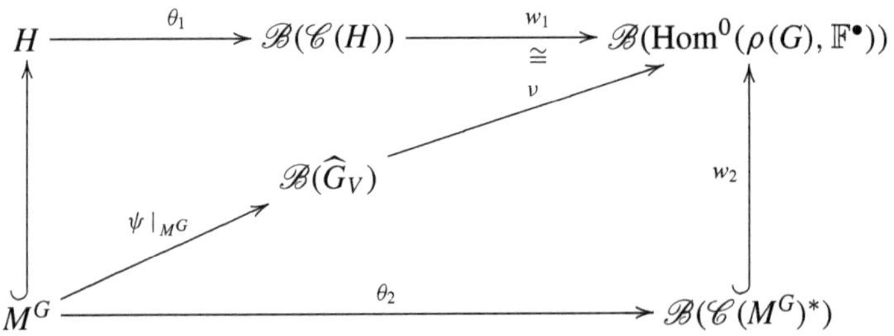

where

- *θ_1 and θ_2 are transfer homomorphisms of Krull monoids as given in Proposition 3.7.*
- *w_1 is the extension to the monoid of product-one sequences of the group isomorphism $\overline{w}\,|_{\mathsf{q}(D_0)/\mathsf{q}(H)}$ given in Theorem 4.5.4*
- *w_2 is the extension to the monoid of product-one sequences of the restriction to $\mathcal{C}(M^G)^*$ of the group isomorphism $\overline{w}\,|_{\mathsf{q}(M_0)/\mathsf{q}(M^G)}$ given in Proposition 4.8*
- *ψ is given in Proposition 4.7.*
- *ν will be defined below (indeed, ν is a transfer homomorphism as given in Proposition 3.8).*

3. *If $\widehat{G}_V = \widehat{G}$, then every class of $\mathcal{C}(M^G)$ contains a prime divisor.*

Proof 1. By Proposition 4.7.1, $\mathbb{F}[V]^G$ is the monoid algebra of M^G over $\mathbb{F}$. Thus, by [7, Theorem 8], every class of $\mathbb{F}[V]^G$ contains a prime divisor.

2. In order to show that the diagram is commutative, we fix an $m \in M^G$. We consider the divisor theory $M^G \hookrightarrow M_0$ from Proposition 4.8 and factorize m in M_0, say $m = \prod_{i=1}^n \left(x_i^{e(x_i)}\right)^{a_i}$ where $a_1, \ldots, a_n \in \mathbb{N}_0$. Since $\overline{w}(x_i^{e(x_i)}) = \chi_i^{e(x_i)}$ for all $i \in [1, n]$, it follows that

$$(w_2 \circ \theta_2)(m) = (\chi_1^{e(x_1)})^{[a_1]} \cdot \ldots \cdot (\chi_n^{e(x_n)})^{[a_n]} \in \mathscr{B}(\mathrm{Hom}^0(\rho(G), \mathbb{F}^\bullet)).$$

Next we view m as an element in H and consider the divisor theory $H \hookrightarrow D_0$. Since $M_0 \subset D_0$ is divisor closed, $m = \prod_{i=1}^n \left(x_i^{e(x_i)}\right)^{a_i}$ is a factorization of m in D_0. Therefore $(w_1 \circ \theta_1)(m) = (w_2 \circ \theta_2)(m)$.

By definition of ψ, we infer that

$$\psi(m) = \chi_1^{[e(x_1)a_1]} \cdot \ldots \cdot \chi_n^{[e(x_n)a_n]}.$$

We define a partition of $\widehat{G}_V = G_1 \cup G_2$, where $G_2 = \{\chi_i \colon \chi_i = \chi_j$ for some distinct $i, j \in [1, n]\}$ and $G_1 = \widehat{G}_V \setminus G_2$. Let $\nu \colon \mathscr{B}(\widehat{G}_V) \to \mathscr{B}(\mathrm{Hom}^0(\rho(G), \mathbb{F}^\bullet))$ be defined as in Proposition 3.8 (with respect to the partition $G_0 = G_1 \uplus G_2$). By Proposition 4.8.2, $e(x_i) = 1$ if $\chi_i \in G_2$, and $e(x_i)$ equals the number $e(\chi_i)$ in Proposition 3.8 if $\chi_i \in G_1$. Therefore it follows that

$$\nu(\psi(m)) = (\chi_1^{e(x_1)})^{[a_1]} \cdot \ldots \cdot (\chi_n^{e(x_n)})^{[a_n]},$$

and hence the diagram commutes.

3. In a finite abelian group all elements are contained in the subgroup generated by the other elements, with the only exception of the generator of a 2-element group. Therefore unless G is the 2-element group and the non-trivial character occurs with multiplicity one in the sequence $\chi_1 \cdot \ldots \cdot \chi_n$, all the $e(x_i) = 1$ by Proposition 4.8.3, and the elements x_i are all prime in M_0, so they represent all the divisor classes, as i varies in $[1, n]$. In the missing case we have $\mathbb{F}[V]^G = \mathbb{F}[x_1, \ldots, x_{n-1}, x_n^2]$ (after a renumbering of the variables if necessary), hence both class groups are trivial, and x_1 and x_2^2 are prime elements in the unique class.

Thus Proposition 4.9.1 gives a partial answer to Problem 1. Using that notation it states that $m_g \geq 1$ for all $g \in \mathscr{C}(\mathbb{F}[V]^G)$.

Example 4.10 The set $\mathscr{C}(M^G)^*$ may be a proper subset of $\mathscr{C}(M^G)$, and consequently the monoid homomorphism $\nu : \mathscr{B}(\widehat{G}_V) \to \mathscr{B}(\mathrm{Hom}^0(\rho(G), \mathbb{F}^\bullet))$ is not surjective in general.

1. Indeed, let G be cyclic of order 3, $g \in G$ with $\mathrm{ord}(g) = 3$, and the action on $\mathbb{F}[x_1, x_2, x_3]$ is given by $g \cdot x_i = \omega x_i$, where ω is a primitive cubic root of 1. Then $\chi_1 = \chi_2 = \chi_3 = \chi$, so $e(x_1) = e(x_2) = e(x_3) = 1$, implying $\overline{w}(\mathscr{C}(M^G)^*) = \{\chi\}$ (each of the x_i is a prime element in the class χ), whereas $\overline{w}(\mathscr{C}(M^G)) = \{\chi, \chi^2, \chi^3 = 1\}$, the 3-element group. Thus $\mathscr{B}(\widehat{G}_V) = \{\chi^{[3k]} \colon k \in \mathbb{N}_0\}$, and $\nu(\mathscr{B}(\widehat{G}_V))$ is the free

abelian monoid $\mathscr{F}(\{\chi^3\})$ generated by $\chi^3 = 1 \in \widehat{G}$. The polynomials $x_1^2 + x_2x_3$ and $x_1^3 + x_2^2x_3$ are irreducible, they are relative invariants of weight χ^2 and χ^3, so they represent prime elements of D_0 in the remaining classes χ^2 and $\chi^3 = 1$.

2. To provide an example with a multiplicity free module, let G be cyclic of order 5, $g \in G$ with $\mathrm{ord}(g) = 5$, and the action on $\mathbb{F}[x_1, x_2, x_3]$ is given by $g \cdot x_1 = \omega x_1$, $g \cdot x_2 = \omega^2 x_2$, $g \cdot x_3 = \omega^3 x_3$, where ω is a primitive fifth root of 1. Then setting $\chi = \chi_1$, we have $\chi_2 = \chi^2$, $\chi_3 = \chi^3$ and $\overline{w}(\mathscr{C}(M^G)) = \langle \chi \rangle$ is the 5-element group, so V is multiplicity free. Still we have $e(x_1) = e(x_2) = e(x_3) = 1$, so $\overline{w}(\mathscr{C}(M^G)^*) = \{\chi, \chi^2, \chi^3\}$ (and x_1, x_2, x_3 are the prime elements of M_0 in these classes). The remaining classes χ^4 and $\chi^5 = 1$ contain the prime elements of D_0 represented by $x_2^2 + x_1x_3$ and $x_1^5 + x_2x_3$.

4.4 A Monoid Associated with G-Modules

Throughout this subsection, suppose that $\mathrm{char}(\mathbb{F}) \nmid |G|$.

In this subsection, we discuss a monoid associated with representations of not necessarily abelian groups which in the case of abelian groups recovers the monoid of G-invariant monomials. Decompose V into the direct sum of G-modules:

$$V = V_1 \oplus \cdots \oplus V_r \tag{4}$$

and denote by $\rho_i : G \to \mathrm{GL}(V_i)$ the corresponding group homomorphisms. Then (4) induces a decomposition of $\mathbb{F}[V]$ into multihomogeneous components as follows. The coordinate ring $\mathbb{F}[V]$ is the symmetric algebra $\mathrm{Sym}(V^*) = \bigoplus_{n=0}^{\infty} \mathrm{Sym}^n(V^*)$. Writing $\mathbb{F}[V]_a = \mathrm{Sym}^{a_1}(V_1^*) \otimes \ldots \otimes \mathrm{Sym}^{a_r}(V_r^*)$ we have $\mathrm{Sym}^n(V^*) = \bigoplus_{|a|=n} \mathbb{F}[V]_a$, and hence $\mathbb{F}[V] = \bigoplus_{a \in \mathbb{N}_0^r} \mathbb{F}[V]_a$. The summands $\mathbb{F}[V]_a$ are G-submodules in $\mathbb{F}[V]$, and $\mathbb{F}[V]_a \mathbb{F}[V]_b \subset \mathbb{F}[V]_{a+b}$, so $\mathbb{F}[V]$ is a $\mathbb{N}_0^r$-graded algebra. Moreover, $\mathbb{F}[V]^G$ is spanned by its multihomogeneous components $\mathbb{F}[V]_a^G = \mathbb{F}[V]^G \cap \mathbb{F}[V]_a$. For $f \in \mathbb{F}[V]_a$ we call a the *multidegree* of f. We are in the position to define

$$\mathscr{B}(G, V) = \{a \in \mathbb{N}_0^r : \mathbb{F}[V]_a^G \neq \{0\}\} \tag{5}$$

the set of multidegrees of multihomogeneous G-invariants. We give precise information on $\mathscr{B}(G, V)$ in terms of quantities associated to the direct summands V_i of V. For $i \in [1, r]$ denote by c_i the greatest common divisor of the elements of $\mathscr{B}(G, V_i)$, and F_i the Frobenius number of the numerical semigroup $\mathscr{B}(G, V_i) \subset \mathbb{N}_0$, so F_i is the minimal positive integer N such that $\mathscr{B}(G, V_i)$ contains $N + kc_i$ for all $k \in \mathbb{N}_0$.

Proposition 4.11

1. *$\mathscr{B}(G, V) \subset \mathbb{N}_0^r$ is a reduced finitely generated* C*-monoid.*
2. *For each $i \in [1, r]$ and all $a \in \mathbb{N}_0^r$ satisfying $a_i \geq b(G, V_i) + F_i$ we have*

$$a \in \mathscr{B}(G, V) \quad \text{if and only if} \quad c_i e_i + a \in \mathscr{B}(G, V). \tag{6}$$

3. *For each* $i \in [1, r]$ *we have* $c_i = |\rho_i(G) \cap \mathbb{F}^\bullet \mathrm{id}_{V_i}|$.

Proof 1. Take $a, b \in \mathscr{B}(G, V)$, so there exist non-zero $f \in \mathbb{F}[V]^G_a$ and $h \in \mathbb{F}[V]^G_b$. Now $0 \neq fh \in \mathbb{F}[V]^G_{a+b}$, hence $a + b \in \mathscr{B}(G, V)$. This shows that $\mathscr{B}(G, V)$ is a submonoid of $\mathbb{N}_0$. Moreover, the multidegrees of a multihomogeneous $\mathbb{F}$-algebra generating system of $\mathbb{F}[V]^G$ clearly generate the monoid $\mathscr{B}(G, V)$. Thus $\mathscr{B}(G, V)$ is finitely generated by Theorem 4.1.

To show that $\mathscr{B}(G, V)$ is also a C-monoid, recall that by Proposition 2.6.3 a finitely generated submonoid H of $\mathbb{N}_0^r$ is a C-monoid if and only if each standard basis element $e_i \in \mathbb{N}_0^r$ has a multiple in H. Now this condition holds for $\mathscr{B}(G, V)$, since by Theorem 4.1.2 $\mathbb{F}[V_i]^G \subset \mathbb{F}[V]^G$ contains a homogeneous element of positive degree for each $i \in [1, r]$.

2. By symmetry it is sufficient to verify (6) in the case $i = 1$. Suppose $a \in \mathscr{B}(G, V)$, so there is a non-zero G-invariant $f \in \mathrm{Sym}^{a_1}(V_1^*) \otimes \cdots \otimes \mathrm{Sym}^{a_r}(V_r^*)$. Decompose $\mathrm{Sym}^{a_1}(V_1^*) = \bigoplus_j W_j$ into a direct sum of irreducible G-modules. This gives a direct sum decomposition $\mathrm{Sym}^{a_1}(V_1^*) \otimes \cdots \otimes \mathrm{Sym}^{a_r}(V_r^*) = \bigoplus_j (W_j \otimes \mathrm{Sym}^{a_2}(V_2^*) \otimes \cdots \otimes \mathrm{Sym}^{a_r}(V_r^*))$. It follows that $\mathrm{Sym}^{a_1}(V_1^*)$ contains an irreducible G-module direct summand W such that $W \otimes \mathrm{Sym}^{a_2}(V_2^*) \otimes \cdots \otimes \mathrm{Sym}^{a_r}(V_r^*)$ contains a non-zero G-invariant. By definition of $b(G, V_1)$ we know that $\mathbb{F}[V_1]$ is generated as an $\mathbb{F}[V_1]^G$ module by its homogeneous components of degree $\leq b(G, V_1)$. Therefore, there exists a $d \leq b(G, V_1)$ such that the degree d homogeneous component of $\mathbb{F}[V]$ contains a G-submodule $U \cong W$, and $a_1 \in d + \mathscr{B}(G, V_1)$. Now for any homogeneous $h \in \mathbb{F}[V_1]^G$ we have $hU \otimes \mathrm{Sym}^{a_2}(V_2^*) \otimes \cdots \otimes \mathrm{Sym}^{a_r}(V_r^*)) \subset \mathbb{F}[V]_{(d+\deg(h), a_2, \ldots, a_r)}$ contains a non-zero G-invariant, since it is isomorphic to $W \otimes \mathrm{Sym}^{a_2}(V_2^*) \otimes \cdots \otimes \mathrm{Sym}^{a_r}(V_r^*))$. It follows that $(k, a_2, \ldots, a_r) \in \mathscr{B}(G, V)$ for all $k \in d + \mathscr{B}(G, V_1)$, in particular, for all $k \in \{d + F_1, d + F_1 + c_1, d + F_1 + 2c_1, \ldots\}$.

3. Let $i \in [1, r]$, and to simplify notation set $W = V_i$, $c = c_i$, and $\phi = \rho_i$. Recall that $\mathbb{F}[W]^A = \mathbb{F}[W]^B$ for some finite subgroups $A, B \subset \mathrm{GL}(W)$ implies that $A = B$. Indeed, the condition implies equality $\mathbb{F}(W)^A = \mathbb{F}(W)^B$ of the corresponding quotient fields, and so both A and B are the Galois groups of the field extension $\mathbb{F}(W)$ over $\mathbb{F}(W)^A = \mathbb{F}(W)^B$, implying $A = B$. Now denote by $Z \subset \mathrm{GL}(W)$ the subgroup of scalar transformations $Z = \{\omega \mathrm{id}_W : \omega^c = 1\}$, so Z is a central cyclic subgroup of $\mathrm{GL}(W)$ of order c. Clearly every homogeneous element of $\mathbb{F}[W]$ whose degree is a multiple of c is invariant under Z. It follows that $\mathbb{F}[W]^G \subset \mathbb{F}[W]^Z$, hence denoting by $\tilde{G}$ the subgroup $\phi(G)Z$ of $\mathrm{GL}(W)$, we have $\mathbb{F}[W]^G = \mathbb{F}[W]^{\tilde{G}}$. It follows that $\phi(G) = \tilde{G}$, i.e. $Z \subset \phi(G)$, and so $c = |Z|$ divides the order of $\phi(G) \cap \mathbb{F}^\bullet \mathrm{id}_W$. Conversely, if $\lambda \mathrm{id}_W$ belongs to $\rho(G)$, then every element of $\mathbb{F}[W]^G$ must be invariant under the scalar transformation $\lambda \mathrm{id}_W$, whence all homogeneous components of $\mathbb{F}[W]^G$ have degree divisible by the order of λ, so the order of the cyclic group $\phi(G) \cap \mathbb{F}^\bullet \mathrm{id}_W$ must divide c.

In general $\mathscr{B}(G, V)$ is not a Krull monoid. To provide an example, consider the two-dimensional irreducible representation V of the symmetric group $S_3 = D_6$. Its ring of polynomial invariants is generated by an element of degree 2 and 3, hence $\mathscr{B}(G, V) = \langle 2, 3 \rangle \subset (\mathbb{N}_0, +)$, which is not Krull.

Proposition 4.12 *For every $k \in \mathbb{N}$ we have $\mathsf{D}_k(\mathscr{B}(G, V)) \leq \beta_k(G, V)$.*

Proof Let $k \in \mathbb{N}$. Take $a \in \mathscr{B}(G, V)$ such that $|a| > \beta_k(G, V)$. By (5) a multihomogeneous invariant $f \in \mathbb{F}[V]^G_a$ exists. As $\deg(f) = |a| > \beta_k(G, V)$ it follows that $f = \sum_{i=1}^N f_{i,1} \dots f_{i,k+1}$ for some non-zero multihomogeneous invariants $f_{i,j}$ of positive degree. Denoting by $a_{i,j} \in \mathbb{N}_0^r$ the multidegree of $f_{i,j}$, we have that $a = a_{i,1} + \dots + a_{i,k+1}$, where $0 \neq a_{i,j} \in \mathscr{B}(G, V)$. This shows that all $a \in \mathscr{B}(G, V)$ with $|a| > \beta_k(G, V)$ factor into the product of more than k atoms, implying the desired inequality.

Remarks 1. Let G be abelian and suppose that $\mathbb{F}$ is algebraically closed. Then we may take in (4) a decomposition of V into the direct sum of 1-dimensional submodules and so V_i^*, is spanned by a variable x_i as in Sect. 4.3. Then $\mathbb{F}[V]_a$ is spanned by the monomial $x_1^{a_1} \cdots x_r^{a_r}$ and $a \in \mathscr{B}(G, V)$ holds if and only if the corresponding monomial is G-invariant. So in this case $\mathscr{B}(G, V)$ can be naturally identified with M^G and the transfer homomorphism $\psi \mid_{M^G}$ of Proposition 4.7 can be thought of as a transfer homomorphism $\mathscr{B}(G, V) \to \mathscr{B}(\widehat{G}_V)$, which is an isomorphism if V is multiplicity free. However, this transfer homomorphism does not seem to have an analogues for non-abelian G (i.e., the study of $\mathscr{B}(G, V)$ can not be reduced to the multiplicity free case), as it is shown by the example below.

2. The binary tetrahedral group $G = \widetilde{A}_4 \cong SL_2(\mathbb{F}_3)$ of order 24 has a two-dimensional complex irreducible representation V such that $\mathbb{F}[V]^G$ is minimally generated by elements of degree 6, 8, 12 (see for example [4, Appendix A]), hence $\mathscr{B}(G, V) = \{0, 6, 8, 12, 14, 16, 18, \dots\}$. On the other hand under this representation G is mapped into the special linear group of V, so on $V \oplus V$ the function maping $((x_1, x_2), (y_1, y_2)) \mapsto \det \begin{pmatrix} x_1 & y_1 \\ x_2 & y_2 \end{pmatrix}$ is a G-invariant of multidegree $(1, 1)$, implying that $(1, 1) \in \mathscr{B}(G, V \oplus V)$. This shows that the transfer homomorphism $\tau : \mathbb{N}_0^2 \to \mathbb{N}_0$, $(a_1, a_2) \mapsto a_1 + a_2$ does not map $\mathscr{B}(G, V \oplus V)$ into $\mathscr{B}(G, V)$, as $\tau(1, 1) = 2 \notin \mathscr{B}(G, V)$.

Recall that the multigraded Hilbert series of $\mathbb{F}[V]^G$ in r indeterminates $T = (T_1, \dots, T_r)$ is

$$H(\mathbb{F}[V]^G, T) = \sum_{a \in \mathbb{N}_0^r} \dim_{\mathbb{F}}(\mathbb{F}[V]^G_a) T_1^{a_1} \cdots T_r^{a_r}, \quad \text{and hence}$$

$$\mathscr{B}(G, V) = \{a \in \mathbb{N}_0^r \colon \text{ the coefficient of } T^a \text{ in } H(\mathbb{F}[V]^G, T) \text{ is nonzero}\}.$$

By this observation Proposition 4.12 can be used for finding lower bounds on the Noether number $\beta(G, V)$, thanks to the following classical result of Molien (see for example [4, Theorem 2.5.2]):

Proposition 4.13 *Given a G-module* $V = V_1 \oplus \ldots \oplus V_r$ *over* $\mathbb{C}$, *let* $\rho_i(g) \in \mathrm{GL}(V_i)$ *be the linear transformation defining the action of* $g \in G$ *on* V_i. *Then we have*

$$H(\mathbb{C}[V]^G, T) = \frac{1}{|G|} \sum_{g \in G} \prod_{i=1}^{r} \frac{1}{\det(\mathrm{id}_{V_i} - \rho_i(g) \cdot T_i)}.$$

Example 4.14 (*see pp. 54–55 in* [62]) Consider the alternating group A_5 and its 3-dimensional representation over $\mathbb{C}^3$ as the group of symmetries of an icosahedron. The Hilbert series then equals

$$\frac{1 + T^{15}}{(1 - T^2)(1 - T^6)(1 - T^{10})}$$

whence it is easily seen that $\mathscr{B}(A_5, \mathbb{C}^3) = \langle 2, 6, 10, 15 \rangle$ and consequently $\beta(A_5) \ge \mathsf{D}(\mathscr{B}(A_5, \mathbb{C}^3)) = 15$. Note that this lower bound is stronger than what we could get from $\beta(G) \ge \max_{H \subsetneq G} \beta(H)$, since $\beta(H) \le |H| \le 12$ for any proper subgroup H of A_5.

5 Constants from Invariant Theory and Their Counterparts in Arithmetic Combinatorics

In Sect. 5.1 we compare known reduction lemmas for the Noether number with reduction lemmas for the Davenport constants achieved in previous sections. We demonstrate how to use them to determine the precise value of Noether numbers and Davenport constants in new examples. In Sect. 5.2 we consider an invariant theoretic analogue of the constant $\eta(G)$ (for the definition of $\eta(G)$ see the discussions before Proposition 2.8 and Lemma 3.1).

Throughout this section, suppose that $\mathrm{char}(\mathbb{F}) \nmid |G|$.

5.1 *The Noether Number Versus the Davenport Constant*

In the non-abelian case no structural connection (like Proposition 4.7) is known between the G-invariant polynomials and the product-one sequences over G. Nevertheless, a variety of features of the kth Noether numbers and the kth Davenport constants are strikingly similar, and we offer a detailed comparison.

Recall that $\beta_k(G) = b_k(G) + 1$ ((3)) and that $\mathsf{d}_k(G) + 1 \le \mathsf{D}_k(G)$ (Proposition 2.8.1).

1. The inequalities

$$(a)\quad \beta_k(G) \le k\beta(G) \quad (b)\quad \mathsf{d}_k(G)+1 \le k(\mathsf{d}(G)+1) \quad (c)\quad \mathsf{D}_k(G) \le k\mathsf{D}(G) \tag{7}$$

2. Reduction lemma for normal subgroups $N \triangleleft G$:

$$(a)\quad \beta_k(G) \le \beta_{\beta_k(G/N)}(N) \qquad (b)\quad \mathsf{d}_k(G) \le \mathsf{d}_{\mathsf{d}_k(N)+1}(G/N) \tag{8}$$

3. Reduction lemma for arbitrary subgroups $H \le G$ with index $l = [G:H]$:

$$(a)\ \beta_k(G) \le \beta_{kl}(H) \le l\beta_k(H) \quad (b)\ \mathsf{d}_k(G)+1 \le l(\mathsf{d}_k(H)+1) \quad (c)\ \mathsf{D}_k(G) \le l\mathsf{D}_k(H) \tag{9}$$

4. Supra-additivity: for a normal subgroup $N \triangleleft G$ we have

$$(a)\quad b_{k+r-1}(G) \ge b_k(N) + b_r(G/N) \text{ if } G/N \text{ is abelian} \tag{10}$$

$$(b)\quad \mathsf{d}_{k+r-1}(G) \ge \mathsf{d}_k(N) + \mathsf{d}_r(G/N)$$

5. Monotonicity: for an arbitrary subgroup $H \le G$ we have

$$(a)\quad \beta_k(G) \ge \beta_k(H) \quad (b)\quad \mathsf{d}_k(G) \ge \mathsf{d}_k(H) \quad (c)\quad \mathsf{D}_k(G) \ge \mathsf{D}_k(H) \tag{11}$$

6. Almost linearity in k: there are positive constants $C, C', C'', k_0, k_0', k_0''$ depending only on G such that

$$(a)\ \beta_k(G) = k\sigma(G) + C \text{ for all } k > k_0 \text{ if } \mathrm{char}(\mathbb{F}) = 0 \quad (b)\ \mathsf{d}_k(G) = k\mathsf{e}(G) + C' \tag{12}$$

for all $k > k_0'$ and (c) $\mathsf{D}_k(G) = k\mathsf{e}(G) + C''$ for all $k > k_0''$

7. The following functions are nonincreasing in k:

$$(a)\quad \beta_k(G)/k \quad \text{if} \quad \mathrm{char}(\mathbb{F}) = 0 \qquad (b)\quad \mathsf{D}_k(G)/k \tag{13}$$

The inequality (7) (a) is observed in [12], (b) is shown in Proposition 3.9.4, whereas (c) is observed in the beginning of Sect. 2.5.

For the proof of (8) (a) see [12, Lemma 1.5] and for part (b) see Proposition 3.9.2. Note that the roles of N and G/N are swapped in the formulas (a) respectively (b), but in the abelian case they amount to the same.

The first inequality in part (a) of (9) is proved in [12, Corollary 1.11] for cases when (i) $\mathrm{char}(\mathbb{F}) = 0$ or $\mathrm{char}(\mathbb{F}) > [G:H]$; (ii) H is normal in G and $\mathrm{char}(\mathbb{F}) \nmid [G:H]$; (iii) $\mathrm{char}(\mathbb{F})$ does not divide $|G|$. It is conjectured, however that it holds in fact whenever $\mathrm{char}(F) \nmid [G:H]$ (see [55]). By [11, Lemma 4.3], we have $\beta_{kl}(H) \le$

$l\beta_k(H)$ for all positive integers k, l, implying the second inequality in part (a). Parts (b) and (c) of (9) appear in Proposition 3.9 (3. and 5.).

Part (a) of (10) appears in [13, Theorem 4.3 and Remark 4.4] while part (b) is proved in Proposition 3.9.1.

Parts (b) and (c) of (11) are immediate from the definitions, while part (a) follows from an argument of B. Schmid [73, Proposition 5.1] which also shows that $\beta_k(G, \mathrm{Ind}_H^G V) \geq \beta_k(H, V)$ for all $k \geq 1$ (see [13, Lemma 4.1]).

Part (a) of (12) is proved in [11, Proposition 4.5] (the constant $\sigma(G)$ will be discussed in Sect. 5.2, and for (12) (b) and (c) we refer to Proposition 2.7.2 and Proposition 2.8.2.

Part (a) of (13) is proved in [11, Sect. 4] and for (13) (b) we refer to Proposition 2.7.3.

Furthermore, for a normal subgroup $N \triangleleft G$ we have

$$(a) \quad \beta(G) \leq \beta(G/N)\beta(N) \qquad\qquad (b) \quad \mathsf{D}(G) \leq \mathsf{D}(N)\mathsf{D}(G/N), \qquad (14)$$

where in (b) we assume that $N \cap G' = \{1\}$. Here part (a) is originally due to B. Schmid [73, Lemma 3.1] and it is an immediate consequence of (7) (a) and (8) (a) while part (b) is proven in [39, Theorem 3.3].

The above reduction lemmas on the Noether numbers are key tools in the proof of the following theorem.

Theorem 5.1 *Let* $k \in \mathbb{N}$.

1. $\beta_k(A_4) = 4k + 2$ *and* $\beta(\tilde{A}_4) = 12$, *where* A_4 *is the alternating group of degree* 4 *and* $\tilde{A}_4$ *is the binary tetrahedral group.*
2. *If G is a non-cyclic group with a cyclic subgroup of index two, then*

$$\beta_k(G) = \frac{1}{2}|G|k + \begin{cases} 2 & \text{if } G = Dic_{4m},\ m > 1; \\ 1 & \text{otherwise.} \end{cases}$$

   where $Dic_{4m} = \langle a, b : a^{2m} = 1, b^2 = a^m, bab^{-1} = a^{-1} \rangle$ *is the dicyclic group.*
3.

$$\beta(G) \geq \frac{1}{2}|G| \quad \text{if and only if} \quad G \text{ has a cyclic subgroup of index at most two or}$$
$$G \text{ is isomorphic to } C_3 \oplus C_3,\ C_2 \oplus C_2 \oplus C_2,\ A_4 \text{ or } \tilde{A}_4$$

Proof For 1. see [12, Theorem 3.4 and Corollary 3.6], for 2. see [13, Theorem 10.3], and 3. can be found in [12, Theorem 1.1].

It is worthwhile to compare Theorem 5.1.3 with the statement from [65] asserting that $\mathsf{d}(G) < \frac{1}{2}|G|$ unless G has a cyclic subgroup of index at most two. If G is abelian, then Lemma 3.13 and Proposition 4.7 imply $\mathsf{d}(G) + 1 = \beta(G) = \mathsf{D}(G)$. Combining Theorems 3.10 and 5.1 we obtain that all groups G having a cyclic subgroup of index at most two satisfy the inequality $\mathsf{d}(G)+1 \le \beta(G) \le \mathsf{D}(G)$. Moreover, for these groups $\beta(G) = \mathsf{d}(G)+1$, except for the dicyclic groups, where $\beta(G) = \mathsf{d}(G)+2$. On the other hand, it was shown in [14] that for the Heisenberg group H_{27} of order 27 we have $\mathsf{D}(H_{27}) < \beta(H_{27})$.

Problem 2 Study the relationship between the invariants $\mathsf{d}(G)$, $\beta(G)$, and $\mathsf{D}(G)$. In particular,

- Characterize the groups G satisfying $\mathsf{d}(G)+1 \le \beta(G)$.
- Characterize the groups G satisfying $\beta(G) \le \mathsf{D}(G)$.

In the following examples we demonstrate how the reduction results presented at the beginning of this section do work. This allows us to determine Noether numbers and Davenport constants of non-abelian groups, for which they were not known before.

Example 5.2 Let p, q be primes such that $q \mid p-1$.

1. Consider the non-abelian semi-direct product $G = C_p \rtimes C_q$. A conjecture attributed to Pawale [81] states that $\beta(C_p \rtimes C_q) = p+q-1$ and many subsequent research was done in this direction [12, 17]. Currently it is fully proved only for the cases $q=2$ in [73] and $q=3$ in [10] whereas for arbitrary q we have only upper bounds in [12], proved using known results related to the Olson constant of the cyclic group of order p. Theorem 3.11.3 implies that $\mathsf{d}(G)+1 = p+q-1$ and hence $\mathsf{d}(G)+1$ coincides with the conjectured value for $\beta(G)$.

2. In view of the great difficulties related to Pawale's conjecture it is quite remarkable that we can determine the exact value of the Noether number for the non-abelian semi-direct product $C_{pq} \rtimes C_q$. Indeed, this group contains an index p subgroup isomorphic to $C_q \oplus C_q$, hence $\beta(C_{pq} \rtimes C_q) \le \beta_p(C_q \oplus C_q)$ by (9). By Proposition 4.7 4. we have $\beta_p(C_q \oplus C_q) = \mathsf{D}_p(C_q \oplus C_q)$, and finally, $\mathsf{D}_p(C_q \oplus C_q) = pq+q-1$ by Theorem 3.14. Thus we have $\beta(C_{pq} \rtimes C_q) \le pq+q-1$. The reverse inequality also holds, since $\beta(C_{pq} \rtimes C_q)$ contains a normal subgroup $N \cong C_{pq}$ with $G/N \cong C_q$, so by (10) and (3) we have $\beta(C_{pq} \rtimes C_q) \ge \beta(C_{pq}) + \beta(C_q) - 1 = pq+q-1$. So we have $\beta(C_{pq} \rtimes C_q) = pq+q-1$.

Next we determine the small Davenport constant of this group. Since C_{pq} is a normal subgroup and the corresponding factor group is C_q, we have by Proposition 3.9.1 that $\mathsf{d}(C_{pq} \rtimes C_q) \ge \mathsf{d}(C_{pq}) + \mathsf{d}(C_q) = p+q-2$. The reverse inequality $\mathsf{d}(C_{pq} \rtimes C_q) \le p+q-2$ follows from Theorem 3.11.4, since $C_{pq} \rtimes C_q$ contains also a normal subgroup $N \cong C_p$ such that $G/N \cong C_q \oplus C_q$. Consequently, by Lemma 3.1.2.(a) we have

$$\mathsf{D}(C_{pq} \rtimes C_q) \ge \mathsf{d}(C_{pq} \rtimes C_q) + 1 = pq+q-1.$$

Example 5.3 The symmetric group S_4 has a normal subgroup $N \cong C_2 \oplus C_2$ such that $S_4/N \cong D_6$. We know that $\beta(D_6) = 4$ (say by Theorem 5.1 2.). Thus by (8) and Theorem 3.14 we have $\beta(S_4) \le \beta_{\beta(D_6)}(C_2 \oplus C_2) = \mathsf{D}_4(C_2 \oplus C_2) = 2 \cdot 4 + 1 = 9$.

Now let V be the standard 4-dimensional permutation representation of S_4 and sign : $S_4 \to \{\pm 1\}$ the sign character. It is not difficult to prove the algebra isomorphism $\mathbb{F}[V \otimes \mathrm{sign}]^{S_4} \cong \mathbb{F}[V]^{S_4}_{even} \oplus \Delta_4 \mathbb{F}[V]^{S_4}_{odd}$ where Δ_4 is the Vandermonde determinant in 4 variables, $\mathbb{F}[V]^{S_4}_{even}$ is the span of the even degree homogeneous components of $\mathbb{F}[V]^{S_4}$, and $\mathbb{F}[V]^{S_4}_{odd}$ is the span of the odd degree homogeneous components of $\mathbb{F}[V]^{S_4}$. Moreover, the algebra $\mathbb{F}[V]^{S_4}_{even} \oplus \Delta_4 \mathbb{F}[V]^{S_4}_{odd}$ is easily seen to be minimally generated by $\sigma_2, \sigma_1^2, \sigma_1\sigma_3, \sigma_4, \sigma_3^2, \sigma_1\Delta_4, \sigma_3\Delta_4$, where σ_i is the ith elementary symmetric polynomial. As a result $\beta(S_4, V \otimes \mathrm{sign}) = \deg(\sigma_3\Delta_4) = 3 + \binom{4}{2} = 9$. So we conclude that $\beta(S_4) = 9$ (and not 10, as it is claimed on p. 14 of [57]).

Example 5.4 Let G be the group generated by the complex Pauli matrices

$$\begin{pmatrix} 0 & 1 \\ 1 & 0 \end{pmatrix}, \begin{pmatrix} 0 & -i \\ i & 0 \end{pmatrix}, \begin{pmatrix} 1 & 0 \\ 0 & -1 \end{pmatrix}.$$

This is a pseudoreflection group, hence the ring of invariants on $V = \mathbb{C}^2$ is generated by two elements, namely $\mathbb{C}[x, y]^G = \mathbb{C}[x^4 + y^4, x^2y^2]$. Moreover, $b(G, V)$ is the sum of the degrees of the generators minus $\dim(V)$ (again because G is a pseudoreflection group, see [9]), so $b(G, V) = 6$. It follows by (3) that $\beta(G) = b(G) + 1 \geq b(G, V) + 1 = 7$.

On the other hand, G is a non-abelian semi-direct product$(C_4 \oplus C_2) \rtimes C_2$. Therefore G has a normal subgroup N such that $N \cong G/N \cong C_2 \oplus C_2$ and thus

$$\beta(G) \leq \beta_{\beta(C_2 \oplus C_2)}(C_2 \oplus C_2) = \mathsf{D}_3(C_2 \oplus C_2) = 7.$$

So we conclude that $\beta(G) = 7$.

5.2 The Constants $\sigma(G, V)$ and $\eta(G, V)$

Definition 5.5 1. Let $\sigma(G, V)$ denote the smallest $d \in \mathbb{N}_0 \cup \{\infty\}$ such that $\mathbb{F}[V]^G$ is a finitely generated module over a subring $\mathbb{F}[f_1, \dots, f_r]$ such that $\max\{\deg(f_i) : i \in [1, r]\} = d$. We define $\sigma(G) = \sup\{\sigma(G, W) : W \text{ is a } G\text{-module}\}$.
2. Let $S \subset \mathbb{F}[V]^G$ be the F-subalgebra of $\mathbb{F}[V]^G$ generated by its elements of degree at most $\sigma(G, V)$. Then $\eta(G, V)$ denotes the maximal degree of generators of $\mathbb{F}[V]^G_+$ as an S-module.

One motivation to study $\sigma(G, V)$ and $\eta(G, V)$ is that by a straightforward induction argument [11, Sect. 4] we have

$$\beta_k(G, V) \leq (k-1)\sigma(G, V) + \eta(G, V).$$

By [11, Proposition 6.2], $\sigma(C_p \rtimes C_q) = p$ (this is also true in characteristic q, see [18, Proposition 4.5]).

If $\mathbb{F}$ is algebraically closed, then, by Hilbert's Nullstellensatz, $\sigma(G, V)$ is the smallest d such that there exist homogeneous invariants of degree at most d whose common zero locus is the origin. It is shown in Lemmas 5.1, 5.4 and 5.6 of [11] (some extensions to the modular case and for linear algebraic groups are given in [18]) that

- $\sigma(G) \le \sigma(G/N)\sigma(N)$ if $N \triangleleft G$;
- $\sigma(H) \le \sigma(G) \le [G : H]\sigma(H)$ if $H \le G$;
- $\sigma(G) = \max\{\sigma(G, V) \colon V$ is an irreducible G-module$\}$.

Proposition 5.6 *Let G be abelian.*

1. $\sigma(G) = \exp(G) = \mathsf{e}(G)$.
2. $\eta(G) = \sup\{\eta(G, W) \colon W$ *is a G-module*$\}$.

Proof For 1. see [11, Corollary 5.3]. To prove 2., let $T \in \mathscr{F}(\widehat{G})$ with $|T| = \eta(G) - 1$ such that T has no product-one subsequence U with $|U| \in [1, \mathsf{e}(G)]$. Let V be the regular representation of G, and denote by S the subalgebra of $\mathbb{F}[V]^G$ generated by its elements of degree at most $\sigma(G) = \mathsf{e}(G)$. Now $\psi \colon M \to \mathscr{F}(\widehat{G})$ is an isomorphism (see the proof of Proposition 4.7.3.). Thus $\psi^{-1}(T) \in M$ is not divisible by a G-invariant monomial of degree smaller than $\mathsf{e}(G)$. Since both S and $\mathbb{F}[V]$ are spanned by monomials, it follows that $\psi^{-1}(T) \in M$ is not contained in the S-submodule of $\mathbb{F}[V]^G_+$ generated by elements of degree less than $\deg(\psi^{-1}(T))$. This shows that for the regular representation V of G we have $\eta(G, V) \ge \eta(\widehat{G})$.

On the other hand let W be an arbitrary G-module, and $m \in M$ a monomial with $\deg(m) > \eta(G)$. Then $\psi(m)$ has a product-one subsequence with length at most $\mathsf{e}(G) = \sigma(G)$, hence m is divisible by a G-invariant monomial of length at most $\sigma(G)$ (see the beginning of the proof of Proposition 4.7.2). This shows the inequality $\eta(G, W) \le \eta(\widehat{G})$. Taking into account the isomorphism $\widehat{G} \cong G$ we are done.

For the state of the art on $\eta(G)$ (in the abelian case) we refer to [22, 23], [40, Theorem 5.8.3]. Proposition 5.6 inspires the following problem.

Problem 3 Let G be a finite non-abelian group. Is $\sup\{\eta(G, W) \colon W$ is a G-module$\}$ finite? Is it related to $\eta(\mathscr{B}(G))$ (see Sects. 2.5 and 3.1)?

Acknowledgments This work was supported by the *Austrian Science Fund FWF* (Project No. P26036-N26) and by OTKA K101515 and PD113138.

References

1. N.R. Baeth, A. Geroldinger, Monoids of modules and arithmetic of direct-sum decompositions. Pacific J. Math. **271**, 257–319 (2014)
2. P. Baginski, S.T. Chapman, Arithmetic congruence, monoids: a survey, *Combinatorial and Additive Number Theory: CANT, and 2012*, Springer Proceedings in Mathematics and Statistics (Springer, Berlin, 2014), pp. 15–38

3. J. Bass, Improving the Erdős-Ginzburg-Ziv theorem for some non-abelian groups. J. Number Theor. **126**, 217 (2007)
4. D.J. Benson, *Polynomial Invariants of Finite Groups*, vol. 190, London Mathematical Society Lecture Notes Series (Cambridge University Press, Cambridge, 1993)
5. G. Bhowmik, J.-C. Schlage-Puchta, Davenport's constant for groups of the form $\mathbb{Z}_3 \oplus \mathbb{Z}_3 \oplus \mathbb{Z}_{3d}$, in *Additive Combinatorics*, vol. 43, CRM Proceedings and Lecture Notes, ed. by A. Granville, M.B. Nathanson, J. Solymosi (American Mathematical Society, 2007), pp. 307–326
6. G. Bhowmik, J.-C. Schlage-Puchta, Davenport's constant for groups with large exponent, *Theory and Applications of Finite Fields*, Contemporary Mathematics (AMS, 2013), pp. 21–31
7. G.W. Chang, Every divisor class of Krull monoid domains contains a prime ideal. J. Algebra **336**, 370–377 (2011)
8. F. Chen, S. Savchev, Long minimal zero-sum sequences in the groups $C_2^{r-1} \oplus C_{2k}$. Integers **14**, Paper A23 (2014)
9. C. Chevalley, Invariants of finite groups generated by reflections. Amer. J. Math. **77**, 778–782 (1955)
10. K. Cziszter, The Noether number of the non-abelian group of order $3p$. Per. Math. Hungarica **68**, 150–159 (2014)
11. K. Cziszter, M. Domokos, On the generalized Davenport constant and the Noether number. Central Eur. J. Math. **11**, 1605–1615 (2013)
12. K. Cziszter, M. Domokos, Groups with large Noether bound. Ann. Inst. Fourier (Grenoble) **64**, 909–944 (2014)
13. K. Cziszter, M. Domokos, The Noether number for the groups with a cyclic subgroup of index two. J. Algebra **399**, 546–560 (2014)
14. K. Cziszter, M. Domokos, I. Szöllősi, The Noether number and the Davenport constants of the groups of order less than 32, manuscript in preparation
15. C. Delorme, O. Ordaz, D. Quiroz, Some remarks on Davenport constant. Discrete Math. **237**, 119–128 (2001)
16. H. Derksen, G. Kemper, Computational invariant theory, *Encyclopedia of Mathematical Sciences*, vol. 130 (Springer, Berlin, 2002)
17. M. Domokos, P. Hegedűs, Noether's bound for polynomial invariants of finite groups. Arch. Math. **74**, 161–167 (2000)
18. J. Elmer, M. Kohls, Zero-separating invariants for finite groups. J. Algebra **411**, 92–113 (2014)
19. A. Facchini, Direct sum decomposition of modules, semilocal endomorphism rings, and Krull monoids. J. Algebra **256**, 280–307 (2002)
20. A. Facchini, Krull monoids and their application in module theory, in *Algebras, Rings and their Representations*, ed. by A. Facchini, K. Fuller, C.M. Ringel, C. Santa-Clara (World Scientific, 2006), pp. 53–71
21. A. Facchini, Direct-sum decompositions of modules with semilocal endomorphism rings. Bull. Math. Sci. **2**, 225–279 (2012)
22. Y. Fan, W. Gao, Q. Zhong, On the Erdős-Ginzburg-Ziv constant of finite abelian groups of high rank. J. Number Theory **131**, 1864–1874 (2011)
23. Y. Fan, W. Gao, L. Wang, Q. Zhong, Two zero-sum invariants on finite abelian groups. Europ. J. Comb. **34**, 1331–1337 (2013)
24. B.W. Finklea, T. Moore, V. Ponomarenko, Z.J. Turner, Invariant polynomials and minimal zero sequences. Involve **1**, 159–165 (2008)
25. P. Fleischmann, The Noether bound in invariant theory of finite groups. Adv. Math. **156**, 23–32 (2000)
26. P. Fleischmann, M. Sezer, R.J. Shank, C.F. Woodcock, The Noether numbers for cyclic groups of prime order. Adv. Math. **207**, 149–155 (2006)
27. P. Fleischmann, C.F. Woodcock, Relative invariants, ideal classes, and quasi-canonical modules of modular rings of invariants. J. Algebra **348**, 110–134 (2011)
28. J. Fogarty, On Noether's bound for polynomial invariants of a finite group. Electron. Res. Announc. Amer. Math. Soc. **7**, 5–7 (2001). (electronic)

29. R.M. Fossum, *The Divisor Class Group of a Krull Domain* (Springer, Berlin, 1973)
30. M. Freeze, W.A. Schmid, Remarks on a generalization of the Davenport constant. Discrete Math. **310**, 3373–3389 (2010)
31. W. Gao, A. Geroldinger, On zero-sum sequences in $\mathbb{Z}/n\mathbb{Z} \oplus \mathbb{Z}/n\mathbb{Z}$. Integers **3**, Paper A08, 45 p (2003)
32. W. Gao, A. Geroldinger, Zero-sum problems in finite abelian groups : a survey. Expo. Math. **24**, 337–369 (2006)
33. W. Gao, Z. Lu, The Erdős-Ginzburg-Ziv theorem for dihedral groups. J. Pure Appl. Algebra **212**, 311–319 (2008)
34. W. Gao, Y. Li, The Erdős-Ginzburg-Ziv theorem for finite solvable groups. J. Pure Appl. Algebra **214**, 898–909 (2010)
35. W. Gao, A. Geroldinger, D.J. Grynkiewicz, Inverse zero-sum problems III. Acta Arith. **141**, 103–152 (2010)
36. W. Gao, Y. Li, J. Peng, An upper bound for the Davenport constant of finite groups. J. Pure Appl. Algebra **218**, 1838–1844 (2014)
37. A. Geroldinger, Additive group theory and non-unique factorizations, in *Combinatorial Number Theory and Additive Group Theory*, Advanced Courses in Mathematics CRM Barcelona, ed. by A. Geroldinger, I. Ruzsa (Birkhäuser, 2009), pp. 1–86
38. A. Geroldinger, Non-commutative Krull monoids: a divisor theoretic approach and their arithmetic. Osaka J. Math. **50**, 503–539 (2013)
39. A. Geroldinger, D.J. Grynkiewicz, The large Davenport constant I: groups with a cyclic index 2 subgroup. J. Pure Appl. Algebra **217**, 863–885 (2013)
40. A. Geroldinger, F. Halter-Koch, *Non-Unique Factorizations. Algebraic, Combinatorial and Analytic Theory*, vol. 278, Pure and Applied Mathematics (Chapman & Hall/CRC, 2006)
41. A. Geroldinger, W. Hassler, Arithmetic of Mori domains and monoids. J. Algebra **319**, 3419–3463 (2008)
42. A. Geroldinger, R. Schneider, On Davenport's constant, J. Comb. Theory, Ser. A **61**, 147–152 (1992)
43. A. Geroldinger, Q. Zhong, *The set of minimal distances in Krull monoids*. Acta Arithmetica **173**, 97–120 (2016)
44. A. Geroldinger, M. Liebmann, A. Philipp, On the Davenport constant and on the structure of extremal sequences. Period. Math. Hung. **64**, 213–225 (2012)
45. A. Geroldinger, S. Ramacher, A. Reinhart, On v -Marot Mori rings and C -rings. J. Korean Math. Soc. **52**, 1–21 (2015)
46. R. Gilmer, Multiplicative Ideal Theory. Queen's Papers **90** (1992)
47. B. Girard, Inverse zero-sum problems and algebraic invariants. Acta Arith. **135**, 231–246 (2008)
48. D.J. Grynkiewicz, The large Davenport constant II: general upper bounds. J. Pure Appl. Algebra **217**, 2221–2246 (2013)
49. D.J. Grynkiewicz, *Structural Additive Theory*, Developments in Mathematics (Springer, Berlin, 2013)
50. F. Halter-Koch, A generalization of Davenport's constant and its arithmetical applications. Colloq. Math. **63**, 203–210 (1992)
51. F. Halter-Koch, *Ideal Systems. An Introduction to Multiplicative Ideal Theory* (Marcel Dekker, 1998)
52. F. Halter-Koch, Multiplicative ideal theory in the context of commutative monoids, in *Commutative Algebra: Noetherian and Non-Noetherian Perspectives*, ed. by M. Fontana, S.-E. Kabbaj, B. Olberding, I. Swanson (Springer, 2011), pp. 203–231
53. D. Han, The Erdős-Ginzburg-Ziv Theorem for finite nilpotent groups. Archiv Math. **104**, 325–332 (2015)
54. P. Hegedűs, L. Pyber, *Bounding the Noether Number of Finite Groups*. In preparation
55. G. Kemper, Separating invariants. J. Symbolic Comput. **44**(9), 1212–1222 (2009)
56. F. Knop, On Noether's and Weyl's bound in positive characteristic, *Invariant Theory in All Characteristics*, vol. 35, CRM Proceedings of Lecture Notes (American Mathematical Society, Providence, 2004), pp. 175–188

57. H. Kraft, C. Procesi, *Classical Invariant Theory, A Primer* Website of H. Kraft (1996)
58. H. Nakajima, Relative invariants of finite groups. J. Algebra **79**, 218–234 (1982)
59. J. Neukirch, *Algebraic Number Theory* (Springer, Berlin, 1999)
60. M.D. Neusel, Degree bounds—an invitation to postmodern invariant theory. Topology Appl. **154**, 792–814 (2007)
61. M.D. Neusel, Invariant Theory, *Student Mathematical Library*, vol. 36 (AMS, 2007)
62. M.D. Neusel, L. Smith, Invariant theory of finite groups, *Mathematical Surveys and Monographs*, vol. 94 (AMS, 2002)
63. E. Noether, Der Endlichkeitssatz der Invarianten endlicher Gruppen. Math. Ann. **77**, 89–92 (1916)
64. E. Noether, Der Endlichkeitssatz der Invarianten endlicher Gruppen der Charakteristik p. Nachr. Ges. Wiss. Göttingen, Math.-Phys. Kl. **1926**, 28–35 (1926)
65. J.E. Olson, E.T. White, Sums from a sequence of group elements, in *Number Theory and Algebra*, ed. by H. Zassenhaus (Academic Press, 1977), pp. 215–222
66. V. Ponomarenko, Minimal zero sequences of finite cyclic groups. Integers **4**, A24, 6 p (2004)
67. M. Radziejewski, On the distribution of algebraic numbers with prescribed factorization properties. Acta Arith. **116**, 153–171 (2005)
68. C. Reiher, A proof of the theorem according to which every prime number possesses property B, Ph.D. thesis, Rostock, 2010
69. A. Reinhart, On integral domains that are C-monoids. Houston J. Math. **39**, 1095–1116 (2013)
70. D.R. Richman, Invariants of finite groups over fields of characteristic p. Adv. Math. **124**, 25–48 (1996)
71. S. Savchev, F. Chen, Long zero-free sequences in finite cyclic groups. Discrete Math. **307**, 2671–2679 (2007)
72. S. Savchev, F. Chen, *Long minimal zero-sum sequences in the group* $C_2 \oplus C_{2k}$. Integers **12**, Paper A51, 18 p (2012)
73. B.J. Schmid, Finite groups and invariant theory, *Topics in Invariant Theory*, vol. 1478, Lecture Notes in Mathematics (Springer, Berlin, 1991), pp. 35–66
74. W.A. Schmid, Higher-order class groups and block monoids of Krull monoids with torsion class group. J. Algebra Appl. **9**, 433–464 (2010)
75. W.A. Schmid, Inverse zero-sum problems II. Acta Arith. **143**, 333–343 (2010)
76. W.A. Schmid, The inverse problem associated to the Davenport constant for $C_2 \oplus C_2 \oplus C_{2n}$, and applications to the arithmetical characterization of class groups. Electron. J. Comb. **18**(1) (2011). Research Paper 33
77. W.A. Schmid, Some recent results and open problems on sets of lengths of Krull monoids with finite class group, in *Multiplicative Ideal Theory and Factorization Theory*, ed. by S.T. Chapman, M. Fontana, A. Geroldinger, B. Olberding (Springer, Berlin, 2016)
78. M. Sezer, Sharpening the generalized Noether bound in the invariant theory of finite groups. J. Algebra **254**, 252–263 (2002)
79. Q. Wang, Y. Qu, On the critical number of finite groups II. Ars Comb. **113**, 321–330 (2014)
80. Q. Wang, J.J. Zhuang, On the critical number of finite groups of order pq. Int. J. Number Theor. **8**, 1271–1279 (2012)
81. D.L. Wehlau, The Noether number in invariant theory. C. R. Math. Acad. Sci. Soc. R. Can. **28**(2), 39–62 (2006)
82. K. Cziszter, M. Domokos, I. Szöllősi, The Noether number and the Davenport constants of the groups of order less than 32, manuscript in preparation

Ring and Semigroup Constructions

Marco D'Anna

Abstract In this paper we present a survey on some ring constructions, recently introduced and studied, and we show how to produce some analogous semigroup constructions. Moreover, we describe how to translate at semigroup level some ring properties of these constructions; in particular, we will focus on the Gorenstein property and to its semigroup counterpart, the symmetry.

MSC 20M14 · 13H10 · 13A30 · 14H20

1 Introduction

Let R be a commutative ring with unity and let M be an R-module; the idealization, also called trivial extension, is a classical construction introduced by Nagata (see [24, page 2], [20, Chap. VI, Sect. 25] and [18]) that produces a new ring containing an ideal isomorphic to M. Recently, D'Anna and Fontana introduced the so-called *amalgamated duplication*:

$$R \bowtie I = \{(r, r + i) \mid r \in R,\ i \in I\} \subset R \times R;$$

(see [8, 9], studied also in, e.g. [2, 10, 23]), that, starting with a ring R and an ideal I, produces a new ring that, if $M = I$, has many properties coinciding with the idealization. On the other hand, while the idealization is never reduced, the duplication can be reduced, but is never an integral domain. Looking for a unified approach to these two constructions, D'Anna and Re in [11] observed that it is possible to present both of them as quotients of the Rees algebra modulo particular ideals. This observation leaded to the joint paper by Barucci, D'Anna, Strazzanti ([4]), where the authors study a more general construction, that produces a ring which,

M. D'Anna (✉)
Dipartimento di Matematica e Informatica, Università di Catania, Viale A. Doria 6, 95125 Catania, Italy
e-mail: mdanna@dmi.unict.it

S. Chapman et al. (eds.), *Multiplicative Ideal Theory and Factorization Theory*, Springer Proceedings in Mathematics & Statistics 170,
DOI 10.1007/978-3-319-38855-7_4

in some cases, is an integral domain. More precisely, given a monic polynomial $t^2 + at + b \in R[t]$ and denoting with $\mathcal{R}_+$ the Rees algebra associated to the ring R with respect to the ideal I, i.e. $\mathcal{R}_+ = \bigoplus_{n \geq 0} I^n t^n$, the authors define and study the quotient ring

$$\mathcal{R}_+/(I^2(t^2 + at + b)),$$

where $(I^2(t^2 + at + b))$ is the contraction to $\mathcal{R}_+$ of the ideal generated by $t^2 + at + b$ in $R[t]$. We will denote such ring by $R(I)_{a,b}$.

Meanwhile D'Anna, Finocchiaro and Fontana proposed another possible generalization of the duplication: let R and U be commutative rings with unity, let J be an ideal of U and let $f : R \longrightarrow U$ be a ring homomorphism. In this setting, we can define the following subring of $R \times U$:

$$R \bowtie^f J = \{(r, f(r) + j) \mid r \in R,\ j \in J\}$$

called *the amalgamation of R with U along J with respect to f* (see [13–15], studied also in, e.g. [17, 29]). This construction is a generalization of the amalgamated duplication and other classical constructions (such as the $A + XB[X]$ construction, the $D + M$ construction) can be studied as particular cases of the amalgamation. On the other hand, the amalgamation $R \bowtie^f J$ is related to a construction proposed by D.D. Anderson in [1] and motivated by a classical construction due to Dorroh [16], concerning the embedding of a ring without identity in a ring with identity.

The level of generality chosen for the amalgamation is due to the fact that it can be studied in the frame of pullback constructions. This point of view allows to provide easily an ample description of the properties of $R \bowtie^f J$, in connection with the properties of R, J and f.

For all these constructions, we can consider the particular cases when the ring we are starting with is an algebroid branch (so it has an associated value semigroup, which is a numerical semigroup). The rings obtained using the constructions cited above are algebroid curves (or branches) whose associated semigroup can be obtained with a corresponding semigroup construction, that can be defined independently by the ring case. Two of these semigroup constructions are described in [8], in the case of amalgamated duplication, and in [12], when the quotient of the Rees algebra is a domain.

In this paper, after presenting in Sect. 1 a survey on the ring constructions cited above, in Sect. 2 we will focus on the semigroup constructions. In particular, after presenting the definitions of semigroup duplication and numerical duplication we will give the new definition of semigroup amalgamation; then we will see how these semigroup constructions correspond to ring constructions if we start from an algebroid branch (Sect. 3.1), providing the proofs of the new results (see Theorem 3.5 and Proposition 3.6). Finally, we will focus on the symmetry of the semigroups we are dealing with, which is the counterpart of the Gorenstein property for rings (see, for the semigroup amalgamation, Proposition 3.10).

2 Survey on Ring Constructions

In this section we intend to present some algebraic properties of the rings obtained with the ring constructions introduced above. First we will study the properties of the elements in the family of quotients of the Rees algebra (obtaining also, as a by-product, the corresponding results for the amalgamated duplication); then we will investigate the same properties for the amalgamation.

Let R be a commutative ring with unity and I a proper ideal of R; the amalgamated duplication (or simply duplication) of R with respect to I is defined as $R \bowtie I = \{(r, r+i) \mid r \in R,\ i \in I\} \subset R \times R$; it is not difficult to see that $R \bowtie I \cong R \oplus I$ endowed with the multiplication $(r,i)(s,j) = (rs, rj + si + ij)$. On the other hand the Nagata's idealization, or simply idealization, of R with respect to an ideal I of R, denoted by $R \ltimes I$ (that could be defined for any R-module M) is defined as the R-module $R \oplus I$ endowed with the multiplication $(r,i)(s,j) = (rs, rj + si)$.

Let now t be an indeterminate. The Rees algebra (also called Blow-up algebra) associated to R and I is defined as the following graded subring of $R[t]$:

$$\mathscr{R}_+ = \bigoplus_{n \geq 0} I^n t^n \subseteq R[t].$$

It is possible to prove that, if $f(t) \in R[t]$ is a monic polynomial of degree $k > 0$, then $f(t)R[t] \cap \mathscr{R}_+ = \{f(t)g(t) : g(t) \in I^k \mathscr{R}_+\}$. Denoting this ideal by $(I^k f(t))$, it is clear that each element of the factor ring $\mathscr{R}_+/(I^k f(t))$ is represented by a unique polynomial of $\mathscr{R}_+$ of degree $< k$.

Now if we choose particular polynomials of degree 2 we can obtain both the idealization and the duplication:

Proposition 2.1 ([4, Proposition 1.4]) *We have the following isomorphisms of rings:*

(1) $\mathscr{R}_+/(I^2t^2) \cong R \ltimes I$

(2) $\mathscr{R}_+/(I^2(t^2 - t)) \cong R \bowtie I$

The previous proposition makes natural to consider the family

$$R(I)_{a,b} = \mathscr{R}_+/(I^2(t^2 + at + b)),$$

where $a, b \in R$. As R-module $R(I)_{a,b} \cong R \oplus I$ and the natural injection $R \hookrightarrow R(I)_{a,b}$ is a ring homomorphism; however, $0 \oplus I$ in general (if $b \neq 0$) is not an ideal of $R(I)_{a,b}$, although this happens for both idealization and duplication.

We recall some relevant properties of the rings of the form $R(I)_{a,b}$ that do not depend on the choice of the polynomial.

Proposition 2.2 ([4, Proposition 1.3]) *The ring extensions* $R \subseteq R(I)_{a,b} \subseteq R[t]/(t^2 + at + b))$ *are both integral and the three rings have the same Krull dimension.*

Using the chain of inclusions $R \subseteq R(I)_{a,b} \subseteq R[t]/(t^2 + at + b)$ and the fact that these extensions are integral, we can get information on $\mathrm{Spec}(R(I)_{a,b})$ with respect to $\mathrm{Spec}(R)$.

Proposition 2.3 ([4, Proposition 1.9]) *For each prime ideal P of R, there are at most two prime ideals of $R(I)_{a,b}$ lying over P. Moreover, if $t^2 + at + b$ is irreducible on $R/\mathfrak{m}$ for any maximal ideal $\mathfrak{m}$ of R, then there is exactly one prime ideal of $R(I)_{a,b}$ lying over P.*

Remark 2.4 (1) The primes of $R[t]/(t^2 + at + b)$ (and so also those of $R(I)_{a,b}$) lying over P depend on the factorization of $t^2 + at + b$ in $Q(R/P)[t]$ (where $Q(R/P)$ denotes the field of fractions of R/P). For particular a and b, this factorization may not depend on P. For example, in the case of the idealization, the equality $t^2 = t \cdot t$, implies that there is only one prime lying over P, both in $R[t]/(t^2)$ and in the idealization. As for the case of the duplication, the equality $t^2 - t = t \cdot (t - 1)$, implies that there are two primes in $R[t]/(t^2 - t)$ lying over P, namely (P, t) and $(P, t - 1)$. Contracting these primes to the duplication we get the same prime if and only if $P \supseteq I$ (see, e.g. [9]).

(2) The proof of the previous proposition implies that a sufficient condition for $R(I)_{a,b}$ to be an integral domain is that R is an integral domain and $t^2 + at + b$ is irreducible in $Q(R)[t]$. Under particular assumptions on R, it is possible to prove the existence of such polynomials (see next Proposition 2.7).

Proposition 2.5 ([4, Proposition 1.11]) *The following conditions are equivalent:*

(i) *R is a Noetherian ring;*
(ii) *$R(I)_{a,b}$ is a Noetherian ring for all $a, b \in R$;*
(iii) *$R(I)_{a,b}$ is a Noetherian ring for some $a, b \in R$.*

Assume that R is local, with maximal ideal $\mathfrak{m}$. Then it is known that both $R \bowtie I$ and $R \ltimes I$ are local with maximal ideals $\mathfrak{m} \oplus I$ (in the first case under the isomorphism $R \bowtie I \cong R \oplus I$). More generally:

Proposition 2.6 ([4, Proposition 2.1]) *R is local if and only if $R(I)_{a,b}$ is local. In this case the maximal ideal of $R(I)_{a,b}$ is $\mathfrak{m} \oplus I$ (as R-module).*

It is also clear that, if $(R, \mathfrak{m})$ is local and if we denote by M the maximal ideal of $R(I)_{a,b}$, then $k = R/\mathfrak{m} \cong R(I)_{a,b}/M$.

If R is a local Noetherian integral domain, we can always find integral domains in the family of rings $R(I)_{a,b}$.

Proposition 2.7 ([4, Proposition 2.5 and Corollary 2.6]) *Let R be a local Noetherian integral domain with* $\dim R \geq 1$, *let $Q(R)$ be its field of fractions and let $I \subset R$ be an ideal. Then there exist infinitely many elements $r \in R$ such that $R(I)_{0,-r}$ is an integral domain.*

For the local Noetherian case we are interested in studying the Cohen–Macaulay (briefly CM) property and the Gorensteinness.

Proposition 2.8 ([4, Proposition 2.7]) *Assume that R is a local ring. Then the following conditions are equivalent:*

(i) *R is a CM ring and I is a maximal CM R-module.*
(ii) *$R(I)_{a,b}$ is a CM R-module.*
(iii) *$R(I)_{a,b}$ is a CM ring.*
(iv) *R is a CM ring and each regular R-sequence of R is also an $R(I)_{a,b}$-regular sequence.*

Theorem 2.9 ([4, Theorem 3.2 and Corollary 3.3]) *Let R be a local ring of dimension $d = 1$ and let I be a proper ideal of R. Then, for every $a, b \in R$, the ring $R(I)_{a,b}$ is Gorenstein if and only if R is a CM ring and I is a canonical ideal of R.*

The previous theorem can be stated and proved in higher dimension (see [5]). Notice also that this result is a generalization of analogous results given for idealization and amalgamated duplication (see [8, 18, 26, 27]).

In the remaining part of this section, after showing why the rings of the form $R \bowtie^f J$ can be obtained as pullbacks, we will investigate for them the same algebraic properties just presented above for the quotients of the Rees algebra.

Let $f : R \longrightarrow U$ be a ring homomorphism and J an ideal of U. As stated in the introduction, the amalgamation of R with U along J with respect to f is defined as the following subring of $R \times U$:

$$R \bowtie^f J = \{(r, f(r) + j) \mid r \in R,\ j \in J\}.$$

We recall that, if $\alpha : R \to C$, $\beta : U \to C$ are ring homomorphisms, the subring $D := \alpha \times_C \beta := \{(r, u) \in R \times U \mid \alpha(r) = \beta(u)\}$ of $R \times U$ is called the *pullback* (or *fiber product*) of α and β.

The fact that D is a pullback can also be described by saying that the triplet (D, p_R, p_U) is a solution of the universal problem of rendering commutative the diagram built on α and β

$$\begin{array}{ccc} D & \xrightarrow{p_R} & R \\ {\scriptstyle p_U}\downarrow & & \downarrow{\scriptstyle \alpha} \\ U & \xrightarrow{\beta} & C \end{array}$$

where p_R (respectively, p_U) is the restriction to $\alpha \times_C \beta$ of the projection of $R \times U$.

Proposition 2.10 ([13, Proposition 4.2]) *Let $f : R \to U$ be a ring homomorphism and J be an ideal of U. If $\pi : U \to U/J$ is the canonical projection and $\breve{f} := \pi \circ f$, then $R \bowtie^f J = \breve{f} \times_{U/J} \pi$.*

It is possible to characterize those pullbacks that give rise to amalgamated algebras. Recall that a ring homomorphism $r : B \to A$ is called *a ring retraction* if there exists a ring homomorphism $\iota : A \to B$, such that $r \circ \iota = \mathrm{id}_A$. In this situation, ι is necessarily injective, r is necessarily surjective, and A is called a *retract* of B. Now, if we consider the amalgamation, we have that R is a retract of $R \bowtie^f J$. More precisely, $p_R : R \bowtie^f J \to R, (r, f(r) + j) \mapsto r$, is a retraction, since the map $\iota : R \to R \bowtie^f J, r \mapsto (r, f(r))$, is a ring embedding such that $p_R \circ \iota = \mathrm{id}_R$.

Proposition 2.11 ([13, Proposition 4.7]) *Let $R, U, C, \alpha, \beta, p_R, p_U$ be as in the above definition of fiber product. Then, the following conditions are equivalent.*

(i) $p_R : \alpha \times_C \beta \to R$ *is a ring retraction.*
(ii) There exist an ideal J of U and a ring homomorphism $f : R \to U$ such that $\alpha \times_C \beta = R \bowtie^f J$.

Let us give some straightforward consequences of the definition of amalgamated algebra $R \bowtie^f J$.

Proposition 2.12 ([13, Proposition 5.1]) *Let R, U, J and f as above.*

(1) The natural ring homomorphism defined by $\iota(r) := (r, f(r))$, for all $r \in R$ is an embedding, making $R \bowtie^f J$ a ring extension of R.
(2) Let I be an ideal of R and set $I \bowtie^f J := \{(i, f(i) + j) \mid i \in I, j \in J\}$. Then $I \bowtie^f J$ is an ideal of $R \bowtie^f J$ and we have the following canonical isomorphism:

$$\frac{R \bowtie^f J}{I \bowtie^f J} \cong \frac{R}{I}.$$

(3) Let $p_R : R \bowtie^f J \to R$ and $p_U : R \bowtie^f J \to U$ be the natura l projections of $R \bowtie^f J \subseteq R \times U$ into R and U, respectively. Then p_R is surjective and $\mathrm{Ker}(p_R) = \{0\} \times J$, *while* $p_U(R \bowtie^f J) = f(R) + J$ *and* $\mathrm{Ker}(p_U) = f^{-1}(J) \times \{0\}$.
(4) Let $\gamma : R \bowtie^f J \to (f(R) + J)/J$ be the natural ring homomorphism, defined by $(r, f(r) + j) \mapsto f(r) + J$. Then γ is surjective and $\mathrm{Ker}(\gamma) = f^{-1}(J) \times J$. *Thus, there exists a natural isomorphism*

$$\frac{R \bowtie^f J}{f^{-1}(J) \times J} \cong \frac{f(R) + J}{J}.$$

It is clear that if $U = R$ and $f = \mathrm{id}_R$, we obtain the duplication. However with this construction we can obtain rings with many different properties. For example, notice that, in general, the embedding $\iota : R \hookrightarrow R \bowtie^f J$, is not an integral extension; in particular, the description of $\mathrm{Spec}(R \bowtie^f J)$ and of $\dim(R \bowtie^f J)$ is quite complicated and we can obtain many different situations depending on the choice of R, U, J, and f (for a complete study of this problem see [14]).

Let us see when the ring $R \bowtie^f J$ is a domain or when it is reduced; the subring $f(R) + J \subseteq U$ plays an important role.

Proposition 2.13 ([13, Proposition 5.2]) *With the notation of Proposition 2.12, assume $J \neq \{0\}$. Then, the following conditions are equivalent.*

(i) $R \bowtie^f J$ is an integral domain.

(ii) $f(R) + J$ *is an integral domain and* $f^{-1}(J) = \{0\}$.

In particular, if U *is an integral domain and* $f^{-1}(J) = \{0\}$, *then* $R \bowtie^f J$ *is an integral domain.*

Note that, if $R \bowtie^f J$ is an integral domain, then R is also an integral domain, by Proposition 2.12(1).

Proposition 2.14 ([13, Proposition 5.4]) *We preserve the notation of Proposition 2.12. The following conditions are equivalent.*

(i) $R \bowtie^f J$ *is a reduced ring.*
(ii) R *is a reduced ring and* $\mathrm{Nilp}(U) \cap J = \{0\}$.

Notice that the previous proposition implies that the property of being reduced for $R \bowtie^f J$ is independent of the nature of f. Moreover, if R and $f(R) + J$ are reduced rings, then $R \bowtie^f J$ is a reduced ring.

The next proposition provides an answer to the question of when $R \bowtie^f J$ is a Noetherian ring.

Proposition 2.15 ([13, Proposition 5.6]) *The following conditions are equivalent.*

(i) $R \bowtie^f J$ *is a Noetherian ring.*
(ii) R *and* $f(R) + J$ *are Noetherian rings.*

This result has a moderate interest because the Noetherianity of $R \bowtie^f J$ is not directly related to the data (i.e. R, U, f and J), but to the ring $f(R) + J$. Therefore, in order to obtain more useful criteria for the Noetherianity of $R \bowtie^f J$, we specialize Proposition 2.15 in some relevant cases.

Proposition 2.16 ([13, Proposition 5.7]) *With the notation of Proposition 2.12, assume that at least one of the following conditions holds:*

(1) J *is a finitely generated* R*-module (with the structure naturally induced by* f*);*
(2) J *is a Noetherian* R*-module (with the structure naturally induced by* f*);*
(3) $f(R) + J$ *is Noetherian as* R*-module (with the structure naturally induced by* f*);*
(4) f *is a finite homomorphism.*

Then $R \bowtie^f J$ *is Noetherian if and only if* R *is Noetherian.*

In particular, if R *is a Noetherian ring and* U *is a Noetherian* R*-module (e.g. if* f *is a finite homomorphism), then* $R \bowtie^f J$ *is a Noetherian ring for all ideals* J *of* U.

We now concentrate on the local case.

Proposition 2.17 ([15, Corollary 2.7]) $R \bowtie^f J$ *is a local ring if and only if* R *is a local ring and* J *is contained in the Jacobson radical of* U. *In particular, if* $\mathfrak{m}$ *is the unique maximal ideal of* R, *then* $\mathfrak{m} \bowtie^f J$ *is the unique maximal ideal of* $R \bowtie^f J$.

Assuming that $R \bowtie^f J$ is local and Noetherian, it is possible to investigate when $R \bowtie^f J$ is a Cohen–Macaulay (briefly CM) ring or a Gorenstein ring. Throughout the rest of this section we are assuming that R is Noetherian, local, with maximal ideal $\mathfrak{m}$, that J is an ideal of U contained in its Jacobson radical and that J is finitely generated as an R-module.

In this situation we know that the amalgamated algebra $R \bowtie^f J$ is a Noetherian local ring, with maximal ideal $\mathfrak{m} \bowtie^f J$. Moreover, the canonical map $\iota : R \to R \bowtie^f J$ is a finite ring embedding, since J is finitely generated as an R-module, and thus $\dim(R) = \dim(R \bowtie^f J)$. Moreover $\mathrm{Ann}(R \bowtie^f J) = (0)$, hence the dimension of $R \bowtie^f J$ as R-module (or, equivalently, $\dim(R/\mathrm{Ann}(R \bowtie^f J))$, since $R \bowtie^f J$ is a finite R-module) equals the Krull dimension of $R \bowtie^f J$.

Remark 2.18 ([15, Remark 5.1]) We observe that, under the previous assumptions, $R \bowtie^f J$ is a CM ring if and only if it is a CM R-module if and only if J is a maximal CM R-module.

In order to study when $R \bowtie^f J$ is a Gorenstein ring, we need to look at R endowed with a natural structure of an $R \bowtie^f J$-module. The next proposition holds in general, without assuming the additional hypotheses on R stated above.

Proposition 2.19 ([15, Proposition 5.3]) *Consider the natural map $\Lambda : f^{-1}(J) \to \mathrm{Hom}_{R \bowtie^f J}(R, R \bowtie^f J)$, where $\Lambda(x) := \lambda_x : R \to R \bowtie^f J$ is the $R \bowtie^f J$-linear map defined by $\lambda_x(r) := (rx, 0)$, for each $r \in R$ and $x \in f^{-1}(J)$. Then, Λ is an R-linear embedding and Λ is surjective if and only if* $\mathrm{Ann}_{f(R)+J}(J) = (0)$.

Now we are able to give a sufficient condition and a necessary condition for the ring $R \bowtie^f J$ to be Gorenstein.

Remark 2.20 ([15, Remark 5.4]) If R is a local Cohen–Macaulay ring, with maximal ideal $\mathfrak{m}$, having a canonical module isomorphic (as an R-module) to J, then $R \bowtie^f J$ is Gorenstein.

Proposition 2.21 ([15, Proposition 5.5]) *Assume that R is a local Cohen–Macaulay ring and that* $\mathrm{Ann}_{f(R)+J}(J) = (0)$. *If $R \bowtie^f J$ is Gorenstein, then R has a canonical module isomorphic to $f^{-1}(J)$.*

With extra assumptions on the ideal $f^{-1}(J)$ and on the ring $f(R) + J$, we can obtain the following characterization of when $R \bowtie^f J$ is Gorenstein as a consequence of [25, Theorem 4].

Proposition 2.22 ([15, Proposition 5.7]) *Assume that R is a CM ring, $f(R) + J$ is (S_1) and equidimensional, $J \neq 0$ and that $f^{-1}(J)$ is a regular ideal of R. Then, the following conditions are equivalent.*

(i) *$R \bowtie^f J$ is Gorenstein.*
(ii) *$f(R) + J$ is a CM ring, J is a canonical module of $f(R) + J$ and $f^{-1}(J)$ is a canonical module of R.*

3 Semigroup Constructions

In this section we present three semigroup constructions and then we show how they correspond to (particular cases of) the ring constructions presented in the previous section.

We always start with a numerical semigroup S, that is a submonoid of $\mathbb{N}$, such that $|\mathbb{N} \setminus S| < \infty$. We will denote by $\mathscr{F} = \mathscr{F}(S)$ its Frobenius number, i.e. the maximum integer not belonging to S.

What we obtain with the first and the third construction will be a subsemigroup of $\mathbb{N}^2$. We will show that the result of these constructions will be a good subsemigroup of $\mathbb{N}^2$, in the sense of [3] (where can be found the general definition of good subsemigroups of $\mathbb{N}^h$). We specialize the definition for the case $h = 2$.

Definition 3.1 Let $W \subseteq \mathbb{N}^2$ be a monoid; we say that W is a *good semigroup* if it satisfies the following properties:

(1) If $\boldsymbol{\alpha}, \boldsymbol{\beta} \in W$, then $\min(\boldsymbol{\alpha}, \boldsymbol{\beta}) = (\min\{\alpha_1, \beta_1\}, \min\{\alpha_2, \beta_2\}) \in W$.
(2) If $\boldsymbol{\alpha}, \boldsymbol{\beta} \in W$, $\boldsymbol{\alpha} \neq \boldsymbol{\beta}$ and $\alpha_i = \beta_i$ for some $i \in \{1, 2\}$, then there exists $\boldsymbol{\varepsilon} \in W$ such that $\varepsilon_i > \alpha_i = \beta_i$ and $\varepsilon_j = \min\{\alpha_j, \beta_j\}$ for $j \neq i$.
(3) There exists $\boldsymbol{\delta} \in \mathbb{N}^2$ such that $W \supseteq \boldsymbol{\delta} + \mathbb{N}^2$.

These properties are always satisfied by value semigroups of algebroid curves and are a consequence of the properties of discrete valuations (see [3]).

The first construction we present was introduced in [8] and, starting with a numerical semigroup S, it produces a subsemigroup of $\mathbb{N}^2$. Let $E \subseteq S$ be a semigroup ideal (i.e. $E \subseteq S$ such that $e + s \in E$, for every $e \in E$ and $s \in S$). Notice that we have $|\mathbb{N} \setminus E| < \infty$ and, denoting by $\mathscr{F}(E)$ the maximum integer not belonging to E (called *Frobenius number* of E), $\mathscr{F}(E) \geq \mathscr{F}$.

We define the *semigroup duplication* (or simply duplication) $S \bowtie E \subseteq \mathbb{N}^2$ of S with respect to E as the following subset of $\mathbb{N}^2$:

$$(u, v) \in S \bowtie E \iff \begin{cases} \bullet \text{ either } u = v \in S \\ \bullet \text{ or } u < v,\ u \in E,\ v \in S \\ \bullet \text{ or } v < u,\ v \in E,\ u \in S. \end{cases}$$

It is easy to see that $S \bowtie E$ is a semigroup. Moreover it is a good semigroup. This fact can be checked directly or it can be deduced from the fact that $S \bowtie E$ is the value semigroup of an algebroid curve with two branches, as it will be shown in the next subsection (see Theorem 3.3). An alternative definition is the following: if we set $D = \{(s, s) : s \in S\}$, then

$$S \bowtie E = D \cup (E \times E) \cup \{\min(\boldsymbol{\alpha}, \boldsymbol{\beta}) : \boldsymbol{\alpha} \in D,\ \boldsymbol{\beta} \in E \times E\}$$

(it is not difficult to check the equivalence of the two definitions).

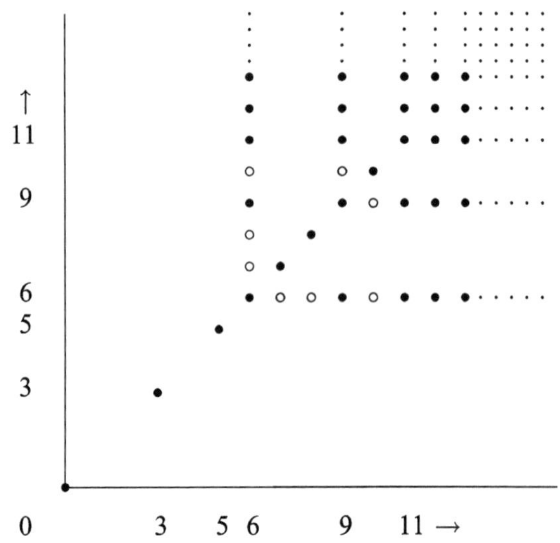

Fig. 1 $S \bowtie E$

In the next subsection we will see that the semigroup duplication corresponds to the amalgamated duplication, when the ring R we are starting with is an algebroid branch.

If $S = \langle 3, 5, 7\rangle = \{0, 3, 5, \rightarrow\}$ and $E = (6) + S = \{6, 9, 11, \rightarrow\}$, the semigroup $S \bowtie E$ is depicted in Fig. 1 (where the elements of $D \cup (E \times E)$ are marked as dots, while the elements obtained taking the minimums are depicted with circles).

The second construction we describe is the so-called *numerical duplication* (introduced and studied in [12]; see also [28]).

Let $E \subseteq S$ be an ideal of S. We set $2 \cdot S := \{2s : s \in S\}$ and $2 \cdot E := \{2t : t \in E\}$ (notice $2 \cdot S \neq 2S = S + S$ and $2 \cdot E \neq 2E = E + E$). Let $b \in S$ be an odd integer. Then we define the *numerical duplication*, $S \bowtie^b E$, of S with respect to E and b as the following subset of $\mathbb{N}$:

$$S \bowtie^b E = 2 \cdot S \cup (2 \cdot E + b).$$

It is straightforward to check that $S \bowtie^b E$ is a numerical semigroup. In fact $0 = 2 \cdot 0 \in S \bowtie^b E$; moreover, since $\mathscr{F}(E) \geq \mathscr{F}(S)$, every integer $n > 2\mathscr{F}(E) + b$ belongs to $S \bowtie^b E$; finally, the conditions $b \in S$ and E ideal of S immediately imply that $S \bowtie^b E$ is closed with respect to the sum.

We will see that this construction correspond to the domain case (for algebroid branches) in the family of quotients of the Rees algebra.

If $S = \langle 3, 5, 7\rangle = \{0, 3, 5, \rightarrow\}$ and $E = (6) + S = \{6, 9, 11, \rightarrow\}$, choosing $b = 3$, we get the following numerical semigroup

$$S \bowtie^3 E = \{0, 6, 10, 12, 14, 15, 16, 18, 20, 21, 22, 24, \rightarrow\}.$$

Finally, we present a new construction that will correspond to the amalgamation. Let S and T be two numerical semigroups and let $g : S \to T$ be a semigroup homomorphism. It is easy to see that g has to be the multiplication by a positive integer c such that $cs \in T$, for every $s \in S$. Let E be an ideal of T; then we set $D = \{(s, cs) : s \in S\}$ and we define the *amalgamation of S with T along E with respect to g* the following subset of $\mathbb{N}^2$:

$$S \bowtie^g E = D \cup (g^{-1}(E) \times E) \cup \{\min(\boldsymbol{\alpha}, \boldsymbol{\beta}) : \boldsymbol{\alpha} \in D,\ \boldsymbol{\beta} \in (g^{-1}(E) \times E)\}.$$

In order to check that it is a good subsemigroup of $\mathbb{N}^2$, it would be necessary a case by case argument; otherwise, this fact can be deduced from the fact that $S\bowtie^g E$ is the value semigroup of an algebroid curve with two branches, as it will be shown in the next subsection.

Notice that the projections of $S\bowtie^g E$ are S and $g(S) \cup E$; this last one is itself a numerical semigroup, as it is easy to check.

If $S = \langle 2, 3\rangle = \{0, 2, 3, \to\}$, $T = \langle 3, 4\rangle = \{0, 3, 4, 6, \to\}$ and $E = (3) + T = \{3, 6, 7, 9, \to\}$, then, if we choose g as the multiplication by 2, $g^{-1}(E) = (3) + S = \{3, 5, \to\}$ and the semigroup $S \bowtie^g E$ is depicted in Fig. 2 (where the elements of $D \cup (g^{-1}(E) \times E)$ are marked as dots, while the elements obtained taking the minimums are depicted with circles).

In order to better handle the elements of $S \bowtie^g E$ we will need a lemma, that gives a description of its elements analogue to the first definition of duplication.

Lemma 3.2 *Let $(s, t) \in S \times (g(S) \cup E)$, with $t \neq cs$. Then $(s, t) \in S \bowtie^g E$ if and only if one of the following condition holds:*

(1) if $t < cs$, then $t \in E$;
(2) if $t > cs$, then $s \in g^{-1}(E)$ and, either $t \in E$, or $t = c\tilde{s}$, for some $\tilde{s} \in S$, $\tilde{s} > s$.

Proof Let $(s, t) \in S \bowtie^g E$. If $(s, t) \in g^{-1}(E) \times E$, then it is clear that one of the two conditions is satisfied. If (s, t) is obtained as a minimum, then, either $(s, t) = \min((s, cs), (s', t))$ (with $(s, cs) \in D$ and $(s', t) \in g^{-1}(E) \times E$) and we get the first condition, or $(s, t) = \min((s, t'), (\tilde{s}, c\tilde{s}))$ (with $(\tilde{s}, c\tilde{s}) \in D$ and $(s', t) \in g^{-1}(E) \times E$) and we get the second condition.

Conversely, if one of the two conditions is satisfied, it is easy to check that either $(s, t) \in g^{-1}(E) \times E$ or it is obtained as a minimum of an element of D and one of $g^{-1}(E) \times E$.

3.1 Algebroid Branches

In this subsection we apply the ring constructions described in the first section to curve singularities and we show how, taking the corresponding value semigroups, we get the semigroup constructions defined in this section.

Fig. 2 $S \bowtie^g E$

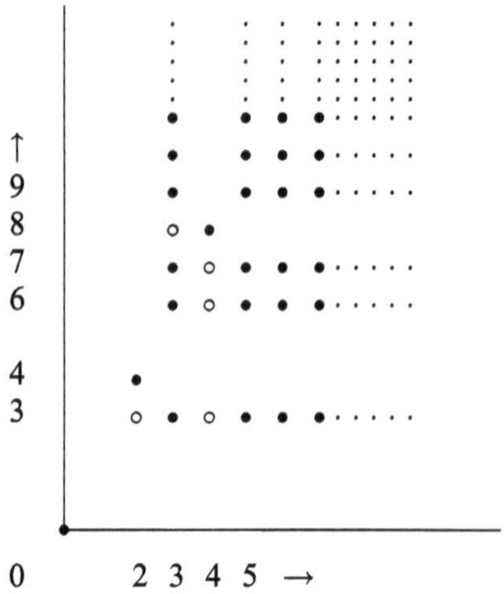

Following Zariski's terminology (see e.g. [30]), by an algebroid curve (with h branches) we mean a one-dimensional reduced ring of the form

$$R = k[[x_1, \ldots, x_n]]/(P_1 \cap \cdots \cap P_h),$$

where $x_1, \ldots, x_n$ are indeterminates over the field k (that we assume to be algebraically closed and of characteristic 0) and $P_1, \ldots, P_n$ are prime ideals of height $n-1$ in $k[[x_1, \ldots, x_n]]$. The ring $R_i = k[[x_1, \ldots, x_n]]/P_i$ is the i-th algebroid branch of the curve. If we consider the completion (with respect to the topology induced by its maximal ideal) of the local ring at a singular point (that we can assume to be the origin) of an algebraic curve over an algebraically closed field, we get an algebroid curve.

Under these hypotheses the quotient field of R_i ($i = 1, \ldots, h$ is isomorphic to the field of formal Laurent series $k((t_i))$ and its integral closure is isomorphic to $k[[t_i]]$ and it is a finite R_i module. The total ring of fractions of R is $Q(R) \cong k((t_1)) \times \cdots \times k((t_h))$ and the integral closure of R in $Q(R)$ is $\overline{R} \cong k[[t_1]] \times \cdots \times k[[t_h]]$ (cf. [3]). Let v_i be the usual valuation on $k((t_i))$, i.e. the order of a series; hence, looking at any element $r \in Q(R)$ as an element of $k((t_1)) \times \cdots \times k((t_h))$, we define $v(r) = (v_1(r_1), \ldots, v_h(r_h))$. If we set $v(R) := \{v(r) : r \in R,\ r \notin Z(R)\}$ (where $Z(R)$ is the set of the zero divisors of R) we get a good subsemigroup of $\mathbb{N}^h$ (cf. [3]). More generally, if I is a regular fractional ideal of R (i.e. I contains a non zero divisor of $Q(R)$) we define its value set as $v(I) := \{v(i) : i \in I,\ i \notin Z(Q(R))\}$ (where $Z(Q(R))$ is the set of the zero divisors of $Q(R)$). If R is an algebroid branch (i.e. $h = 1$), $S = v(R)$ is a numerical semigroup.

A particular case of algebroid branches is given by the semigroup rings of the form $R = k[[t^s : s \in S]]$ (where S is a numerical semigroup). In this case the correspondence between ring properties and semigroup properties is more strict than the general case. For example, if E is a semigroup ideal, the monomial ideal $I = (t^e : e \in E)$ is an ideal of R such that $v(I) = E$.

In order to consider the value of any element of R we can define $v_i(0) = \infty$, with the following conventions: $m < \infty$ and $m + \infty = \infty$, for every integer m. With these assumptions, the value semigroup of an algebroid curve with two branches will be a subsemigroup W of $(\mathbb{N} \cup \{\infty\})^2$ and the same properties of Definition 3.1 hold, with the obvious generalizations if one component of the elements $\boldsymbol{\alpha}, \boldsymbol{\beta} \in W$ equals ∞.

Now consider an algebroid branch R and a regular ideal I. If we consider the ring $R \bowtie I$, it is again a one-dimensional reduced local ring. More precisely we have:

Theorem 3.3 ([8, Theorem 4.1, Corollary 4.2, Proposition 4.5]) *Let R be an algebroid branch and let $I \neq (0)$ be a proper ideal of R. Then $R \bowtie I$ is an algebroid curve with 2 branches both isomorphic to R. Moreover the value semigroup of $R \bowtie I$ is $v(R \bowtie I) = v(R) \bowtie v(I)$.*

In particular, since for every numerical semigroup S and every ideal $E \subset S$ we can find an algebroid branch R (e.g. the numerical semigroup ring) such that $v(R) = S$ and an ideal $I \subset R$ such that $v(I) = E$, we get that the construction $S \bowtie E$ always produces a good subsemigroup of $\mathbb{N}^2$.

Starting again from an algebroid branch R, we consider now $R(I)_{0,-r}$, choosing r in such a way that $R(I)_{0,-r}$ is an integral domain.

Theorem 3.4 ([4, Theorem 3.6]) *Let R be an algebroid branch and let $I \neq (0)$ be a proper ideal of R; let $r \in R$, such that $b = v(r)$ is odd. Then $R(I)_{0,-r}$ is an algebroid branch and its value semigroup is $v(R) \bowtie^b v(I)$.*

Finally, we consider the case of amalgamation. Let $R = k[[x_1, \dots, x_n]]/P$ and $U = k[[y_1, \dots, y_m]]/Q$ be algebroid branches and $f : R \to U$ be an homomorphism. Let $J \neq (0)$ be a proper ideal of U. In this case, since R and U are both integral domains, it follows immediately, by Proposition 2.13 that $R \bowtie^f J$ is reduced. Moreover, since they are both local we also get by Proposition 2.17, that $R \bowtie^f J$ is local. Consider now R as subring of $k[[t]]$ and U as subring of $k[[u]]$; let $g(u)$ be the image of any non-zero element in R; then $k[[g(u)]] \subset f(R) \subset f(R) + J \subset U \subset k[[u]]$; since the inclusion $k[[g(u)]] \subset k[[u]]$ is finite, then also the inclusion $f(R) \subset U$ is finite. Hence, in particular, by Proposition 2.16, $R \bowtie^f J$ is Noetherian; moreover, J is a finite generated R-module, hence $\dim(R \bowtie^f J) = \dim(R) = 1$.

Theorem 3.5 *Under the assumptions stated above, $R \bowtie^f J$ is an algebroid curve with two branches.*

Proof By the discussion above we know that $R \bowtie^f J$ is one-dimensional, local, Noetherian and reduced. So it is enough to show that it can be presented as a quotient of a power series ring over k, modulo the intersection of two prime ideals of coheight 1.

Let $\varphi_1 : k[[x_1, \dots, x_n]] \longrightarrow k[[t]]$ be the composition of the homomorphisms $k[[x_1, \dots, x_n]] \longrightarrow R$ and $R \hookrightarrow \overline{R} \cong k[[t]]$.

Analogously, let $\varphi_2 : k[[y_1, \dots, y_m]] \longrightarrow k[[u]]$ be the composition of the homomorphisms $k[[y_1, \dots, y_m]] \longrightarrow U$ and $U \hookrightarrow \overline{U} \cong k[[u]]$.

Let $\bar{g}_i = f(\bar{x}_i)$ (for $i = 1, \dots, n$) be the images of $\bar{x}_i$ in U and let $\{\bar{h}_j : j = 1, \dots, r\}$ be a minimal set of generators of J as R-module, with g_i and h_j elements of $k[[y_1, \dots, y_m]]$. If we identify R with its image in $k[[t]]$ and $f(R)$ and J with their images in $k[[u]]$, respectively, then $f(R) = k[[\varphi_2(g_1), \dots, \varphi_2(g_n)]]$, since, if $\overline{F} \in R$, with $F \in k[[x_1, \dots, x_n]]$, then $f(\varphi_1(F)) = f(\overline{F}) = F(\varphi_2(g_1), \dots, \varphi_2(g_n))$, and $J = f(R)\varphi_2(h_1) + \dots + f(R)\varphi_2(h_r)$ as $f(R)$-module. Let $z_1, \dots, z_r$ be new indeterminates and define the following homomorphisms:

$$\begin{aligned} \psi_1 : k[[x_1, \dots, x_n, z_1, \dots, z_r]] &\longrightarrow \overline{R} \\ x_i &\longmapsto \varphi_1(x_i) \\ z_j &\longmapsto 0 \end{aligned}$$

$$\begin{aligned} \psi_2 : k[[x_1, \dots, x_n, z_1, \dots, z_r]] &\longrightarrow \overline{U} \\ x_i &\longmapsto \varphi_2(g_i) \\ z_j &\longmapsto \varphi_2(h_j) \end{aligned}$$

We have that $\mathrm{Im}\psi_1 = R$ and, since J is an ideal of U, $\mathrm{Im}\psi_2 = f(R) + J$ (in fact, if $F \in k[[x_1, \dots, x_n, z_1, \dots, z_r]]$, we can write uniquely $F = F_1 + F_2$, where $F_1 \in k[[x_1, \dots, x_n]]$ and F_2 contains only terms in which some z_j appears; by definition we obtain: $\psi_2(F_1) = F_1(\phi_2(g_1), \dots, \psi_2(g_n)) = f(\overline{F}_1) \in f(R)$ and $\psi_2(F_2) \in (\psi_2(z_1), \dots, \psi_2(z_r)) = (\varphi_2(h_1), \dots, \varphi_2(h_r)) \subseteq J$; the other inclusion is trivial). In particular, both the ideals $\mathrm{Ker}\psi_i$ are prime ideals of co-height 1. Now define the following homomorphism:

$$\begin{aligned} \Omega : k[[x_1, \dots, x_n, z_1, \dots, z_r]] &\longrightarrow \overline{R} \times \overline{U} \\ F(x_1, \dots, x_n, z_1, \dots, z_r) &\longmapsto (\psi_1(F), \psi_2(F)) \end{aligned}$$

We have that $\mathrm{Ker}\Omega = \mathrm{Ker}\psi_1 \cap \mathrm{Ker}\psi_2$ and, therefore,

$$\mathrm{Im}\Omega \cong k[[x_1, \dots, x_n, z_1, \dots, z_r]]/\mathrm{Ker}\Omega$$

is an algebroid curve with 2 branches. Hence to conclude the proof we need to show that $R \bowtie^f J \cong \mathrm{Im}\Omega$.

Since $R \bowtie^f J = \{(r, f(r) + j \mid r \in R, j \in J\}$, as sub-$R$-module of $R \times U \subseteq \overline{R} \times \overline{U}$, it is generated by $(1, 1), (0, \varphi_2(h_1)), \dots, (0, \varphi_2(h_r))$; hence $R \bowtie^f J \subseteq \mathrm{Im}\Omega$.

Conversely, we know that $\mathrm{Im}\Omega \subseteq \mathrm{Im}\psi_1 \times \mathrm{Im}\psi_2 = (R \times f(R) + J)$; let $(r, s) \in R \times (f(R) + J)$ be an element of $\mathrm{Im}\Omega$; hence there exists $F \in k[[x_1, \dots, x_n, z_1, \dots, z_r]]$ such that $\psi_1(F) = r$ and $\psi_2(F) = s$. We need to show that $s - f(r) \in J$; as above, we write uniquely $F = F_1 + F_2$, where $F_1 \in k[[x_1, \dots, x_n]]$ and F_2 contains only terms in which some z_j appears. By definition we obtain: $\psi_1(F_1) = \varphi_1(F_1)$, $\psi_1(F_2) = 0$ and, as above,

$$\psi_2(F_1) = F_1(\phi_2(g_1), \ldots, \psi_2(g_n)) = f(\overline{F}_1) = f(\varphi_1(F_1))$$

and $\psi_2(F_2) \in (\psi_2(y_1), \ldots, \psi_2(y_r)) = (\varphi_2(h_1), \ldots, \varphi_2(h_r)) \subseteq J$.

It follows that $s - f(r) = \psi_2(F) - f(\psi_1(F)) = \psi_2(F_1) + \psi_2(F_2) - f(\psi_1(F_1)) = f(\varphi_1(F_1)) + \psi_2(F_2) - f(\varphi_1(F_1)) = \psi_2(F_2) \in J$, that is $\mathrm{Im}\Omega \subseteq R \bowtie^f J$.

Proposition 3.6 *Let $R = k[[t^s : s \in S]]$, $U = k[[u^t : t \in T]]$ and $J = (u^e : e \in E)$; assume that $cs \in T$, for all $s \in S$ and let $f : R \longrightarrow U$ be the homomorphism defined by $f(F(t)) = F(u^c)$. Then $v(R \bowtie^f J) = S \bowtie^g E$, where $g : S \longrightarrow T$ is the multiplication by c.*

In particular, the construction $S \bowtie^g E$ always produces a good subsemigroup of $\mathbb{N}^2$.

Proof Let v_1 be the usual discrete valuation on $k((t))$ and v_2 be the usual discrete valuation on $k((u))$. Let $(r, f(r) + j) \in R \bowtie^f J$ and set $v_1(r) = a$ and $v_2(j) = b \in E$. If $v_2(f(r)) = ca < b$, then $v(r, f(r) + j) = (a, ca) \in S \bowtie^g E$.

If $v_2(f(r)) = ca > b$, then $v(r, f(r) + j) = (a, b)$. By Lemma 3.2 we need to check condition (1). If $a \notin g^{-1}(E)$, pick $a' \in S$, $a' > a$, such that ca' is bigger than $\mathscr{F}(E)$. Hence $a' \in g^{-1}(E)$ and condition (1) of Lemma 3.2 is satisfied.

Finally, we consider the case $ca = b$. If $v_2(f(r) + j) = ca$ there is nothing to prove. So assume that $d = v_2(f(r) + j) > ca$. Again by Lemma 3.2, to show that $(a, d) \in S \bowtie^g E$ we need to check condition (2). Since $ca = b, a \in g^{-1}(E)$; if $d \notin E$, then the expression of $f(r) + j$ as a power series in u has, as summand of lower order, a monomial appearing in $f(r)$; therefore its order has to be of the form $d = c\tilde{a}$, with $a < \tilde{a} \in S$; hence condition (2) of Lemma 3.2 is satisfied and we have proved the inclusion $v(R \bowtie^f J) \subseteq S \bowtie^g E$.

To prove the reverse inclusion we use again Lemma 3.2. Consider an element $(a, b) \in S \bowtie^g E$. If $b = ca$ then $(a, b) = v(t^a, f(t^a)) = v(t^a, u^{ca})$. If $b \neq ca$, one of the two conditions of the lemma are satisfied.

In the first case $b < ca$; then $(a, b) = v(t^a, f(t^a) + u^b) = v(t^a, u^b + u^{ca})$.

In the second case, $b > ca$ and $a \in g^{-1}(E)$, i.e. $ca \in E$ or equivalently $u^{ca} \in J$; this implies that $(t^a, 0) \in da$. If $b \in E$, then $u^b \in J$, $(0, u^b) \in R \bowtie^f J$ and $(a, b) = v((t^a, 0) + (0, u^b)) \in v(R \bowtie^f J)$.

Otherwise, if $b = c\tilde{a}$, for some $\tilde{a} > a$, $(t^{\tilde{a}}, u^b) \in R \bowtie^f J$ and $(a, b) = v((t^a, 0) + (t^{\tilde{a}}, u^b)) \in R \bowtie^f J$.

The proof is complete.

3.2 Symmetry

In this subsection we study the symmetry of the semigroups obtained with the three constructions previously described. One possible strategy is to study the Gorensteinnes of the corresponding algebroid curve (or branch), while the other is to prove it directly.

As a matter of fact, let us recall that, if R is an algebroid curve, it is a one-dimensional reduced ring, then it is CM. Moreover, since R is a local complete reduced ring, by [19, Satz 6.21], R always has a canonical module ω_R which can be identified with a fractional ideal in $Q(R)$; moreover, since the invertible fractional ideals of R are principal, by [19, Satz 2.8], we have that, if ω_R is a canonical ideal of R, for each nonzero divisor $z \in Q$, $z\omega_R$ is a canonical ideal and, if ω_R and ω'_R are two canonical ideals of R, then there exists a non-zero divisor $z \in Q(R)$ such that $\omega_R = z\omega'_R$. In particular, we can always assume that $\omega_R \subseteq R$ or, when it is needed, that $R \subseteq \omega_R \subseteq \overline{R}$.

For a numerical semigroup S, with Frobenius number $\mathscr{F}$, if $s \in S$, then $\mathscr{F} - s \notin S$. When the converse is true, that is

$$x \in S \Longleftrightarrow \mathscr{F} - x \notin S,$$

the semigroup S is said to be *symmetric*; by [22] the Gorenstein algebroid branches are characterized as those algebroid branches that have a symmetric value semigroup. Moreover, if we set $K(S) := \{x \in \mathbb{Z} : f(S) - x \notin S\}$ (notice that $K(S) = S$ if and only if S is symmetric), then it is proved in [21, Satz 5] that a fractional ideal I of R, such that $R \subseteq I \subseteq \overline{R}$, is a canonical module for R if and only if $v(I) = K(S)$. More generally, a canonical ideal of S is an ideal of the form $x + K(S)$ (where $x \in \mathbb{N}$) and a proper ideal of R is a canonical ideal of R if and only if its value set $v(I)$ is a canonical ideal of $v(R)$. Both these results can be generalized to the case of algebroid curves with more than one branch, giving proper definitions of symmetric semigroup and of $K(S)$ (see [6, 7]).

As for the first two constructions, since the rings $R \bowtie I$ and $R(I)_{0,-r}$ are members of the same family of quotients of the Rees algebra, by Theorem 2.9, it is enough to prove the Gorensteinness of one of them. In the second case (that is for the algebroid branch $R(I)_{0,-r}$), this is equivalent to the fact that $v(R) \bowtie^b v(I)$ is symmetric.

The symmetry for numerical duplication has been characterized in [12]:

Proposition 3.7 ([12, Proposition 3.1]) *The numerical semigroup $S \bowtie^b E$ is symmetric if and only if E is a canonical ideal of S.*

The next corollaries are now straightforward.

Corollary 3.8 [4, Corollary 3.3] *The rings $R(I)_{a,b}$ are Gorenstein if and only if I is a canonical ideal of R. In particular, this statement holds for the amalgamated duplication.*

Corollary 3.9 *The subsemigroup $S \bowtie E$ of $\mathbb{N}^2$ is symmetric if and only if E is a canonical ideal of S.*

We finish the paper studying the symmetry for the amalgamation $S \bowtie^g E$. Using Propositions 2.22 and 3.6, we get immediately the following characterization.

Proposition 3.10 *The semigroup $S \bowtie^g E$ is symmetric if and only if $g^{-1}(E)$ is a canonical ideal of S and E is a canonical ideal of $g(S) \cup E$.*

The previous proposition can be proved also directly. Since in this paper we defined the semigroup amalgamation $S \bowtie^g E$ independently by the ring amalgamation, for the sake of completeness we give also the numerical proof.

Recall that $g : S \longrightarrow T$ is the multiplication by c and that $g(S) \cup E$ is a subsemigroup of T. We need also to recall the definition and a characterization of symmetry for good subsemigroups of $\mathbb{N}^2$.

Let $W \subseteq \mathbb{N}^2$ be a good semigroup; let $\boldsymbol{\delta} = \boldsymbol{\delta}(W)$ be the minimum element of $\mathbb{N}^2$ such that $W \supseteq \boldsymbol{\delta} + \mathbb{N}^2$ and define $\boldsymbol{\gamma} = \boldsymbol{\delta} - (1, 1)$. We also set $\Delta(\boldsymbol{\alpha}) = \{(\alpha_1, b) : b > \alpha_2\} \cup \{(a, \alpha_2) : a > \alpha_1\}$. Then W is said to be *symmetric* if

$$\boldsymbol{\alpha} \in W \Longleftrightarrow \Delta(\boldsymbol{\gamma} - \boldsymbol{\alpha}) \cap W = \emptyset$$

(it is clear that the vector $\boldsymbol{\gamma}$ plays the role of the Frobenius number). Let W_1 and W_2 be the two projections of W; we have that W is symmetric if and only if

$$\begin{aligned} &|W_1 \cap \{0, 1, \ldots, \delta_1 - 1\}| + |\{b < \delta_2 : (\delta_1, b) \in W\}| \\ &+|W_2 \cap \{0, 1, \ldots, \delta_2 - 1\}| + |\{a < \delta_1 : (a, \delta_2) \in W\}| \\ &= \delta_1 + \delta_2 \end{aligned}$$

(for more details see [6, 7]).

We also recall that, for an ideal E of a numerical semigroup S, we have the inequality $|\{e \in E : e < \mathscr{F}(E)\}| \leq |\mathbb{N} \setminus S|$ and that the equality holds if and only if E is a canonical ideal of S (see [12]).

Now we can give the proof of Proposition 3.10.

Proof Set $W = S \bowtie^g E$; we know that $W_1 = S$ and $W_2 = g(S) \cup E$. Moreover $\boldsymbol{\delta} = \boldsymbol{\delta}(W) = (\mathscr{F}(g^{-1}(E)) + 1, \mathscr{F}(E) + 1)$.

Then W is symmetric if and only if

$$\begin{aligned} &|S \cap \{0, 1, \ldots, \mathscr{F}(g^{-1}(E))\}|+ \\ &|\{b < \mathscr{F}(E) + 1 : (\mathscr{F}(g^{-1}(E)) + 1, b) \in W\}|+ \\ &|(g(S) \cup E) \cap \{0, 1, \ldots, \mathscr{F}(E)\}|+ \\ &|\{a < \mathscr{F}(g^{-1}(E)) + 1 : (a, \mathscr{F}(E) + 1) \in W\}| = \\ &\mathscr{F}(g^{-1}(E)) + 1 + \mathscr{F}(E) + 1\,. \end{aligned}$$

By definition of $S \bowtie^g E$, we have

$$|\{b < \mathscr{F}(E) + 1 : (\mathscr{F}(g^{-1}(E)) + 1, b) \in W\}| = |\{b \in E : b < \mathscr{F}(E) + 1\}|$$

and

$$|\{a < \mathcal{F}(g^{-1}(E)) + 1 : (a, \mathcal{F}(E) + 1) \in W\}| \\ = |\{a \in g^{-1}(E) : a < \mathcal{F}(g^{-1}(E)) + 1\}|.$$

It follows that

$$|S \cap \{0, \ldots, \mathcal{F}(g^{-1}(E))\}| + |\{a < \mathcal{F}(g^{-1}(E)) + 1 : (a, \mathcal{F}(E) + 1) \in W\}| \\ = |S \cap \{0, 1, \ldots, \mathcal{F}(g^{-1}(E))\}| + |\{a \in g^{-1}(E) : a < \mathcal{F}(g^{-1}(E)) + 1\}| \\ \leq |S \cap \{0, 1, \ldots, \mathcal{F}(g^{-1}(E))\}| + |\mathbb{N} \setminus S| = \mathcal{F}(g^{-1}(E)) + 1$$

and the equality holds if and only if $g^{-1}(E)$ is a canonical ideal of S.

Analogously,

$$|(g(S) \cup E) \cap \{0, \ldots, \mathcal{F}(E)\}| + \\ |\{b < \mathcal{F}(E) + 1 : (\mathcal{F}(g^{-1}(E)) + 1, b) \in W\}| \leq \mathcal{F}(E) + 1$$

and the equality holds if and only if E is a canonical ideal of $g(S) \cup E$.

Now the thesis follows immediately.

References

1. D.D. Anderson, Commutative rings, in *Multiplicative Ideal Theory in Commutative Algebra: A Tribute to the Work of Robert Gilmer*, ed. by J. Brewer, S. Glaz, W. Heinzer, B. Olberding (Springer, New York, 2006), pp. 1–20
2. A. Bagheri, M. Salimi, E. Tavasoli, S. Yassemi, A construction of quasi-Gorenstein rings. J. Algebra Appl. **11**(1) (2012). doi:10.1142/s0219498811005361
3. V. Barucci, M. D'Anna, R. Fröberg, Analytically unramified one-dimensional semilocal rings and their value semigroups. J. Pure Appl. Algebra **147**, 215–254 (2000)
4. V. Barucci, M. D'Anna, F. Strazzanti, A family of quotients of the Rees algebra. Commun. Algebra **43**(1), 130–142 (2015)
5. V. Barucci, M. D'Anna, F. Strazzanti, *Gorenstein and almost Gorenstein property for a family of quotients of the Rees algebra,* preprint
6. A. Campillo, F. Delgado, K. Kiyek, Gorenstein property and symmetry for one-dimensional local Cohen-Macaulay rings. Manuscripta Math. **83**, 405–423 (1994)
7. M. D'Anna, The canonical module of a one-dimensional reduced local ring. Commun. Alg. **25**, 2939–2965 (1997)
8. M. D'Anna, A construction of Gorenstein rings. J. Algebra **306**(2), 507–519 (2006)
9. M. D'Anna, M. Fontana, An amalgamated duplication of a ring along an ideal: basic properties. J. Algebra Appl. **6**(3), 443–459 (2007)
10. M. D'Anna, M. Fontana, The amalgamated duplication of a ring along a multiplicative-canonical ideal. Arkiv Mat. **45**, 241–252 (2007)
11. M. D'Anna, R. Re, *On the amalgamated duplication of a curve singularity along an ideal*, private communication
12. M. D'Anna, F. Strazzanti, The numerical duplication of a numerical semigroup. Semigroup forum **87**, 149–160 (2013)

13. M. D'Anna, C.A. Finocchiaro, M. Fontana, Amalgamated algebras along an ideal, in *Commutative Algebra and Applications*, ed. by M. Fontana, S. Kabbaj, O. Olberding, I. Swanson (de Gruyter, Berlin, 2009)
14. M. D'Anna, C.A. Finocchiaro, M. Fontana, Properties of chains of prime ideals in an amalgamated algebra along an ideal. J. Pure Appl. Algebra **214**, 1633–1641 (2010)
15. M. D'Anna, C.A. Finocchiaro, M. Fontana, *New algebraic properties of an amalgamated algebra along an ideal*, to appear on Commun. Algebra
16. J.L. Dorroh, Concerning adjunctions to algebras. Bull. Am. Math. Soc. **38**, 85–88 (1932)
17. C. Finocchiaro, Prüfer-like conditions on an amalgamated algebra along an ideal. Houst. J. Math. **40**(1), 63–79 (2013)
18. R. Fossum, Commutative extensions by canonical modules are Gorenstein rings. Proc. Am. Math. Soc. **40**, 395–400 (1973)
19. J. Herzog, E. Kunz, *Kanonische Modul eines Cohen-Macaulay Rings*, vol. 238, Lecture Notes in Mathematics (Springer, Berlin, 1971)
20. J. Huckaba, *Commutative Rings with Zero Divisors* (M. Dekker, New York, 1988)
21. J. Jäger, Längeberechnungen und kanonische Ideale in eindimensionalen Ringen. Arch. Math. **29**, 504–512 (1977)
22. E. Kunz, The value-semigroup of a one-dimensional Gorenstein ring. Proc. Am. Math. Soc. **25**, 748–751 (1970)
23. H.R. Maimani, S. Yassemi, Zero-divisor graphs of amalgamated duplication of a ring along an ideal. J. Pure Appl. Algebra **212**, 168–174 (2008)
24. M. Nagata, *Local Rings* (Interscience, New York, 1962)
25. T. Ogoma, *Fiber products of Noetherian rings*. Adv. Stud. Pure Math. **11**, 173–182 (1987). Commutative Algebra and Combinatorics, ed. by M. Nagata, H. Matsumura
26. I. Reiten, The converse of a theorem of Sharp on Gorenstein modules. Proc. Am. Math. Soc. **32**, 417–420 (1972)
27. J. Shapiro, On a construction of Gorenstein rings proposed by M. D'Anna. J. Algebra **323**, 1155–1158 (2010)
28. F. Strazzanti, *One half of almost symmetric numerical semigroups*, to appear on Semigroup Forum
29. J. Wook Lim, D.Y. Oh, S-Noetherian properties on amalgamated algebras along an ideal. J. Pure Appl. Algebra **218**(6), 10751080 (2014)
30. O. Zariski, Studies in equisingularity. Am. J. Math. **87**, 507–536 (1965)

New Distinguished Classes of Spectral Spaces: A Survey

Carmelo A. Finocchiaro, Marco Fontana and Dario Spirito

Abstract In the present survey paper, we present several new classes of Hochster's spectral spaces "occurring in nature," actually in multiplicative ideal theory, and not linked to or realized in an explicit way by prime spectra of rings. The general setting is the space of the semistar operations (of finite type), endowed with a Zariski-like topology, which turns out to be a natural topological extension of the space of the overrings of an integral domain, endowed with a topology introduced by Zariski. One of the key tool is a recent characterization of spectral spaces, based on the ultrafilter topology, given in Finocchiaro, Commun Algebra, 42:1496–1508, 2014, [15]. Several applications are also discussed.

Keywords Spectral space and spectral map · Star and semistar operations · Zariski, constructible, inverse and ultrafilter topologies · Gabriel-Popescu localizing system · Riemann–Zariski space of valuation domains · Kronecker function ring

MSC(2010) 13A15 · 13G05 · 13B10 · 13E99 · 13C11 · 14A05

The authors gratefully acknowledge partial support from *INdAM, Istituto Nazionale di Alta Matematica.*

M. Fontana (✉) · D. Spirito
Dipartimento di Matematica e Fisica, Università degli Studi "Roma Tre",
00146 Rome, Italy
e-mail: fontana@mat.uniroma3.it

D. Spirito
e-mail: spirito@mat.uniroma3.it

C.A. Finocchiaro
Institute of Analysis and Number Theory, University of Technology, Kopernikusgasse 24,
8010 Graz, Austria
e-mail: finocchiaro@math.tugraz.at

S. Chapman et al. (eds.), *Multiplicative Ideal Theory and Factorization Theory*,
Springer Proceedings in Mathematics & Statistics 170,
DOI 10.1007/978-3-319-38855-7_5

1 Introduction and Preliminaries

Let X be a topological space. According to [35], X is called a *spectral space* if there exists a ring R such that $\mathtt{Spec}(R)$, with the Zariski topology, is homeomorphic to X. Spectral spaces can be characterized in a purely topological way: a topological space X is spectral if and only if X is T_0 (this means that for every pair of distinct points of X, at least one of them has an open neighborhood not containing the other), quasi-compact, admits a basis of quasi-compact open subspaces that is closed under finite intersections, and every irreducible closed subspace C of X has a (unique) generic point (i.e., there exists one point $x_C \in C$ such that C coincides with the closure of this point) [35, Proposition 4].

In the present survey paper, we present several new classes of spectral spaces occurring naturally in multiplicative ideal theory. Before doing this, we introduce, for convenience of the reader, some background material.

1.1 Semistar Operations

Let D be an integral domain with quotient field K. Let $\overline{\boldsymbol{F}}(D)$ [respectively, $\boldsymbol{F}(D)$; $\boldsymbol{f}(D)$] be the set of all nonzero D-submodules of K [respectively, nonzero fractional ideals; nonzero finitely generated fractional ideals] of D (thus, $\boldsymbol{f}(D) \subseteq \boldsymbol{F}(D) \subseteq \overline{\boldsymbol{F}}(D)$).

A *semistar operation* on D is a map $\star : \overline{\boldsymbol{F}}(D) \to \overline{\boldsymbol{F}}(D)$, $E \mapsto E^\star$, such that, for every $z \in K$, $z \neq 0$, and for every $E, F \in \overline{\boldsymbol{F}}(D)$, the following properties hold: $(\star_1)$ $E \subseteq E^\star$; $(\star_2)$ $E \subseteq F$ implies $E^\star \subseteq F^\star$; $(\star_3)$ $(E^\star)^\star = E^\star$; $(\star_4)$ $(zE)^\star = z \cdot E^\star$. If $D = D^\star$, then the map $\star|_{\boldsymbol{F}(D)} : \boldsymbol{F}(D) \to \boldsymbol{F}(D)$ is called a *star operation* on D.

Semistar operations were introduced by Okabe and Matsuda in 1994 [46] (although this kind of operations were considered by J. Huckaba in 1988, in the setting of rings with zero divisors [38, Sect. 20]), producing a more general and flexible concept than the earlier notion of a star operations which in turn were defined by Krull [40–42] and used, among others, by Gilmer [32, Sect. 32].

A star operation, in Krull's original terminology, was called "prime operation" (*Strich-Operation* or *′-Operation*, in German [40, 41]). The notion of semiprime operation and the relation with that of semistar operation has been investigated in [14] (see also [51]). Semiprime operations include various examples of specific closures, used mainly in the Noetherian setting, the most important of which is probably tight closure, originally defined in [36]. (See [13] for a survey on closure operations.)

1.2 Riemann–Zariski Spaces

Let K be a field and let A be any subring of K. Let $\mathtt{Zar}(K|A)$ denote the set of all the valuation domains of K that contain A as a subring. In the special case where $A := D$ is an integral domain with quotient field K, we simply set

$$\mathtt{Zar}(D) := \mathtt{Zar}(K|D) = \{V \mid V \text{ is a valuation domain overring of } D\}.$$

O. Zariski [52] introduced a topological structure on the set $Z := \mathtt{Zar}(K|A)$ by taking, as a basis for the open sets, the subsets $\mathtt{B}_F := \{V \in Z \mid V \supseteq A[F]\}$, for F varying in the family of all finite subsets of K (see also [53, Chap. VI, Sect. 17, p. 110]). This topology is called *the Zariski topology on* Z and the set Z, equipped with this topology (denoted also by $Z^{\mathtt{zar}}$), is usually called *the Riemann–Zariski space of* $K|A$ (sometimes also called abstract Riemann surface or generalized Riemann manifold of $K|A$).

In 1944, Zariski [52] proved a general result that implies the quasi-compactness of $Z^{\mathtt{zar}}$, and later it was proven that $Z^{\mathtt{zar}}$ is a spectral space, in the sense of M. Hochster [35] (for the case of the space $\mathtt{Zar}(D)$ see [12, Theorem 4.1]). More precisely, in [11, Theorem 2] (respectively, in [17, Corollary 3.4]) the authors provide explicitly a ring R_D (respectively, $R_{K|A}$) having the property that $\mathtt{Spec}(R_D)$ (respectively, $\mathtt{Spec}(R_{K|A})$) is canonically homeomorphic to $\mathtt{Zar}(D)$ (respectively, to $\mathtt{Zar}(K|A)$), both endowed with the Zariski topology (see also [37]).

Recently in [21] the Zariski topology on $\mathtt{Zar}(D)$ was explicitly extended on the larger space $\mathtt{Overr}(D)$ of all overrings of D, by taking, as a basis of open sets the collection of the sets of the type $\mathtt{Overr}(D[F])$, for F varying in the family of all finite subsets of K (see also [53, p. 115]). Clearly, in this way, $\mathtt{Zar}(D)$ becomes a subspace of $\mathtt{Overr}(D)$.

1.3 The Inverse Topology on a Spectral Space

Let X be a topological space and let Y be any subset of X. We denote by $\mathtt{Cl}(Y)$ the closure of Y in the topological space X. Recall that the topology on X induces a natural preorder $\leq_X$ on X (simply denoted by $\leq$, if no confusion can arise), defined by setting $x \leq_X y$ if $y \in \mathtt{Cl}(\{x\})$. It is straightforward that $\leq_X$ is a partial order if and only if X is a T_0 space (e.g., this holds when X is spectral). The set $Y^{\mathtt{gen}} := \{x \in X \mid y \in \mathtt{Cl}(\{x\}), \text{ for some } y \in Y\}$ is called *closure under generizations of* Y. Similarly, using the opposite order, the set $Y^{\mathtt{sp}} := \{x \in X \mid x \in \mathtt{Cl}(\{y\}), \text{ for some } y \in Y\}$ is called *closure under specializations of* Y. We say that Y is *closed under generizations* (respectively, *closed under specializations*) if $Y = Y^{\mathtt{gen}}$ (respectively, $Y = Y^{\mathtt{sp}}$). For two elements x, y in a spectral space X, we have:

$$x \leq y \quad \Leftrightarrow \quad \{x\}^{\mathtt{gen}} \subseteq \{y\}^{\mathtt{gen}} \quad \Leftrightarrow \quad \{x\}^{\mathtt{sp}} \supseteq \{y\}^{\mathtt{sp}}.$$

Suppose that X is a spectral space; then, X can be endowed with another topology, introduced by Hochster [35, Proposition 8], whose basis of closed sets is the collection of all the quasi-compact open subspaces of X. This topology is called *the inverse topology on* X. For a subset Y of X, let $\mathtt{Cl}^{\mathtt{inv}}(Y)$ be the closure of Y, in the inverse topology of X; we denote by $X^{\mathtt{inv}}$ the set X, equipped with the inverse topology. The name given to this new topology is due to the fact that, given $x, y \in X$, $x \in \mathtt{Cl}^{\mathtt{inv}}(\{y\})$ if and only if $y \in \mathtt{Cl}(\{x\})$, i.e., the partial order induced by the inverse topology is the opposite order of the partial order induced by the given spectral topology [35, Proposition 8].

By definition, for any subset Y of X, we have

$$\mathtt{Cl}^{\mathtt{inv}}(Y) = \bigcap\{U \mid U \text{ open and quasi-compact in } X,\ U \supseteq Y\}.$$

In particular, keeping in mind that the inverse topology reverses the order of the given spectral topology, it follows [35, Proposition 8] that the closure under generizations $\{x\}^{\mathtt{gen}}$ of a singleton is closed in the inverse topology of X, since

$$\{x\}^{\mathtt{gen}} = \mathtt{Cl}^{\mathtt{inv}}(\{x\}) = \bigcap\{U \mid U \subseteq X \text{ quasi-compact and open},\ x \in U\}.$$

On the other hand, it is trivial, by the definition, that the closure under specializations of a singleton $\{x\}^{\mathtt{sp}}$ is closed in the given topology of X, since $\{x\}^{\mathtt{sp}} = \mathtt{Cl}(\{x\})$.

2 Ultrafilter Topology and Spectral Spaces

The characterization of spectral spaces given in [35, Proposition 4] is often not easy to handle. In particular, it might be arduous to verify that a space is spectral using direct arguments involving irreducible closed subspaces.

The main result of the present section (Theorem 2.8) provides a criterion for deciding when a topological space is spectral, based on the use of ultrafilters. To introduce this statement, we need some basic and preliminary results on various topological structures that can be considered on the prime spectrum of a ring.

It is well known that the prime spectrum of a commutative ring endowed with the Zariski topology is always T_0, but almost never T_2 nor T_1 (it is T_2 or Hausdorff only in the zero-dimensional case, cf. for instance [45, Théorème 1.3]). Thus, in the general case, it is natural to look for a Hausdorff topology $\mathcal{T}$ on $\mathtt{Spec}(R)$ such that the following properties are satisfied at the same time:

- $\mathcal{T}$ is finer than the Zariski topology;
- $(\mathtt{Spec}(R), \mathcal{T})$ is compact (i.e., quasi-compact and T_2, using the terminology of [33]).

A classical answer to the previous question is given in [33, (7.2.11)], even in the more general setting of the underlying topological space of a scheme, by considering the *constructible topology* (see [10], [4, Chap. 3, Exercises 27, 28 and 30]) or the *patch topology* [35].

As in [49], we introduce the *constructible topology* by a Kuratowski closure operator: if X is a spectral space, we set, for each subset Y of X,

$$\mathtt{Cl}^{\mathtt{cons}}(Y) := \bigcap \{U \cup (X \setminus V) \mid U \text{ and } V \text{ open and quasi-compact in } X,\ U \cup (X \setminus V) \supseteq Y\}.$$

We denote by $X^{\mathtt{cons}}$ the set X, equipped with the constructible topology. For Noetherian spectral spaces, the clopen subsets of the constructible topology are precisely the constructible subsets after C. Chevalley [10], i.e., the finite unions of locally closed subspaces. It is straightforward that the constructible topology is a refinement of the given topology (it is the coarsest topology on X for which the quasi-compact open subspaces are clopen) and it is always Hausdorff. Finally, by [17, Remark 2.2], we have $\mathtt{Cl}^{\mathtt{inv}}(Y) = (\mathtt{Cl}^{\mathtt{cons}}(Y))^{\mathtt{gen}}$. It follows that each closed set in the inverse topology is closed under generizations and, from [17, Proposition 2.6], that a quasi-compact subspace Y of X closed for generizations is inverse closed. On the other hand, the closure of a subset Y in the given topology of X, $\mathtt{Cl}(Y)$, coincides with $(\mathtt{Cl}^{\mathtt{cons}}(Y))^{\mathtt{sp}}$ [17, Remark 2.2].

In the following result we collect some well-known classical properties of $\mathtt{Spec}(R)$, equipped with the constructible topology.

Theorem 2.1 (cf. [4, Chap. 3, Exercises 27, 28 and 30], [26, Proposition 5], [45, Théorème 2.2], [47, Proposition 5] and [48]) *Let R be a ring. We denote by $\mathtt{Spec}(R)^{\mathtt{zar}}$ (respectively, $\mathtt{Spec}(R)^{\mathtt{cons}}$) the set $\mathtt{Spec}(R)$, endowed with the Zariski topology (respectively, the constructible topology). The following properties hold.*

(1) $\mathtt{Spec}(R)^{\mathtt{cons}}$ is compact, Hausdorff and totally disconnected (and, by definition, the topology is finer than the Zariski topology).
(2) $\mathtt{Spec}(R)^{\mathtt{cons}} = \mathtt{Spec}(R)^{\mathtt{zar}}$ if and only if R is zero-dimensional.
(3) Assume that $\mathtt{Spec}(R)^{\mathtt{zar}}$ is a Noetherian space. Then, a subset of $\mathtt{Spec}(R)$ is clopen in $\mathtt{Spec}(R)^{\mathtt{cons}}$ if and only if it is constructible, *according to Chevalley (see [9, 10] and [33, (2.3.11) and (2.4.1)]) (i.e., it is a finite union of locally closed subsets of $\mathtt{Spec}(R)^{\mathtt{zar}}$).*
(4) Let $\{\mathbb{X}_f \mid f \in R\}$ be a collection of algebraically independent indeterminates over R, let I be the ideal of the polynomial ring $R[\{\mathbb{X}_f \mid f \in R\}]$ generated by the set $\{f^2\mathbb{X}_f - f;\ f\mathbb{X}_f^2 - \mathbb{X}_f \mid f \in R\}$, and consider the ring $\mathrm{T}(R) := R[\{\mathbb{X}_f \mid f \in R\}]/I$. Then, the following statements hold.

  (4.a) $\mathrm{T}(R)$ is absolutely flat (or, von Neumann regular, *i.e., for each $a \in \mathrm{T}(R)$ there exists $x \in \mathrm{T}(R)$ such that $ax^2 = a$), called the* absolutely flat cover *of R.*
  (4.b) The canonical embedding $\iota : R \to \mathrm{T}(R)$ is an epimorphism in the category of rings. Furthermore, ι is an isomorphism if and only if R is absolutely flat.

(4.c) The canonical continuous map $\iota^a : \mathtt{Spec}(\mathrm{T}(R))^{\mathtt{zar}} \to \mathtt{Spec}(R)^{\mathtt{cons}}$, *induced by* ι, *is an homeomorphism. In particular, the topological space* $\mathtt{Spec}(R)^{\mathtt{cons}}$ *is spectral.*

In [26] a new description of $\mathtt{Spec}(R)^{\mathtt{cons}}$ is presented, by using a new tool: *convergence by ultrafilters.*

For the reader's convenience, we recall now some basic facts about ultrafilters (for further properties see, for example, [43]). Let $\mathbf{X}$ be a nonempty set. A nonempty collection $\mathscr{U}$ of nonempty subsets of $\mathbf{X}$ is called *an ultrafilter on* $\mathbf{X}$ if the following axioms hold:

- If $Y, Z \in \mathscr{U}$, then $Y \cap Z \in \mathscr{U}$.
- If $Y \in \mathscr{U}$ and $Y \subseteq Z \subseteq \mathbf{X}$, then $Z \in \mathscr{U}$.
- If $Y \subseteq \mathbf{X}$ then either $Y \in \mathscr{U}$ or $\mathbf{X} \setminus Y \in \mathscr{U}$.

It is easy to see that, for each $x \in \mathbf{X}$, the collection $\mathscr{U}_x := \{Y \subseteq \mathbf{X} \mid x \in Y\}$ is an ultrafilter on $\mathbf{X}$, called *the trivial (or principal) ultrafilter generated by* x. Every finite set admits only trivial ultrafilters. The existence of nontrivial ultrafilters on infinite sets is guaranteed by the Axiom of Choice. Precisely, it is proved under ZFC that, if $\mathcal{F}$ is a nonempty collection of subsets of $\mathbf{X}$ with the finite intersection property, then there exists an ultrafilter $\mathscr{U}$ on $\mathbf{X}$ such that $\mathcal{F} \subseteq \mathscr{U}$.

Now, let R be a ring, let Y be a nonempty subset of $\mathtt{Spec}(R)$ and let $\mathscr{U}$ be an ultrafilter on Y. For each $f \in R$ we set $\mathtt{V}(f) := \{P \in \mathtt{Spec}(R) \mid f \in P\}$. It is easy to show that the set $P_{Y,\mathscr{U}} := P_{\mathscr{U}} := \{f \in R \mid \mathtt{V}(f) \cap Y \in \mathscr{U}\}$ is a prime ideal of R [8, Lemma 2.4], called *the ultrafilter limit point of* Y*, with respect to* $\mathscr{U}$. According to [26, Definition 1], a nonempty subset Y of $\mathtt{Spec}(R)$ is *ultrafilter closed* if, for any ultrafilter $\mathscr{U}$ on Y, we have $P_{\mathscr{U}} \in Y$. We assume that the empty set is ultrafilter closed. The following result relates the constructible topology and the convergence by ultrafilters.

Theorem 2.2 (cf. [26, Theorem 8]) *Let* R *be a ring and let* $Y \subseteq \mathtt{Spec}(R)$. *Then, the following conditions are equivalent.*

(i) Y *is closed, with respect to the constructible topology.*
(ii) Y *is ultrafilter closed.*

In [15, Sect. 2], the convergence by ultrafilters, presented in [26], is extended in a more general setting. Precisely, let $\mathbf{X}$ be a nonempty set and $\mathcal{F}$ be a nonempty collection of subsets of $\mathbf{X}$. If Y is a nonempty subset of $\mathbf{X}$ and $\mathscr{U}$ is an ultrafilter on Y, we define

$$Y_{\mathcal{F}}(\mathscr{U}) := \{x \in \mathbf{X} \mid [\ \forall F \in \mathcal{F}, x \in F \iff F \cap Y \in \mathscr{U}\ ]\}$$

and call it *the* $\mathcal{F}$*-ultrafilter limit set of* Y*, with respect to* $\mathscr{U}$.

Example 2.3 (cf. [15, Example 2.1(2)]) Let R be a ring, let $\boldsymbol{\mathcal{P}}$ denote the collection of the principal open subset of $\mathtt{Spec}(R)$, i.e.,

$$\boldsymbol{\mathcal{P}} := \{\mathrm{D}(f) := \{P \in \mathtt{Spec}(R) \mid f \notin P\} \mid f \in R\}.$$

If $\mathscr{U}$ is an ultrafilter on a subset Y of $\mathtt{Spec}(R)$, then $Y_{\boldsymbol{\mathcal{P}}}(\mathscr{U}) = \{P_{\mathscr{U}}\}$, where $P_{\mathscr{U}}$ denotes, as before, the ultrafilter limit point of Y, with respect to $\mathscr{U}$.

Example 2.4 Let K be a field and let A be any subring of K. In the space $\mathtt{Zar}(K|A)$, let

$$\boldsymbol{\mathcal{B}} := \{\mathrm{B}_F := \mathtt{Zar}(K|A[F]) \mid F \subseteq K,\ F \text{ finite}\},$$

denote the standard basis for the open sets for the Zariski topology on $\mathtt{Zar}(K|A)$. If Z is a nonempty subset of $\mathtt{Zar}(K|D)$ and $\mathscr{U}$ is an ultrafilter on Z, it is easy to show that the subset

$$Z_{\mathscr{U}} := \{x \in K \mid \mathtt{Zar}(K|A[x]) \cap Z \in \mathscr{U}\}$$

is still a valuation domain of K (cf. [8, Lemma 2.9] and [16, Proposition 3.1]), called *the ultrafilter limit point of* Z, *with respect to* $\mathscr{U}$. Then we have $Z_{\boldsymbol{\mathcal{B}}}(\mathscr{U}) = \{Z_{\mathscr{U}}\}$.

The next goal is to extend the notion of ultrafilter closure given for the prime spectrum of a ring in a general setting.

Let $\mathbf{X}$ be a nonempty set, $\mathcal{F}$ a nonempty collection of subsets of $\mathbf{X}$, and fix a nonempty subset Y of $\mathbf{X}$. We say that Y is *$\mathcal{F}$-stable under ultrafilters* if, for any ultrafilter $\mathscr{U}$ on Y, we have $Y_{\mathcal{F}}(\mathscr{U}) \subseteq Y$.

Let $\boldsymbol{\mathcal{P}}$ be as in Example 2.3. It is easily seen that a subset of the prime spectrum of a ring is $\boldsymbol{\mathcal{P}}$-stable under ultrafilters if and only if it is ultrafilter closed, that is, it is closed in the constructible topology (by Theorem 2.2).

Proposition 2.5 (cf. [15, Propositions 2.6, 2.11, 2.13 and Theorem 2.14]) *Let* $\mathbf{X}$ *be a nonempty set,* $\mathcal{F}$ *be a nonempty collection of subsets of* $\mathbf{X}$*. Then, the following properties hold.*

(1) The collection of all the subsets of $\mathbf{X}$ *that are stable under ultrafilters is the family of the closed sets for a topology on* $\mathbf{X}$*, called the* $\mathcal{F}$-ultrafilter topology. *We will denote by* $\mathbf{X}^{\mathcal{F}\text{-}\mathtt{ultra}}$ *the set* $\mathbf{X}$*, equipped with the* $\mathcal{F}$*-ultrafilter topology.*
(2) If $\mathcal{B}$ *is the Boolean subalgebra of the power set of* $\mathbf{X}$ *generated by* $\mathcal{F}$*, then* $\mathcal{B}$ *is a collection of clopen subsets of* $\mathbf{X}^{\mathcal{F}\text{-}\mathtt{ultra}}$.
(3) For each subset Y *of* $\mathbf{X}$*, the closure of* Y *in* $\mathbf{X}^{\mathcal{F}\text{-}\mathtt{ultra}}$ *is the set*

$$\bigcup\{Y_{\mathcal{F}}(\mathscr{U}) \mid \mathscr{U} \text{ ultrafilter on } Y\}.$$

(4) The following conditions are equivalent.

(i) $\mathbf{X}^{\mathcal{F}\text{-}\mathtt{ultra}}$ *is quasi-compact.*
(ii) For any ultrafilter $\mathscr{U}$ *on* $\mathbf{X}$*, the ultrafilter limit set* $\mathbf{X}_{\mathcal{F}}(\mathscr{U})$ *is nonempty.*

Example 2.6 (cf. [15, Remark 2.7]) Let $\mathbf{X}$ be a nonempty set.

(1) If $\mathcal{B}(\mathbf{X})$ denotes the power set of $\mathbf{X}$, the $\mathcal{B}(\mathbf{X})$-ultrafilter topology is the discrete topology.
(2) The $\{\mathbf{X}\}$-ultrafilter topology is the chaotic topology (i.e., the open sets are just X and $\emptyset$).
(3) Let R be a ring, $\mathbf{X} := \mathtt{Spec}(R)$ and $\boldsymbol{\mathcal{P}}$ be as in Example 2.3. Then, the $\boldsymbol{\mathcal{P}}$-ultrafilter topology is the constructible topology on $\mathbf{X}$ by [15, Corollary 2.17].

We apply the previous construction when the given set is a topological space and the collection of subsets $\mathcal{F}$ is a basis for the topology.

Proposition 2.7 (cf. [15, Proposition 3.1]) *Let $(X, \mathcal{T})$ be a nonempty topological space and $\boldsymbol{\mathcal{B}}$ be a basis of open sets of X. Then, the following statements hold.*

(1) The $\boldsymbol{\mathcal{B}}$-ultrafilter topology is finer than or equal to the topology $\mathcal{T}$.
(2) If $(X, \mathcal{T})$ is a T_0 space, then $X^{\boldsymbol{\mathcal{B}}-\mathrm{ultra}}$ is a Hausdorff and totally disconnected space.
(3) Assume now that $(X, \mathcal{T})$ is T_0 and that $X^{\boldsymbol{\mathcal{B}}-\mathrm{ultra}}$ is compact. Then, the $\boldsymbol{\mathcal{B}}$-ultrafilter topology is the coarsest topology for which $\boldsymbol{\mathcal{B}}$ is a family of clopen sets. Moreover, $(X, \mathcal{T})$ is a spectral space and the constructible topology on $(X, \mathcal{T})$ is precisely the $\boldsymbol{\mathcal{B}}$-ultrafilter topology.

Note that part (3) of the previous proposition generalizes [26, Theorem 8] and [16, Theorem 3.4].

By using Propositions 2.5(4), 2.7(3) and keeping in mind [35, Corollary to Proposition 7], we can deduce new characterizations of spectral spaces and hence new criteria, based on ultrafilters, to decide if a given topological space is spectral.

Theorem 2.8 (cf. [15, Corollary 3.3]) *For a nonempty topological space X, the following conditions are equivalent:*

(i) X is a spectral space.
(ii) There exists a basis $\boldsymbol{\mathcal{B}}$ for the open sets of X such that $X^{\boldsymbol{\mathcal{B}}-\mathrm{ultra}}$ is a compact and Hausdorff space.
(iii) X is a T_0 space and there is a basis $\boldsymbol{\mathcal{B}}$ for the open sets of X such that, for any ultrafilter $\mathscr{U}$ on X, the ultrafilter limit set $X_{\boldsymbol{\mathcal{B}}}(\mathscr{U})$ is nonempty.
(iv) X is a T_0 space and there is a subbasis $\boldsymbol{\mathcal{S}}$ for the open sets of X such that, for any ultrafilter $\mathscr{U}$ on X, the ultrafilter limit set $X_{\boldsymbol{\mathcal{S}}}(\mathscr{U})$ is nonempty.

The proof of Theorem 2.8 is not constructive, since it is based on the Axiom of Choice and some of its consequences.

As an application of Theorem 2.8, we now determine some new classes of spectral spaces. The key point of the proofs resides on the existence of ultrafilter limit points.

Example 2.9 (cf. [15, Proposition 3.5]) Let $A \subseteq B$ be a ring extension, and let $X := \mathbf{R}(B|A)$ denote the collection of all the intermediate rings between A and B. We can make X a topological space, by generalizing the Zariski topology introduced on the space of the overrings on an integral domain (see Sect. 1.2) and taking as a subbasis of open sets the collection

$$\mathcal{S} := \{\mathbf{R}(B|A[x]) \mid x \in B\}.$$

We claim that X is a spectral space. It is easily seen that X is T_0 because, if $C \neq D \in X$, we can assume, without loss of generality, that there is an element $c \in C \setminus D$, and then the open set $\mathbf{R}(B|A[c])$ contains C and does not contain D. By Theorem 2.8, we have to show that, if $\mathscr{U}$ is an ultrafilter on X, then the ultrafilter limit set $X_{\mathcal{S}}(\mathscr{U})$ is nonempty. Consider the subset

$$A_{\mathscr{U}} := \{x \in B \mid \mathbf{R}(B|A[x]) \in \mathscr{U}\}$$

of B. We claim that $A_{\mathscr{U}}$ is a subring of B.

This follows immediately from the definition of an ultrafilter, since, if $x, y \in A_{\mathscr{U}}$ then each of the sets $\mathbf{R}(B|A[x-y])$, $\mathbf{R}(B|A[xy])$ contain $\mathbf{R}(B|A[x]) \cap \mathbf{R}(B|A[y]) \in \mathscr{U}$, and thus $\mathbf{R}(B|A[x-y])$, $\mathbf{R}(B|A[xy]) \in \mathscr{U}$, that is, $x-y,\ xy \in A_{\mathscr{U}}$. Furthermore, $A_{\mathscr{U}}$ contains A because, for each $a \in A$, $\mathbf{R}(B|A[a]) = X \in \mathscr{U}$. Therefore, $A_{\mathscr{U}}$ is an element of X. The fact that $A_{\mathscr{U}} \in X_{\mathcal{S}}(\mathscr{U})$ follows immediately from the definition of $A_{\mathscr{U}}$ and thus, by Theorem 2.8, X is a spectral space.

In particular, if $A := D$ is an integral domain and $B := K$ is the quotient field of D, we deduce from the previous example that:

Corollary 2.10 *The space* $\texttt{Overr}(D)$ *of the overrings of an integral domain D, endowed with the Zariski topology, is a spectral space.*

Example 2.11 (cf. [15, Proposition 3.6]) Let A, B and X be as in the previous example, and let $X' := \mathbf{R}'(B|A)$ be the subset of X consisting of all the subrings of B that are integrally closed in B. We claim that, with the subspace topology induced by that of X, the topological space X' is spectral.

It is obvious that a subbasis of open sets for the topology of X' is given by the family $\mathcal{S}' := \{\mathbf{R}'(B|A[x]) \mid x \in B\}$. As in the previous example, the key fact is the existence in X' of ultrafilter limit points, with respect to every ultrafilter $\mathscr{U}$ on X'. Indeed, it is not difficult to show that

$$A'_{\mathscr{U}} := \{x \in B \mid \mathbf{R}'(B|A[x]) \in \mathscr{U}\}$$

is a subring of B containing A that is integrally closed in B. Thus, again by definition, the ultrafilter limit set $X'_{\mathcal{S}'}(\mathscr{U})$ is nonempty, containing $A'_{\mathscr{U}}$. Again, by Theorem 2.8, we conclude that X' is a spectral space.

In particular, if $A := D$ is an integral domain and $B := K$ is the quotient field of D, we deduce from the previous example that:

Corollary 2.12 *The subspace* $\texttt{Overr}_{ic}(D)$ *of* $\texttt{Overr}(D)$, *consisting of the integrally closed overrings of an integral domain D, endowed with the Zariski topology, is a spectral space.*

Example 2.13 We preserve the notation of Example 2.9, and let $X'' := \mathbf{L}(B|A)$ be the (possibly empty) subspace of $\mathbf{R}(B|A)$ consisting of all the local rings T such that $A \subseteq T \subseteq B$. A subbasis for the open sets of X'' is clearly the family

$$\mathcal{S}'' := \{\mathbf{L}(B|A[x]) \mid x \in B\}$$

We claim that, if X'' is nonempty, then it is spectral. Again, we need to prove that, for any ultrafilter $\mathscr{U}$ on X'' the ultrafilter limit set $X''_{\mathcal{S}''}(\mathscr{U})$ is nonempty. As before, it is easy to infer that $A''_{\mathscr{U}} := \{x \in B \mid \mathbf{L}(B|A[x]) \in \mathscr{U}\} \in \mathbf{R}(B|A)$. It will be immediate to conclude that $A''_{\mathscr{U}} \in X''_{\mathcal{S}''}(\mathscr{U})$ if we show that $A''_{\mathscr{U}}$ is a local ring. We claim that the unique maximal ideal of $A''_{\mathscr{U}}$ is

$$M := \{x \in B \mid \{T \in X'' \mid x \in T \backslash U(T)\} \in \mathscr{U}\}$$

where, as usual, $U(T)$ denotes the set of units of a ring T. Thus it suffices to note that $U(A''_{\mathscr{U}}) = A''(\mathscr{U})\backslash M$ (this follows easily from definitions).

In particular, if $A := D$ is an integral domain and $B := K$ is the quotient field of D, we deduce from the previous example that:

Corollary 2.14 *The subspace* $\mathtt{Overr}_{\mathtt{loc}}(D)$ *of* $\mathtt{Overr}(D)$*, consisting of the local overrings of an integral domain D, endowed with the Zariski topology, is a spectral space.*

3 Spaces of Semistar Operations

Let D be an integral domain with quotient field K. As in the star operation setting, to each semistar operation $\star$ can be associated a map $\star_f : \overline{\boldsymbol{F}}(D) \to \overline{\boldsymbol{F}}(D)$ defined by

$$E^{\star_f} := \bigcup \{F^\star \mid F \subseteq E,\ F \in \boldsymbol{f}(D)\},$$

for every $E \in \overline{\boldsymbol{F}}(D)$. The map $\star_f$ is again a semistar operation, which coincides with $\star$ on finitely generated modules; moreover, $(\star_f)_f = \star_f$. If $\star = \star_f$, we say that $\star$ is a *semistar operation of finite type*. We call $\star_f$ the *finite-type semistar operation associated to* $\star$.

For each $T \in \mathtt{Overr}(D)$, the map $\wedge_{\{T\}} : \overline{\boldsymbol{F}}(D) \to \overline{\boldsymbol{F}}(D)$, defined by $E^{\wedge_{\{T\}}} := ET$, for each $E \in \overline{\boldsymbol{F}}(D)$, is an example of semistar operation of finite type on D, called the *semistar extension to* T.

We denote by $\mathtt{SStar}(D)$ (respectively, $\mathtt{SStar}_f(D)$) the set of all semistar operations (respectively, semistar operations of finite type) on D. The set $\mathtt{SStar}(D)$ can be endowed with a natural partial order $\preceq$ which turns it into a complete lattice: if $\star_1, \star_2$ are two semistar operations, say that $\star_1 \preceq \star_2$ if $E^{\star_1} \subseteq E^{\star_2}$ for every

$E \in \overline{\boldsymbol{F}}(D)$. In particular, $\star_f \preceq \star$, and $\star_f$ is the biggest semistar operation of finite type smaller than $\star$.

The infimum $\wedge_{\mathscr{S}}$ of a nonempty family $\mathscr{S}$ of semistar operations can be written explicitly as follows:

$$E^{\wedge_{\mathscr{S}}} = \bigcap \{E^{\star} \mid \star \in \mathscr{S}\}, \qquad \text{for each } E \in \overline{\boldsymbol{F}}(D).$$

In particular, if $\mathcal{T}$ is a nonempty family of overrings of D, then the infimum of the family of semistar operations $\{\wedge_{\{T\}} \mid T \in \mathcal{T}\}$ is denoted by $\wedge_{\mathcal{T}}$.

On the other hand, there is not a general explicit formula for the supremum $\vee_{\mathscr{S}} := \bigwedge\{\sigma \in \mathtt{SStar}(D) \mid \star \preceq \sigma \text{ for all } \star \in \mathscr{S}\}$, although, if $\mathscr{S} \subseteq \mathtt{SStar}_f(D)$, then

$$E^{\vee_{\mathscr{S}}} = \bigcup \{E^{\star_1 \circ \star_2 \circ \cdots \circ \star_n} \mid \star_1, \ldots, \star_n \in \mathscr{S}\} \tag{1}$$

where $\star_1 \circ \star_2 \circ \cdots \circ \star_n$ denotes the usual composition of functions (see [3, p.1628] and [21, Lemma 2.12]).

A nonzero ideal I of D is called a *quasi-$\star$-ideal* if $I = I^{\star} \cap D$. A *quasi-$\star$-prime* is a quasi-$\star$-ideal which is also a prime ideal; the set of all quasi-$\star$-prime ideals of D is denoted by $\mathtt{QSpec}^{\star}(D)$. The set of maximal elements in the set of proper quasi-$\star$-ideals of D (ordered by set-theoretic inclusion) is denoted by $\mathtt{QMax}^{\star}(D)$, and it is a subset of $\mathtt{QSpec}^{\star}(D)$. By Zorn's Lemma, it is easy to show that if $\star$ is a semistar operation of finite type then $\mathtt{QMax}^{\star}(D) \neq \emptyset$. If every quasi-$\star$-ideal is contained in a quasi-$\star$-prime, then $\star$ is said to be *quasi-spectral* or *semifinite*. Every operation of finite type is not only quasi-spectral, but it has the stronger property that every quasi-$\star$-ideal is contained in a maximal quasi-$\star$-ideal. Note that a semistar operation $\star$ may be quasi-spectral even if $\mathtt{QMax}^{\star}(D)$ is empty (see [21, Remark 5.6] for an example).

A semistar operation $\star$ is called *spectral* if there is a nonempty subset $Y \subseteq \mathtt{Spec}(D)$ such that $\star = \wedge_{\mathcal{L}(Y)}$, where $\mathcal{L}(Y) := \{D_P \mid P \in Y\}$. We set $s_Y := \wedge_{\mathcal{L}(Y)}$ and we call s_Y *the spectral semistar operation associated to* $Y \subseteq \mathtt{Spec}(D)$.

A *semistar operation* $\star$ is called *stable* if $(E \cap F)^{\star} = E^{\star} \cap F^{\star}$ for every pair $E, F \in \overline{\boldsymbol{F}}(D)$.

Remark 3.1 Every spectral semistar operation is quasi-spectral (or semifinite) by [22, Lemma 1.4(5)] and every spectral semistar operation, or more generally every operation induced by a family of D-flat overrings of D, is stable. However, the converse does not hold in general [34, Sect. 3, p. 441], but if $\star$ is a stable semistar operation then $\star$ is spectral if and only if it is quasi-spectral (see [1, Theorem 4] and [22, Theorem 4.12(3)]). In particular, a stable semistar operation of finite type is spectral.

In [21], the set $\mathtt{SStar}(D)$ was endowed with a topology (called the *Zariski topology*) by declaring open the sets of the form

$$\mathtt{V}_E := \{\star \in \mathtt{SStar}(D) \mid 1 \in E^{\star}\},$$

for $E \in \overline{\boldsymbol{F}}(D)$. This topology makes $\mathtt{SStar}(D)$ into a quasi-compact, T_0 space with a unique closed point (the identity semistar operation d_D) and a generic point (the trivial semistar extension $\wedge_{\{K\}}$). In particular, $\mathtt{SStar}(D)$ is never T_1 (nor T_2) unless $D = K$.

Proposition 3.2 *Let D be an integral domain, let $\mathtt{Overr}(D)$ and $\mathtt{SStar}_{f}(D)$ be endowed with their Zariski topologies, and let $\iota : \mathtt{Overr}(D) \to \mathtt{SStar}_{f}(D)$ be the injective map defined by $\iota(T) := \wedge_{\{T\}}$, for each $T \in \mathtt{Overr}(D)$. Then, the following statements hold.*

(1) The map ι is a topological embedding [21, Proposition 2.5].
(2) The mapping $\pi : \mathtt{SStar}_{f}(D) \to \mathtt{Overr}(D)$, defined by $\pi(\star) := D^{\star}$, for each $\star \in \mathtt{SStar}_{f}(D)$, is a continuous surjection such that $\pi \circ \iota$ is the identity of $\mathtt{Overr}(D)$. In other words, π is a topological retraction.

Note that part (2) of the previous proposition follows from the fact that, for each subbasic open set $\mathtt{B}_x := \mathtt{Overr}(D[x])$ of $\mathtt{Overr}(D)$, we have $\pi^{-1}(\mathtt{B}_x) = \{\star \in \mathtt{SStar}_{f}(D) \mid D[x] \subseteq D^{\star}\} = \{\star \in \mathtt{SStar}_{f}(D) \mid 1 \subseteq (x^{-1}D)^{\star}\} = \mathtt{V}_{x^{-1}D}$.

The following result relates the quasi-compactness of a collection of semistar operations on the same integral domain with the finite type property of their infimum.

Proposition 3.3 (cf. [21, Proposition 2.7]) Let D be an integral domain and let $\mathscr{S}$ be a quasi-compact subspace of $\mathtt{SStar}_{f}(D)$. Then, $\wedge_{\mathscr{S}}$ is of finite type.

Remark 3.4 Let $\mathscr{S}$ be a subset of $\mathtt{SStar}(D)$ and set $\mathscr{S}_{f} := \{\star_{f} \mid \star \in \mathscr{S}\}$. Consider the following properties:

(a) $\mathscr{S}$ is quasi-compact in $\mathtt{SStar}(D)$;
(b) $\mathscr{S}_{f}$ is quasi-compact in $\mathtt{SStar}_{f}(D)$;
(c) $\wedge_{\mathscr{S}_{f}}$ is a semistar operation of finite type;
(d) $\wedge_{\mathscr{S}_{f}} = (\wedge_{\mathscr{S}})_{f}$.

Then (a) $\Rightarrow$ (b) $\Rightarrow$ (c) $\Leftrightarrow$ (d).

In fact, it is straightforward that (a) $\Rightarrow$ (b) (see also Proposition 3.10). By Proposition 3.3, (b) $\Rightarrow$ (c). For (c) $\Rightarrow$ (d), note that in general $\wedge_{\mathscr{S}_{f}} \leq \wedge_{\mathscr{S}}$ and $(\wedge_{\mathscr{S}})_{f} \leq \wedge_{\mathscr{S}_{f}}$. The conclusion follows from the fact that, under (c), $(\wedge_{\mathscr{S}_{f}})_{f} = \wedge_{\mathscr{S}_{f}}$. Finally, (d) $\Rightarrow$ (c) is trivial.

Since, for each overring T of an integral domain D, the semistar operation $\wedge_{\{T\}}$ is of finite type, we get the following result, just by applying Propositions 3.2 and 3.3.

Corollary 3.5 (cf. [21, Corollary 2.8]) *Let D be an integral domain and let $\boldsymbol{\mathcal{T}}$ be a quasi-compact subspace of $\mathtt{Overr}(D)$. Then $\wedge_{\boldsymbol{\mathcal{T}}}$ is of finite type.*

In particular, the previous corollary applies when $\boldsymbol{\mathcal{T}}$ is *locally finite*, i.e., if every nonzero element of D is nonunit in finitely many overrings of the family $\boldsymbol{\mathcal{T}}$ [21, Corollary 2.10]. However, the finite type property of a semistar operation $\wedge_{\boldsymbol{\mathcal{T}}}$, induced by a collection $\mathcal{T}$ of overrings, does not imply the quasi-compactness of $\mathcal{T}$, as the following example shows. This example provides a negative answer to the Conjecture in [21, p. 214].

Example 3.6 Let k be a field, let $\mathbb{X}$ be an indeterminate over k, let $D := k[\![\mathbb{X}^4, \mathbb{X}^5, \mathbb{X}^6, \mathbb{X}^7]\!] = k + \mathbb{X}^4 k[\![\mathbb{X}]\!]$ and let $K := k(\!(\mathbb{X})\!)$. Since D is Noetherian and a conductive domain (i.e., $(D : T) \neq (0)$ for each $T \in \texttt{Overr}(D)$ with $T \neq K$, see [5, Theorem 1]), $\overline{\boldsymbol{F}}(D) = \boldsymbol{F}(D) \cup \{K\} = \boldsymbol{f}(D) \cup \{K\}$, and thus every semistar operation on D is of finite type. For every $\alpha \in K$, consider the ring $T_\alpha := D[\mathbb{X}^2 + \alpha\mathbb{X}^3] = k + (\mathbb{X}^2 + \alpha\mathbb{X}^3)k + \mathbb{X}^4 k[[\mathbb{X}]]$, and, for every $A \subseteq k$, let $\mathcal{T}_A := \{T_\alpha \mid \alpha \in A\}$. Then, as observed above, the semistar operation $\wedge_{\mathcal{T}_A}$ is of finite type. However, if A is infinite (so, for example, if k is infinite and $A = k$), then $\mathcal{T}_A$ is not quasi-compact. Indeed, the open cover $\{\texttt{Overr}(T_\alpha) \mid \alpha \in A\}$ of $\mathcal{T}_A$ in $\texttt{Overr}(D)$ has no finite subcovers, since $\texttt{Overr}(T_\alpha) \cap \mathcal{T}_A = \{T_\alpha\}$.

The following example shows how to use Corollary 3.5 for establishing the failure of quasi-compactness for some distinguished subspaces of $\texttt{Overr}(D)$.

Example 3.7 Let D be a Noetherian domain of dimension ≥ 2, and let $\mathcal{D}$ be the set of Noetherian valuation overrings of D, i.e., the union of $\{K\}$ with the set of discrete valuation overrings of D. If I is a proper ideal of D, then $I^{\wedge_{\mathcal{D}}} = I^b$, where $b := \wedge_{\texttt{Zar}(D)}$ (see, for example, [39, Proposition 6.8.4], after noting that the terminology used therein is slightly different). In particular, the same holds for every $F \in \boldsymbol{f}(D)$, so that $(\wedge_{\mathcal{D}})_f = b$. However, if $W \in \texttt{Zar}(D) \setminus \mathcal{D}$ (for example, if $\dim(W) \geq 2$, where the existence of such a W is guaranteed by [32, Corollary 19.7]), then W is contained in (at most) one element V of $\mathcal{D}$, so that $WV = V$, while $WV' = K$ for each $V' \in \mathcal{D}$, $V' \neq V$. Hence, $W^{\wedge_{\mathcal{D}}} \neq W$, while $W^b = W$ and thus, $\wedge_{\mathcal{D}} \neq b$. Therefore, $\wedge_{\mathcal{D}}$ is not of finite type, and so $\mathcal{D}$ is not a quasi-compact subset of $\texttt{Overr}(D)$ (or of $\texttt{Zar}(D)$).

Theorem 3.8 (cf. [21, Theorem 2.13]) *Let D be an integral domain. Then, $\texttt{SStar}_f(D)$ is a spectral space.*

The proof uses Theorem 2.8, so it is not constructive. However, if A is a ring such that $\texttt{Spec}(A) \simeq \texttt{SStar}_f(D)$, we can assume that:

(a) $A_{\texttt{red}}$ (the reduced ring associated to A) is an integral domain (since $\texttt{SStar}_f(D)$ has a unique generic point),
(b) $A_{\texttt{red}}$ (and A) is local (since $\texttt{SStar}_f(D)$ has a unique closed point), and
(c) $\dim(A) = \dim(A_{\texttt{red}}) \geq |\texttt{Spec}(D)| - 1$ (see the following Propositions 4.3 and 4.6).

On the other hand, since the proof of Theorem 2.8 uses in a crucial way the characterization (1) of the supremum of a family of finite-type semistar operations, it cannot readily be adapted to $\texttt{SStar}(D)$ and so, up to now, we do not know whether $\texttt{SStar}(D)$ is a spectral space.

We denote by $\overline{\texttt{SStar}}(D)$ (respectively, $\widetilde{\texttt{SStar}}(D)$) the subset of $\texttt{SStar}(D)$ consisting of all stable semistar operations (respectively, all stable semistar operations of finite type).

Remark 3.9 (a) If we set $\texttt{SStar}_{sp}(D) := \{\star \in \texttt{SStar}(D) \mid \star \text{ is spectral}\}$ (respectively, $\texttt{SStar}_{f,sp}(D) := \{\star \in \texttt{SStar}_f(D) \mid \star \text{ is spectral}\}$), then by Remark 3.1 $\texttt{SStar}_{sp}(D) \subseteq \overline{\texttt{SStar}}(D)$, and the inclusion might be proper. However, in the finite type case, we have equality [22, Proposition 4.23(2)], i.e.,

$$\texttt{SStar}_{f,sp}(D) = \overline{\texttt{SStar}}(D) \cap \texttt{SStar}_f(D) = \widetilde{\texttt{SStar}}(D).$$

(b) Let $\texttt{Loc}(D)$ and $\texttt{Overr}_{flat}(D)$ be, respectively, the set of localizations of D and the set of D-flat overrings of D (and so $\texttt{Loc}(D) \subseteq \texttt{Overr}_{flat}(D)$). We observe that the topological embedding $\iota : \texttt{Overr}(D) \hookrightarrow \texttt{SStar}_f(D)$, considered in Proposition 3.2(1), restricts to a topological embedding $\iota_{Loc} : \texttt{Loc}(D) \hookrightarrow \widetilde{\texttt{SStar}}(D)$ (or to a topological embedding $\iota_{flat} : \texttt{Overr}_{flat}(D) \hookrightarrow \widetilde{\texttt{SStar}}(D)$).

On the opposite side, the map $\pi : \texttt{SStar}_f(D) \to \texttt{Overr}(D)$ (Proposition 3.2(2)) does not always restrict to a map $\widetilde{\texttt{SStar}}(D) \to \texttt{Overr}_{flat}(D)$, since not all intersection of localizations of D are D-flat (see for instance [34, Sect. 3, p. 441]).

Given a semistar operation $\star$ on D, we can always associate to $\star$ two semistar operations $\overline{\star}$ and $\widetilde{\star}$ on D defined as follows: for each $E \in \overline{\boldsymbol{F}}(D)$,

$$\begin{aligned} E^{\overline{\star}} &:= \bigcup\{(E : I) \mid I \text{ nonzero ideal of } D \text{ such that } I^\star = D^\star\}, \\ E^{\widetilde{\star}} &:= \bigcup\{(E : J) \mid J \text{ nonzero finitely generated ideal of } D \\ &\qquad \text{such that } J^\star = D^\star\}. \end{aligned}$$

It is easy to see that $\widetilde{\star} \preceq \overline{\star} \preceq \star$ and, moreover, that $\overline{\star}$ (respectively, $\widetilde{\star}$) is the largest stable (respectively, stable of finite type) semistar operation that precedes $\star$, called *the stable* (respectively, the *the finite type stable*) *semistar operation associated to* $\star$. Therefore, $\star$ is stable (respectively, stable of finite type) if and only if $\star = \overline{\star}$ (respectively, $\star = \widetilde{\star}$) [22, Proposition 3.7, Corollary 3.9]. Note that, for each semistar operation $\star$, we always have $\widetilde{\star} = s_Y$, where $Y = \texttt{QMax}^{\star_f}(D)$ (cf. [22, p. 182,Proposition 4.3], [24, Proposition 3.4(4)], [25, Remark 10] and, for the star operation case, [2, Corollary 2.10]).

Proposition 3.10 (cf. [18, Proposition 4.1] and [21, Proposition 2.4]) *Let* $\Phi_f : \texttt{SStar}(D) \to \texttt{SStar}(D)$ *(respectively,* $\overline{\Phi} : \texttt{SStar}(D) \to \texttt{SStar}(D)$*;* $\widetilde{\Phi} : \texttt{SStar}(D) \to \texttt{SStar}(D)$*) be the map defined by* $\star \mapsto \star_f$ *(respectively,* $\star \mapsto \overline{\star}$*;* $\star \mapsto \widetilde{\star}$*). Then:*

(1) The images of Φ_f*,* $\overline{\Phi}$ *and* $\widetilde{\Phi}$ *are, respectively,* $\texttt{SStar}_f(D)$*,* $\overline{\texttt{SStar}}(D)$ *and* $\widetilde{\texttt{SStar}}(D)$*.*
(2) The maps Φ_f*,* $\overline{\Phi}$ *and* $\widetilde{\Phi}$ *are continuous in the Zariski topology.*
(3) The maps Φ_f*,* $\overline{\Phi}$ *and* $\widetilde{\Phi}$ *are topological retractions of* $\texttt{SStar}(D)$ *onto their respective images.*

Another point of similarity between finite type, stable, and spectral operations is given by the open sets needed to generate the Zariski topology, induced by the Zariski topology of $\texttt{Overr}(D)$. Indeed, if $\star$ is of finite type, let $E \in \overline{\boldsymbol{F}}(D)$, and let $\star \in \texttt{V}_E$, that is, $1 \in E^\star$, then there is a finitely generated submodule $F \subseteq E$ such that $1 \in F^\star$, so that $\star \in \texttt{V}_F$; it follows that

$$\texttt{V}_E \cap \texttt{SStar}_{\texttt{f}}(D) = \bigcup \{\texttt{V}_F \cap \texttt{SStar}_{\texttt{f}}(D) \mid F \subseteq E,\ F \in \boldsymbol{f}(D)\}$$

and thus $\{\texttt{V}_F \cap \texttt{SStar}_{\texttt{f}}(D) \mid F \in \boldsymbol{f}(D)\}$ is a subbasis for the Zariski topology on $\texttt{SStar}_{\texttt{f}}(D)$. Similarly, if $\star$ is stable, then $1 \in E^\star$ if and only if $1 \in E^\star \cap D^\star = (E \cap D)^\star$. Therefore, the Zariski topology on $\overline{\texttt{SStar}}(D)$ is generated by the $\texttt{V}_I \cap \overline{\texttt{SStar}}(D)$, as I ranges among the integral ideals of D. The same reasoning shows that $\{V_J \cap \widetilde{\texttt{SStar}}(D) \mid J \subseteq D,\ J \in \boldsymbol{f}(D)\}$ is a subbasis for the Zariski topology on $\widetilde{\texttt{SStar}}(D)$. This implies that stable semistar operations are completely determined by their action inside the ring. In particular, if $* : \boldsymbol{F}(D) \to \boldsymbol{F}(D)$ is a stable *star* operation, then there is a unique stable *semistar* operation $\hat{*} : \overline{\boldsymbol{F}}(D) \to \overline{\boldsymbol{F}}(D)$ such that $\hat{*}|_{\boldsymbol{F}(D)} = *$.

Remark 3.11 Note that the subbasic open sets $\overline{\texttt{U}}_I := \texttt{V}_I \cap \overline{\texttt{SStar}}(D) = \{\star \in \texttt{SStar}(D) \mid 1 \in I^\star\} \cap \overline{\texttt{SStar}}(D)$ (respectively, $\widetilde{\texttt{U}}_I := \texttt{V}_I \cap \widetilde{\texttt{SStar}}(D) = \{\star \in \texttt{SStar}(D) \mid 1 \in I^\star\} \cap \widetilde{\texttt{SStar}}(D)$) of $\overline{\texttt{SStar}}(D)$ (respectively, of $\widetilde{\texttt{SStar}}(D)$), where I is an ideal of D, form a basis of $\overline{\texttt{SStar}}(D)$ (respectively, $\widetilde{\texttt{SStar}}(D)$), since $\overline{\texttt{U}}_{I'} \cap \overline{\texttt{U}}_{I''} = \overline{\texttt{U}}_{I' \cap I''}$ (respectively, $\widetilde{\texttt{U}}_{I'} \cap \widetilde{\texttt{U}}_{I''} = \widetilde{\texttt{U}}_{I' \cap I''}$), for all I' and I'' ideals of D.

On the other hand, when considering finitely generated ideals J of D, in general the $\widetilde{\texttt{U}}_J$'s do not form a basis for the open sets in $\widetilde{\texttt{SStar}}(D)$, since $\widetilde{\texttt{U}}_{J'} \cap \widetilde{\texttt{U}}_{J''} = \widetilde{\texttt{U}}_{J' \cap J''}$, and $J' \cap J''$ is not necessarily finitely generated, even if J' and J'' are finitely generated ideals of D.

Besides the Zariski topology, we can also endow $\texttt{SStar}(D)$ with possibly weaker topologies induced by the sets considered in the above paragraph.

Proposition 3.12 (cf. [21, Proposition 2.1 and Remark 2.2]) *Preserve the notation of Proposition 3.10, and endow* $\texttt{SStar}(D)$ *with the topology generated by* $\{V_F \mid F \in \boldsymbol{f}(D)\}$ *(respectively,* $\{V_I \mid I$ *ideal in* $D\}$*;* $\{V_J \mid J \subseteq D,\ J \in \boldsymbol{f}(D)\}$*). Then,* $\Phi_{\texttt{f}}$ *(respectively,* $\overline{\Phi}$*;* $\widetilde{\Phi}$*) is the Kolmogoroff quotient of* $\texttt{SStar}(D)$ *onto* $\texttt{SStar}_{\texttt{f}}(D)$ *(respectively,* $\overline{\texttt{SStar}}(D)$*;* $\widetilde{\texttt{SStar}}(D)$*), i.e., it is the canonical map to the quotient by the equivalence relation of "topological indistinguishability" (where two points of a topological space are topologically indistinguishable if they have exactly the same neighborhoods).*

Let $Y \subseteq \texttt{Spec}(D)$ be a nonempty set defining a spectral semistar operation. Then its closure, in the inverse topology (denoted by $\texttt{Cl}^{\texttt{inv}}(Y)$, see Sect. 1.3), provides some useful information about s_Y.

Proposition 3.13 (cf. [21, Corollaries 4.4 and 5.2, Proposition 5.1] and [22, Lemma 4.2 and Remark 4.5]) *Let D be an integral domain and let Y and Z be two nonempty subsets of* $\texttt{Spec}(D)$*. The following statements hold.*

(1) $s_Y = s_Z$ *if and only if* $Y^{\mathtt{gen}} = Z^{\mathtt{gen}}$.
(2) s_Y *is of finite type if and only if* Y *is quasi-compact.*
(3) $\widetilde{s_Y} = \widetilde{s_Z}$ *if and only if* $\mathtt{Cl}^{\mathtt{inv}}(Y) = \mathtt{Cl}^{\mathtt{inv}}(Z)$.
(4) $\widetilde{s_Y} = s_{\mathtt{Cl}^{\mathtt{inv}}(Y)}$.

Note that, in general, $(s_Y)_f$ is quasi-spectral but not spectral, and it is spectral if and only if $(s_Y)_f$ is stable [22, Proposition 4.23(2)]. In other words, it is possible that $\widetilde{s_Y} \lneq (s_Y)_f$ (see [21, Remark 5.3] and [2, p. 2466]) and thus it is not true in general that $(s_Y)_f = s_{\mathtt{Cl}^{\mathtt{inv}}(Y)}$.

The following result provides control of the infimum and the supremum of a family of spectral operations:

Lemma 3.14 (cf. [18, Lemma 4.3]) *Let $\mathscr{D}$ be a nonempty set of spectral semistar operations on an integral domain D. For each spectral semistar operation $\star$, set $\Delta(\star) := \mathtt{QSpec}^\star(D)$. Then, the following statements hold.*

(1) $\wedge_{\mathscr{D}}$ *is spectral with* $\Delta(\wedge_{\mathscr{D}}) = \bigcup\{\Delta(\star) \mid \star \in \mathscr{D}\}$.
(2) If $\vee_{\mathscr{D}}$ *is quasi-spectral, then it is spectral with* $\Delta(\vee_{\mathscr{D}}) = \bigcap\{\Delta(\star) \mid \star \in \mathscr{D}\}$.

Note that the hypothesis that $\vee_{\mathscr{D}}$ be quasi-spectral in point (2) is necessary: for example, if $\mathbb{A}$ is the ring of all algebraic integers, $\star_P := s_{\mathrm{Max}(\mathbb{A})\setminus\{P\}}$ and $\mathscr{D} := \{\star_P \mid P \in \mathrm{Max}(\mathbb{A})\} \subseteq \mathtt{SStar}_{sp}(\mathbb{A})$, then $\vee_{\mathscr{D}}$ is a semistar operation that closes $\mathbb{A}$ and thus closes every principal ideal of $\mathbb{A}$, while $\mathtt{QSpec}^{\vee_{\mathscr{D}}}(D) = \{(0)\}$, hence $\vee_{\mathscr{D}}$ is not quasi-spectral. (See [18, Example 4.4] for more details.)

Lemma 3.14(2) provides useful information on the supremum of a family of spectral semistar operations of finite type, allowing one to prove that the space of all stable semistar operations of finite type is spectral. The proof of the following theorem follows closely the one of Theorem 3.8.

Theorem 3.15 (cf. [18, Theorem 4.5]) *Let D be an integral domain. Then, $\widetilde{\mathtt{SStar}}(D)$ is a spectral space.*

Stable semistar operations are closely related to the concept of *localizing systems*, in the sense of Gabriel-Popescu (cf. for instance [6, Chap. II], [7, 30, 44, 50]). Recall that a localizing system on D is a subset $\mathcal{F}$ of ideals of D such that:

- if $I \in \mathcal{F}$ and J is an ideal of D such that $I \subseteq J$, then $J \in \mathcal{F}$;
- if $I \in \mathcal{F}$ and J is an ideal of D such that, for each $i \in I$, $(J :_D iD) \in \mathcal{F}$, then $J \in \mathcal{F}$.

A localizing system $\mathcal{F}$ is said to be *of finite type* if for each $I \in \mathcal{F}$ there exists a nonzero finitely generated ideal $J \in \mathcal{F}$ such that $J \subseteq I$. For instance, if T is an overring of R, $\mathcal{F}(T) := \{I \mid I \text{ ideal of } D, IT = T\}$ is a localizing system of finite type, while, if V is a valuation domain with a nonzero idempotent prime ideal P, then $\hat{\mathcal{F}}(P) := \{I \mid I \text{ ideal of } V \text{ and } I \supseteq P\}$ is a localizing system of V which is not of finite type [28, Proposition 5.1.12 and Remark 5.1.13]. We denote by $\mathtt{LS}(D)$ (respectively, $\mathtt{LS}_f(D)$) the set of all localizing systems (respectively, localizing systems of finite type) on D. We can introduce on these sets a natural topology, that we still call the *Zariski topology*, whose subbasic open sets are the $\mathtt{W}_I := \{\mathcal{F} \in \mathtt{LS}(D) \mid I \in \mathcal{F}\}$, as I varies among the ideals in D.

Theorem 3.16 (cf. [18, Proposition 3.5,Proposition 4.1(5) and Corollary 4.6]) *Let D be an integral domain. The map $\boldsymbol{\lambda} : \mathtt{LS}(D) \to \overline{\mathtt{SStar}}(D)$ (respectively, the map $\boldsymbol{\lambda}_f : \mathtt{LS}_f(D) \to \widetilde{\mathtt{SStar}}(D)$), defined by $\mathcal{F} \mapsto \star_{\mathcal{F}}$, establishes a homeomorphism between spaces endowed with the Zariski topologies (respectively, the induced topologies from the Zariski topologies). In particular, by Theorem 3.15, $\mathtt{LS}_f(D)$ is a spectral space.*

4 The Space of Inverse-Closed Subsets of a Spectral Space

Let D be an integral domain. By the results in the previous sections, the spaces $\mathtt{Overr}(D)$, $\widetilde{\mathtt{SStar}}(D)$ and $\mathtt{SStar}_f(D)$ are spectral spaces. Since $\mathtt{Spec}(D)$ can be embedded in each of these spaces, they can be seen as peculiar "spectral extensions" of $\mathtt{Spec}(D)$.

In particular, in this section we focus on the canonical embedding $\mathtt{Spec}(D) \hookrightarrow \widetilde{\mathtt{SStar}}(D)$, in order to generalize this spectral extension to arbitrary rings or to arbitrary spectral spaces. For this purpose, we need some preliminaries, including the notions and properties of Sect. 1.3.

We start by observing that the natural injection $s : \mathtt{Spec}(D) \to \widetilde{\mathtt{SStar}}(D)$, defined by $s(P) := s_{\{P\}} = \wedge_{\{D_P\}}$, is a topological embedding of topological (spectral) spaces (both endowed with the Zariski topology). Indeed, if J is a finitely generated ideal of D and $\widetilde{\mathtt{U}}_J := \mathtt{V}_J \cap \widetilde{\mathtt{SStar}}(D) = \{\star \in \mathtt{SStar}(D) \mid 1 \in J^\star\} \cap \widetilde{\mathtt{SStar}}(D)$ is a generic subbasic open set of $\widetilde{\mathtt{SStar}}(D)$, then

$$s^{-1}(\widetilde{\mathtt{U}}_J) = \{P \in \mathtt{Spec}(D) \mid 1 \in JD_P\} = \mathtt{D}(J).$$

Remark 4.1 The map $s : \mathtt{Spec}(D) \to \widetilde{\mathtt{SStar}}(D)$ is the composition of the homeomorphism $\ell : \mathtt{Spec}(D) \to \mathtt{Loc}(D)$, defined by $\ell(P) := D_P$, for each $P \in \mathtt{Spec}(D)$ and the topological embedding $\iota_{Loc} : \mathtt{Loc}(D) \hookrightarrow \widetilde{\mathtt{SStar}}(D)$ (defined in Remark 3.9(b)). Note also that the homeomorphism ℓ induces an isomorphism of partially ordered sets (with the ordering induced by the topologies), however the ordering in $\mathtt{Loc}(D)$, induced by the Zariski topology, is the opposite order of the set-theoretic inclusion.

Given a spectral space X, let $\boldsymbol{\mathcal{X}}(X) := \{Y \subseteq X \mid Y \neq \emptyset,\ Y = \mathtt{Cl}^{\mathrm{inv}}(Y)\}$. If $X = \mathtt{Spec}(R)$ for some ring R, we write for short $\boldsymbol{\mathcal{X}}(R)$ instead of $\boldsymbol{\mathcal{X}}(\mathtt{Spec}(R))$.

We define a *Zariski topology on* $\boldsymbol{\mathcal{X}}(X)$ by taking, as subbasis of open sets, the sets of the form

$$\boldsymbol{\mathcal{U}}(\Omega) := \{Y \in \boldsymbol{\mathcal{X}}(X) \mid Y \subseteq \Omega\},$$

where Ω varies among the quasi-compact open subspaces of X. Note that the previous subbasis is in fact a basis, since $\boldsymbol{\mathcal{U}}(\Omega) \cap \boldsymbol{\mathcal{U}}(\Omega') = \boldsymbol{\mathcal{U}}(\Omega \cap \Omega')$ and $\Omega \cap \Omega'$ is a quasi-compact open subspace of X, for any pair Ω, Ω' of quasi-compact open subspaces of X. Moreover, $\Omega \in \boldsymbol{\mathcal{U}}(\Omega)$, since a quasi-compact open subset Ω of X is a closed set in the inverse topology of X. Note also that, when $X = \texttt{Spec}(R)$, for some ring R, a generic basic open set of the Zariski topology on $\boldsymbol{\mathcal{X}}(R)$ is of the form

$$\boldsymbol{\mathcal{U}}(\mathrm{D}(J)) = \{Y \in \boldsymbol{\mathcal{X}}(R) \mid Y \subseteq \mathrm{D}(J)\}$$

where J is any finitely generated ideal of R.

The main result in this setting is the following, which provides a description of the space $\boldsymbol{\mathcal{X}}(X)$ (see [19]).

Theorem 4.2 *Let X be a spectral space.*

(1) The space $\boldsymbol{\mathcal{X}}(X)$, endowed with the Zariski topology, is a spectral space.
(2) Let $Y_1,\ Y_2 \in \boldsymbol{\mathcal{X}}(X)$. Then, $Y_1 \subseteq Y_2$ if and only if $Y_1 \leq_{\boldsymbol{\mathcal{X}}(X)} Y_2$.
(3) The canonical map $\varphi : X \to \boldsymbol{\mathcal{X}}(X)$, defined by $\varphi(x) := \{x\}^{\mathrm{gen}}$, for each $x \in X$, is a spectral embedding (which is also an order-preserving embedding between ordered sets, with the ordering induced by the Zariski topologies).
(4) $\boldsymbol{\mathcal{X}}(X)$ has a unique maximal point (i.e., X).
(5) Let Z be another spectral space and let $\varphi : X \to \boldsymbol{\mathcal{X}}(X)$ be the spectral embedding defined in (3). Consider a spectral map $\lambda : X \to Z$ satisfying the following condition:

(`sup-completion`) *For each nonempty quasi-compact subspace Y of X, there exists $z_Y := \sup\{\lambda(y) \mid y \in Y\}$ (where* sup *is taken with respect to the ordering induced by the topology of Z) and if Y' is another nonempty quasi-compact subspace of X, with $\texttt{Cl}_X^{\texttt{inv}}(Y') \neq \texttt{Cl}_X^{\texttt{inv}}(Y)$, then $z_{Y'} \neq z_Y$. Moreover, if $\mathcal{W}$ denotes the set of all nonempty quasi-compact open subspaces Ω of X, then $\mathscr{B} := \{\{z_\Omega\}^{\mathrm{gen}} \mid \Omega \in \mathcal{W}\}$ is a subbasis for the open sets of Z.*

Then, the following properties hold.

(5.a) There exists a spectral embedding $\boldsymbol{\lambda}^\sharp : \boldsymbol{\mathcal{X}}(X) \to Z$ such that $\boldsymbol{\lambda}^\sharp \circ \varphi = \lambda$.
(5.b) If, furthermore, $z = \sup_Z\{\lambda(x) \mid x \in \lambda^{-1}(\{z\}^{\mathrm{gen}})\}$ for each $z \in Z$, then $\boldsymbol{\lambda}^\sharp : \boldsymbol{\mathcal{X}}(X) \to Z$ is the unique spectral embedding (in fact, homeomorphism) such that $\boldsymbol{\lambda}^\sharp \circ \varphi = \lambda$.

Let X be a spectral space and let $\hat{\boldsymbol{\mathcal{X}}}(X) := \{Y \subseteq X \mid Y = \texttt{Cl}^{\texttt{inv}}(Y)\} = \boldsymbol{\mathcal{X}}(X) \cup \{\emptyset\}$. The techniques used for proving Theorem 4.2(1) allow also to show that $\hat{\boldsymbol{\mathcal{X}}}(X)$ (endowed with an obvious extension of the topology of $\boldsymbol{\mathcal{X}}(X)$) is a spectral space. Moreover, since $\boldsymbol{\mathcal{U}}(\emptyset) = \{\emptyset\}$ is open in $\hat{\boldsymbol{\mathcal{X}}}(X)$, then we deduce that $\boldsymbol{\mathcal{X}}(X)$ is a closed (spectral) subspace of $\hat{\boldsymbol{\mathcal{X}}}(X)$.

As a consequence of the previous theorem, it is possible to compare the dimensions of X and $\boldsymbol{\mathcal{X}}(X)$ with the cardinality $|X|$ of the spectral space X (see [19]).

Proposition 4.3 *Let X be a spectral space and let* $\varphi : X \to \mathcal{X}(X)$ *be the topological embedding defined in Theorem 4.2(2). Then,*

(1) $\varphi(X) = \mathcal{X}(X)$ *if and only if* $(X, \leq)$ *is linearly ordered.*
(2) $\dim(\mathcal{X}(X)) = |X| - 1 \geq \dim(X)$. *Moreover, in the finite dimensional case,* $\dim(\mathcal{X}(X)) = \dim(X)$ *if and only if X is linearly ordered.*

While the inequality $|X| - 1 \geq \dim(X)$ is sharp, the more noncomparable elements the set X contains, the smaller $\dim(X)$ is with respect to $|X|$. For example, if X is homeomorphic to the prime spectrum of the direct product of $n + 1$ fields, $n \geq 1$, then $\dim(X) = 0$ while $|X| - 1 = n$.

Furthermore, if $\dim(X)$ is not finite, then clearly $\dim(\mathcal{X}(X)) = \dim(X)$, but we can easily choose X to be not totally ordered.

We also note that, if $\phi : X \longrightarrow Y$ is a spectral map of spectral space, the map $\mathcal{X}(\phi) : \mathcal{X}(X) \longrightarrow \mathcal{X}(Y)$ defined by $\mathcal{X}(\phi)(C) := \phi(C)^{\text{gen}}$ for every inverse-closed subset C of X is again a spectral map. It follows that the assignment $X \mapsto \mathcal{X}(X)$, $\phi \mapsto \mathcal{X}(\phi)$ is a (covariant) functor from the category of spectral spaces into itself (see [19] for details).

We show next that the map $\boldsymbol{\lambda}^{\#} : \mathcal{X}(X) \to Z$ (Theorem 4.2(5.a)) is not unique. The following example shows in fact that it is possible that there exist two different spectral maps (with at most one non-injective) $\boldsymbol{\Lambda}_1, \boldsymbol{\Lambda}_2 : \mathcal{X}(X) \to Z$, $\boldsymbol{\Lambda}_1 \neq \boldsymbol{\Lambda}_2$, such that $\boldsymbol{\Lambda}_1 \circ \varphi = \lambda = \boldsymbol{\Lambda}_2 \circ \varphi$.

Example 4.4 Consider the spectral space $X := \{0, a, b, c\}$, with $0 < a, b, c$ and a, b, c not comparable. Let $\boldsymbol{\Lambda} : \mathcal{X}(X) \to \mathcal{X}(X)$ be the function defined by

$$\boldsymbol{\Lambda}(C) := \begin{cases} C & \text{if } C \neq \{a, b\}^{\mathtt{gen}}, \\ X & \text{if } C = \{a, b\}^{\mathtt{gen}}. \end{cases}$$

The unique basic open set of $\mathcal{X}(X)$ containing $\{a, b\}^{\mathtt{gen}}$ is $\mathcal{U}(\{a, b\}^{\mathtt{gen}})$, and clearly we have $\boldsymbol{\Lambda}^{-1}(\mathcal{U}(\{a, b\}^{\mathtt{gen}})) = \mathcal{U}(\{a\}^{\mathtt{gen}}) \cup \mathcal{U}(\{b\}^{\mathtt{gen}})$. For any other basic open set $\mathcal{U}$ of $\mathcal{X}(X)$, we have $\boldsymbol{\Lambda}^{-1}(\mathcal{U}) = \mathcal{U}$. This shows that $\boldsymbol{\Lambda}$ is a nontrivial spectral map, $\boldsymbol{\Lambda} \neq \mathtt{id}_{\mathcal{X}(X)}$, such that $\boldsymbol{\Lambda}(\{x\}^{\mathtt{gen}}) = \{x\}^{\mathtt{gen}}$, for each $x \in X$.

The following statement provides an explicit characterization of the space $\mathcal{X}(X)$ and follows immediately from Theorem 4.2(5).

Corollary 4.5 *Let* $\lambda : X \to Z$ *be a spectral embedding of spectral spaces. Then, the following conditions are equivalent.*

(i) Z is a partially ordered set (under the ordering induced by the topology), for each $z \in Z$, $z = \sup_Z\{\lambda(x) \mid x \in \lambda^{-1}(\{z\}^{\mathtt{gen}})\}$, *and* λ *satisfies the condition* (`sup-completion`).
(ii) Z is homeomorphic to $\mathcal{X}(X)$, *via a unique homeomorphism* $\boldsymbol{\Lambda} : \mathcal{X}(X) \to Z$ *such that* $\boldsymbol{\Lambda} \circ \varphi = \lambda$.

In the special case where $X = \mathtt{Spec}(D)$ for some integral domain D, the spectral space $\boldsymbol{\mathcal{X}}(D) := \{Y \subseteq \mathtt{Spec}(D) \mid \emptyset \neq Y = \mathtt{Cl}^{\mathtt{inv}}(Y)\}$ can be interpretated in terms of stable semistar operations of finite type (see [19]).

Proposition 4.6 *Let D be an integral domain and let $\boldsymbol{\mathcal{X}}(D) := \{Y \subseteq \mathtt{Spec(D)} \mid \emptyset \neq Y = \mathtt{Cl}^{\mathtt{inv}}(Y)\}$. The map $s^\sharp : \boldsymbol{\mathcal{X}}(D) \to \widetilde{\mathtt{SStar}}(D)$, defined by $s^\sharp(Y) := s_Y$ for each $Y \in \boldsymbol{\mathcal{X}}(D)$, is a homeomorphism with inverse map $\Delta : \widetilde{\mathtt{SStar}}(D) \to \boldsymbol{\mathcal{X}}(D)$, defined by $\Delta(\star) := \mathtt{QSpec}^{\star}(D)$ for each $\star$ stable semistar operation of finite type on D. Moreover, if $\varphi : \mathtt{Spec}(D) \to \boldsymbol{\mathcal{X}}(D)$ is the topological embedding defined in Theorem 4.2(3) and $s : \mathtt{Spec}(D) \to \widetilde{\mathtt{SStar}}(D)$ is the topological embedding defined by $P \mapsto s_{\{P\}}$, for each prime ideal P of D, then $s^\sharp \circ \varphi = s$.*

As a consequence of the previous proposition and Theorem 4.2(1) we reobtain immediately Theorem 3.15, that is, the space of all stable semistar operations of finite type on an integral domain is a spectral space.

5 A Topological Version of Hilbert's Nullstellensatz

As a first application of the general construction considered in the previous section, we give now a topological version of Hilbert's Nullstellensatz.

Given a ring R, consider the set $\mathtt{Rd}(R) := \{I \mid I \text{ ideal of } R \text{ and } I = \mathrm{rad}(I)\}$ of radical ideals of R and, more generally, the set $\mathtt{Id}(R) := \{I \mid I \text{ ideal of } R\}$, endowed with the *hull–kernel topology*, defined by taking as a basis for the open sets the subsets

$$U(x_1, x_2, \ldots, x_n) := \{I \in \mathtt{Id}(R) \mid x_i \notin I \text{ for some } i,\ 1 \leq i \leq n\},$$

where $x_1, x_2, \ldots, x_n \in R$. We denote by $\mathtt{Id}(R)^{\mathrm{hk}}$ (respectively, $\mathtt{Rd}(R)^{\mathrm{hk}}$) the set of all the ideals of R (respectively, of all the radical ideals of R), endowed with the hull–kernel topology (respectively, with the induced topology from the hull–kernel topology of $\mathtt{Id}(R)$). In this situation, the inclusion maps $\mathtt{Spec}(R) \subseteq \mathtt{Rd}(R) \subseteq \mathtt{Id}(R)$ become topological embeddings; in other words the hull–kernel topology induced on $\mathtt{Spec}(R)$ coincides with the Zariski topology.

For deepening the study of the topological space $\mathtt{Rd}(R)^{\mathrm{hk}}$ we introduce an analogue, in the inverse topology, of the space $\boldsymbol{\mathcal{X}}(R)$ (Sect. 4).

Let X be a spectral space and let $\mathtt{Cl}(Y)$ denote the closure of a subspace Y in the given topology of X. For the sake of simplicity, we denote by X' the spectral space $X^{\mathtt{inv}}$, i.e., the set X endowed with the inverse topology [35, Proposition 8]. We set $\boldsymbol{\mathcal{X}}'(X) := \{Y \subseteq X \mid Y \neq \emptyset,\ Y = \mathtt{Cl}(Y)\}$ and, for each quasi-compact open subspaces Ω of X, we set $\boldsymbol{\mathcal{U}}'(\Omega) := \{Y \in \boldsymbol{\mathcal{X}}'(X) \mid Y \cap \Omega = \emptyset\} = \boldsymbol{\mathcal{U}}(\Omega')$, where $\Omega' := X \setminus \Omega$.

It is well known that $(X^{\mathtt{inv}})^{\mathtt{inv}}$ coincides with X (with the given spectral topology) [35, Proposition 8] hence, *mutatis mutandis*, we can now apply Theorem 4.2, since $\boldsymbol{\mathcal{X}}'(X) = \boldsymbol{\mathcal{X}}(X')$, and we easily get the following.

Proposition 5.1 *Let X be a spectral space and let $X' := X^{\mathtt{inv}}$.*

(1) The space $\boldsymbol{\mathcal{X}}'(X) := \{Y \subseteq X \mid Y \neq \emptyset,\ Y = \mathtt{Cl}(Y)\}$ is a spectral space, when endowed with the topology, called the Zariski topology, *having as a basis of open sets, the sets of the form $\boldsymbol{\mathcal{U}}'(\Omega)$, where Ω varies among the quasi-compact open subspaces of X.*

(2) The canonical map $\varphi' : X' \to \boldsymbol{\mathcal{X}}'(X)$, defined by $\varphi'(x) := \{x\}^{\mathtt{sp}}$, for each $x \in X$, is a spectral embedding between spectral spaces.

Suppose now that $X := \mathtt{Spec}(R)$ is the prime spectrum of a commutative ring R, endowed with the Zariski topology. We recall that a basis of open sets of $X^{\mathtt{inv}}$ is the collection of sets $\{\mathtt{V}(J) \mid J$ is a finitely generated ideal of $R\}$ which makes $X^{\mathtt{inv}}$ a spectral space [35, Proposition 8].

Remark 5.2 With the notation introduced above, let $\varphi' : X' = \mathtt{Spec}(R)^{\mathtt{inv}} \hookrightarrow \boldsymbol{\mathcal{X}}(X')^{\mathtt{zar}} = \boldsymbol{\mathcal{X}}'(X)^{\mathtt{zar}}$ be the canonical topological embedding defined by $\varphi'(x) := \{x\}^{\mathtt{sp}}$. Then, it is easy to see that the map $\psi := (\varphi')^{\mathtt{inv}} : X = (\mathtt{Spec}(R)^{\mathtt{inv}})^{\mathtt{inv}} \hookrightarrow \boldsymbol{\mathcal{X}}'(X)^{\mathtt{inv}}$ defined by $\psi(x) := \{x\}^{\mathtt{gen}}$ is a topological embedding (acting like φ as a set-theoretic map).

The next result provides a topological version of Hilbert Nullstellensatz (see [20]).

Theorem 5.3 *Let R be a ring and let $\boldsymbol{\mathcal{X}}'(R) := \boldsymbol{\mathcal{X}}'(\mathtt{Spec}(R))$ be the spectral space of the nonempty Zariski closed subspaces of $\mathtt{Spec}(R)$ (Proposition 5.1). We can also consider the space $\boldsymbol{\mathcal{X}}'(R)$ as a spectral space endowed with the inverse topology [35, Proposition 8]. Then, for each $C \in \boldsymbol{\mathcal{X}}'(R)$, the map:*

$$\mathscr{J} : \boldsymbol{\mathcal{X}}'(R)^{inv} \to \mathtt{Rd}(R)^{hk} \text{ defined by } \mathscr{J}(C) := \bigcap\{P \in \mathtt{Spec}(R) \mid P \in C\},$$

is a homeomorphism.

Related to the previous Theorem 5.3, it is possible to prove, with a standard argument based on Theorem 2.8, that the set of all ideals of a ring is also a spectral space. More precisely:

Proposition 5.4 (cf. [20]) *Let $\mathtt{Id}(R)$ be the space of all ideals of a ring R, endowed with the hull–kernel topology. Then, $\mathtt{Id}(R)$ is a spectral space, having $\mathtt{Rd}(R)$ (endowed with the hull–kernel topology) as a spectral subspace.*

The following Hasse diagram summarizes some of the results proved above.

$$\begin{array}{ccccc}
 & & \mathtt{Id}(R)^{\mathtt{hk}} & & \\
 & & \uparrow & & \\
\boldsymbol{\mathcal{X}}'(\mathtt{Spec}(R))^{\mathtt{inv}} & \simeq & \mathtt{Rd}(R)^{\mathtt{hk}} & & \boldsymbol{\mathcal{X}}(\mathtt{Spec}(R))^{\mathtt{zar}} \\
\uparrow & & \uparrow & & \uparrow \\
\mathtt{Spec}(R)^{\mathtt{zar}} & \Longleftrightarrow & \mathtt{Spec}(R)^{\mathtt{hk}} & \Longleftrightarrow & \mathtt{Spec}(R)^{\mathtt{zar}}
\end{array}$$

6 The Space of eab Semistar Operations of Finite Type

In the present section, we give another application of Theorem 4.2. More precisely, we apply the construction of the space $\mathcal{X}(X)$ to the case of the Riemann–Zariski spectral space $X := \mathtt{Zar}(D)$ of all valuation overrings of an integral domain D (endowed with the Zariski topology, see Sect. 1.2).

Let $\star$ be a semistar operation on an integral domain D. We say that $\star$ is an eab *semistar operation* (respectively, an ab *semistar operation*) if, for every $F, G, H \in \boldsymbol{f}(D)$ (respectively, for every $F \in \boldsymbol{f}(D)$, $G, H \in \overline{\boldsymbol{F}}(D)$) the inclusion $(FG)^\star \subseteq (FH)^\star$ implies $G^\star \subseteq H^\star$. Note that, if $\star$ is eab, then $\star_{_f}$ is also eab, since $\star$ and $\star_{_f}$ agree on finitely generated fractional ideals. The concepts of eab and ab operations coincide on finite-type operations, but not in general [27, 29].

It is easy to see that a *valuative semistar operation*, i.e., a semistar operation of the type $\wedge_{\mathcal{W}}$, where $\mathcal{W} \subseteq \mathtt{Zar}(D)$, is an eab semistar operation. In particular, the b-operation, where $b := \wedge_{\mathtt{Zar}(D)}$, is an eab semistar operation of finite type, since $\mathtt{Zar}(D)$ is quasi-compact (Corollary 3.5).

To every semistar operation $\star \in \mathtt{SStar}(D)$ we can associate a map $\star_a$ defined by

$$F^{\star_a} := \bigcup\{((FG)^\star : G^\star) \mid G \in \boldsymbol{f}(D)\}$$

for every $F \in \boldsymbol{f}(D)$, and then we can extend it to arbitrary D-modules $E \in \overline{\boldsymbol{F}}(D)$ by setting $E^{\star_a} := \bigcup\{F^{\star_a} \mid F \subseteq E,\ F \in \boldsymbol{f}(D)\}$. The map $\star_a$ is always an eab semistar operation of finite type on D. Moreover, $\star = \star_a$ if and only if $\star$ is an eab semistar operation of finite type and, if $\star$ is an eab semistar operation, then $\star_a = \star_{_f}$ [23, Proposition 4.5].

Remark 6.1 (a) Let T be an overring of D, and let $\star_T$ be a semistar operation on T. Then, we can define a semistar operation $\star$ on D by $\star := \star_T \circ \wedge_{\{T\}}$, i.e., $E^\star := (ET)^{\star_T}$ for every $E \in \overline{\boldsymbol{F}}(D)$. If now $F \in \boldsymbol{f}(T)$, then

$$\begin{aligned} F^{\star_a} &= \bigcup\{((FG)^\star : G^\star) \mid G \in \boldsymbol{f}(D)\} = \bigcup\{((FGT)^{\star_T} : (GT)^{\star_T}) \mid G \in \boldsymbol{f}(D)\} = \\ &= \bigcup\{((FTH)^{\star_T} : H^{\star_T}) \mid H \in \boldsymbol{f}(T)\} = (FT)^{(\star_T)_a} = F^{(\star_T)_a}. \end{aligned}$$

Hence, for every $E \in \overline{\boldsymbol{F}}(D)$, $E^{\star_a} = (ET)^{(\star_T)_a}$, that is, $\star_a = (\star_T)_a \circ \wedge_{\{T\}}$.

(b) W. Krull only considered the concept of an "arithmetisch brauchbar" operation(for short ab-operation, as above) [41]. He did not consider the concept of "endlich arithmetisch brauchbar" operation (or, more simply, eab-operation as above). This concept stems from the original version of Gilmer's book [31].

(c) Denote by $\mathtt{SStar}_{val}(D)$ (respectively, $\mathtt{SStar}_{eab}(D)$; $\mathtt{SStar}_{f,eab}(D)$) the set of valutative (respectively, eab; eab of finite type) semistar operations on D. Every valutative operation is eab, but not every eab operation is valutative; however, the two definitions agree on finite-type operations, i.e.,

$$\mathtt{SStar}_{eab}(D) \cap \mathtt{SStar}_f(D) =: \mathtt{SStar}_{f,eab}(D) = \mathtt{SStar}_{val}(D) \cap \mathtt{SStar}_f(D),$$

(see, for instance, [23, Corollary 5.2]). A similar relationship holds between spectral and stable semistar operations, with the valutative operations corresponding to the spectral ones and the $\mathtt{eab}$ operations to the stable ones, i.e., every spectral semistar operation is stable but not every stable semistar operation is spectral, however $\mathtt{SStar}_{f,sp}(D) = \overline{\mathtt{SStar}}(D) \cap \mathtt{SStar}_f(D) = \widetilde{\mathtt{SStar}}(D)$ (Remark 3.9(a)).

Recall also that there are examples of $\mathtt{eab}$ semistar operations which are quasi-spectral but not valutative [27, Example 15].

It is not hard to prove the following statement, which is a companion to Proposition 3.10.

Proposition 6.2 (cf. [18, Proposition 5.2]) *Let D be an integral domain and let $\Phi_a : \mathtt{SStar}(D) \to \mathtt{SStar}(D)$ be the map defined by $\star \mapsto \star_a$. Then:*

(1) The image of Φ_a coincides with $\mathtt{SStar}_{f,eab}(D)$.
(2) The map Φ_a is continuous in the Zariski topology.
(3) The map Φ_a is a topological retraction of $\mathtt{SStar}(D)$ onto $\mathtt{SStar}_{f,eab}(D)$.

The relation between valutative operations and subsets of $\mathtt{Zar}(D)$ behaves very similarly to the relation between spectral operations and subsets of $\mathtt{Spec}(D)$ (Proposition 3.13).

Proposition 6.3 *Let D be an integral domain and let Y and Z be two nonempty subsets of $\mathtt{Zar}(D)$. Then, the following statements hold.*

(1) $\wedge_Y = \wedge_Z$ if and only if $Y^{\mathrm{gen}} = Z^{\mathrm{gen}}$.
(2) $\wedge_Y$ is of finite type if and only if Y is quasi-compact [21, Proposition 4.5].
(3) $(\wedge_Y)_f = (\wedge_Z)_f$ if and only if $\mathtt{Cl}^{\mathrm{inv}}(Y) = \mathtt{Cl}^{\mathrm{inv}}(Z)$ [17, Theorem 4.9].
(4) $(\wedge_Y)_f = \wedge_{\mathtt{Cl}^{\mathrm{inv}}(Y)}$ [17, Corollary 4.17].

Note that $Y^{\mathrm{gen}} = \{V \in \mathtt{Zar}(D) \mid V \supseteq V_0, \text{ for some } V_0 \in Y\}$. For the statement (1), assume first that $\wedge_Y = \wedge_Z$. Let V be a valuation domain such that $V \in Y^{\mathrm{gen}} \setminus Z^{\mathrm{gen}}$. Then, for any $W \in Z$, we can pick an element $x_W \in W \setminus V$. It follows that $I := (x_W^{-1} \mid W \in Z) \subseteq M_V$, where M_V is the maximal ideal of V. Thus, if $V_0 \in Y$ is such that $V_0 \subseteq V$ (such a V_0 exists since $V \in Y^{\mathrm{gen}}$), we have $IV_0 \subseteq M_{V_0}$ and, in particular, $1 \notin I^{\wedge_Y}$. On the other hand, clearly $1 \in I^{\wedge_Z}$, a contradiction. The converse it is straightforward since, for each $Y \subseteq \mathtt{Zar}(D)$, $\wedge_Y = \wedge_{Y^{\mathrm{gen}}}$.

Remark 6.4 Since $b = \wedge_{\mathtt{Zar}(D)}$ is a semistar operation of finite type (and this can be proved completely independently from the topological point of view, see [39, Proposition 6.8.2] and [21, Remark 4.6]), from Proposition 6.3 we get a new proof of the fact that $\mathtt{Zar}(D)$ is a quasi-compact space (this is a special case of Zariski's theorem [53, Theorem 40, p. 113]).

The embedding $\iota : \mathtt{Overr}(D) \to \mathtt{SStar}_f(D)$ (Proposition 3.2) restricts to an embedding $\mathtt{Zar}(D) \hookrightarrow \mathtt{SStar}_{f,eab}(D)$, while the image of the restriction $\pi|_{\mathtt{SStar}_{f,eab}(D)}$ of the canonical map $\pi : \mathtt{SStar}_f(D) \to \mathtt{Overr}(D)$ (defined by

$\star \mapsto D^\star$) coincides with $\texttt{Overr}_{ic}(D)$, i.e., with the space of the overrings of D that are integrally closed in K (since, by a well known Krull's theorem, every integrally closed ring can be represented as an intersection of valuation rings [53, Theorem 6, p. 15]).

Using the b-operation, we can introduce a general version of the classical Kronecker function ring, introduced by L. Kronecker in the case of Dedekind domains. Let $\mathbb{X}$ be is an indeterminate over D and let $\boldsymbol{c}(h)$ be the content of a polynomial $h \in D[\mathbb{X}]$ (i.e., the ideal of D generated by the coefficients of h). Then, we set:

$$\begin{aligned}\mathrm{Kr}(D) := \mathrm{Kr}(D, b) &:= \{f/g \mid f, g \in D[\mathbb{X}],\ g \neq 0,\ \text{with } \boldsymbol{c}(f)^b \subseteq \boldsymbol{c}(g)^b\} \\ &= \bigcap\{V(\mathbb{X}) \mid V \in \texttt{Zar}(D)\},\end{aligned}$$

where $V(\mathbb{X})$ denotes the Gaussian (or trivial) extension of V to $K(\mathbb{X})$, i.e., $V(\mathbb{X}) := V[\mathbb{X}]_{(MV[\mathbb{X}])}$. This is a Bézout domain with quotient field $K(\mathbb{X})$, called *the b-Kronecker function ring of D* (see [23, Definition 3.2, Corollary 3.4(2) and Theorem 5.1], [25, Theorem 14] and [32, Theorem 32.11]). It follows immediately that the localization map $\texttt{Spec}(\mathrm{Kr}(D)) \longrightarrow \texttt{Zar}(\mathrm{Kr}(D))$ (defined by $P \mapsto \mathrm{Kr}(D)_P$) is actually an homeomorphism. Moreover, the map $\Psi : \texttt{Zar}(D) \longrightarrow \texttt{Zar}(\mathrm{Kr}(D))$ (defined by $V \mapsto V(\mathbb{X})$) is a homeomorphism [17, Propositions 3.1 and 3.3], so that $\texttt{Spec}(\mathrm{Kr}(D))$ realizes $\texttt{Zar}(D)$ as a spectral space [11, Theorem 2].

In particular, the homeomorphism (and so the isomorphism of partially ordered sets) that we denote by θ, from $\texttt{Spec}(\mathrm{Kr}(D))$ to $\texttt{Zar}(D)$ induces a 1-1 correspondence $\boldsymbol{\Theta}_0$ between the set $\{Y \subseteq \texttt{Spec}(\mathrm{Kr}(D)) \mid Y = Y^{\downarrow}\}$ (where $Y^{\downarrow} := \{z \in \texttt{Spec}(\mathrm{Kr}(D)) \mid z \leq y, \text{ for some } y \in Y\} = Y^{\texttt{gen}}$) and the set $\{\mathcal{W} \subseteq \texttt{Zar}(D) \mid \mathcal{W} = \mathcal{W}^{\uparrow}\}$ (where $\mathcal{W}^{\uparrow} := \{W' \in \texttt{Zar}(D) \mid W' \supseteq W, \text{ for some } W \in \mathcal{W}\} = \mathcal{W}^{\texttt{gen}}$). Therefore $\boldsymbol{\Theta}_0$ induces a bijection $\boldsymbol{\Theta} : \texttt{SStar}_{sp}(\mathrm{Kr}(D)) \rightarrow \texttt{SStar}_{val}(D)$ defined by $\boldsymbol{\Theta}(s_Y) := \wedge_{\boldsymbol{\Theta}_0(Y)}$, where $\boldsymbol{\Theta}_0(Y) = \{V \in \texttt{Zar}(D) \mid M(\mathbb{X}) \cap \mathrm{Kr}(D) \in Y\} =: \mathcal{V}(Y)$ and $M(\mathbb{X})$ is the maximal ideal of $V(\mathbb{X})$.

Theorem 6.5 (cf. [18, Theorem 5.11]) *Let D be an integral domain. Then, the bijection $\boldsymbol{\Theta}$, restricted to $\texttt{SStar}_f(D)$, induces a homeomorphism between $\widetilde{\texttt{SStar}}(\mathrm{Kr}(D))$ and $\texttt{SStar}_{f,eab}(D)$. In particular, $\texttt{SStar}_{f,eab}(D)$ is a spectral space.*

Another interpretation of the previous theorem can be given by considering the spectral space $\mathcal{X}(X)$, when X coincides with $\texttt{Zar}(D)$. This point of view sheds new light on the analogies between the spectral spaces $\widetilde{\texttt{SStar}}(D)$ ($=\texttt{SStar}_{f,sp}(D)$, by Remark 3.9(a)) and $\texttt{SStar}_{f,eab}(D)$, after recalling that $\mathcal{X}(D) := \mathcal{X}(\texttt{Spec}(D))$ is canonically homeomorphic to $\widetilde{\texttt{SStar}}(D)$ (Proposition 4.6).

Corollary 6.6 *Let D be an integral domain. The map*

$$\Lambda : \mathcal{X}(\texttt{Zar}(D)) \rightarrow \texttt{SStar}_{f,eab}(D),\ \textit{defined by } \Lambda(\mathcal{Y}) := \wedge_{\mathcal{Y}},$$

for each inverse-closed subset $\mathcal{Y}$ of $\texttt{Zar}(D)$, is a homeomorphism.

Proof (Sketch) The proof is based on the following key facts. The space $\mathcal{X}(\mathtt{Zar}(D))$ is canonically homeomorphic to $\mathcal{X}(\mathrm{Kr}(D))$ [19]. By Proposition 4.6, $\mathcal{X}(\mathrm{Kr}(D)) \simeq \widetilde{\mathtt{SStar}}(\mathrm{Kr}(D))$ $(= \mathtt{SStar}_{f,sp}(\mathrm{Kr}(D)))$ and finally that the map Θ_f, restriction of Θ to $\mathtt{SStar}_{f,sp}(\mathrm{Kr}(D))$, from $\mathtt{SStar}_{f,sp}(\mathrm{Kr}(D))$ onto $\mathtt{SStar}_{f,eab}(D)$, is a homeomorphism (for more details [18, Theorem 5.11(2)]). □

The following Hasse diagram summarizes the topological embeddings of some of the spaces considered in the present paper. All spaces are spectral except possibly the three spaces denoted by $^{(\blacklozenge)}$.

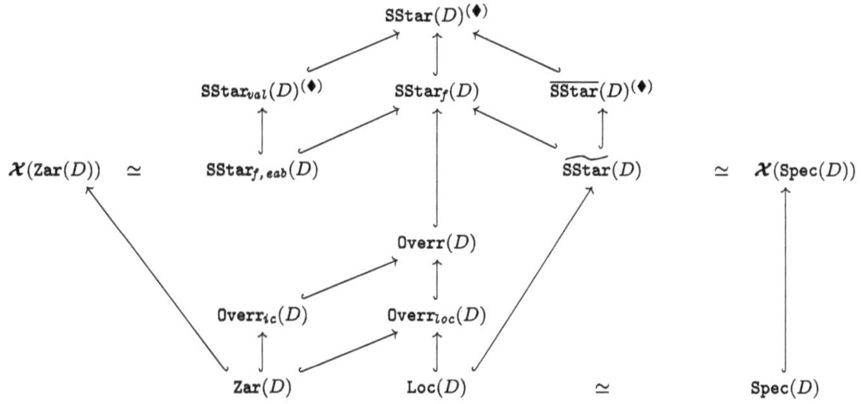

References

1. D.D. Anderson, Star-operations induced by overrings. Commun. Algebra **16**, 2535–2553 (1988)
2. D.D. Anderson, S.J. Cook, Two star operations and their induced lattices. Commun. Algebra **28**, 2461–2475 (2000)
3. D.F. Anderson, D.D. Anderson, Examples of star operations on integral domains. Commun. Algebra **18**, 1621–1643 (1990)
4. M.F. Atiyah, I.G. Macdonald, *Introduction to Commutative Algebra* (Addison-Wesley, Reading, 1969)
5. V. Barucci, D. Dobbs, M. Fontana, Conducive integral domains as pullbacks. Manuscr. Math. **54**, 261–277 (1986)
6. N. Bourbaki, *Algèbre Commutative, Chap. 1–2* (Hermann, Paris, 1961)
7. P.-J. Cahen, Commutative torsion theory. Trans. Am. Math. Soc. **184**, 73–85 (1973)
8. P.-J. Cahen, K.A. Loper, F. Tartarone, Integer-valued polynomials and Prüfer v-multiplication domains. J. Algebra **226**, 765–787 (2000)
9. C. Chevalley, Sur la théorie des variétés algébriques. Nagoya Math. J. **8**, 1–43 (1955)
10. C. Chevalley, H. Cartan, Schémas normaux; morphismes; ensembles constructibles. Séminaire Henri Cartan **8**, Exp. No. 7, 1–10 (1955–1956)
11. D. Dobbs, M. Fontana, Kronecker function rings and abstract Riemann surfaces. J. Algebra **99**, 263–274 (1986)
12. D. Dobbs, R. Fedder, M. Fontana, Abstract Riemann surfaces of integral domains and spectral spaces. Ann. Mat. Pura Appl. **148**, 101–115 (1987)

13. N. Epstein, A guide to closure operations in commutative algebra, *Progress in Commutative Algebra*, vol. 2 (Walter de Gruyter, Berlin, 2012), pp. 1–37
14. N. Epstein, Semistar operations and standard closure operations. Commun. Algebra **43**, 325–336 (2015)
15. C.A. Finocchiaro, Spectral spaces and ultrafilters. Commun. Algebra **42**, 1496–1508 (2014)
16. C.A. Finocchiaro, M. Fontana, K.A. Loper, Ultrafilter and constructible topologies on spaces of valuation domains. Commun. Algebra **41**, 1825–1835 (2013)
17. C.A. Finocchiaro, M. Fontana, K.A. Loper, The constructible topology on spaces of valuation domains. Trans. Am. Math. Soc. **365**, 6199–6216 (2013)
18. C.A. Finocchiaro, M. Fontana, D. Spirito, Spectral Spaces of Semistar Operations . J. Pure Appl. Algebra. **220**, 2897–2913 (2016)
19. C.A. Finocchiaro, M. Fontana, D. Spirito, The Space of Inverse-Closed Subsets of a Spectral Space. (2016) (submitted)
20. C.A. Finocchiaro, M. Fontana, D. Spirito, On a Topological Version of Hilbert's Nullstellensatz. J. Algebra (to appear)
21. C.A. Finocchiaro, D. Spirito, Some topological considerations on semistar operations. J. Algebra **409**, 199–218 (2014)
22. M. Fontana, J. Huckaba, Localizing systems and semistar operations, in *Non-Noetherian Commutative Ring Theory*, ed. by Scott T. Chapman, Sarah Glaz (Kluwer Academic Publishers, Dordrecht, 2000), pp. 169–198
23. M. Fontana, K.A. Loper, Kronecker function rings: a general approach, in *Ideal theoretic methods in commutative algebra* (Columbia, MO, 1999), pp. 189–205. (Lecture Notes in Pure and Applied Mathematics, vol. 220 (Dekker, New York, 2001))
24. M. Fontana, K.A. Loper, Nagata rings, Kronecker function rings and related semistar operations. Commun. Algebra **31**, 4775–4805 (2003)
25. M. Fontana, K.A. Loper, An historical overview of Kronecker function rings, Nagata rings, and related star and semistar operations, in *Multiplicative Ideal Theory in Commutative Algebra: A Tribute to the Work of Robert Gilmer*, ed. by J.W. Brewer, S. Glaz, W. Heinzer, B. Olberding (Springer, New York, 2006), pp. 169–187
26. M. Fontana, K.A. Loper, The patch topology and the ultrafilter topology on the prime spectrum of a commutative ring. Commun. Algebra **36**, 2917–2922 (2008)
27. M. Fontana, K.A. Loper, Cancellation properties in ideal systems: a classification of e.a.b. semistar operations. J. Pure Appl. Algebra **213**, 2095–2103 (2009)
28. M. Fontana, J. Huckaba, I. Papick, *Prüfer domains* (M. Dekker, New York, 1997)
29. M. Fontana, K.A. Loper, R. Matsuda, Cancellation properties in ideal systems: an e.a.b. not a.b. star operation. AJSE (Arabian Journal for Science and Engineering)–Mathematics **35**, 45–49 (2010)
30. P. Gabriel, La localisation dans les anneaux non commutatifs, in *Séminaire Dubreil (sous la direction de P. Dubreil, M.-L. Dubreil-Jacotin, C. Pisot). Algèbre et théorie des nombres*, 13 no. 1, Exposé No. 2 (1959–1960), 35 p
31. R. Gilmer, *Multiplicative Ideal Theory*. Queen's Papers in Pure and Applied Mathematics, vol. I & II (Kingston, Ontario, Canada, 1968)
32. R. Gilmer, *Multiplicative Ideal Theory* (M. Dekker, New York, 1972)
33. A. Grothendieck, J. Dieudonné, *Éléments de Géométrie Algébrique I, IHES 1960* (Springer, Berlin, 1970)
34. W. Heinzer, M. Roitman, Well-centered overrings of an integral domain. J. Algebra **272**(2), 435–455 (2004)
35. M. Hochster, Prime ideal structure in commutative rings. Trans. Am. Math. Soc. **142**, 43–60 (1969)
36. M. Hochster, C. Huneke, Tight closure, invariant theory, and the Briançon-Skoda theorem. J. Am. Math. Soc. **3**(1), 31–116 (1990)
37. O. Heubo-Kwegna, Kronecker function rings of transcendental field extensions. Commun. Algebra **38**, 2701–2719 (2010)
38. J. Huckaba, *Commutative Rings with Zero Divisors* (M. Dekker, New York, 1988)

39. C. Huneke, I. Swanson, *Integral Closure of Ideals, Rings, and Modules*, vol. 336, London Mathematical Society Lecture Note Series (Cambridge University Press, Cambridge, 2006)
40. W. Krull, *Idealtheorie* (Springer, Berlin, 1935). (2nd edn. 1968)
41. W. Krull, Beiträge zur Arithmetik kommutativer Integritätsbereiche, I - II. Math. Z. **41**, 545–577, 665–679 (1936)
42. W. Krull, *Gesammelte Abhandlungen/Collected Papers, Hrsg. v. Paulo Ribenboim* (Walter de Gruyter, Berlin, 1999)
43. T. Jech, *Set Theory* (Springer, New York, 1997). (1st edn. Academic Press, 1978)
44. J. Lambek, *Torsion Theories, Additive Semantics, and Rings of Quotients*. Lecture Notes in Mathematics, vol. 177 (Springer, Berlin, 1971)
45. P. Maroscia, Sur les annéaux de dimension zéro. Rend. Acc. Naz. Lincei **56**, 451–459 (1974)
46. A. Okabe, R. Matsuda, Semistar operations on integral domains. Math. J. Toyama Univ. **17**, 1–21 (1994)
47. J.-P. Olivier, Anneaux absolument plats universels et épimorphismes à buts réduits, *Sém P. Samuel, Algèbre Commutative, Année*, Ex. N. 6 (1967/68)
48. J.-P. Olivier, Anneaux absolument plats universels et épimorphismes d'anneaux. C.R. Acad. Sci. Paris **266**, 317–318 (1968)
49. N. Schwartz, M. Tressl, Elementary properties of minimal and maximal points in Zariski spectra. J. Algebra **323**, 698–728 (2010)
50. B. Stenström, *Rings and Modules of Quotients*. Lecture Notes in Math, vol. 237 (Springer, Berlin, 1971)
51. J.C. Vassilev, Structure on the set of closure operations of a commutative ring. J. Algebra **321**, 2737–2753 (2009)
52. O. Zariski, The compactness of the Riemann manifold of an abstract field of algebraic functions. Bull. Am. Math. Soc **50**, 683–691 (1944)
53. O. Zariski, P. Samuel, *Commutative Algebra*, vol. II (Van Nostrand, Princeton, 1960)

Relative Polynomial Closure and Monadically Krull Monoids of Integer-Valued Polynomials

Sophie Frisch

Dedicated to Franz Halter-Koch on the occasion of his 70th birthday.

Abstract Let D be a Krull domain and $\mathrm{Int}(D)$ the ring of integer-valued polynomials on D. For any $f \in \mathrm{Int}(D)$, we explicitly construct a divisor homomorphism from $[\![f]\!]$, the divisor-closed submonoid of $\mathrm{Int}(D)$ generated by f, to a finite sum of copies of $(\mathbb{N}_0, +)$. This implies that $[\![f]\!]$ is a Krull monoid. For V a discrete valuation domain, we give explicit divisor theories of various submonoids of $\mathrm{Int}(V)$. In the process, we modify the concept of polynomial closure in such a way that every subset of D has a finite polynomially dense subset. The results generalize to $\mathrm{Int}(S, V)$, the ring of integer-valued polynomials on a subset, provided S does not have isolated points in v-adic topology.

Keywords Monoid · Factorization · Krull monoids · Divisor homomorphism · Divisor theory · Integer-valued polynomial · Polynomial closure

Mathematics Subject Classification 2010 Primary 13F20 · Secondary 20M13 · 13A05 · 13B25 · 11C08 · 11R09

1 Introduction

The ring of integer-valued polynomials $\mathrm{Int}(\mathbb{Z})$ enjoys quite chaotic nonunique factorization: given any finite list of natural numbers $1 < n_1 \le n_2 \le \cdots \le n_k$, one can

S. Frisch (✉)
Department of Mathematics, Graz University of Technology, Steyrergasse 30, 8010 Graz, Austria
e-mail: frisch@blah.math.tugraz.at

S. Chapman et al. (eds.), *Multiplicative Ideal Theory and Factorization Theory*, Springer Proceedings in Mathematics & Statistics 170,
DOI 10.1007/978-3-319-38855-7_6

find a polynomial $f \in \mathrm{Int}(\mathbb{Z})$ that has exactly k essentially different factorizations into irreducible elements of $\mathrm{Int}(\mathbb{Z})$, namely, one with n_1 irreducible factors, one with n_2, etc. [4]. In contrast to this, A. Reinhart [9] has shown for any unique factorization domain D that $\mathrm{Int}(D)$ is monadically Krull, i.e., that the divisor-closed submonoid $[\![f]\!]$ generated by any single polynomial $f \in \mathrm{Int}(D)$ (the monoid consisting of all divisors in $\mathrm{Int}(D)$ of powers of f) is a Krull monoid. So, we have here an interesting case of Krull monoids with rather wild factorization properties.

In this paper, we find divisor homomorphisms and, in some cases, divisor theories for the divisor- closed submonoids generated by single polynomials $f \in \mathrm{Int}(S, D)$, the ring of integer-valued polynomials on a subset of a Krull domain. If S does not have any isolated points in any of the topologies given by essential valuations of D, we can construct a divisor homomorphism from $[\![f]\!]$ to a finite direct sum of copies of $(\mathbb{N}_0, +)$ [Theorem 5.4]. This implies that $[\![f]\!]$ is a Krull monoid, and hence, that $\mathrm{Int}(S, D)$ is monadically Krull.

In the special case of D being a discrete valuation domain, we can determine explicitly the divisor theories of certain submonoids of $\mathrm{Int}(S, D)$ [Theorems 4.2 and 5.3].

As a tool for constructing divisor homomorphisms on monoids of integer-valued polynomials, we introduce "relative" polynomial closure, that is, polynomial closure with respect to a subset of $K[x]$, in Sect. 2. This modification of the concept of polynomial closure makes it possible to find finite polynomially dense subsets of arbitrary sets in Sect. 3. Equipped with these finite polynomially dense sets, we construct the actual divisor homomorphisms and, in some cases, divisor theories, to finite sums of copies of $(\mathbb{N}_0, +)$ in Sects. 4 and 5.

The remainder of this introduction contains a short review of concepts and notation related to integer-valued polynomials.

Definition 1.1 Let D be a domain with quotient field K and $f \in K[x]$. f is called integer-valued if $f(D) \subseteq D$. For a subset $S \subseteq K$, $f \in K[x]$ is called integer-valued on S if $f(S) \subseteq D$. When there are several possibilities for D, we say D-valued on S instead of integer-valued on S.

The ring of integer-valued polynomials on D is written $\mathrm{Int}(D)$, and the ring of integer-valued polynomials on a subset S of the quotient field of D is denoted by $\mathrm{Int}(S, D)$:

$$\mathrm{Int}(S, D) = \{f \in K[x] \mid f(S) \subseteq D\}, \quad \mathrm{Int}(D) = \mathrm{Int}(D, D).$$

Definition 1.2 Let D be a domain with quotient field K, $S \subseteq D$ and $f \in \mathrm{Int}(S, D)$. The divisor-closed submonoid of $\mathrm{Int}(S, D)$ generated by f, which we write $[\![f]\!]$, is the multiplicative monoid consisting of all $g \in \mathrm{Int}(S, D)$ for which there exists $m \in \mathbb{N}$ and $h \in \mathrm{Int}(S, D)$, such that $g \cdot h = f^m$.

Keep in mind that an element of $[\![f]\!]$ is not just a polynomial $g \in \mathrm{Int}(S, D)$ that divides some power of f in $K[x]$. The cofactor $h = f^m/g$ is also required to be in $\mathrm{Int}(S, D)$. Take for example $\binom{x}{2}$ in $\mathrm{Int}(\mathbb{Z})$. Here x divides f in $K[x]$, but $x \notin [\![f]\!]$.

We will frequently use the following divisibility criterion for $[\![f]\!]$.

Remark 1.3 Let $[\![f]\!]$ be the divisor-closed submonoid of $\mathrm{Int}(S, D)$ as in Definition 1.2 and $g, h \in [\![f]\!]$. Then g divides h in $[\![f]\!]$ if and only if g divides h in $K[x]$ and the cofactor h/g is in $\mathrm{Int}(S, D)$.

Multiplying a polynomial in $[\![f]\!]$ by a constant in D does not in general result in an element of $[\![f]\!]$. We can multiply elements of $[\![f]\!]$ by some suitable constants, though, see Lemma 1.4.

Regarding valuation terminology: we use additive valuations, that is, a valuation is a map $v\colon K\backslash\{0\} \to \Gamma$, where $(\Gamma, +)$ is a totally ordered group, satisfying

1. $v(ab) = v(a) + v(b)$
2. $v(a + b) \geq \min(v(a), v(b))$

and we set $v(0) = \infty$. The valuation group of v is the image of v in Γ. The valuation domain of a valuation v on a field K is $V = \{k \in K \mid v(k) \geq 0\}$.

Lemma 1.4 *Let V be the valuation domain of a valuation v on K, $S \subseteq V$, $f \in \mathrm{Int}(S, V)$ and $[\![f]\!]$ the divisor-closed submonoid of* $\mathrm{Int}(S, V)$ *generated by f. Let $g \in [\![f]\!]$ and $a \in K$. If $-\min_{s\in S} v(g(s)) \leq v(a) \leq 0$ then $ag \in [\![f]\!]$.*

Proof Let $g, h \in \mathrm{Int}(S, V)$ and $m \in \mathbb{N}$ such that $gh = f^m$. Then both ag and $a^{-1}h$ are in $\mathrm{Int}(S, V)$, and $ag \cdot a^{-1}h = f^m$.

We recall the definitions of ideal content and fixed divisor, whose interplay will be an important ingredient of proofs. Let R be a domain and $f \in R[x]$. The content of f, denoted $c(f)$, is the fractional ideal generated by the coefficients of f. If R is a principal ideal domain, we identify, by abuse of notation, ideals by their generators and say that $c(f)$ is the gcd of the coefficients of f. A polynomial $f \in R[x]$ is called primitive if $c(f) = R$, that is, in the case of a PID, if $c(f) = 1$.

Definition 1.5 Let D be a domain with quotient field K, $S \subseteq D$ and $f \in K[x]\backslash\{0\}$. The *fixed divisor of f on S*, denoted $\mathrm{d_S}(f)$, is the D-submodule of K generated by the image $f(S)$. Note that $\mathrm{d_S}(f)$ is a fractional ideal. If $S = D$, we write $\mathrm{d}(f)$ for $\mathrm{d_D}(f)$. If D is a PID, we will, by abuse of notation, sometimes write a generator to stand for the ideal, e.g., $\mathrm{d_S}(f) = 1$ for $\mathrm{d_S}(f) = D$. A polynomial $f \in \mathrm{Int}(S, D)$ is called *image-primitive* if $\mathrm{d_S}(f) = D$.

For polynomials in $D[x]$, image-primitive implies primitive, but not vice versa. One difference between ideal content and fixed divisor is that the ideal content is multiplicative for sufficiently nice rings—called Gaussian rings—including principal ideal rings, whereas the fixed divisor is not multiplicative. $\mathrm{d_S}(f)\,\mathrm{d_S}(g)$ contains $\mathrm{d_S}(fg)$, but the containment is often strict.

Remark 1.6 Two easy but useful facts:

1. If $f \in \mathrm{Int}(S, D)$ is image-primitive then f^n is image-primitive for all $n \in \mathbb{N}$.
2. If $f \in \mathrm{Int}(S, D)$ is image-primitive then all divisors in $\mathrm{Int}(S, D)$ of f are also image-primitive.

Remark 1.7 In case D is an intersection of valuation rings, then every $f \in \text{Int}(S, D)$ is also in $\text{Int}(S, V)$ for all these valuation rings, and f may be image-primitive as an element of $\text{Int}(S, V)$, but not as an element of $\text{Int}(S, D)$. In this case, we write

$$v(f(S)) := \min_{s \in S} v(f(s))$$

and write $v(f(S)) = 0$ to express that f is image-primitive when regarded as an element of $\text{Int}(S, V)$.

2 Relative Polynomial Closure

Definition 2.1 (*relative polynomial closure*) Fix a domain D with quotient field K. Let $T \subseteq K$ and $\mathcal{F} \subseteq K[x]$.

The polynomial closure of T relative to $\mathcal{F}$ is

$$C_{\mathcal{F}}(T) = \{s \in K \mid \forall f \in \mathcal{F} \cap \text{Int}(T, D) : \quad f(s) \in D\}.$$

If $T \subseteq S \subseteq K$, and $C_{\mathcal{F}}(T) \supseteq S$ we call T polynomially dense in S relative to $\mathcal{F}$.

The definition of polynomial closure and polynomial density depends on the choice of D. If there is any doubt about D, we say D-polynomial closure and D-polynomially dense.

Polynomial closure relative to $K[x]$ is the "usual" polynomial closure, introduced by Gilmer [6] and studied by McQuillan [7], the present author [3], Cahen [1], Park and Tartarone [8] and Chabert [2], among others. The reason why we generalize the well-known concept of polynomial closure will become apparent in the next section: when we consider polynomial closure relative to a set of polynomials whose irreducible factors are restricted to a finite set, it becomes possible to find finite polynomially dense subsets of any fractional set.

Remark 2.2 The following properties of polynomial closure relative to a subset $\mathcal{F}$ of $K[x]$ are easy to check.

1. $C_{\mathcal{F}}(T) = \bigcap_{f \in \mathcal{F} \cap \text{Int}(T,D)} f^{-1}(D)$
2. Polynomial closure relative to $\mathcal{F}$ is a closure operator, in the sense that
   a. $T \subseteq C_{\mathcal{F}}(T)$
   b. $C_{\mathcal{F}}(C_{\mathcal{F}}(T)) = C_{\mathcal{F}}(T)$
   c. $T \subseteq S \Longrightarrow C_{\mathcal{F}}(T) \subseteq C_{\mathcal{F}}(S)$
3. Polynomial closure relative to $\mathcal{F}$ is the closure given by a Galois correspondence that maps every subset T of K to a subset of $\mathcal{F}$, and every subset G of $\mathcal{F}$ to a subset of K, namely,

$$T \mapsto \mathscr{F} \cap \mathrm{Int}(T, D) \quad \text{and} \quad G \mapsto \bigcap_{f \in G} f^{-1}(D).$$

4. If $\mathscr{F}_0 \subseteq \mathscr{F}_1 \subseteq K[x]$ then $C_{\mathscr{F}_1}(T) \subseteq C_{\mathscr{F}_0}(T)$.
5. If T is polynomially dense in S relative to $\mathscr{F}_1$, and $\mathscr{F}_0 \subseteq \mathscr{F}_1$, then T is polynomially dense in S relative to $\mathscr{F}_0$.

When D is a valuation domain, then polynomially dense subsets of S relative to $\mathscr{F}$ are easily characterized (subject to a weak condition on $\mathscr{F}$): they are the subsets T such that, for each $f \in \mathscr{F}$, $\min_{s \in S} v(f(s))$ is attained by some $s \in T$.

Lemma 2.3 *Let v be a valuation on a field K, V its valuation ring, $T \subseteq S \subseteq K$ and $\mathscr{F} \subseteq K[x]$. Consider*

1. $\forall f \in \mathscr{F}\ \min_{s \in S} v(f(s)) = \min_{t \in T} v(f(t))$
2. T is V-polynomially dense in S relative to $\mathscr{F}$.

(1) *implies* (2). *If $\mathscr{F}$ is closed under multiplication by nonzero constants in K then* (2) *implies* (1).

Proof (1 $\Rightarrow$ 2) For every polynomial $f \in \mathscr{F} \cap \mathrm{Int}(T, V)$, $\min_{t \in T} v(f(t)) \geq 0$. Therefore, by (1), $\min_{s \in S} v(f(s)) \geq 0$ and hence $f \in \mathrm{Int}(S, V)$.

(2 $\Rightarrow$ 1) For every $f \in \mathscr{F}$, $\min_{t \in T} v(f(t)) \geq \min_{s \in S} v(f(s))$, since $T \subseteq S$. If $f \in \mathscr{F}$ and $\alpha \in \mathbb{Z}$ are such that $\min_{t \in T} v(f(t)) \geq \alpha > \min_{s \in S} v(f(s))$, pick $a \in K$ with $v(a) = -\alpha$. Then $af \in \mathscr{F} \cap \mathrm{Int}(T, V)$, but $af \notin \mathrm{Int}(S, V)$, so T is not V-polynomially dense in S relative to $\mathscr{F}$.

3 Finite Polynomially Dense Subsets

Let F be a finite set of irreducible polynomials in $K[x]$ and $\mathscr{F}$ the multiplicative submonoid of $K[x]$ generated by F and the nonzero constants of K. That is, $\mathscr{F}$ consists of all nonzero polynomials in $K[x]$ whose irreducible factors in $K[x]$ are (up to multiplication by nonzero constants) in F.

We will now construct, for every subset S of a discrete valuation ring V, a finite polynomially dense subset of S relative to $\mathscr{F}$. It is possible to admit fractional subsets of K, but for simplicity's sake we restrict ourselves to subsets of V.

By discrete valuation, we mean, more precisely, a discrete rank 1 valuation, that is, a valuation v whose value group is isomorphic to $\mathbb{Z}$. A normalized discrete valuation is one whose value group is actually equal to $\mathbb{Z}$. The valuation ring of a discrete valuation is called discrete valuation ring, abbreviated DVR. As we all know, a DVR is a local principal ideal domain.

Remark 3.1 Let v be a discrete valuation on K with valuation ring V, $f \in K[x]$, and $L \supseteq K$ a finite-dimensional field extension over which f splits. Let w be an extension of v to L ($w \mid_K = v$), W the valuation ring of w and P its maximal ideal. Say f splits as $f(x) = c \prod_{j=1}^{k}(x - b_j) \prod_{j=1}^{m}(x - a_j)$ with $w(b_j) < 0$ for $1 \leq j \leq k$ and $w(a_j) \geq 0$ for $1 \leq j \leq m$ over L.

Then for all $s \in V$,

$$v(f(s)) = w(c) + \sum_{j=1}^{k} w(b_j) + \sum_{j=1}^{m} w(s - a_j)$$

Proof This follows from the fact that $w(s \pm b) = w(b)$ whenever $w(b) < w(s)$.

Definition 3.2 Let X be a topological space and $S \subseteq X$. An isolated point of S is an element $s \in X$ having a neighborhood U such that $U \cap S = \{s\}$.

Proposition 3.3 *Let v be a discrete valuation on K and V its valuation ring. Let $F \neq \emptyset$ be a finite set of monic irreducible polynomials in $K[x]$ and $\mathscr{F}$ the set of those polynomials in $K[x]$ whose monic irreducible factors are all in F. Let $S \subseteq V$.*

1. *Then there exists a finite subset $T \subseteq S$ such that*

$$\forall f \in \mathscr{F} \ \min_{t \in T} v(f(t)) = \min_{s \in S}(v(f(s)))$$

and every such $T \subseteq S$ is, in particular, a finite set that is polynomially dense in S relative to $\mathscr{F}$.

2. *If no root of any $f \in F$ is an isolated point of S in v-adic topology, then the above set T can be chosen such as not to contain any root of any $f \in F$.*
3. *Let L be the splitting field of F over K, w an extension of v to L and W the valuation ring of w. Let A be the set of distinct roots of polynomials of F in W. Then T in (1) and (2) can be chosen with $|T| \leq \max(1, |A|)$.*

Proof Let L, w, W, and A as in (3). Let P be the maximal ideal of W. We call the elements of A "the roots". We may assume $S \neq \emptyset$ and $A \neq \emptyset$ (otherwise the claimed facts are trivial). In view of Remark 3.1, to show (1) it suffices to construct a finite set $T \subseteq S$ such that, for every finite sequence $(a_i)_{i=1}^{m}$ in A,

$$\min_{t \in T} \sum_{i=1}^{m} w(t - a_i) = \min_{s \in S} \sum_{i=1}^{m} w(s - a_i)$$

We will do this by constructing a finite covering $\mathscr{C}$ of S by disjoint sets $C \subseteq W$ and for each $C \in \mathscr{C}$ choosing a representative $t \in C \cap S$ such that $w(t - a) \leq w(s - a)$ for every $a \in A$ and every $s \in C \cap S$. This representative $t \in C \cap S$ then satisfies $\forall f \in \mathscr{F} \ v(f(t)) = \min_{s \in C \cap S} v(f(s))$, by Remark 3.1. If we take T to be the set of representatives of covering sets $C \in \mathscr{C}$ then for every $f \in \mathscr{F}$, $\min_{s \in S} v(f(s))$ is realized by some $s \in T$. By Lemma 2.3, this makes T polynomially dense in S relative to $\mathscr{F}$.

For any ideal I of W, we call a residue class $r + I$ "relevant" if $S \cap (r + I) \neq \emptyset$.

We construct $\mathscr{C}$, $\mathscr{C}_n$ $(n \geq 0)$ and T inductively. Before step 0, initialize $T = \emptyset$, $\mathscr{C} = \emptyset$, $\mathscr{C}_0 = \{W\}$.

At the beginning of step n, $\mathscr{C}$ is a finite set of relevant residue classes of various P^k with $k < n$ while $\mathscr{C}_n$ is a finite set of relevant residue classes of P^n each containing at least one root. In step n, initialize $\mathscr{C}_{n+1} = \emptyset$; then go through each $C \in \mathscr{C}_n$ and process it as follows:

1. If $C \cap S = \{c\}$ with $c \in A$ then put c in T and C in $\mathscr{C}$. Note that in this case $C \cap V$ is a v-adic neighborhood of c whose intersection with S is $\{c\}$, and that therefore $c \in A$ is an isolated point of S.
2. Else, if C contains a relevant residue class D of P^{n+1} which does not contain a root, pick such a D, add a representative of $D \cap S$ to T; then put C in $\mathscr{C}$.
3. Else place all relevant residue classes of P^{n+1} contained in C (each containing a root, by construction) in $\mathscr{C}_{n+1}$.

If $\mathscr{C}_{n+1}$ is empty at the end of step n, stop. Otherwise proceed to step $n + 1$.

Note that after each step n, $\mathscr{C} \cup \mathscr{C}_{n+1}$ is a covering of S. When the algorithm terminates with $\mathscr{C}_{n+1} = \emptyset$, then $\mathscr{C}$ is a covering of S and T contains for each $C \in \mathscr{C}$ a representative $t \in C \cap S$ satisfying $w(t - a) = \min_{s \in C \cap S} w(s - a)$ for all $a \in A$. Therefore $v(f(t)) = \min_{s \in C \cap S} v(f(s))$ for all $f \in \mathscr{F}$ by Remark 3.1.

The algorithm terminates when no root is left in $\bigcup \mathscr{C}_{n+1}$. For each root $a \in A$, one can give an upper bound on n such that a is no longer in $\mathscr{C}_{n+1}$. Namely, let n such that $w(a - a') < n$ for all roots $a \neq a'$. If $(a + P^{n+1}) \cap S = \emptyset$ then a residue class containing a has been dropped as not relevant at or before step n, so $a + P^{n+1} \notin \mathscr{C}_{n+1}$. If $(a + P^{n+1}) \cap S = \{a\}$, then a residue class containing a is placed in $\mathscr{C}$ at step $n + 1$ or earlier. Otherwise, $a + P^{n+1}$ contains an element of S other than a. Let $s \in (a + P^{n+1}) \cap S$, with $w(s - a) = m$ minimal. Then $a + P^m$ will be placed in $\mathscr{C}$ by step m.

This shows (1). For (2), note that the set T thus constructed contains no root of any $f \in F$ except such as are isolated points of S in v-adic topology. For (3), note that every time an element is added to T, a set containing at least one root is transferred from $\mathscr{C}_n$ to $\mathscr{C}$ and the number of roots in $\bigcup_{C \in \mathscr{C}_n} C$ decreases.

Remark 3.4 Thanks to the anonymous referee for pointing out that parts (1) and (2) of Proposition 3.3 can be shown more quickly by applying Dickson's theorem [5, Theorem 1.5.3], which says that the set of minimal elements of any subset N of $\mathbb{N}_0^m$ is finite and that for every $a \in N$ there exists a minimal element $b \in N$ with $b \leq a$, to the subset $N = \{(w(s - a))_{a \in A} \mid s \in S\}$ of $\mathbb{N}_0^A$.

4 Divisor Theories for Monoids of Integer-Valued Polynomials on Discrete Valuation Rings

We are going to construct divisor homomorphisms from submonoids of $\mathrm{Int}(S, D)$, where D is a Krull domain, to finite sums of copies of $(\mathbb{N}_0, +)$. The idea is to gain insight into divisibility in $\mathrm{Int}(S, D)$ by relating it to divisibility in a finitely generated free commutative monoid. In this section, we assume V to be a discrete valuation domain and determine the divisor theory of the submonoid consisting of all elements of $\mathrm{Int}(S, V)$ whose irreducible factors in $K[x]$ come from a fixed finite set.

By *monoid* we mean a semigroup that has a neutral element. All monoids that we examine here are cancellative, that is, whenever $ab = cb$ or $ba = bc$, it follows that $a = c$. Also, all our monoids will be commutative.

A short review of divisibility terminology, in the perhaps less familiar additive form: Let $(M, +)$ be a commutative monoid, written additively, and $a, b \in M$.

1. We say that *a divides b* in M and write $a \mid b$, whenever there exists $c \in M$ such that $a + c = b$.
2. We call an element $d \in M$ a *greatest common divisor*, abbreviated gcd, of a subset $A \subseteq M$, if
   a. $d \mid a$ for all $a \in A$
   b. for all $c \in M$: if $c \mid a$ for all $a \in A$ then $c \mid d$.

If $(M, +)$ is a direct sum of k copies of $(\mathbb{N}_0, +)$, then the divisibility relation in M is just the partial order given by the order relations on each component: Let $a, b \in M = \sum_{i=1}^{k}(\mathbb{N}_0, +)$ with $a = (a_1, \ldots, a_k)$ and $b = (b_1, \ldots, b_k)$. Then $a \mid b$ in M is equivalent to $a_i \leq b_i$ for all $1 \leq i \leq k$. Therefore, any set $\{(m_{i1}, m_{i2}, \ldots, m_{ik}) \mid i \in I\}$ of elements of M has a unique gcd, namely, $d = (\min_i(m_{i1}), \min_i(m_{i2}), \ldots, \min_i(m_{ik}))$.

Definition 4.1 A monoid homomorphism $\varphi\colon G \to H$ is called a divisor homomorphism if $\varphi(a) \mid \varphi(b)$ in H implies $a \mid b$ in G. (Note that the reverse implication holds for every monoid homomorphism.)

A divisor homomorphism $\varphi\colon G \to \sum_{i=1}^{n}(\mathbb{N}_0, +)$ is called a divisor theory if each of the unit vectors e_i (having 1 in the ith coordinate and zeros elsewhere) occurs as gcd of a finite set of images of elements of G.

In what follows, we denote the normalized discrete valuation on $K(x)$ corresponding to an irreducible polynomial $h \in K[x]$ by v_h; that is, for $g \in K[x]$, $v_h(g)$ is the exponent to which h occurs in the essentially unique factorization of g in $K[x]$ into irreducible polynomials, and for $g_1/g_2 \in K(x)$, $v_h(g_1/g_2) = v_h(g_1) - v_h(g_2)$.

In this section we examine the special case $\mathrm{Int}(S, V)$, where V is a discrete valuation ring (DVR).

Theorem 4.2 *Let v be a normalized discrete valuation on K and V its valuation ring. Let H be a finite set of pairwise nonassociated irreducible polynomials in $K[x]$ and $\mathscr{H}$ the multiplicative submonoid of $K[x]$ generated by H and the nonzero constants in K. Let $S \subseteq V$ such that no root of any $h \in H$ is an isolated point of S in v-adic topology. Let $\mathscr{F} = \mathscr{H} \cap \mathrm{Int}(S, V)$.*

There exists a finite subset T of S that is polynomially dense in S relative to $\mathscr{H}$ and contains no root of any $h \in H$; and for every such T

$$\varphi\colon \mathscr{F} \to \sum_{h\in H}(\mathbb{N}_0,+) \oplus \sum_{t\in T}(\mathbb{N}_0,+), \quad \varphi(g) = ((v_h(g) \mid h \in H), (v(g(t)) \mid t \in T)),$$

is a divisor homomorphism. If T is chosen minimal, φ is a divisor theory.

Proof The existence of a finite polynomially dense subset T containing no root of any $h \in H$ is Proposition 3.3. Once we have a finite dense set, a minimal dense set can be obtained by removing redundant elements.

φ is well defined, because T contains no root of any $h \in H$. Once φ is a well-defined function, it clearly is a monoid homomorphism. Now suppose $a, b \in \mathscr{F}$ such that $\varphi(a) \mid \varphi(b)$, and set $c = b/a$. We must show $c \in \mathrm{Int}(S, V)$.

$\varphi(a) \mid \varphi(b)$ means $v_h(a) \le v_h(b)$ for all $h \in H$ and $v(a(t)) \le v(b(t))$ for all $t \in T$. The first shows $c \in K[x]$, and therefore $c \in \mathscr{H}$, and the second shows that $c(t) \in V$ for all $t \in T$. Since T is polynomially dense in S relative to $\mathscr{H}$, it follows that $c \in \mathrm{Int}(S, V)$. We have shown φ to be a divisor homomorphism.

It remains to show that every e_h for any $h \in H$ and every e_t for any $t \in T$ occurs as the gcd of a finite set of images of elements of $\mathscr{F}$, provided T is minimal.

We may assume, without changing $\mathscr{H}$, $\mathscr{F}$ or φ in any way, that the elements of H are in $V[x]$ and primitive.

First, let p be a generator of the maximal ideal of V. The constant polynomial p is an element of $\mathscr{F}$ satisfying $v_h(p) = 0$ for all $h \in H$ and $v(p(t)) = 1$ for all $t \in T$.

Second, we note that every polynomial $h \in H$ is an element of $\mathscr{F}$ satisfying $v_h(h) = 1$ and $v_l(h) = 0$ for every $l \in H\backslash\{h\}$.

Third, we show that for every $t \in T$, there exists $g_t \in \mathscr{F}$ such that $v(g_t(t)) = 0$ and $v(g_t(r)) > 0$ for all $r \in T\backslash\{t\}$. We use the minimality of T and Lemma 2.3: Since T is polynomially dense in S relative to $\mathscr{H}$, but $T\backslash\{t\}$ is not, there exists a polynomial $k \in \mathscr{H}$ with $v(k(t)) = \min_{s\in S} v(k(s))$ and $v(k(r)) > \min_{s\in S} v(k(s))$ for all $r \in T\backslash\{t\}$. Let k be such a polynomial and $\alpha = v(k(t))$. Then $g_t(x) = p^{-\alpha}k(x)$ has the desired properties.

Fourth, we show that for every $t \in T$ and $h \in H$ there exists $g_{th} \in \mathscr{F}$ such that $v(g_{th}(t)) = 0$ and $v_h(g_{th}) > 0$. Let k be any polynomial in $\mathscr{F}$ with $v_h(k) > 0$. If $v(k(t)) = \alpha > 0$, set $g_{th}(x) = p^{-\alpha}k(x)g_t(x)^\alpha$.

Now for any $h \in H$ and $t \in T$,

$$e_h = \gcd\left(\{\varphi(g_{th}) \mid t \in T\} \cup \{\varphi(h)\}\right) \quad \text{and} \quad e_t = \gcd\left(\{\varphi(g_r) \mid r \neq t\} \cup \{\varphi(p)\}\right).$$

5 Divisor Homomorphisms on Monadic Monoids of Integer-Valued Polynomials

What we have found out about the submonoid of $\mathrm{Int}(S, V)$ consisting of polynomials whose irreducible factors in $K[x]$ come from a fixed finite set, we now apply to the divisor-closed submonoid of $\mathrm{Int}(S, V)$ generated by a single polynomial. We consider discrete valuation domains first and afterwards generalize to Krull domains.

Recall from Definition 1.2 that $[\![f]\!]$, the divisor-closed submonoid of $\mathrm{Int}(S, D)$ generated by f, is the multiplicative monoid consisting of all those $g \in \mathrm{Int}(S, D)$ which divide some power of f in $\mathrm{Int}(S, D)$. Also, recall the definition of image-primitive, and of $\mathrm{d}_S(f)$, the fixed divisor of f on S from Definition 1.5.

First, let us get a trivial case out of the way

Lemma 5.1 *Let V be a DVR, $S \subseteq V$ and $f \in V[x]$ with $\mathrm{d}_S(f) = V$. Let $F \subseteq V[x]$ be a set of primitive polynomials in $V[x]$ representing the different irreducible factors of f in $K[x]$. Let $\mathscr{F}_0$ be the multiplicative submonoid of $V[x]$ generated by F and the units of V. Then*

1. $[\![f]\!] = \mathscr{F}_0$
2. *Every element g of $[\![f]\!]$ is in $V[x]$, is primitive, and satisfies $\mathrm{d}_S(g) = V$.*
3. *If $g, h \in [\![f]\!]$, then g divides h in $[\![f]\!]$ if and only if g divides h in $K[x]$.*
4. $\varphi\colon [\![f]\!] \to \sum_{h\in F}(\mathbb{N}_0, +)$, $\quad \varphi(g) = (v_h(g) \mid h \in F)$, *is a divisor theory.*

Proof We will show (1) and (2). The remaining statements follow.

$f \in V[x]$ is image-primitive on S and hence primitive. The same holds for all powers of f and for all divisors in $V[x]$ of any power of f by Remark 1.6.

Clearly, every element of $\mathscr{F}_0$ divides in $V[x]$ some power of f. Therefore $\mathscr{F}_0 \subseteq [\![f]\!]$, and every element of $\mathscr{F}_0$ is image-primitive on S.

Now let $g \in [\![f]\!]$. Let $m \in \mathbb{N}$ and $h \in \mathrm{Int}(S, V)$ with $hg = f^m$. Then $h = c\tilde{h}$ and $g = d\tilde{g}$ with $\tilde{g}, \tilde{h} \in \mathscr{F}_0$ and $c, d \in K$. Since $\tilde{g}$ and $\tilde{h}$ are image-primitive on S, we must have $v(c) \geq 0$ and $v(d) \geq 0$. Since f^m is primitive, $v(c) = -v(d)$. It follows that $v(c) = v(d) = 0$ and therefore $g, h \in \mathscr{F}_0$.

Let D be a domain with quotient field K, S a subset of D, and $f \in \mathrm{Int}(S, D)$. Let H be a set of representatives (up to multiplication by a nonzero constant) of the irreducible factors of f in $K[x]$. For instance, H could be the set of monic irreducible factors of f in $K[x]$. Or, in case that D is a principal ideal domain, such as, for instance, a discrete valuation domain, H can be chosen to be a set of primitive irreducible polynomials in $D[x]$. By $\mathscr{H}$ we denote the multiplicative submonoid of $K[x]\backslash\{0\}$ generated by H and the constants in $K\backslash\{0\}$. (Note that $\mathscr{H}$ depends only on f, not on the choice of H). Obviously $[\![f]\!] \subseteq \mathscr{H} \cap \mathrm{Int}(S, D)$. We now examine when equality holds. In this case, we can give a divisor theory of $[\![f]\!]$ [Theorem 5.3]. Otherwise, we have to be content with a divisor homomorphism [Theorem 5.4].

Theorem 5.2 *Let V be a discrete valuation domain with quotient field K, $S \subseteq V$ and $f \in \mathrm{Int}(S, V)\backslash\{0\}$. Let $\mathscr{H}$ be the multiplicative submonoid of $K[x]$ generated* *by the monic irreducible factors of* f *in* $K[x]$ *and the nonzero constants in* K. *If* $\mathrm{d}_S(f) \neq V$ *then*

$$[\![f]\!] = \mathscr{H} \cap \mathrm{Int}(S, V).$$

Proof Clearly, $[\![f]\!] \subseteq \mathscr{H} \cap \mathrm{Int}(S, V)$. For the reverse inclusion, let $f = c\tilde{f}$ with $c \in K\backslash\{0\}$ and $\tilde{f} \in V[x]$ primitive. We will first show that $b\tilde{f} \in [\![f]\!]$, for arbitrary $b \in V\backslash\{0\}$:

Since $\mathrm{d}_S(f) \neq V, v(\mathrm{d}_S(f)) > 0$, and we may apply the Archimedean axiom. Let $m \in \mathbb{N}$ such that $mv(\mathrm{d}_S(f)) \geq v(b) - v(c)$. Then $f^{m+1} = (f^m cb^{-1})b\tilde{f}$, and both $(f^m cb^{-1})$ and $b\tilde{f}$ are in $\mathrm{Int}(S, V)$. Therefore $b\tilde{f} \in [\![f]\!]$.

Furthermore, for arbitrary $b \in V\backslash\{0\}$, all divisors in $V[x]$ of $b\tilde{f} \in [\![f]\!]$ are also in $[\![f]\!]$. Therefore, all primitive irreducible factors of f and all nonzero constants of V, as well as all products of such elements, are in $[\![f]\!]$. Finally, by Lemma 1.4, we can multiply elements of $[\![f]\!]$ by any constant $a \in K$ with $v(a) < 0$, as long as the result is integer-valued on S. Therefore, $\mathscr{H} \cap \mathrm{Int}(S, V) \subseteq [\![f]\!]$.

Theorem 5.3 *Let* v *be a normalized discrete valuation on* K *and* V *its valuation ring. Let* $S \subseteq V$ *and* $f \in \mathrm{Int}(S, V)$, *such that no root of* f *is an isolated point of* S *in* v*-adic topology. Let* H *be the set of different monic irreducible factors of* f *in* $K[x]$ *and* $\mathscr{H}$ *the multiplicative submonoid of* $K[x]$ *generated by* H *and the nonzero constants in* K. *By* $[\![f]\!]$ *denote the divisor-closed submonoid of* $\mathrm{Int}(S, V)$ *generated by* f.

There exists a finite polynomially dense subset T *of* S *relative to* $\mathscr{H}$ *that does not contain any root of* f*; and for every such* T

$$\varphi\colon [\![f]\!] \to \sum_{h\in H}(\mathbb{N}_0, +) \oplus \sum_{t\in T}(\mathbb{N}_0, +) \quad \varphi(g) = ((v_h(g) \mid h \in H), (v(g(t)) \mid t \in T)),$$

is a divisor homomorphism.

If $\mathrm{d}_S(f) \neq V$ *and* T *is chosen minimal then* φ *is a divisor theory.*

Proof $[\![f]\!]$ is a submonoid of $\mathscr{H} \cap \mathrm{Int}(S, V)$. The monoid homomorphism φ in the theorem is the restriction of the divisor homomorphism of Theorem 4.2 to $[\![f]\!]$ and is therefore itself a divisor homomorphism. If $\mathrm{d}_S(f) \neq V$ then $[\![f]\!] = \mathscr{H} \cap \mathrm{Int}(S, V)$ by Theorem 5.2. In this case, φ is a divisor theory by Theorem 4.2, provided T is minimal.

Recall that a Krull domain D is a domain satisfying the following conditions with respect to $\mathrm{Spec}^1(D)$, the set of prime ideals of height 1:

1. For every $P \in \mathrm{Spec}^1(D)$, the localization D_P is a DVR.
2. $D = \bigcap_{P\in\mathrm{Spec}^1(D)} D_P$
3. Each nonzero $r \in D$ lies in only finitely many $P \in \mathrm{Spec}^1(D)$.

If D is a Krull domain, we denote the normalized discrete valuation on the quotient field of D whose valuation ring is D_P, where $P \in \mathrm{Spec}^1(D)$, by v_P. Such a valuation is called an essential valuation of the Krull domain D.

Theorem 5.4 *Let D be a Krull domain with quotient field K and $S \subseteq D$. Let $f \in \text{Int}(S, D)\backslash\{0\}$, and $[\![f]\!]$ the divisor-closed multiplicative submonoid of $\text{Int}(S, D)$ generated by f. Let H be the finite set of different monic irreducible factors of f in $K[x]$ and $\mathcal{H}$ the multiplicative submonoid of $K[x]$ generated by H and the nonzero constants. Let $\mathcal{P}$ be the finite set of primes P of height 1 of D such that either $f \notin D_P[x]$ or $f \in D_P[x]$ and $\text{v}_P(f(S)) > 0$.*

If S does not contain any isolated points in v_P-adic topology for any $P \in \mathcal{P}$, then for each $P \in \mathcal{P}$, there exists a finite subset T_P of S that is D_P-polynomially dense relative to $\mathcal{H}$ in S and contains no root of f. For any such choice of sets T_P, let

$$(M, +) = \sum_{h \in H} (\mathbb{N}_0, +) \oplus \sum_{P \in \mathcal{P}} \sum_{t \in T_P} (\mathbb{N}_0, +).$$

Then

$$\varphi \colon [\![f]\!] \to M, \qquad \varphi(g) = ((v_h(g) \mid h \in H), ((\text{v}_P(g(t)) \mid t \in T_P) \mid P \in \mathcal{P})),$$

is a divisor homomorphism.

Proof The existence of the sets T_P is guaranteed by Proposition 3.3. Since no element of any T_P contains a root of any polynomial in $[\![f]\!]$, φ is a well-defined monoid homomorphism.

Now assume $a, b \in [\![f]\!]$ with $\varphi(a) \mid \varphi(b)$; we need to show $a \mid b$ in $[\![f]\!]$. By Remark 1.3, it suffices to show that a divides b in $K[x]$ and that the cofactor $c = b/a$ is in $\text{Int}(S, D_P)$ for every $P \in \text{Spec}^1(D)$.

Let $c = b/a$. That c is in $K[x]$ follows from $v_h(a) \le v_h(b)$ for all irreducible factors h of a and b in $K[x]$.

Consider a prime P of height 1 of D that is not in $\mathcal{P}$. For such a prime, $f \in D_P[x]$ and f is image-primitive in $\text{Int}(S, D_P)$. We may apply Lemma 5.1 (3) and deduce that $c \in \text{Int}(S, D_P)$.

Now for $P \in \mathcal{P}$, let ψ_P be the projection of M onto $\sum_{h \in H} (\mathbb{N}_0, +) \oplus \sum_{t \in T_P} (\mathbb{N}_0, +)$, and call the latter monoid $M(P)$. From $\varphi(a) \mid \varphi(b)$ it follows that $\psi_P(\varphi(a))$ divides $\psi_P(\varphi(b))$. Let $[\![f]\!]_P$ be the divisor-closed submonoid of $\text{Int}(S, D_P)$ generated by f. Then $[\![f]\!]$ is a submonoid of $[\![f]\!]_P$, and $\psi_P \circ \varphi$ is the restriction to $[\![f]\!]$ of the divisor homomorphism in Theorem 4.2. Now the fact that $\psi_P(\varphi(a))$ divides $\psi_P(\varphi(b))$ implies $c \in \text{Int}(S, D_P)$, by Theorem 4.2.

Corollary 5.5 *Let D be a Krull domain and S a subset that does not have any isolated points in any of the topologies given by essential valuations of D. Let $f \in \text{Int}(S, D)\backslash\{0\}$. Then $[\![f]\!]$, the divisor-closed submonoid of $\text{Int}(S, D)$ generated by f, is a Krull monoid.*

In particular, for every Krull domain D and every $f \in \text{Int}(D)\backslash\{0\}$, the divisor-closed submonoid $[\![f]\!]$ of $\text{Int}(D)$ generated by f is a Krull monoid.

Proof Indeed, the existence of a divisor homomorphism from $[\![f]\!]$ to a finite sum of copies of $(\mathbb{N}_0, +)$ in Theorem 5.4 ensures that $[\![f]\!]$ is a Krull monoid, see [5, Theorem 2.4.8].

Monoids with the property that the divisor-closed submonoid generated by any single element is a Krull monoid have been called *monadically Krull* by A. Reinhart. Without using divisor homomorphisms, through an approach completely different from ours, Reinhart showed that $\text{Int}(D)$ is monadically Krull whenever D is a principal ideal domain [9, Theorem 5.2].

Corollary 5.5 generalizes Reinhart's result to Krull domains, and also to integer-valued polynomials on (sufficiently nice) subsets. The explicit divisor homomorphisms of Theorems 4.2, 5.3 and 5.4 give additional information on the arithmetic of submonoids of $\text{Int}(D)$.

It remains an open problem to find the precise divisor theories (cf. Definition 4.1) of those monoids of integer-valued polynomials for which Theorems 5.3 and 5.4 provide divisor homomorphisms.

Acknowledgments This publication was supported by the Austrian Science Fund FWF, grant P23245-N18. Enthusiastic thanks go out to the anonymous referee for his/her meticulous reading of the paper and the resulting corrections.

References

1. P.-J. Cahen, Polynomial closure. J. Number Theory **61**(2), 226–247 (1996)
2. J.-L. Chabert, On the polynomial closure in a valued field. J. Number Theory **130**(2), 458–468 (2010). (MR 2564907 (2011d:13032))
3. S. Frisch, Substitution and closure of sets under integer-valued polynomials. J. Number Theory **56**(2), 396–403 (1996)
4. S. Frisch, A construction of integer-valued polynomials with prescribed sets of lengths of factorizations, Monatsh. Math. **171**(3–4), 341–350 (2013). (MR 3090795)
5. A. Geroldinger, F. Halter-Koch, Non-unique factorizations, *Pure and Applied Mathematics (Boca Raton)*, vol. 278 (Chapman & Hall/CRC, Boca Raton, 2006). (MR 2194494 (2006k:20001))
6. R.W. Gilmer, Sets that determine integer-valued polynomials. J. Number Theory **33**, 95–100 (1989)
7. D.L. McQuillan, On a theorem of R. Gilmer. J. Number Theory **39**, 245–250 (1991)
8. M.H. Park, F. Tartarone, Polynomial closure in essential domains. Manuscr. Math. **117**, 29–41 (2005)
9. A. Reinhart, On monoids and domains whose monadic submonoids are Krull, in *Commutative Algebra – Recent Advances in Commutative Rings, Integer-valued Polynomials, and Polynomial Functions*, ed. by M. Fontana, S. Frisch, S. Glaz (Springer, Berlin, 2014)

An Overview of the Computational Aspects of Nonunique Factorization Invariants

P.A. García-Sánchez

Abstract We give an overview of the existing algorithms to compute nonunique factorization invariants in finitely generated monoids.

1 Introduction

In this manuscript, we give a general overview of the existing procedures to compute nonunique factorization invariants. These methods have gained importance since they provide batteries of examples that can be used to understand how to prove theoretical results (or disprove ideas that we initially thought would hold). The algorithms improve when we obtain new theoretical results, and in many cases from advances in integer linear programming (and in particular in the study of systems of linear homogeneous Diophantie equalities and inequations; since factorizations can be seen as nonnegative integer solutions of systems of this form). Thus in a sense, this is a wheel: theory produces algorithms that can be used to test new ideas, and these yield new results.

A *semigroup* is a set with a binary associative operation. If a semigroup S has an identity element (an element e such that $ex = xe = x$ for all $x \in S$), then we say that the semigroup is a *monoid*. Let $(M, \cdot)$ be a monoid. An element $m \in M$ is a *unit* if there exists $m' \in M$ such that $m \cdot m' = e = m' \cdot m$, where e is the identity element of M. A monoid is *reduced* if the only unit is the identity element. We are concerned with factorizations up to units, so we can at the very beginning remove the units from our monoid and suppose that it is reduced. If $\mathscr{U}(M)$ denote the set of units, then M_{red} is defined as $M_{\text{red}} = \{m + \mathscr{U}(M) \mid m \in M\}$, which is the *reduced* monoid associated to M.

P.A. García-Sánchez (✉)
Departamento de Álgebra and IEMath-GR, Universidad de Granada,
18071 Granada, Spain
e-mail: pedro@ugr.es

S. Chapman et al. (eds.), *Multiplicative Ideal Theory and Factorization Theory*,
Springer Proceedings in Mathematics & Statistics 170,
DOI 10.1007/978-3-319-38855-7_7

A monoid M is *commutative* if $m \cdot m' = m' \cdot m$ for all $m, m' \in M$. All monoids in this paper are commutative, and thus we will adopt additive notation, and will use 0 to denote the identity element.

A monoid M is *cancellative* if whenever $m + m' = m + m''$ for some $m, m', m'' \in M$, we have $m' = m''$. If $(R, +, \cdot)$ is a domain, then the underlying monoid $(R, \cdot)$ is commutative and cancellative. As with commutativity, we will also assume that our monoids are cancellative.

Thus in what follows a monoid M is meant to be commutative, cancellative, and reduced. We denote $M^* = M \setminus \{0\}$.

Since we are assuming that our monoids are cancellative, we can consider their quotient groups. Let M be a monoid, the *quotient group* of M, denoted by $\mathfrak{G}(M)$, is the set $(M \times M)/\sim$, where $\sim$ is the congruence defined as $(x, y) \sim (x', y')$ if $x + y' = x' + y$. We use $[(x, y)]$ to denote the equivalence class of (x, y) modulo this relation. Addition in $\mathfrak{G}(M)$ is defined by the rule $[(x, y)] + [(x', y')] = [(x + x', y + y')]$. It is easy to show that $(\mathfrak{G}(M), +)$ is a group, and that the natural embedding $i : M \to \mathfrak{G}(M)$, $m \mapsto [(m, 0)]$ is a monoid homomorphism. We can represent $\mathfrak{G}(M)$ as the set $\{x - y \mid x, y \in M\}$ via this embedding.

Assume that M is a submonoid of a free monoid F. Then we say that M is *saturated* if $\mathfrak{G}(M) \cap F = M$. A *Krull* monoid is a monoid M such that M_{red} is a saturated submonoid of a free monoid (this is just one of the many possible definitions; see [27]).

An element m in M^* is said to be an *atom* or *irreducible* if whenever $m = m' + m''$ for some $m', m'' \in M$, then either $m' = 0$ or $m'' = 0$ (recall that we are assuming that M is reduced). Let $\mathscr{A}(M)$ denote the set of atoms of M. We say that M is *atomic* if every element $m \in M$ can be expressed as a sum of finitely many atoms.

For a given set X, let $\mathscr{F}(X)$ be the free monoid on X, that is, the expressions of the form $\sum_{x \in X} \lambda_x x$ with $\lambda_x \in \mathbb{N}$ ($\mathbb{N}$ denotes the set of nonnegative integers), and all but finitely many λ_x are zero. For M an atomic monoid, denote by $\mathsf{Z}(M) = \mathscr{F}(\mathscr{A}(M))$. There is a natural monoid epimorphism

$$\varphi : \mathsf{Z}(M) \to M, \quad \varphi\Big(\sum_{a \in \mathscr{A}(M)} \lambda_a a\Big) = \sum_{a \in \mathscr{A}(M)} \lambda_a a.$$

Observe that many expressions of the form $\sum_{a \in \mathscr{A}(M)} \lambda_a a$ may correspond to the same element in M. For $m \in M$, we define $\mathsf{Z}(m) = \varphi^{-1}(m)$. Every element in $\mathsf{Z}(m)$ is a *factorization* of m. For $N \subseteq M$, we will write $\mathsf{Z}(N) = \bigcup_{m \in N} \mathsf{Z}(m)$.

It may happen that the cardinality of $\mathsf{Z}(m)$ is one for all m (and consequently φ is an isomorphism and M is a free monoid); in this case M is said to be a *factorial monoid*. It also may happen that there are finitely many factorizations for every element in the monoid M, and then we say that M is a *FF-monoid* (which stands for finite factorization monoid). The *length* of a factorization $\sum_{a \in \mathscr{A}(M)} \lambda_a a$ is $\sum_{a \in \mathscr{A}(M)} \lambda_a$. If for every element $m \in M$, all the lengths of its factorizations coincide, then we say that M is a *half-factorial* monoid; and if the set of possible lengths of factorizations are finite for every element, the monoid is a *BF-monoid* (BF stands for bounded factorizations; see [27] for more details and properties of these monoids).

Observe that from a computational point of view it is desirable that M can be described in a "finite" way, and this happens in the case M is an atomic monoid with finitely many atoms. In this setting, if the cardinality of $\mathscr{A}(M)$ is e, we can identify $\mathsf{Z}(M)$ with $\mathbb{N}^e$. As we are assuming M is cancellative and reduced, this implies, that any two factorizations are incomparable with respect to the usual partial ordering in $\mathbb{N}^e$. Dickson's lemma implies that $\mathsf{Z}(m)$ will have finitely many elements for any $m \in M$.

A monoid morphism $f : M \to M'$ is a *transfer homomorphism* if

(T1) $M' = \mathrm{im}(f) + \mathscr{U}(M')$ and $f^{-1}(\mathscr{U}(M')) = \mathscr{U}(M)$,
(T2) if $u \in M$ and $f(u) = b + c$ for some $b, c \in M'$, then there exist v, w such that $u = v + w$, $f(v) \in b + \mathscr{U}(M')$ and $f(w) \in c + \mathscr{U}(M')$.

Transfer homomorphisms allow to study the arithmetical invariants (such as sets of lengths and catenary degree) of Krull and weakly Krull monoids in associated auxiliary monoids. In many cases these auxiliary monoids are finitely generated (see [27]). So, in these cases we will have FF-monoids, and we will be able to determine some properties using a computer.

Notice also that if we are assuming that M is finitely generated, then according to [37, Proposition 3.1], we can assume that M "lives" in $\mathbb{Z}^k \times \mathbb{Z}_{d_1} \times \cdots \times \mathbb{Z}_{d_r}$. If $A = \{m_1, \ldots, m_e\}$ is the set of atoms of M, then $M = \langle A \rangle = \left\{ \sum_{i=1}^{e} n_i m_i \mid n_1, \ldots, n_t \in \mathbb{N} \right\}$. For $m \in M$ the set of factorizations of m corresponds with the set of nonnegative integer solutions of the system of equations

$$(m_1 \mid \cdots \mid m_e)(x_1 \ldots x_e)^T = m,$$

where the m_i's are written in columns, and the last r equations are in congruences modulo $d_1, \ldots, d_r$, respectively. In order to deal with these equations in congruences, we can introduce auxiliary variables and then project to the original ones (see for instance [37, Chap. 7]). The software `Normaliz` [6] can handle these kinds of systems of equations.

By removing equations in congruences, we then have a monoid that is *torsion free*, that is, whenever $km = km'$ for k a positive integer and $m, m' \in M$, we have $m = m'$. Every finitely generated commutative, cancellative, reduced, and torsion free monoid is isomorphic to a submonoid of $\mathbb{N}^k$ for some positive integer k (this is known in the literature as Grillet's Theorem, see for instance [37, Theorem 3.11]). A monoid with all these conditions is called an *affine semigroup*. The set of atoms of an affine semigroup M is $M^* \setminus (M^* + M^*)$, and it is the unique minimal generating system of M. So here minimal generators correspond with atoms (irreducibles).

We will give the definition of arithmetic invariants in the scope of affine semigroups. This does not mean that some of the methods reviewed can be used in a more general scope (even in an noncomputatonal framework), see for instance [8, 32–34].

Recall that the *kernel congruence* of a monoid morphism $f : M \to M'$ is defined as

$$\ker(f) = \{(x, y) \in M \times M \mid f(x) = f(y)\}.$$

Observe that this definition is slightly different from that of kernel of a group morphism (or ring morphism or linear map), because we do not have inverses, and from $f(x) = f(y)$ we cannot write $f(x - y) = 0$.

If z and z' are two factorizations of $m \in M$, then the pair (z, z') is in the kernel of the morphism φ defined above. The map φ, in the setting of affine semigroups with atoms $\{m_1, \ldots, m_e\}$, can be written as

$$\varphi : \mathbb{N}^e \to M,\ \varphi(n_1, \ldots, n_e) = n_1 m_1 + \cdots + n_e m_e.$$

A *presentation* σ of M is a generating system of $\ker \varphi$, that is, $\ker \varphi$ is the minimal congruence containing σ.

Remark 1 Notice that from the definition of presentation, if σ is a presentation for M and z, z' are two factorizations of $m \in M$, then there exists a chain of factorizations $z_1, \ldots, z_r$ of m such that

- $z_1 = z$, $z_r = z'$,
- for every $i \in \{1, \ldots, r - 1\}$ there exists $a_i, b_i, c_i \in \mathbb{N}^e$ such that $(z_i, z_{i+1}) = (a_i + c_i, b_i + c_i)$ with either $(a_i, b_i) \in \sigma$ or $(b_i, a_i) \in \sigma$.

This idea actually catches the fact that $\ker \varphi$ is the least congruence containing σ, or in other words, it is the reflexive-symmetric-transitive closure of σ compatible with addition.

Hence knowing a presentation of M (a generating set of $\ker \varphi$) allows us to know how to move from z to z', and consequently it will be a fundamental tool in the study factorizations of elements in affine semigroups. This is the case of catenary degree and Delta sets.

Recently it has been shown that some invariants are related to the calculation of the set of factorizations of a principal ideal (this occurs with the tame degree and the ω-primality).

An affine semigroup $M \subseteq \mathbb{N}^k$ is *full* if $\mathfrak{G}(M) \cap \mathbb{N}^k = M$, that is, it is a finitely generated saturated submonoid of $\mathbb{N}^k$ for some positive integer k. Clearly, a monoid is full affine if and only if it is a reduced finitely generated Krull monoid (see [27, Theorem 2.7.14]). For full affine semigroups, there are specific procedures that significantly speed up the process of computing factorizations of principal ideals.

For numerical semigroups there are methods, based mainly on computing with Apéry sets, which avoid the use of linear integer programming, and work well for small generators. We will describe them when applicable.

This manuscript is meant to give a state of art of the implementations existing for the calculation of nonunique factorization invariants. We will simply explain the theory that supports these procedures, but will not describe deeply the functions used. We have implemented everything that is described here in the `GAP` [20] package `numericalsgps` ([18]; see the manual of the package for a description of the functions, examples and mode of operation). The reader interested in a full description and implementation of the algorithms can have a look at the source code available either on the GAP web page, or for the development version in https://bitbucket.org/gap-system/numericalsgps (the files containing the functions described here for numerical semigroups are in `catenary-tame.gi` and `contributions.gi`; those for affine semigroups are in `affine.gi`, both in the folder `gap`). The package tests availability of other packages [2, 16, 26, 28, 29] that interact with `4ti2` [1], `Normaliz` [5, 6] and `Singular` [17]. Depending on this availability, the package will use an specific method for the calculations. So, in some cases, we wrote up to four different implementations for computing the same invariant (this is why there are several files with prefix `affine-extra` in the `gap` folder).

2 Presentations

Rédei proved in [36] that every finitely generated commutative monoid is finitely presented. In our setting, this means that every affine semigroup admits a presentation with finitely many elements. Since then, many alternative and shorter proofs have been published. We recall here one of these approaches.

Let t be a symbol and let $\mathbb{K}$ be a field. For M an affine semigroup, define the *semigroup ring* $\mathbb{K}[M] = \bigoplus_{m\in M} \mathbb{K}t^m$, where addition is performed component-wise and multiplication follows the rule $t^m t^{m'} = t^{m+m'}$. Observe that we can think of $\mathbb{K}[M]$ as the set of polynomials in the variable t but with exponents in the monoid M (so this is not necessarily a subring of $\mathbb{K}[t]$, the ring of polynomials in t, since M does not have to be a submonoid of $\mathbb{N}$).

Assume that $\{m_1, \ldots, m_e\}$ is a generating system of M. Herzog in [30] proves that σ is a presentation of M if and only if the ideal $I_M = (X^a - X^b \mid (a, b) \in \sigma)$, where I_M is the kernel of the ring homomorphism induced by

$$\mathbb{K}[x_1, \ldots, x_t] \to \mathbb{K}[M],\ x_i \mapsto t^{m_i}.$$

Observe that for $n = (n_1, \ldots, n_k)$, we can write t^m as $t_1^{n_1} \ldots t_k^{n_k}$ and in this way we can see $\mathbb{K}[M]$ as a subring of $\mathbb{K}[t_1, \ldots, t_k]$. In particular, we can compute a presentation of M by using elimination: we start with the ideal $\big(x_1 - t^{m_1}, \ldots, x_e - t^{m_e}\big) \subseteq \mathbb{K}[x_1, \ldots, x_e, t_1, \ldots, t_k]$, and then eliminate the variables $t_1, \ldots, t_k$ to obtain I_M.

Example 2.1 Let us compute a presentation of $M = \langle (2, 0), (0, 2), (1, 1), (2, 1)\rangle$ with `singular`, [17].

```
> ring r=0,(x,y,z,t,u,v),lp;
> ideal i = (x-u^2,y-v^2,z-u*v,t-u*v^2);
> eliminate(i,u*v);
_[1]=yz2-t2
_[2]=xt2-z4
_[3]=xy-z2
```

This means that $I_M = \left(yz^2 - t^2, xt^2 - z^4, xy - z^2\right)$, and in light of Herzog's correspondence, the set

$$\left\{((0,1,2,0),(0,0,0,2)),((1,0,0,2),(0,0,4,0)),((1,1,0,0),(0,0,2,0))\right\}$$

is a presentation for M.

A *minimal presentation* of M is a presentation that cannot be refined to another presentation of M, that is, it is minimal with respect to set inclusion (it turns out that it is also minimal with respect to cardinality; see [37, Corollary 9.5]).

Example 2.2 The presentation in Example 2.1 is not minimal. If we want to obtain a minimal presentation with `singular` additional work is needed.

```
> ring r=0,(x,y,z,t,u,v),(wp(2,2,2,3),lp(2));
// ** redefining r **
> ideal i = (x-u**2,y-v**2,z-u*v,t-u*v**2);
> ideal j=eliminate(i,u*v);
> minbase(j);
_[1]=xy-z2
_[2]=yz2-t2
```

Given $m \in M$, we define ∇_m as the graph with vertices $\mathsf{Z}(m)$ and zz' is an edge if $z \cdot z' \neq 0$ (dot product). An element m is a *Betti element* of M if the graph ∇_m is not connected. We will denote by $\operatorname{Betti}(M)$ the set of Betti elements of M.

The sets of vertices of the connected components of ∇_m are also known as $\mathcal{R}$-classes of $\mathsf{Z}(m)$. The following method can be used to produce all minimal presentations (up to arrangement of the pairs and symmetry) of M; see for instance [37, Chap 9].

- For all $m \in M$, if ∇_m is connected, then set $\sigma_m = \emptyset$. If not, let $R_1, \ldots, R_r$ be the different $\mathcal{R}$-classes of $\mathsf{Z}(m)$. Consider any tree T with vertices $R_1, \ldots, R_r$. For each $i \in \{1, \ldots, r\}$ take $r_i \in R_i$. Set $\sigma_m = \{(z_i, z_j) \mid R_i R_j \text{ is an edge of } T\}$ (for instance, one might take $\sigma_m = \{(z_1, z_2), (z_1, z_3), \ldots, (z_1, z_r)\}$).
- The set $\sigma = \bigcup_{m \in M} \sigma_m$ is a minimal presentation of M.

It follows that the set of Betti elements of M has finite cardinality and that the cardinality of a (any) minimal presentation is $\sum_{b \in \operatorname{Betti}(M)} (\operatorname{ncc}(\nabla_b) - 1)$, where $\operatorname{ncc}(\nabla_b)$ stands for the number of connected components of ∇_b. This formula holds for every atomic monoid having the ascending chain on principal ideals [8, Corollary 1].

Example 2.3 Let M be as in Example 2.1. Since any presentation contains a minimal presentation, we have that $\operatorname{Betti}(M) \subseteq \{(2,4),(2,2),(4,4)\}$. We use the GAP [20] package `numericalsgps` [18] to calculate the $\mathcal{R}$-classes of each of these elements.

```
gap> RClassesOfSetOfFactorizations(
 FactorizationsVectorWRTList([4,4],[[2,0],[0,2],[1,1],[1,2]]));
[ [ [ 0, 0, 4, 0 ], [ 1, 0, 0, 2 ], [ 1, 1, 2, 0 ], [ 2, 2, 0, 0 ] ] ]
gap> RClassesOfSetOfFactorizations(
 FactorizationsVectorWRTList([2,4],[[2,0],[0,2],[1,1],[1,2]]));
```

```
[ [ [ 0, 1, 2, 0 ], [ 1, 2, 0, 0 ] ], [ [ 0, 0, 0, 2 ] ] ]
gap> RClassesOfSetOfFactorizations(
 FactorizationsVectorWRTList([2,2],[[2,0],[0,2],[1,1],[1,2]]));
[ [ [ 1, 1, 0, 0 ] ], [ [ 0, 0, 2, 0 ] ] ]
```

It follows that $\mathrm{Betti}(M) = \{(2, 2), (2, 4)\}$ (this also follows from Example 2.2).

The function `FactorizationsVectorWRTList` either uses [15], or if available [6] or [1] through the packages `NormalizInterface` [29] or either `4ti2gap` [26] or `4ti2Interface` [28].

We will see that knowing a minimal presentation is of great help for the calculation of catenary degree, and it also provides relevant information on the Delta sets.

3 Apéry Sets

Let M be an affine semigroup generated by $\{m_1, \dots, m_e\}$. Let $m \in M$. The *Apéry set* of m in M is the set

$$\mathrm{Ap}(M, m) = \{m' \in M \mid m' - m \notin M\}.$$

Apéry sets can be defined in a more general setting. If our monoid fulfills the ascending chain condition on principal ideals, then every for every $m' \in M$ there exits unique $(w, k) \in \mathrm{Ap}(M, m) \times \mathbb{N}$ such that $m' = km + w$ (see [8]).

If M is a numerical semigroup, then the cardinality of $\mathrm{Ap}(M, m)$ has exactly m elements. Moreover, if $b \in \mathrm{Betti}(M)$, then b is can be expressed as $b = m_i + w$ for some $i \in \{2, \dots, e\}$ and $w \in \mathrm{Ap}(M, m_1) \setminus \{0\}$ depending on b (see for instance [38, Proposition 8.19]). As minimal presentations are crucial for studying factorizations, this implies that Apéry sets are also important in our study specialized to the numerical semigroup setting.

4 Graver Bases

Let M be an affine semigroup, $M \subseteq \mathbb{N}^k$ generated by $\{m_1, \dots, m_e\}$.

We have seen that a minimal presentation is a minimal generating system of $\ker \varphi$ as a congruence. It turns out that $\ker \varphi$ is not only a congruence, but an affine semigroup itself, and thus it admits a unique minimal generating system, which we denote by $\mathscr{I}(M)$. It follows easily that $\mathscr{I}(M)$ corresponds with the pairs $(x, y) = ((x_1, \dots, x_e), (y_1, \dots, y_e)) \in \mathbb{N}^e \times \mathbb{N}^e \setminus \{(0, 0)\}$ that are minimal (with respect to the usual product order, that is, $(x, y) \le (x', y')$ if $x \le x'$ and $y \le y'$; and now in $\mathbb{N}^e$, $x \le x'$ if $x_i \le x_i'$ for all $i \in \{1, \dots, e\}$) solutions of

$$(m_1 | \cdots | m_e | -m_1 | \cdots | -m_e)(x \mid y)^T = 0,$$

because if $(x, y) \in \ker \varphi$, then $x_1 m_1 + \cdots + x_e m_e = y_1 m_1 + \cdots + y_e m_e$. Moreover, there exists $a_1, \ldots, a_s \in \mathscr{I}(M)$ such that $(x, y) = a_1 + \cdots + a_s$ (each a_i is a pair of factorizations of the same element). That is, every pair of factorizations of the same element can be expressed as a sum of pairs of factorizations of some specific elements. Indeed, we will say that $m \in M$ is *primitive* if there exist $x, y \in \mathsf{Z}(m)$ such that $(x, y) \in \mathscr{I}(M)$.

In particular, $\mathscr{I}(M)$ is a presentation of M, though in general with a lot of redundancy. This is because it is a minimal generating system of $\ker \varphi$ as a monoid, and not as a congruence. For instance, if $\mathbf{e}_i$ is the ith row of the identity $e \times e$ matrix, then $(\mathbf{e}_i, \mathbf{e}_i) \in \mathscr{I}(M)$ for all $i \in \{1, \ldots, e\}$; and these elements are not needed in a presentation, since they are just a consequence of the reflexive property of congruences. Also if $(a, b) \in \mathscr{I}(M)$, then $(b, a) \in \mathscr{I}(M)$; and if we have (a, b) in a presentation, we no longer need (b, a), because this last pair follows by symmetry.

An expression of the form $(x, y) = a_1 + \cdots + a_s$ cannot be achieved by taking the a_i in a minimal presentation. Thus factorizations of primitive elements give more information than minimal presentations, and for some invariants this extra information will come into scene.

Example 4.1 Let M be the numerical semigroup minimally generated by $\{3, 5, 7\}$. Then a minimal presentation of M is given by

$$\{((0, 2, 0), (1, 0, 1)), ((3, 1, 0), (0, 0, 2)), ((4, 0, 0), (0, 1, 1))\}.$$

While

$$\begin{aligned} \mathscr{I}(M) = \{&(\mathbf{e}_1, \mathbf{e}_1), (\mathbf{e}_2, \mathbf{e}_2), (\mathbf{e}_3, \mathbf{e}_3), ((3, 1, 0), (0, 0, 2)), ((0, 0, 2), (3, 1, 0)), \\ &((4, 0, 0), (0, 1, 1)), ((0, 1, 1), (4, 0, 0)), ((1, 0, 1), (0, 2, 0)), \\ &((0, 2, 0), (1, 0, 1)), ((2, 3, 0), (0, 0, 3)), ((0, 0, 3), (2, 3, 0)), \\ &((1, 5, 0), (0, 0, 4)), ((0, 0, 4), (1, 5, 0)), ((0, 7, 0), (0, 0, 5)), \\ &((0, 0, 5), (0, 7, 0)), ((5, 0, 0), (0, 3, 0)), ((0, 3, 0), (5, 0, 0))\}. \end{aligned}$$

On $\mathbb{Z}^e$ define the order $(x_1, \ldots, x_e) \sqsubseteq (y_1, \ldots, y_e)$ if for all $i \in \{1, \ldots, e\}$, $x_i y_i \geq 0$ and $|x_i| \leq |y_i|$. Also, for $x \in \mathbb{Z}^e$ set x^+ and x^- to be the unique elements in $\mathbb{N}^e$ such that $x = x^+ - x^-$ and $x^+ \cdot x^- = 0$. It turns out that $x \sqsubseteq y$ if and only if $(x^+, x^-) \leq (y^+, y^-)$ (usual partial ordering).

Let H be a subgroup of $\mathbb{Z}^e$. A *Graver basis* of H is a set of minimal nonzero elements of H with respect to $\sqsubseteq$.

Notice that the set of integer solutions of

$$(m_1 \mid \cdots \mid m_e) x^T = 0$$

defines a subgroup H_M of $\mathbb{Z}^e$. In fact $(x, y) \in \ker \varphi$ if and only if $x - y \in H_M$ (this is a rephrasing of the necessity condition in [37, Proposition 1.4]). From a Graver basis G of H_M we can easily compute

$$\mathscr{I}(M) = \{(x^+, x^-) \mid x \in G\} \cup \{(\mathbf{e}_i, \mathbf{e}_i) \mid i \in \{1, \ldots, e\}\}.$$

Example 4.2 Let us go back to M in Examples 2.1 and 2.2.

```
gap> GraverBasis4ti2(["mat",TransposedMat([[2,0],[0,2],[1,1],[1,2]])]);
[ [ 1, 0, -4, 2 ], [ 0, 1, 2, -2 ], [ 1, 1, -2, 0 ], [ 1, 2, 0, -2 ] ]
```

The output of `4ti2` does not print an element and its negation. Hence, a Graver basis of H_M consists in 8 elements and $\mathscr{I}(M)$ has $8+4$ elements.

We will see that some nonunique factorization invariants depend on the factorizations of the primitive elements of M.

5 Block Monoids

Let G be an Abelian group. And let $g_1, \ldots, g_k \in G$. A *zero-sum sequence* is an expression of the form $n_1g_1 + \cdots + n_kg_k = 0$ with $(n_1, \ldots, n_k) \in \mathbb{N}^k$. The *length* of this sequence is $n_1 + \cdots + n_k$. We say that a zero-sum sequence is *minimal* if there is no other zero-sum sequence $n_1'g_1 + \cdots + n_k'g_k = 0$ such that $0 \neq (n_1', \ldots, n_k') \lneq (n_1, \ldots, n_k)$. The set of zero-sum sequences is clearly a monoid, actually it can be identified as a submonoid of $\mathbb{N}^k$ and it is generated by the minimal zero-sum sequences (indeed it is a full affine semigroup). We will denote the set of zero-sum sequences in $g_1, \ldots, g_k$ by $\mathscr{B}(\{g_1, \ldots, g_k\})$.

Since G is an Abelian group, it is then isomorphic to $\mathbb{Z}_{d_1} \times \cdots \times \mathbb{Z}_{d_r} \times \mathbb{Z}^l$ for some $d_1, \ldots, d_r, l \in \mathbb{N}$. Hence we can identify the elements $g_1, \ldots, g_k$ with elements in $\mathbb{Z}_{d_1} \times \cdots \times \mathbb{Z}_{d_r} \times \mathbb{Z}^l$. Hence $\mathscr{B}(\{g_1, \ldots, g_k\})$ corresponds with the set of nonnegative integer solutions of the system of $r + l$ equations and k unknowns

$$(g_1 \mid \cdots \mid g_k)x = 0 \in \mathbb{Z}_{d_1} \times \cdots \times \mathbb{Z}_{d_r} \times \mathbb{Z}^l$$

(the first r equations are in congruences modulo $d_1, \ldots, d_r$, respectively). The set of solutions of this system of equations can be computed via `Normaliz` ([6]).

The *Davenport constant* is the supremum (in this setting maximum) of the lengths of minimal zero-sum sequences.

Example 5.1 We can compute the block monoid associated to $\mathbb{Z}_2^2$ in the following way using `numericalsgps`.

```
gap> m2:=[[0,1],[1,0],[1,1]];;
gap> a:=AffineSemigroup("equations",[TransposedMat(m2),[2,2]]);;
gap> GeneratorsOfAffineSemigroup(a);
[ [ 0, 0, 2 ], [ 0, 2, 0 ], [ 1, 1, 1 ], [ 2, 0, 0 ] ]
```

Observe that we are omitting $(0, 0)$ and that the second argument of `Affine Semigroup` is a matrix whose columns are the elements in $(\mathbb{Z}_2^2)^*$ and a list indicating the equations that are congruences with the respective modules. The Davenport constant in this case is 3.

Many factorization properties of monoids can be derived (or bounded in some cases) from the factorization properties of the block monoid of their class groups (see [27]). This is why these affine semigroups are relevant in the study of nonunique factorization invariants.

6 Denumerant and Maximal Denumerant

We have already mentioned that for an affine semigroup M and $m \in M$, the set $\mathsf{Z}(m)$ has finitely many elements. The *denumerant* of m is precisely the cardinality of $\mathsf{Z}(m)$. There is a wide amount of the literature devoted to the study of denumerants of elements in numerical semigroups, indeed few formulas are known, and just for some particular families of monoids ([35] is a nice reference for the reader interested in this topic).

Of course the bigger an integer in a numerical semigroup M is, the larger its denumerant is, and thus it is not bounded. This is not the case if we just count factorizations with maximal length. The *maximal denumerant* of m in M is the number of elements in $\mathsf{Z}(m)$ with maximal length, which is a positive integer, since $\mathsf{Z}(m)$ has finitely many elements. If M is a numerical semigroup, set the maximal denumerant of M as the supremum of the maximal denumerants of elements of M. What is astonishing is that this supremum is indeed a maximum, and thus a positive integer. Bryant and Hamblin give in [7] a procedure to compute the maximal denumerant of any numerical semigroup.

Example 6.1 The semigroup $\langle 3, 5, 7\rangle$ has maximal denumerant 2.

```
gap> s:=NumericalSemigroup(3,5,7);;
gap> MaximalDenumerantOfNumericalSemigroup(s);
2
gap> List(Intersection([0..100],s),
> x->Length(FactorizationsElementWRTNumericalSemigroup(x,s)));
[ 1, 1, 1, 1, 1, 1, 1, 2, 1, 2, 2, 2, 3, 2, 3, 3, 3, 4, 4, 4, 4, 5, 5, 5, 6,
  6, 6, 7, 7, 7, 8, 8, 9, 9, 9, 10, 10, 11, 11, 12, 12, 12, 14, 13, 14, 15,
  15, 16, 16, 17, 17, 18, 19, 19, 20, 20, 21, 22, 22, 23, 24, 24, 25, 26, 26,
  27, 28, 29, 29, 30, 31, 31, 33, 33, 34, 35, 35, 37, 37, 38, 39, 40, 41, 41,
  43, 43, 44, 46, 46, 47, 48, 49, 50, 51, 52, 53, 54, 55 ]
```

7 Length-Based Invariants

Let M be an affine semigroup generated by $\{m_1, \ldots, m_e\}$. Take $m \in M$ and $x = (x_1, \ldots, x_e) \in \mathsf{Z}(m)$. Recall that the length of x is defined as

$$|x| = x_1 + \cdots + x_e.$$

The *set of lengths of factorizations* of m is

$$\mathsf{L}(m) = \{|x| \mid x \in \mathsf{Z}(m)\}.$$

Since $\mathsf{Z}(m)$ has finitely many elements, so has $\mathsf{L}(m)$. This means that affine semigroups are BF-monoids.

Recall that a monoid is half factorial if the cardinality of $\mathsf{L}(s)$ is one for all $s \in S$. This concept was introduced for domains in [43].

From Remark 1 it easily follows that M is half factorial if and only if for every (a, b) in a minimal presentation of M we have $|a| = |b|$ (see [40]). Thus, we can determine whether or not an affine semigroup is half factorial.

Example 7.1 In Example 2.2, since $((1, 2, 0, 0), (0, 0, 0, 2))$ belongs to a minimal presentation of M, we deduce that M is not half factorial.

7.1 *Elasticity*

One of the first nonunique factorization invariants that appeared in the literature was the elasticity (introduced in [42]). It was meant to measure how far a monoid is from being half factorial.

Take m in an affine semigroup M. The *elasticity* of m, $\rho(m)$, is defined as

$$\rho(m) = \frac{\sup \mathsf{L}(m)}{\min \mathsf{L}(m)}.$$

Since $\mathsf{L}(m)$ has finitely many elements, the supremum in the numerator is indeed a maximum. The elasticity of M is defined as

$$\rho(M) = \sup\{\rho(m) \mid m \in M\}.$$

It is not hard to show (see [40]) that

$$\rho(M) = \max\left\{\frac{|a|}{|b|} \;\middle|\; (a, b) \in \mathscr{I}(M)\right\}.$$

Hence, by computing a Graver basis of H_M we can calculate the elasticity of M. However computing a Graver basis, can be highly time consuming. Philipp in his thesis, and published later in [33], provided an alternative method for the computation of the elasticity: he showed that we only have to consider elements $(a, b) \in \mathscr{I}(M)$ with $a \neq b$ and with minimal support (indices of nonzero coordinates). These elements are known in the literature as circuits, and we can use [19, Lemma 8.8] to calculate them by means of determinants.

$$\rho(M) = \max\left\{\frac{|a|}{|b|} \;\middle|\; (a,b) \text{ circuit of } \ker\varphi\right\}.$$

Example 7.2 Let us compute $\rho\left(\mathscr{B}\left(\mathbb{Z}_2^3\right)\right)$.

```
gap> m:=[[0,0,1],[0,1,0],[0,1,1],[1,0,0],[1,0,1],[1,1,0],[1,1,1]];;
gap> a:=AffineSemigroup("equations",[TransposedMat(m),[2,2,2]]);;
gap> ElasticityOfAffineSemigroup(a);
2
```

Example 7.3 We see now with an easy example that the elasticity of the Betti elements of a monoid is not enough to compute the elasticity of the monoid.

```
gap> s:=NumericalSemigroup(3,5,7);;
gap> BettiElementsOfNumericalSemigroup(s);
[ 10, 12, 14 ]
gap> List(last, b-> ElasticityOfFactorizationsElementWRTNumericalSemigroup(b,s));
[ 1, 2, 2 ]
```

We see that the maximum is 2; while it is well known that for numerical semigroups the elasticity of the monoid is the quotient of the largest minimal generator by the multiplicity of the semigroup (the least positive integer in the semigroup). So in this case it should be $7/3$.

```
gap> PrimitiveElementsOfNumericalSemigroup(s);
[ 3, 5, 7, 10, 12, 14, 15, 21, 28, 35 ]
gap> List(last, b-> ElasticityOfFactorizationsElementWRTNumericalSemigroup(b,s));
[ 1, 1, 1, 1, 2, 2, 5/3, 7/3, 2, 11/5 ]
gap> Maximum(last);
7/3
gap> ElasticityOfNumericalSemigroup(s);
7/3
```

7.2 Delta Sets

Another way to measure how far we are from half factoriality, is to determine how distant are the different lengths of factorizations. This is the motivation for the following definition.

Let as above m be an element in the affine semigroup M. Assume that $\mathsf{L}(m) = \{l_1 < \cdots < l_r\}$. Define the *Delta set* of m as

$$\Delta(s) = \{l_2 - l_1, \ldots, l_r - l_{r-1}\},$$

and if $r = 1$, $\Delta(m) = \emptyset$. The Delta set of M is defined as

$$\Delta(M) = \bigcup_{m \in M} \Delta(m).$$

So, the bigger $\Delta(M)$ is, the farther is M from being half factorial.

Recall that $(x, y) \in \ker \varphi$ if and only if $x - y \in H_M$. Indeed, it is not hard to show that H_M is generated as a group by the differences of the pairs in a presentation of M. From this, one can prove that

$$\min \Delta(M) = \gcd \Delta(M)$$

([27, Proposition 1.4.4]).

By using the idea expressed in Remark 1, it can be shown that the maximum of the distances between lengths of factorizations is reached in a Betti element of M ([13, Theorem 2.5]):

$$\max \Delta(M) = \max \{ \max \Delta(b) \mid b \in \mathrm{Betti}(M) \}.$$

This gives us an interval where the elements in $\Delta(M)$ must be, but it is far from being a procedure to compute the whole set $\Delta(M)$.

For numerical semigroups, it is known that the sets of distances between consecutive lengths of factorizations are eventually periodic [14] and a bound for this periodicity is given. This bound was improved in [22]. Hence, we can compute the Delta sets of the elements up to this bound (a dynamic version of this procedure is presented in [3]). The problem is that this bound can be huge.

```
gap> s:=NumericalSemigroup(701,902,1041);
<Numerical semigroup with 3 generators>
gap> DeltaSetOfNumericalSemigroup(s);
[ 1, 2, 3, 4, 5, 6, 11, 17 ]
gap> DeltaSetPeriodicityBoundForNumericalSemigroup(s);
313436
```

Recently in [24] a procedure that runs as fast as Euclid's extended algorithm has been presented for numerical semigroups with embedding dimension three (and not symmetric, though the algorithm seems to work also for symmetric numerical semigroups).

O'Neill in [31] gives new theoretical tools for the computation of $\Delta(M)$ for an arbitrary affine semigroup M. We now have a preliminary implementation of them, and we are currently working on proving the correctness of our algorithm.

8 Distance-Based Invariants

Observe that length-based invariants cannot describe the behavior of factorizations in half-factorial monoids. To measure how spread are the factorizations, we first need a distance.

For $x = (x_1, \ldots, x_e)$, $y = (y_1, \ldots, y_e) \in \mathbb{N}^e$, define the infimum of x and y as

$$x \wedge y = (\min\{x_1, y_1\}, \ldots, \min\{x_p, y_p\})$$

(if we think in multiplicative notation and x and y are factorizations of an element, then $x \wedge y$ translates to greatest common divisor).

The *distance* between x and y is defined as

$$\mathrm{d}(x, y) = \max\{|x - (x \wedge y)|, |y - (x \wedge y)|\}$$

(equivalently $\mathrm{d}(x, y) = \max\{|x|, |y|\} - |x \wedge y|$).

8.1 Catenary Degree

We start with an example that illustrates the idea of catenary degree.

Example 8.1 The factorizations of $66 \in \langle 6, 9, 11\rangle$ are

$$\mathsf{Z}(66) = \big\{(0,0,6), (1,3,3), (2,6,0), (4,1,3), (5,4,0), (8,2,0), (11,0,0)\big\}.$$

The distance between $(11, 0, 0)$ and $(0, 0, 6)$ is 11.

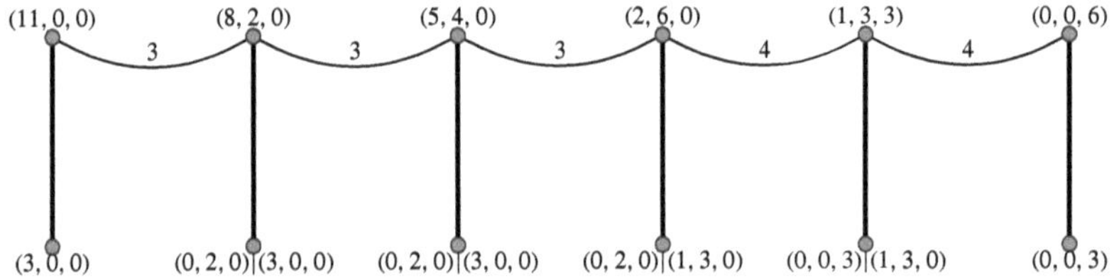

In the above picture the factorizations are depicted in the top of a post, and they are linked by a "catenary" labeled with the distance between two consecutive sticks. On the bottom we have drawn the factorizations removing the common part with the one on the left and that of the right, respectively. So we have linked $(11, 0, 0)$ and $(0, 0, 6)$ with a chain of factorizations, and every two consecutive nodes in the chain are at most at distance 4. This is in fact the best we can do in this example. We do not care about the length of the sequence, but about how closer are two consecutive elements in the chain.

Let M be an affine semigroup, and take $m \in M$. Let $x, y \in \mathsf{Z}(m)$ and let N be a nonnegative integer. An N-*chain* joining x and y is a sequence $x_1, \ldots, x_k \in \mathsf{Z}(m)$ such that

- $x_1 = x$, $x_k = y$,
- for all $i \in \{1, \ldots, k-1\}$, $\mathrm{d}(x_i, x_{i+1}) \le N$.

The *catenary degree* of m, denoted $\mathsf{c}(m)$, is the least N such that for any two factorizations $x, y \in \mathsf{Z}(m)$, there is an N-chain joining them. The catenary degree of M, $\mathsf{c}(M)$, is defined as

$$\mathsf{c}(M) = \sup\left\{\mathsf{c}(m) \mid m \in M\right\}.$$

The calculation of $\mathsf{c}(m)$ can be performed in the following way. We consider the complete graph with vertices the factorizations of m, and edges labeled with the distances between their ends. Then we pick an edge with the largest label, and if it is not a bridge, then we remove it. We keep doing so, until we arrive to a bridge. The label of this bridge is $\mathsf{c}(m)$.

Example 8.2 As an illustration of the above procedure, consider $77 \in S = \langle 10, 11, 23, 35\rangle$. In the following figure, we see that we can remove the edge with label 6, meaning that in order to go from (0, 7, 0, 0) to (2, 1, 2, 0) we can first go to (2, 2, 0, 1) and then to (2, 1, 2, 0), and the distances in this walk between two consecutive nodes are less than 6. Then we remove the edge labeled with 5. But we cannot remove the edge joining (1, 4, 1, 0) and (0, 7, 0, 0) since it is a bridge (we can remove the other labeled with 3).

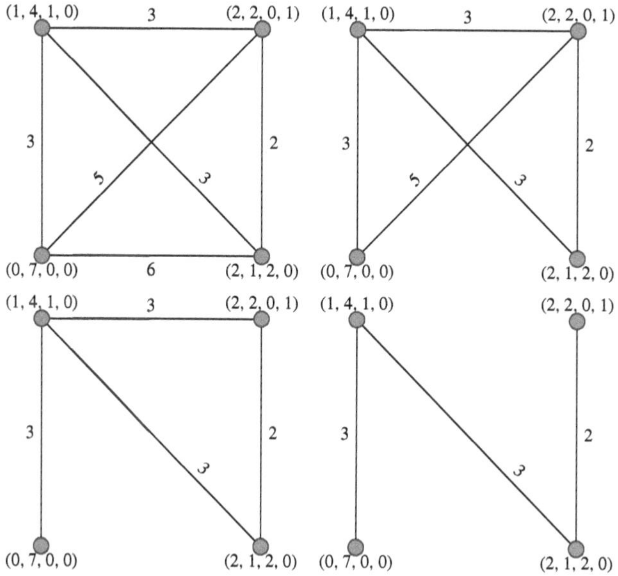

Thus the catenary degree of 77 is 3.

Observe that in Remark 1, we obtained chains joining any two factorizations of the same element, just using translations of elements in a presentation. Since distances are not translation-sensitive, our only concern is how to find a chain joining the first component with the second in a relation in a presentation. It follows (see [12]) that

$$\mathsf{c}(M) = \max\left\{\mathsf{c}(b) \mid b \in \mathrm{Betti}(M)\right\}.$$

This gives a computational procedure to compute the catenary degree of any affine semigroup M.

Example 8.3 Let us recover Example 2.2, $M = \langle(2,0),(0,2),(1,1),(2,1)\rangle$. We already know that $\mathrm{Betti}(M) = \{(2,2),(2,4)\}$.

```
gap> a:=AffineSemigroup([2,0],[0,2],[1,1],[1,2]);;
gap> gens:=GeneratorsOfAffineSemigroup(a);
[ [ 0, 2 ], [ 1, 1 ], [ 1, 2 ], [ 2, 0 ] ]
gap> betti:=BettiElementsOfAffineSemigroup(a);
[ [ 2, 2 ], [ 2, 4 ] ]
gap> List(betti,b->FactorizationsVectorWRTList(b,gens));
[ [ [ 1, 0, 0, 1 ], [ 0, 2, 0, 0 ] ],
  [ [ 2, 0, 0, 1 ], [ 1, 2, 0, 0 ], [ 0, 0, 2, 0 ] ] ]
gap> List(last,CatenaryDegreeOfSetOfFactorizations);
[ 2, 3 ]
gap> CatenaryDegreeOfAffineSemigroup(a);
3
```

So far we do not know of a procedure to compute the (finite) set $\{\mathsf{c}(m) \mid m \in M\}$. It is known that for numerical semigroups, the catenary degree is also eventually periodic, but unfortunately no bounds for this periodicity are known ([9]). For half-factorial monoids it can be shown (see [25, Theorem 2.3]) that

$$\{\mathsf{c}(m) \mid m \in M\} = \{\mathsf{c}(m) \mid m \in \mathrm{Betti}(M)\}.$$

For numerical semigroups, in light of Sect. 3 (see also [10, Corollary 3]),

$$\mathsf{c}(M) = \max\left\{\mathsf{c}(m) \mid m \in \{m_2,\ldots,m_e\} + (\mathrm{Ap}(M,m_1)\setminus\{0\})\right\},$$

and so in this setting it is not needed to compute $\mathrm{Betti}(M)$.

8.2 Monotone, Equal, and Homogeneous Catenary Degrees

In the definition of catenary degree, we are not assuming any restrictions on the shape of the N-chains nor on their lengths. In an attempt to better understand the structure of these chains, new catenary degrees have been introduced in the literature. For instance if we enforce the chain of factorizations to have nondecreasing lengths we obtain the definition of *monotone catenary degree*. This slight change makes its computation much more complicated than the usual catenary degree as we see later.

We can also ask the lengths to be all equal, and then we have *equal catenary degree*. Thus in order to calculate the equal catenary degree of an element we have to arrange the factorizations of this element in layers of factorizations with the same length; and then take the maximum of the "classical" catenary degree in each of the layers. In particular, if all factorizations have different lengths, the equal catenary degree for this element is zero.

Recall that studying the set of factorizations of an element m in a monoid M generated by the columns of a matrix A is equivalent to studying the set of nonnegative integer solutions of the system of linear Diophantine equations $Ax = m$. If we want to study those factorizations with a given length, the standard trick (see for instance [11]) is to add a new row of ones in the matrix A, and the desired length as last component of m. If $M \subseteq \mathbb{N}^k$, and $m \in M$, we write $(m, l) \in \mathbb{N}^{k+1}$ for the element with the first coordinates the coordinates of m and last coordinate equal to l (we have appended the integer l at the "end" of m). Assume that M is minimally generated by $\{m_1, \ldots, m_e\}$. Set

$$M^{eq} = \langle (m_1, 1), \ldots, (m_e, 1) \rangle.$$

Then studying factorizations of an element m in M with length l is the same as studying factorizations of (m, l) in M^{eq} (indeed $(m, l) \in M^{eq}$ if and only if $m \in M$ and $l \in \mathsf{L}(m)$). Consequently, the equal catenary degree of M corresponds with the catenary degree of M^{eq} [25].

Finally, we can also impose that the lengths in the chain are not larger than the maximum of the lengths of the ends of the chain, obtaining in this way the *homogeneous catenary degree*. We mentioned above that for half-factorial monoids, all possible catenary degrees in the monoid arise as catenary degrees of some Betti element. If a monoid M is not half factorial, this is because at least one binomial in the ideal I_M is not homogeneous. One can then homogenize these binomials in the usual way (we choose a new variable z and to each binomial of the form $X^\alpha - X^\beta$ with $|\alpha| > |\beta|$ we associate the binomial $X^\alpha - X^\beta z^{|\alpha|-|\beta|}$; and analogously if $|\alpha| \leq |\beta|$). The resulting binomial ideal is precisely the ideal associated to the monoid

$$M^{hom} = \langle (m_1, 1), \ldots, (m_e, 1), (0, 1) \rangle.$$

This is why we called in [25] this catenary degree homogeneous catenary degree. We proved in that paper that this new catenary degree corresponds with the catenary degree of M^{hom}, and it is between the "classic" catenary degree and the monotone catenary degree.

In order to compute the monotone catenary degree of M, it can be derived from [33] that we have to look at the projections in the first k coordinates of the primitive elements of M^{hom} (see [41, Chap. 3]), and then take the maximum of the monotone catenary degrees of these elements. The monotone catenary degree of m is the maximum of the equal and adjacent catenary degree of m, where the adjacent catenary degree of m is defined as follows: let $\mathsf{L}(m) = \{l_1 < \cdots < l_r\}$, and for every $i \in \{1, \ldots, r\}$ denote by $\mathsf{Z}_{l_i}(m)$ the set of factorizations of m with length l_i; the *adjacent catenary degree* of m is the maximum of the distances $\mathrm{d}(Z_{l_i}, Z_{l_{i+1}})$, $i \in \{1, \ldots, r-1\}$.

Example 8.4 Let us use `numericalsgps` to compute the catenary degrees of $\langle 10, 17, 24, 31, 43 \rangle$.

```
gap> s:=NumericalSemigroup(10,17,24,31,43);
<Numerical semigroup with 5 generators>
gap> MinimalGeneratingSystem(s);
[ 10, 17, 24, 31, 43 ]
gap> CatenaryDegreeOfNumericalSemigroup(s);
6
gap> HomogeneousCatenaryDegreeOfNumericalSemigroup(s);
```

```
11
gap> MonotoneCatenaryDegreeOfNumericalSemigroup(s);
11
gap> EqualCatenaryDegreeOfNumericalSemigroup(s);
11
```

8.3 Tame Degree

Assume that M is an affine semigroup generated by $\{m_1, \ldots, m_e\}$, and let m in M and $x \in \mathsf{Z}(m)$. If there exists $n_1, \ldots, n_e \in \mathbb{N}$ such that $m - \left(\sum_{i=1}^{e} n_i m_i\right) \in M$, then there must be $y = (y_1, \ldots, y_e) \in \mathsf{Z}(m)$ such that $y - (n_1, \ldots, n_e) \in \mathbb{N}^e$. We want to know the smallest possible distance at which we can find such a y. This is the idea of tame degree. We are mostly interested in the case $\sum_{i=1}^{e} n_i m_i = m_j$ for some $j \in \{1, \ldots, e\}$.

Assume that $m - m_i \in M$ for some $i \in \{1, \ldots, e\}$. There is at least an expression of m of the form $m = \lambda_1 m_1 + \cdots + \lambda_e m_e$ with $(\lambda_1, \ldots, \lambda_e) \in \mathbb{N}^e$ and $\lambda_i > 0$, that is, $\lambda = (\lambda_1, \ldots, \lambda_e)$ is a factorization of m with $\lambda_i \neq 0$ (equivalently $\lambda - \mathbf{e}_i \in \mathbb{N}^e$).

The *tame degree* of m with respect to m_i, $\mathsf{t}(m, m_i)$, is the least nonnegative integer t such that for every $z \in \mathsf{Z}(m)$, there exists $z' \in \mathsf{Z}(m)$ with $z' - \mathbf{e}_i \in \mathbb{N}^e$ and $\mathrm{d}(z, z') \leq t$. The tame degree of M with respect to m_i, $\mathsf{t}(M, m_i)$, is the supremum (maximum in this setting, [12]) of all the tame degrees of the elements of $m_i + M$ with respect to m_i.

The tame degree of M, $\mathsf{t}(M)$, is the maximum of the tame degrees of S with respect to all the atoms (affine semigroups are tame and locally tame, [27]). The tame degree of M can be computed by means of the tame degrees of the primitive elements of M ([12]). Recently, a faster approach has been described in [23]. Set $\mathscr{M}_i = \mathrm{Minimals}_{\leq} \mathsf{Z}(m_i + M)$ and $M_i = \{\varphi(z) \mid z \in \mathscr{M}_i\}$. By Dickson's lemma, $\mathscr{M}_i$ and M_i have finitely many elements. Moreover,

$$\mathsf{t}(M, m_i) = \max\left\{\mathsf{t}(m, m_i) \mid m \in M_i\right\}.$$

In [39] there is a procedure to compute M_i (indeed the set of expressions of any ideal of M, not just principal ideals). By using [6] or [1] (or any integer linear programming package) we can also compute this directly as in the following example.

Example 8.5 Let us compute the set M_1 for $M = \langle (2, 0), (0, 2), (1, 1), (1, 2) \rangle$. We need to find the expressions in $(2, 0) + M$. This corresponds with the $(x, y, z, t) \in \mathbb{N}^4$ such that

$$\begin{pmatrix} 2 & 0 & 1 & 1 \\ 0 & 2 & 1 & 2 \end{pmatrix} \begin{pmatrix} x \\ y \\ z \\ t \end{pmatrix} = (2, 0) + \begin{pmatrix} 2 & 0 & 1 & 1 \\ 0 & 2 & 1 & 2 \end{pmatrix} \begin{pmatrix} x' \\ y' \\ z' \\ t' \end{pmatrix}$$

for some $(x', y', z', t') \in \mathbb{N}^4$. This is a system of two equations and eight unknowns. We use the package `4ti2gap` to solve this.

```
gap> m:=[[2,0,1,1,-2,0,-1,-1],[0,2,1,2,0,-2,-1,-2]];
[ [ 2, 0, 1, 1, -2, 0, -1, -1 ], [ 0, 2, 1, 2, 0, -2, -1, -2 ] ]
gap> problem:=["mat",m,"rhs",[[2,0]],"sign",[[1,1,1,1,1,1,1,1]]];
[ "mat", [ [ 2, 0, 1, 1, -2, 0, -1, -1 ], [ 0, 2, 1, 2, 0, -2, -1, -2 ] ],
"rhs", [ [ 2, 0 ] ], "sign", [ [ 1, 1, 1, 1, 1, 1, 1, 1 ] ] ]
gap> ZSolve4ti2(problem);
rec( zhom := [ [ 1, 0, 0, 2, 0, 0, 4, 0 ], [ 0, 0, 4, 0, 1, 0, 0, 2 ],
[ 1, 2, 0, 0, 0, 0, 0, 2 ], [ 0, 0, 1, 0, 0, 0, 1, 0 ],
[ 0, 0, 0, 1, 0, 0, 0, 1 ], [ 0, 1, 2, 0, 0, 0, 0, 2 ],
[ 0, 0, 2, 0, 1, 1, 0, 0 ], [ 0, 1, 0, 0, 0, 1, 0, 0 ],
[ 0, 0, 0, 2, 1, 2, 0, 0 ], [ 1, 0, 0, 0, 1, 0, 0, 0 ],
[ 1, 1, 0, 0, 0, 0, 2, 0 ], [ 0, 0, 0, 2, 0, 1, 2, 0 ] ],
zinhom := [ [ 0, 0, 4, 0, 0, 0, 0, 2 ], [ 0, 0, 2, 0, 0, 1, 0, 0 ],
[ 1, 0, 0, 0, 0, 0, 0, 0 ], [ 0, 0, 0, 2, 0, 2, 0, 0 ] ] )
```

This in particular means that

$$\begin{aligned}\mathsf{Z}((2,0)+M) = \{&(0,0,4,0),(0,0,2,0),(1,0,0,0),(0,0,0,2)\}\\ &+\langle(1,0,0,1),(0,0,4,0),(1,2,0,0),(0,0,1,0),(0,0,2,0),\\ &(0,1,0,0),(0,0,0,2),(1,0,0,0),(1,1,0,0)\rangle\end{aligned}$$

And thus $\mathrm{Minimals}_{\le}\mathsf{Z}((2,0)+M) = \{(0,0,2,0),(1,0,0,0),(0,0,0,2)\}$. Hence $M_1 = \{(2,2),(2,0),(2,4)\}$.

If M is a full affine semigroup (for instance in the case of block monoids), then the elements in $\mathscr{M}_i$ can be computed using [4, Corollary 3.5]. In this case M_i is the set of minimal nonnegative integer solutions of

$$(m_1 \mid \cdots \mid m_e)x^T \ge m_i.$$

Example 8.6 Let us compute as explained in [23] the tame degree of $\mathscr{B}\left(\mathbb{Z}_2^3\right)$.

```
gap> c:=[ [ 0, 0, 1 ], [ 0, 1, 0 ], [ 0, 1, 1 ], [ 1, 0, 0 ], [ 1, 0, 1 ],
[ 1, 1, 0 ], [ 1, 1, 1 ] ];
gap> a:=AffineSemigroup("equations",[TransposedMat(m),[2,2,2]]);;
gap> TameDegreeOfAffineSemigroup(a);
4
```

For numerical semigroups, we have a similar behavior as in the catenary degree. The tame degree is reached in an element that has to do with Apéry sets ([10, Theorem 16])):

$$\mathsf{t}(M) = \max\left\{\mathsf{t}(m) \,\middle|\, m \in \{m_1,\ldots,m_e\} + \left(\bigcup_{i=1}^{e} \mathrm{Ap}(M,m_i)\setminus\{0\}\right)\right\}.$$

For small generators, the above formula is faster than computing minimal factorizations in principal ideals (or if we do not have software to solve linear Diophantine equations over the set of nonnegative integers at hand).

9 ω-Primality

Let M be an affine semigroup. Define on M the following binary relation: $m \leq_M m'$ if $m' - m \in M$. This relation is an order relation (the translation of divisibility to additive notation). We say that $m \in M$ is *prime* if whenever $m \leq_M m' + m''$ for some $m', m'' \in M$, either $m \leq_M m'$ or $m \leq_M m''$. Any prime element must be an atom. But it may happen that no atom is prime (this holds in any nontrivial numerical semigroup). The ω-primality is meant to determine how far an element is from being prime.

The *ω-primality* of m in M, denoted $\omega(m)$, is the least positive integer N such that whenever $m \leq_M a_1 + \cdots + a_n$ for some $a_1, \ldots, a_n \in M$, then $m \leq_M a_{i_1} + \cdots + a_{i_N}$ for some $\{i_1, \ldots, i_N\} \subseteq \{1, \ldots, n\}$.

According to this definition an element is prime provided that its ω-primality is one.

Notice that by definition, $m \leq_M m'$ if and only if m' is in the principal ideal $m + M$. Hence, principal ideals play a fundamental role in the computation of ω-primality (as in the calculation of the tame degree). Indeed in [4, Proposition 3.3] it is shown that

$$\omega(m) = \max\left\{|x| \mid x \in \text{Minimals}_{\leq}(\mathsf{Z}(m + M))\right\}.$$

In [21] the above formula together with the algorithm presented in [39] is used to compute the ω-primality of an element in an affine semigroup. One can also proceed as in Example 8.5 and use for instance `Normaliz` or `4ti2`.

The omega primality of M, if M is minimally generated by $\{m_1, \ldots, m_e\}$, is defined as $\omega(M)$ as the maximum of $\{\omega(m_1), \ldots, \omega(m_e)\}$. Note that the sequence $\{\omega(m)\}_{m \in M}$ is not upper bounded in general.

Example 9.1 According to Example 8.5, for $M = \langle (2, 0), (0, 2), (1, 1), (1, 2) \rangle$, we have $\omega((2, 0)) = 2$. Let us double check it with the `numericalsgps` package.

```
gap> a:=AffineSemigroup([2,0],[0,2],[1,1],[1,2]);;
gap> OmegaPrimalityOfElementInAffineSemigroup([2,0],a);
2
gap> OmegaPrimalityOfAffineSemigroup(a);
4
```

For numerical semigroups, we obtain a similar construction as for the tame degree (as expected, since we are using roughly the same elements in the calculations). In [4, Remarks 5.9] it is shown that if we are looking for minimal factorizations in $\mathsf{Z}(m + M)$, then we only have to search for factorizations of the elements of the form $m + w$ with $w \in \mathrm{Ap}(M, m_i)$ for some $i \in \{1, \dots, e\}$. In [3] an improved method that also uses Apéry sets is given (this is actually the procedure implemented in the package `numericalsgps`; see `contributions.gi` in the package `gap` folder).

Example 9.2 Let us compare the timings for $S = \langle 10, 17, 24, 31, 43 \rangle$.

```
gap> s:=NumericalSemigroup(10,17,24,31,43);;
gap> OmegaPrimalityOfNumericalSemigroup(s);time;
11
13
gap> a:=AsAffineSemigroup(s);;
gap> OmegaPrimalityOfAffineSemigroup(a);time;
11
3654
```

(The timings are in milliseconds.)

If the generators are larger, then the principal ideal approach is better.

```
gap> s:=NumericalSemigroup(201,223,357);;
gap> OmegaPrimalityOfNumericalSemigroup(s);time;
75
32245
gap> a:=AsAffineSemigroup(s);;
gap> OmegaPrimalityOfAffineSemigroup(a);time;
75
1934
```

Acknowledgments The author is supported by the projects MTM2014-55367-P, FQM-343, FQM-5849 and FEDER funds. Thanks to Alfred Gerlondiger for his comments and suggestions, and for encouraging me to write this overview. Also thanks to Alfredo Sánchez-R.-Navarro for his comments. The author also thanks the Centro de Servicios de Informtica y Redes de Comunicaciones (CSIRC), Universidad de Granada, for providing the computing time.

References

1. 4ti2 team, 4ti2–a software package for algebraic, geometric and combinatorial problems on linear spaces. www.4ti2.de
2. M. Barakat, S. Gutsche, S. Jambor, M. Lange-Hegermann, A. Lorenz, O. Motsak, GradedModules, A homalg based package for the Abelian category of finitely presented graded modules over computable graded rings, Version 2014.09.17, (GAP package) (2014). http://homalg.math.rwth-aachen.de
3. T. Barron, C. O'Neill, R. Pelayo, On the computation of delta sets and ω-primality in numerical monoids, preprint, (2014)
4. V. Blanco, P.A. García-Sánchez, A. Geroldinger, Semigroup-theoretical characterizations of arithmetical invariants with applications to numerical monoids and Krull monoids. Illinois J. Math. **55**, 1385–1414 (2011)
5. W. Bruns, B. Ichim, C. Söger, The power of pyramid decompositions in Normaliz. J. Symb. Comput. **74**, 513–536 (2016)
6. W. Bruns, B. Ichim, T. Römer, C. Söger, Normaliz, algorithms for rational cones and affine monoids (2014). http://www.math.uos.de/normaliz

7. L. Bryant, J. Hamblin, The maximal denumerant of a numerical semigroup. Semigroup Forum **86**, 571–582 (2013)
8. M. Bullejos, P.A. García-Sánchez, Minimal presentations for monoids with the ascending chain condition on principal ideals. Semigroup Forum **85**, 185–190 (2012)
9. S.T. Chapman, M. Corrales, A. Miller, C. Miller, D. Phatel, The catenary and tame degrees on a numerical monoid are eventually periodic. J. Aust. Math. Soc. **97**, 289–300 (2014)
10. S.T. Chapman, P.A. García-Sánchez, D. Llena, The catenary and tame degree of a numerical semigroup. Forum Math. **21**, 117–129 (2009)
11. S.T. Chapman, P.A. García-Sánchez, D. Llena, J. Marshall, Elements in a numerical semigroup with factorizations with the same length. Can. Math. Bull. **54**, 39–43 (2011)
12. S.T. Chapman, P.A. García-Sánchez, D. Llena, V. Ponomarenko, J.C. Rosales, The catenary and tame degree in finitely generated commutative cancellative monoids. Manuscripta Math. **120**, 253–264 (2006)
13. S.T. Chapman, P.A. García-Sánchez, D. Llena, A. Malyshev, D. Steinberg, On the delta set and the Betti elements of a BF-monoid. Arab J. Math. **1**, 53–61 (2012)
14. S.T. Chapman, R. Hoyer, N. Kaplan, Delta sets of numerical monoids are eventually periodic. Aequ. math. **77**, 273–279 (2009)
15. E. Contejean, H. Devie, An efficient incremental algorithm for solving systems of linear diophantine equations. Inform. Comput. **113**, 143–172 (1994)
16. M. Costantini, W. de Graaf, Gap package singular; the gap interface to singular (2012). http://gap-system.org/Packages/singular.html
17. W. Decker, G.-M. Greuel, G. Pfister, H. Schönemann, SINGULAR 4.0.0 – A computer algebra system for polynomial computations (2012). http://www.singular.uni-kl.de
18. M. Delgado, P. A. García-Sánchez, J. Morais, "NumericalSgps", a GAP package for numerical semigroups. http://www.gap-system.org
19. D. Eisenbud, B. Sturmfels, Binomial ideals. Duke Math. J. **84**, 1–45 (1996)
20. The GAP Group, GAP – Groups, Algorithms, and Programming, Version 4.7.5 (2014). http://www.gap-system.org
21. J. I. García-García, M.A. Moreno-Frías, A. Vigneron-Tenorio, Computation of the w-primality and asymptotic w-primality with applications to numerical semigroups. arXiv:1307.5807
22. J. I. García-García, M.A. Moreno-Frías, A. Vigneron-Tenorio, Computation of Delta sets of numerical monoids. Israel J. Math. **206**, 395–411 (2016)
23. P. A. García-Sánchez, A new approach for the computation of the tame degree. arXiv:1504.02998
24. P. A. García-Sánchez, D. Llena, A. Moscariello, Delta sets for numerical semigroups with embedding dimension three. arXiv:1504.02116
25. P.A. García-Sánchez, I. Ojeda, A. Sánchez-R.-Navarro, Factorization invariants in half-factorial affine semigroups. Int. J. Algebra Comput. **23**, 111–122 (2013)
26. P. A. García-Sánchez, A. Sánchez-R.-Navarro, 4ti2gap, GAP wraper for 4ti2. https://bitbucket.org/gap-system/4ti2gap
27. A. Geroldinger, F. Halter-Koch, *Non-Unique Factorizations. Algebraic, Combinatorial and Analytic Theory*, vol. 278, Pure and Applied Mathematics (Chapman & Hall/CRC, 2006)
28. S. Gutsche, 4ti2interface, a link to 4ti2 (2013). http://www.gap-system.org/Packages/4ti2interface.html
29. S. Gutsche, M. Horn, C. Söger, Normalizinterface for gap (2014). https://github.com/fingolfin/NormalizInterface
30. J. Herzog, Generators and relations of abelian semigroups and semigroup rings. Manuscripta Math. **3**, 175–193 (1970)
31. C. O'Neill, On factorization invariants and Hilbert functions, preprint
32. A. Philipp, A characterization of arithmetical invariants by the monoid of relations. Semigroup Forum **81**(3), 424–434 (2010)
33. A. Philipp, A characterization of arithmetical invariants by the monoid of relations II: the monotone catenary degree and applications to semigroup rings. Semigroup Forum **90**, 220–250 (2015)

34. A. Philipp, A precise result on the arithmetic of non-principal orders in algebraic number fields. J. Algebra Appl. **11**(1250087), 42 (2012)
35. J. L. Ramírez Alfonsín, *The Diophantine Frobenius Problem* (Oxford University Press, Oxford, 2005)
36. L. Rédei, *The Theory of Finitely Generated Commutative Semigroups* (Pergamon, Oxford, 1965)
37. J.C. Rosales, P.A. García-Sánchez, *Finitely Generated Commutative Monoids* (Nova Science Publishers Inc, New York, 1999)
38. J.C. Rosales, P.A. García-Sánchez, *Numerical Semigroups, Developments in Mathematics 20* (Springer, New York, 2009)
39. J.C. Rosales, P.A. García-Sánchez, J.I. García-García, Irreducible ideals of finitely generated commutative monoids. J. Algebra **238**, 328–344 (2001)
40. J.C. Rosales, P.A. García-Sánchez, J.I. García-García, Atomic commutative monoids and their elasticity. Semigroup Forum **68**, 64–86 (2004)
41. A. Sánchez-R.-Navarro, Linear Diophantine equations and applications, Ph.D. Thesis, in preparation
42. R.J. Valenza, Elasticity of factorizations in number fields. J. Number Theory **39**, 212–218 (1990)
43. A. Zaks, Half-factorial domains. Bull. Amer. Math. Soc. **82**, 721–724 (1976)

Arithmetic of Mori Domains and Monoids: The Global Case

Florian Kainrath

Dedicated to Franz Halter-Koch on the occasion of his 70th birthday

Abstract In (Geroldinger and Hassler, J Algebr 319:3419–3463, 2008) [7] the class of weakly C-monoid has been introduced. We continue to study their arithmetic and give more examples of such monoids.

Keywords Class semigroup · Quasi-finite semigroup · Weakly C-monoid · Mori domain · Tame degree · Local tameness · Elasticity · Catenary degree

1 Introduction

The aim of this paper is to continue the work begun in [7]. There the authors introduced the notion of a weakly C-monoid as a multiplicative model for certain semilocal Mori domains (and other classes of monoids). They used this model to show that such domains are locally tame and have finite catenary degree.

We have two goals here. First, we will study further arithmetical invariants (tame degree and elasticity) of weakly C-monoids (Theorem 6.2 for monoids and Theorem 7.2 for domains). Further we will construct Noetherian domains that are weakly C-monoids but not necessarily semilocal.

My approach to weakly C-monoids is as follows. I will consider a condition (**C**) for a monoid (see Sect. 5). I will show that any monoid satisfying (**C**) is a weakly C-monoid. There are two reasons for introducing the new condition (**C**). First, in

---

F. Kainrath (✉)
Institute for Mathematics and Scientific Computing, University of Graz,
NAWI Graz, Heinrichstraße 36, 8010 Graz, Austria
e-mail: florian.kainrath@uni-graz.at

S. Chapman et al. (eds.), *Multiplicative Ideal Theory and Factorization Theory*,
Springer Proceedings in Mathematics & Statistics 170,
DOI 10.1007/978-3-319-38855-7_8

order to show, that certain weakly C-monoids are locally tame, the authors of [7] need two further conditions. I will show that both of these conditions are implied by **(C)**. Second, all examples of Mori domains I am aware of, which are weakly C-monoids, also satisfy condition **(C)**. Maybe the stronger condition **(C)** will also be useful for a deeper study of the arithmetic of Mori domains considered here (for example set of lengths).

I want to present another view on the results of this paper. Let R be a Noetherian integral domain with complete (hence integral) closure $\widehat{R}$. Assume that the arithmetic of $\widehat{R}$ is in some sense nice. How large can the distance from R to $\widehat{R}$ be, in order that the arithmetic of R behaves nicely? It seems natural to measure this distance by the size of some quotient of $\widehat{R}$ by R. Since arithmetic is a purely multiplicative theory, this quotient should not depend on addition.

In this paper, we construct such a multiplicative quotient (the reduced class semigroup). We show that if this quotient is quasi-finite (see Sect. 4) and if the class group of $\widehat{R}$ is finite (forcing the arithmetic of $\widehat{R}$ to be nice) then the arithmetic of R behaves as expected from experience from orders in algebraic number fields.

2 Generalities on Semigroups and Monoids

We denote by $\mathbb{N}$ the set of nonnegative integers and by $\mathbb{N}^+$ the set of strictly positive integers. For a finite set X let $|X|$ be its cardinality. We use the following notational convention: if we have defined a mathematical object $P(X)$, for all subsets X of some set Y, we set $P(y) = P(\{y\})$ for $y \in Y$.

In this paper a semigroup will be a commutative semigroup having an identity. Accordingly, a ring is a commutative ring with identity, and a subsemigroup of a semigroup S is always supposed to contain the identity of S. A monoid is a cancellative semigroup. We will always use multiplicative notation for a semigroup.

If R is a commutative ring, we denote its underlying multiplicative semigroup again by R. If R is a domain, we denote by $R^\bullet$ the subsemigroup of all nonzero elements of R. It is a monoid.

If S is a semigroup let $S^\times$ be the subgroup of all invertible elements of S. If $U \subset S^\times$ is a subgroup, let S/U be the quotient semigroup of S by the congruence relation $\sim$ defined by $s \sim t \iff s = tu$ for some $u \in U$. We set $S_{\text{red}} = S/S^\times$.

Our references for the theory of semigroups are [10, 11]. But note that our semigroups are their commutative monoids.

In the following let S be a semigroup.

For $n \in \mathbb{N}^+$ and $X_1, \ldots, X_n \subset S$ we set

$$X_1 \ldots X_n = \{x_1 \ldots x_n \mid x_i \in X_i,\ i = 1, \ldots, n\}.$$

Then the power set $\mathbb{P}(S)$ of S together with the law of composition $(X, Y) \mapsto XY$ is again a semigroup. For $X \subset S$ and $n \in \mathbb{N}$ we denote by X^n the nth power of X in

$\mathbb{P}(S)$. For $X, Y \subset S$ we set

$$(X : Y) = \{s \in S \mid sY \subset X\}.$$

If it is necessary to indicate S, we denote this set also by $(X :_S Y)$.

A subset I of S is called an s-ideal, if $IS = I$. Note that as in [11] (but in contrast to [10]) $\emptyset$ is an s-ideal of S. Every union of s-ideals is again an s-ideal. If I is an s-ideal of S then

$$\sqrt{I} = \{s \in S \mid s^n \in I \text{ for some } n \in \mathbb{N}\}$$

is again an s-ideal of S. I is called a radical ideal if $I = \sqrt{I}$.

An s-ideal I of S is called prime if $I \neq S$ and $ab \in I$ implies $a \in I$ or $b \in I$ for all $a, b \in S$. Any union of prime s-ideals is again prime. We denote by s-spec(S) the set of all prime s-ideals of S and by s-spec$(S)^\bullet$ the set of all nonempty prime s-ideals. As in the case of rings we have

$$\sqrt{I} = \bigcap_{\substack{p \in s\text{-spec}(S) \\ I \subset p}} p$$

for any s-ideal I of S [10, Theorem 1.1]. In particular, if $s \in S \setminus S^\times$ then there exists some $p \in s$-spec(S) containing s. We set

$$N(S) = \bigcap_{p \in s\text{-spec}(S)^\bullet} p.$$

(with the usual convention $N(S) = S$, if s-spec$(S)^\bullet$ is empty, i.e., if S is a group). This is an s-ideal of S, which may be empty. Note that, if $N(S) \neq \emptyset$, then $N(S)$ is the smallest nonempty, radical s-ideal of S. If s-spec(S) is finite, then clearly $N(S) \neq \emptyset$.

For $X \subset S$ we denote by

$$[X] = \bigcup_{n \in \mathbb{N}} X^n$$

the semigroup generated by X.

A subsemigroup T of S is called

divisor closed, if $ss' \in T$ implies $s, s' \in T$, for all $s, s' \in S$;
cofinal, if for all $s \in S$ there is some $s' \in S$, such that $ss' \in T$;
saturated, if $st \in T$ implies $s \in T$, for all $s \in S$ and all $t \in T$.

Every divisor-closed semigroup is saturated. If $T' \subset T$ are subsemigroups of S, and if $T' \subset T$ and $T \subset S$ are divisor closed (resp. cofinal, resp. saturated) then $T' \subset S$ has the same property. For a subset $X \subset S$ we denote by $[[X]]_S = [[X]]$ the smallest divisor-closed subsemigroup of S containing X. Then $[[X]]$ consists of those $s \in S$, such that $ss' \in [X]$ for some $s' \in S$.

Note, that $p \subset S$ is a prime s-ideal iff $S \backslash p$ is a divisor-closed subsemigroup. Hence, denoting by $\mathrm{Div}(S)$ the set of all divisor-closed subsemigroups of S, we have a bijective map

$$s\text{-spec}(S) \to \mathrm{Div}(S), \quad p \mapsto S \backslash p.$$

In particular, we have for $s \in S$: $s \in N(S) \iff [[s]] = S$.

Let U be subsemigroup of S. Then we have maps (referred to as natural maps in the following)

$$\mathrm{Div}(S) \to \mathrm{Div}(U), \quad T \mapsto T \cap U \qquad s\text{-spec}(S) \to s\text{-spec}(U), \quad p \mapsto p \cap U.$$

By the above observation, one of them is injective (resp. surjective, resp. bijective) if and only if the other is so.

Let T be a saturated subsemigroup of S. Then the inclusion $T \subset S$ induces an injective homomorphism $T_{\mathrm{red}} \to S_{\mathrm{red}}$, by which we identify T_{red} with a subsemigroup of S_{red}. In particular, we have $T^\times = T \cap S^\times$. Moreover, T_{red} is saturated in S_{red}.

Let $T \subset S$ be a subsemigroup. We denote by $T^{-1}S$ the corresponding semigroup of fractions [11, Chap. II, Sect. 1]. It comes equipped with a canonical homomorphism $j_T \colon S \to T^{-1}S$ such that $j(T) \subset (T^{-1}S)^\times$. Any element $x \in T^{-1}S$ has the form $x = j_T(s)j_T(t)^{-1}$ for some $s \in S$ and some $t \in T$. As usual we set $j_T(s)j_T(t)^{-1} = s/t$. Then, if $s, s' \in S$, $t, t' \in T$, $s/t = s'/t'$ iff $t_0t's = t_0ts'$ for some $t_0 \in T$. In particular, if $T = S$, then $S^{-1}S$ is a group. In case S is a monoid, this group is the quotient group of S, which will be denoted by $\mathsf{q}(S)$. In this case $j_S \colon S \to \mathsf{q}(S) = S^{-1}S$ is injective. Hence we will always suppose $S \subset \mathsf{q}(S)$, if S is a monoid. If T is a submonoid of the monoid S, we always suppose $\mathsf{q}(T) \subset \mathsf{q}(S)$.

If S' is a subsemigroup of S such that $T \subset S'$, then the inclusion $S' \subset S$ induces an injective homomorphism $T^{-1}S' \to T^{-1}S$, by means of which we regard $T^{-1}S'$ as a subsemigroup of $T^{-1}S$.

In particular, we have $T^{-1}T \subset T^{-1}S$ and hence $T^{-1}T \subset (T^{-1}S)^\times$. If T is divisor closed in S, then $T^{-1}T = (T^{-1}S)^\times$.

Now let $\bar{T} = [[T]]_S$. Then the identity of S induces a homomorphism $T^{-1}S \to \bar{T}^{-1}S$, which is easily seen to be an isomorphism. We identify therefore $T^{-1}S = \bar{T}^{-1}S$.

Let R be a commutative ring and $T \in \mathrm{Div}(R)$. If $j \colon R \to T^{-1}R$ is the canonical homomorphism we have

$$T = j^{-1}(T^{-1}R^\times) = R \backslash \bigcup_{\substack{p \in \mathrm{spec}(R) \\ p \cap T = \emptyset}} p.$$

It follows that any $p \in s\text{-spec}(R)$ is a union of prime ideals of R.

Let H be a monoid. We denote by $\widehat{H}$ the set of all those $x \in \mathsf{q}(H)$, for which there exists some $h \in H$ such that $hx^n \in H$, for all $n \in \mathbb{N}$. $\widehat{H}$ is a submonoid of $\mathsf{q}(H)$

containing H, and is called the complete integral closure of H. Let D be a submonoid of H. Then $\widehat{D} \subset \widehat{H}$, since $\mathsf{q}(D) \subset \mathsf{q}(H)$. The s-ideal $(H : \widehat{H}) = (H :_{\mathsf{q}(H)} \widehat{H})$ is called the conductor of H.

For $X \subset \mathsf{q}(H)$ we set $X_v = (H : (H : X))$ (here $(:) = (:_{\mathsf{q}(H)})$). A subset J of $\mathsf{q}(H)$ is called a fractional v-ideal of H, if $J = J_v$ and $(H : J) \neq \emptyset$. A v-ideal of H is a fractional v-ideal of H, that is contained in H.

Together with the multiplication $I * J = (IJ)_v$ the set of all fractional v-ideals of H is a semigroup $\mathcal{F}_v(H)$. Then $x \mapsto xH$ defines an injective homomorphism of groups $\mathsf{q}(H) \to \mathcal{F}_v(H)^\times$, whose cokernel $\mathcal{C}_v(H)$ is called the v-class group of H.

H is called v-Noetherian, if the set of all v-ideals of H satisfies the Ascending Chain Condition. H is called a Krull monoid, if H is v-Noetherian and $H = \widehat{H}$.

Now suppose that R is a domain and $H = R^\bullet$. Then $\widehat{R}^\bullet = \widehat{H}$, where $\widehat{R}$ is the complete integral closure of R. Moreover H is v-Noetherian if and only if R is a Mori domain. Hence H is a Krull monoid if and only if R is a Krull domain. This is all explained in detail in Sect. 2.10 in [6]. For the theory of Mori domains see for example [1–4, 14].

Recall that a monoid F is factorial, if any $x \in F \setminus F^\times$ is a product of prime elements. Suppose that F is factorial. We call a set of prime elements P representative, if any prime element of F is associated to exactly one prime of P. Then any $x \in F$ has a unique representation

$$x = \varepsilon \prod_{p \in P} p^{a_p}$$

with $a_p \in \mathbb{N}$ and $a_p = 0$ for almost all $p \in P$. We set $v_p(x) = a_p$.

3 Class Semigroups

As in the last section S is a semigroup.

For a subset $X \subset S$ and $s, t \in S$ set $s \sim_X t \iff (X : s) = (X : t)$. Since for all s, $t \in S$ we have $(X : st) = ((X : s) : t)$, $\sim_X$ is a congruence relation on S (see also Sect. 2.8 in [6] for the case, that S is a monoid). We denote by $\mathcal{C}(X, S)$ the quotient semigroup of S by $\sim_X$. It is called the class semigroup of (X, S). Let

$$[\cdot]_X^S \colon S \to \mathcal{C}(X, S), \quad s \mapsto [s]_X^S$$

be the canonical homomorphism. For any $Y \subset S$ let $[Y]_X^S$ be the image of Y under $[\cdot]_X^S$. Note the following important property:

$$s \in S,\ x \in X,\ [s]_X^S = [x]_X^S \Rightarrow s \in X.$$

Indeed, $1 \in (X : x) = (X : s)$ and hence $s \in X$.

$U := [S^\times]_X^S \subset \mathscr{C}(X, S)^\times$ is a subgroup. We set $\mathscr{C}(X, S)/U = \mathscr{C}_r(X, S)$ and denote the canonical homomorphism $S \to \mathscr{C}(X, S) \to \mathscr{C}_r(X, S)$ by

$$\{\cdot\}_X^S : S \to \mathscr{C}_r(X, S), \quad s \mapsto \{s\}_X^S.$$

Then by definition we have for all $s, t \in S$: $\{s\}_X^S = \{t\}_X^S$ iff $[s]_X^S = [\varepsilon t]_X^S$ for some $\varepsilon \in S^\times$. For $Y \subset X$ let $\{Y\}_X^S$ be the image of Y under $\{\cdot\}_X^S$. We call $\mathscr{C}_r(X, S)$ the reduced class semigroup of $X \subset S$.

We apply these constructions almost always only in the case of an X that is a subsemigroup of S. Only in the proof of Theorem 7.2 we need the case of an X, that is not a subsemigroup of S.

In the following we state and prove the properties of class semigroups, that we use later on.

Lemma 3.1 *Let $T \subset T'$ be subsemigroups of S.*

1. *Let $\varepsilon \in S^\times$. Then $[\varepsilon]_T^S = 1$ if and only if $\varepsilon \in T^\times$.*
2. *Let U be a subsemigroup of S and $u, v \in U$. Then $[u]_T^S = [v]_T^S$ implies $[u]_{U\cap T}^U = [u]_{U\cap T}^U$.*
3. *Suppose that T is cofinal in S. Then the following are equivalent:*
   a. *T is saturated in S.*
   b. *If $s \in S$, then $[s]_T^S = 1$ if and only if $s \in T$.*
   c. *If $s_1, s_2 \in S$, then $[s_1]_T^S = [s_2]_T^S$ if and only if $t_1 s_1 = s_2 t_2$ for some $t_1, t_2 \in T$.*
4. *Suppose that T' is saturated in S. If $t_1', t_2' \in T'$, then $[t_1']_T^{T'} = [t_2']_T^{T'}$ if and only if $[t_1']_T^S = [t_2']_T^S$.*
5. *Suppose that T' is saturated in S and that T is cofinal in T'. If $s_1, s_2 \in S$ and $[s_1]_T^S = [s_2]_T^S$, then $[s_1]_{T'}^S = [s_2]_{T'}^S$*

Proof 1. Let $\varepsilon \in S^\times$. Then $(T : \varepsilon) = \varepsilon^{-1}T$. Hence $[\varepsilon]_T^S = 1$ if and only if $\varepsilon^{-1}T = T$, which is equivalent to $\varepsilon \in T^\times$.

2. If $u \in U$ then $(T \cap U :_U u) = U \cap (T :_S u)$, from which the assertion follows.

3. Suppose that T is cofinal in S.

a $\Rightarrow$ b. We assume that T is saturated in S. We consider some $s \in S$ and suppose first $[s]_T^S = 1$, so that $(T : s) = T$. Since T is cofinal in S there exists some $s' \in S$ such that $ss' \in T$. Then $s' \in (T : s) = T$. Since T is saturated in S we obtain $s \in T$.

Conversely assume $s \in T$. Since T is saturated in S we have $(T : s) \subset T$. The other inclusion $T \subset (T : s)$ follows from $s \in T$. Hence $(T : s) = T$ which implies $[s]_T^S = 1$.

b $\Rightarrow$ c. So let us assume that b is satisfied, and consider $s_1, s_2 \in S$. If $s_1 t_1 = s_2 t_2$ for some $t_1, t_2 \in T$, then b immediately implies $[s_1]_T^S = [s_2]_T^S$.

Conversely, assume $[s_1]_T^S = [s_2]_T^S$. Since T is cofinal in S we may choose some $x \in S$ such that $t_2 := s_1 x \in T$. From $[s_1]_T^S = [s_2]_T^S$ we obtain $t_1 := s_2 x \in T$, too. Then we have $t_1 s_1 = t_2 s_2$.

c ⇒ a. So assume c and let $t, t' \in T$ and $s \in S$ be such that $ts = t'$. Then c implies $[s]_T^S = [1]_T^S$. Since $1 \in T$ we also have $s \in T$.

4. Since T' is saturated in S we have $(T :_{T'} t') = (T :_S t')$ for any $t' \in T'$, from which the assertion follows.

5. Let $s_1, s_2 \in S$ be such that $[s_1]_T^S = [s_2]_T^S$. By symmetry it is enough to show $(T' : s_1) \subset (T' : s_2)$. So let $u \in (T' : s_1)$. Then $us_1 \in T'$ and since T is cofinal in T' we have $t'us_1 \in T$ for some $t' \in T'$. From $[s_1]_T^S = [s_2]_T^S$ we obtain $t'us_2 \in T \subset T'$. Since T' is saturated in S we conclude $u \in (T' : s_2)$.

From Lemma 3.1.3 we obtain the following corollary, which is well known for monoids [6, Proposition 2.8.7.3].

Corollary 3.2 *Suppose that T is a cofinal and saturated subsemigroup of S. Then $\mathscr{C}(T, S)$ and $\mathscr{C}_r(T, S)$ are groups, and there is an exact sequence*

$$1 \longrightarrow T^{-1}T \stackrel{\iota}{\longrightarrow} S^{-1}S \stackrel{\phi}{\longrightarrow} \mathscr{C}(T, S) \longrightarrow 1,$$

where ι is induced by the inclusion $T \subset S$ and $\phi(s/1) = [s]_T^S$, for all $s \in S$.

Proof Since T is cofinal in S, we have $[[T]]_S = S$. Hence $S^{-1}S = T^{-1}S$. In particular, ι is injective.

Let $s \in S$. Then $ss' \in T$ for some $s' \in S$, since T is cofinal in S. By Lemma 3.1.3 we obtain $[ss']_T^S = 1$. Hence $\mathscr{C}(T, S)$ is a group. Since $\mathscr{C}_r(T, S)$ is a homomorphic image of $\mathscr{C}(T, S)$ it is a group, too.

Since $\mathscr{C}(T, S)$ is a group, there is an unique homomorphism $\phi\colon S^{-1}S \to \mathscr{C}(T, S)$ such that $\phi(s/1) = [s]_T^S$, for all $s \in S$. It is clearly onto. It remains to show, that the kernel of ϕ equals the image of ι. So let $s_1, s_2 \in S$. Then $\phi(s_1/s_2) = 1$ iff $[s_1]_T^S = [s_2]_T^S$. By Lemma 3.1.3 this is equivalent to $t_1 s_1 = t_2 s_2$ for some $t_1, t_2 \in T$. Clearly, this is equivalent to $s_1/s_2 \in \iota(T^{-1}T)$.

Suppose that $T \subset S$ is a cofinal and saturated subsemigroup. Then we call $\mathscr{C}_r(T, S)$ the class group of $T \subset S$. If S is a monoid, then it follows from Corollary 3.2 that $\mathscr{C}_r(T, S) \cong \mathsf{q}(S)/(S^\times \mathsf{q}(T))$. Hence our terminology is consistent with that in [7] in the case where T is a saturated and cofinal submonoid of S.

Lemma 3.3 *Let $U \subset T$ be subsemigroups of S. Then there is a surjective homomorphism $f\colon \mathscr{C}(T, S) \to \mathscr{C}(U^{-1}T, U^{-1}S)$ such that $f([s]_T^S) = [s/1]_{U^{-1}T}^{U^{-1}S}$, for all $s \in S$.*

Proof Let $s \in S$. A standard calculation shows that

$$\left(U^{-1}T : \frac{s}{1}\right) = \left\{\frac{x}{u} \mid x \in (T : s),\, u \in U\right\}.$$

Hence the existence of f follows:

If $u \in U$ then $1/u \in (U^{-1}T)^\times$ and hence

$$[1/u]_{U^{-1}T}^{U^{-1}S} = 1$$

by Lemma 3.1.1. Therefore, f is surjective.

The next result is a kind of a third isomorphism theorem for reduced class semigroups.

Proposition 3.4 *Let $T \subset T'$ be subsemigroups of S and assume, that T is cofinal in S and that T' is saturated in S. Then $\{T'\}_T^S$ is cofinal and saturated in $\mathscr{C}_r(T, S)$ and there are isomorphisms*

$$\{T'\}_T^S \cong \mathscr{C}_r(T, T') \quad \textit{and} \quad \mathscr{C}(\{T'\}_T^S, \mathscr{C}_r(T, S)) \cong \mathscr{C}_r(T', S).$$

Proof Note first that T' is cofinal in S (since T is so) and that T is cofinal in T'. Indeed, if $t' \in T'$ then $t's \in T \subset T'$ for some $s \in S$. Since T' is saturated in S, it follows $s \in T'$.

For notational convenience we set $S_0 = \mathscr{C}_r(T, S)$ and $T_0' = \{T'\}_T^S$. Since T' is cofinal in S, T_0' is cofinal in S_0. Now let $s \in S$ and $t_1', t_2' \in T'$ be such that $\{st_1'\}_T^S = \{t_2'\}_T^S$. Then $[st_1']_T^S = [\varepsilon t_2']_T^S$ for some $\varepsilon \in S^\times$. Using Lemma 3.1.5 we obtain $[st_1']_{T'}^S = [\varepsilon t_2']_{T'}^S$. Using part 3 of that Lemma we obtain $[\varepsilon^{-1}s]_{T'}^S = 1$ and therefore $\varepsilon^{-1}s \in T'$. Hence $\{s\}_T^S = \{\varepsilon^{-1}s\}_T^S \in T_0'$, and T_0' is saturated in S_0.

Next, let $t_1', t_2' \in T'$. We show $\{t_1'\}_T^S = \{t_2'\}_T^S$ if and only if $\{t_1'\}_T^{T'} = \{t_2'\}_T^{T'}$. From this we obtain the desired isomorphism $T_0' \cong \mathscr{C}_r(T, T')$. So, assume first $\{t_1'\}_T^S = \{t_2'\}_T^S$. Then

$$[t_1']_T^S = [\varepsilon t_2']_T^S \tag{1}$$

for some $\varepsilon \in S^\times$. Applying first Lemma 3.1.5 and then Lemma 3.1.3 we obtain $\varepsilon \in T' \cap S^\times = T'^\times$. From Eq. (1) and Lemma 3.1.4 we obtain now $[t_1']_T^{T'} = [\varepsilon t_2']_T^{T'}$ and hence $\{t_1'\}_T^{T'} = \{t_2'\}_T^{T'}$.

Conversely, suppose that this equation holds. Then $[t_1']_T^{T'} = [\varepsilon t_2']_T^{T'}$ for some $\varepsilon \in T'^\times$. Using Lemma 3.1.4 again, we obtain $[t_1']_T^S = [\varepsilon t_2']_T^S$ and therefore $\{t_1'\}_T^S = \{t_2'\}_T^S$.

To construct the second isomorphism, it is enough to show

$$\{s_1\}_{T'}^S = \{s_2\}_{T'}^S \iff [\{s_1\}_T^S]_{T_0'}^{S_0} = [\{s_2\}_T^S]_{T_0'}^{S_0},$$

for all $s_1, s_2 \in S$.

So let $s_1, s_2 \in S$ and suppose first $\{s_1\}_{T'}^S = \{s_2\}_{T'}^S$. Then $[s_1]_{T'}^S = [\varepsilon s_2]_{T'}^S$ for some $\varepsilon \in S^\times$ and by Lemma 3.1.3 we obtain $s_1 t_1' = \varepsilon s_2 t_2'$ for some $t_1', t_2' \in T'$. Using, that T_0' is saturated in S_0, and using Lemma 3.1.3 again we obtain $[\{s_1\}_T^S]_{T_0'}^{S_0} = [\{s_2\}_T^S]_{T_0'}^{S_0}$.

Conversely, assume that this equation holds. Then $\{s_1\}_T^S\{t_1'\}_T^S = \{s_2\}_T^S\{t_2'\}_T^S$ for some $t_1', t_2' \in T'$ (again since T_0' is saturated in S_0 and Lemma 3.1.3). Hence $[s_1 t_1']_T^S =$

$[\varepsilon s_2 t_2']_T^S$ for some $\varepsilon \in S^\times$. Using parts 3 and 5 of Lemma 3.1 we obtain $[s_1]_{T'}^S = [\varepsilon s_2]_{T'}^S$ and hence $\{s_1\}_{T'}^S = \{s_2\}_{T'}^S$.

We investigate now when the natural map $\mathrm{Div}(S) \to \mathrm{Div}(T)$ is surjective or injective. For that we introduce the notion of a zero complete semigroup: a semigroup S is called to be zero complete, if any subsemigroup T of S has a zero, i.e., there exists a (necessarily unique) $0 \in T$ such that $0t = 0$, for all $t \in T$.

Lemma 3.5 *Let $T \subset S$ be semigroups and $f\colon \mathrm{Div}(S) \to \mathrm{Div}(T)$, $g\colon s\text{-}spec(S) \to s\text{-}spec(U)$ be the natural maps.*

1. *Suppose that, for all $s \in S$ we have $\varepsilon s^n \in T$ for some $\varepsilon \in S^\times$ and some $n \in \mathbb{N}^+$. Then $V = [[V \cap T]]_S$, for all $V \in \mathrm{Div}(S)$. Hence f and g are injective.*
2. *Suppose that $[T]_T^S$ is zero complete. Then $U = [[U]]_S \cap T$ for any $U \in \mathrm{Div}(T)$. In particular, f and g are surjective.*
3. *Suppose that T is cofinal and saturated in S and that $\mathscr{C}_r(T, S)$ is finite. Then f and g are bijective. Moreover $T \subset S$ satisfies the assumption of 1.*

Proof 1 is obvious. 2. Let U be a divisor-closed subsemigroup of T. The inclusion $U \subset T \cap [[U]]_S$ is clear. Conversely, suppose that $t \in T \cap [[U]]_S$. Then $ts \in U$ for some $s \in S$. Let $n \in \mathbb{N}$ be such that $[t^n]_T^S$ is a zero of the subsemigroup of $[T]_T^S$ generated by $[t]_T^S$. Then

$$[t^{n+1}]_T^S = [t^n]_T^S [t]_T^S = [t^n]_T^S.$$

Hence $t^{n+1}s^{n+1} \in U \subset T$ implies $t_1 = t^n s^{n+1} \in T$. From $tt_1 = (ts)^{n+1} \in U$ we deduce now $t \in U$.

3. Since $\mathscr{C}_r(T, S)$ is finite, it follows from Lemma 3.1.3, that $T \subset S$ satisfies the assumption of 1. Hence f and g are injective.

Again by Lemma 3.1.3, we have $[T]_T^S = 1$. Therefore $[T]_T^S$ is zero complete. By 2 f and g are surjective.

We close this section by showing how zero complete semigroups enter the scene in the theory to be developed later on.

Lemma 3.6 *Let H be a v-Noetherian monoid such that $(H : \widehat{H}) \neq \emptyset$. Then the semigroup $[H]_H^{\widehat{H}}$ is zero complete.*

Proof Let $\bar{T}$ be a subsemigroup of $[H]_H^{\widehat{H}}$ and let T be its inverse image in H. We consider the set I of all fractional v-ideals of the form $(H :_{\widehat{H}} t)$ with $t \in T$. They are all contained in $\widehat{H}$, and $\widehat{H}$ is H-fractional. Since H is v-Noetherian, we may choose $t \in T$ such that $(H :_{\widehat{H}} t)$ is maximal in I.

Now let $s \in T$ be arbitrary. Then $(H :_{\widehat{H}} t) \subset (H :_{\widehat{H}} st)$ implies $(H :_{\widehat{H}} t) = (H :_{\widehat{H}} st)$ by the maximal choice of t. Hence $[st]_H^{\widehat{H}} = [t]_H^{\widehat{H}}$, for all $s \in T$. Therefore $[t]_H^{\widehat{H}}$ is a zero of $\bar{T}$.

4 Quasi-Finite Semigroups

Again S denotes a semigroup.

Some $s \in S$ is called periodic if the sequence $(s^n)_{n\in\mathbb{N}}$ is (ultimately) periodic. This is equivalent to the following

1. $[s]$ is finite.
2. $s^n = s^{2n}$ for some $n \in \mathbb{N}^+$.
3. $s^k = s^l$ for some distinct $k, l \in \mathbb{N}$.

(for the equivalence of 1 and 3 see [10, Theorem 2.1]; the implication $2 \Rightarrow 3$ is trivial; if 3 holds with $k < l$ then $s^n = s^{2n}$ for $n = a(l-k)$ where $a \in \mathbb{N}$ is such that $a(l-k) \geq k$). By that Theorem we also have: if $|[s]| = k$, then

$$[s] = \{1, s, s^2, \dots, s^{k-1}\}.$$

We call S a bounded periodic semigroup, if there is some $N \in \mathbb{N}^+$ such that, for all $s \in S$ there exists $0 \leq k < l \leq N$ such that $s^k = s^l$, or equivalently there exists some $N \in \mathbb{N}^+$ such that $s^N = s^{2N}$, for all $s \in S$.

We call a subset X of S periodic, if it is a periodic element of the semigroup $\mathbb{P}(S)$. Let I be an s-ideal of S. Then for $m > n$ we have $I^m \subset I^{n+1} \subset I^n$. Hence I is periodic if and only if $I^{n+1} = I^n$ for some $n \in \mathbb{N}$.

Lemma 4.1 *Let X and Y be periodic subsets of S. Then $X \cup Y$ is periodic.*

Proof Since X and Y are periodic, there is some $n \in \mathbb{N}^+$ such that $X^n = X^{2n}$ and $Y^n = Y^{2n}$. Let $k \in \mathbb{N}$. If $0 \leq k \leq 2n-1$, then $4n - k \geq 2n$, and hence $Y^{4n-k} = Y^{3n-k}$. If $2n \leq k \leq 4n$, then $X^k = X^{k-n}$. Using this we obtain

$$\begin{aligned}(X \cup Y)^{4n} &= \bigcup_{0\leq k\leq 4n} X^k Y^{4n-k} = \bigcup_{0\leq k\leq 2n-1} X^k Y^{4n-k} \cup \bigcup_{2n\leq k\leq 4n} X^k Y^{4n-k}\\ &= \bigcup_{0\leq k\leq 2n-1} X^k Y^{3n-k} \cup \bigcup_{2n\leq k\leq 4n} X^{k-n} Y^{4n-k}\\ &= \bigcup_{0\leq k\leq 2n-1} X^k Y^{3n-k} \cup \bigcup_{n\leq k\leq 3n} X^k Y^{3n-k}\\ &= \bigcup_{0\leq k\leq 3n} X^k Y^{3n-k} = (X \cup Y)^{3n}.\end{aligned}$$

Hence $X \cup Y$ is periodic.

Definition 4.2 Let $n \in \mathbb{N}^+$. A n-cover of S is a finite cover $S = S_1 \cup \dots \cup S_r$ of S, such that S_i^n is a singleton, for all $i = 1, \dots, r$. We call S quasi-finite, if it has a n-cover for some $n \in \mathbb{N}^+$.

Remarks 4.3 It follows immediately from this definition that:

1. Every finite semigroup is quasi-finite.
2. Every subsemigroup and every homomorphic image of a quasi-finite semigroup is quasi-finite.
3. Any finite product of quasi-finite semigroups is quasi-finite.
4. Every semigroup of fractions of a quasi-finite semigroup is quasi-finite.
5. If S is quasi-finite, then S possesses a n-cover $S = S_1 \cup \cdots \cup S_r$, such that the S_i are pairwise disjoint.

It is easy to construct infinite, quasi-finite semigroups. For an example of such a semigroup, that occurs in the arithmetical theory of Noetherian domains to be developed later on, we refer to Example 5.7.

Next we list some properties of quasi-finite semigroups.

Lemma 4.4 *Let S be a quasi-finite semigroup.*

1. *S is a bounded periodic semigroup.*
2. *Each subset of S is periodic.*
3. *If I is an s-ideal of S, then there exists $m \in \mathbb{N}$ and a finite $E \subset S$ such that $I^{m+1} = I^m \subset ES \subset I$.*
4. *If S is a group, then S is finite. More precisely: if $S = S_1 \cup \cdots \cup S_r$ is a n-cover, then each S_i is a singleton.*

Proof Let $S = S_1 \cup \cdots \cup S_r$ be a n-cover of S. For $1 \le i \le r$ let $s_i \in S$ be such that $S_i^n = \{s_i\}$.

1. Let $s \in S$. Then there exists $0 \le k < l \le r$ and some $1 \le i \le r$ such that s^k, $s^l \in S_i$. Hence $s^{kn} = s^{ln}$ and $0 \le kn < ln \le rn$.

2. Let $X \subset S$ and set $X_i = X \cap S_i$ for $i = 1, \ldots, r$. By Lemma 4.1 it suffices to show that every X_i is periodic. For that we may assume $X_i \ne \emptyset$. Then $X_i^n = S_i^n = \{s_i\}$. By 1 $\{s_i\}$ is periodic and hence X_i is periodic, too.

3. Let I be an s-ideal of S. Then I is periodic by 2. Hence there is some $n_1 \in \mathbb{N}$ such that $I^{n_1+1} = I^{n_1}$. We set $E = I \cap \{s_1, \ldots, s_r\}$, and $n_2 = r(n-1)+1$. If $x_1, \ldots, x_{n_2} \in I$, then there is some $N \subset \{1, \ldots, n_2\}$ and some $i \in \{1, \ldots, r\}$ such $|N| = n$ and $x_j \in S_i$, for all $j \in N$. Then $x_1 \ldots x_{n_2} \in \prod_{j \in N} x_j S = s_i S \subset ES$. We obtain $I^{n_2} \subset ES \subset I$. Now $m = \max\{n_1, n_2\}$ has the requested property.

4. Suppose that S is a group. Let $1 \le i \le r$ and $s, t \in S_i$. Then $s^{n-1}t$, $s^n \in S_i^n$ and hence $s^{n-1}t = s_i = s^n$. Since S is a group we obtain $s = t$.

Our next goal is an ideal–theoretical description of quasi-finite semigroups (Theorem 4.6). One idea used in its proof will be needed later again, so we separate it into a lemma.

Lemma 4.5 *Let S be a semigroup, $\emptyset \ne X \subset N(S)$, $E \subset S$ a finite set and $n \in \mathbb{N}$. Suppose that $N(S)^{n+1} = N(S)^n \subset ES \subset N(S)$ and that $x/1 = y/1$ in $S^{-1}S$ for all x, $y \in X$. Then X^{n+1} is a singleton.*

Proof Let $x, y \in X$. Since $x/1 = y/1$ in $S^{-1}S$, the s-ideal $I_{x,y} = \{s \in S \mid sx = sy\}$ is not empty. Since by assumption $N(S) \neq \emptyset$ we obtain $ES \subset N(S) \subset \sqrt{I_{x,y}}$. Since E is finite, there is some $m \in \mathbb{N}^+$ such that $(ES)^m \subset I_{x,y}$. From $nm \geq n$ and $N(S)^{n+1} = N(S)^n$ we obtain $N(S)^n = N(S)^{nm} \subset (ES)^m \subset I_{x,y}$.

Now let $x_1, \ldots, x_{n+1}, y_1, \ldots, y_{n+1} \in X$. We show $x_1 \ldots x_{n+1} = y_1 \ldots y_{n+1}$, which will finish the proof. To do this, we show by induction on $i \in \{0, \ldots, n+1\}$ that $x_1 \ldots x_{n+1} = y_1 \ldots y_i x_{i+1} \ldots x_{n+1}$. If $i = 0$ this is clear. So suppose that $i \in \{0, \ldots, n\}$ and that $x_1 \ldots x_{n+1} = y_1 \ldots y_i x_{i+1} \ldots x_{n+1}$. Then by above we obtain

$$y_1 \ldots y_i x_{i+2} \ldots x_{n+1} \in N(S)^n \subset I_{x_{i+1}, y_{i+1}}.$$

Hence $y_1 \ldots y_i x_{i+1} \ldots x_{n+1} = y_1 \ldots y_{i+1} x_{i+2} \ldots x_{n+1}$.

Theorem 4.6 *Let S be a semigroup. Then S is quasi-finite if and only if S has the following properties:*

1. *s-spec(S) is finite.*
2. *For every divisor-closed subsemigroup T of S the group $T^{-1}T$ is finite.*
3. *For every divisor-closed subsemigroup T of S, there exists some $n \in \mathbb{N}$ and some finite $E \subset T$, such that $N(T)^{n+1} = N(T)^n \subset ET \subset N(T)$.*

Proof We show first that, if S is quasi-finite, then S has properties 1–3. We choose a n-cover $S = S_1 \cup \cdots \cup S_r$ and let for $i = 1, \ldots, r$ $s_i \in S$ be such that $S_i^n = \{s_i\}$.

1. Let $p \in s$-spec(S) and let E be the set of those s_i that are contained in p. We show $p = \sqrt{ES}$. The inclusion $\supset$ is clear. Conversely, let $s \in p$. Then $s \in S_i$ for some $i \in [1, r]$. Then $s^n = s_i$ and $s_i \in E$. Hence $s^n \in ES$ and $s \in \sqrt{ES}$.

2. This follows from Remarks 4.3.2, 4.3.4 and Lemma 4.4.4.

3. Let T be a divisor-closed subsemigroup of S. Then T is quasi-finite (Remark 4.3.2). Hence $N(T)$ has the requested property by Lemma 4.4.3.

We now assume, that S has properties 1–3 and show that S is quasi-finite. If $s \in S$ then $s \in N([[s]])$ and hence

$$S = \bigcup_{T \in \mathrm{Div}(S)} N(T).$$

Now let $T \in \mathrm{Div}(S)$. We construct a cover $N(T) = N(T)_1 \cup \cdots \cup N(T)_r$ and some $n(T) \in \mathbb{N}^+$ such that $N(T)_i^{n(T)}$ is a singleton for each $i = 1, \ldots, r$. Since $\mathrm{Div}(S)$ is finite by 1, it then follows S has a n-cover, where

$$n = \prod_{T \in \mathrm{Div}(S)} n(T).$$

By assumption there exists some $m \in \mathbb{N}^+$ and some finite $E \subset T$ such that $N(T)^{m+1} = N(T)^m \subset ET \subset N(T)$.

Let $j \colon T \to T^{-1}T$ be the canonical homomorphism. For $z \in T^{-1}T$ let $N(T)_z$ be the set of those $x \in N(T)$ such that $j(x) = z$. By 2. it suffices to show for $z \in T^{-1}T$, that $N(T)_z^{n+1}$ is a singleton, if $N(T)_z \neq \emptyset$. But this follows from Lemma 4.5.

We draw Corollaries from that theorem, which will prove useful later on.

Corollary 4.7 *Let S be a semigroup and U a subsemigroup of S.*

1. *If U is a finite subgroup of $S^\times$, such that S/U is quasi-finite, then S is quasi-finite.*
2. *Suppose that U is quasi-finite, cofinal and saturated in S and that $\mathscr{C}(U, S)$ is finite. Then S is quasi-finite.*

Proof We show in both cases that S has properties 1–3 of Theorem 4.6.

1. We use that S/U has properties 1–3 of Theorem 4.6. We let $\pi : S \to S/U$ be the canonical homomorphism. Since the ideal theories in S and S/U are essentially the same, S has properties 1 and 3 of Theorem 4.6.

Next let T be a divisor-closed subsemigroup of S and set $\pi(T) = \bar{T}$. Then $\bar{T}$ is a divisor-closed subsemigroup of S/U. A routine calculation shows, that there is a surjective homomorphism $\pi' : T^{-1}T \to \bar{T}^{-1}\bar{T}$ such that $\pi'(t_1/t_2) = \pi(t_1)/\pi(t_2)$ for all $t_1, t_2 \in T$. Let $j\colon U \to T^{-1}T$ be defined by $j(u) = u/1$, for all $u \in U$. Again, a standard calculation shows that the sequence of abelian groups

$$U \xrightarrow{j} T^{-1}T \xrightarrow{\pi'} \bar{T}^{-1}\bar{T} \to 1$$

is exact. Hence $T^{-1}T$ is finite.

2. We use, that U has the properties of Theorem 4.6. Applying Lemma 3.5.3 (since $\mathscr{C}_r(U, S)$ is a homomorphic image of $\mathscr{C}(U, S)$ it is finite, too), we see that the natural maps $\mathrm{Div}(S) \to \mathrm{Div}(U)$, $s\text{-spec}(S) \to s\text{-spec}(U)$ are bijective. In particular, $s\text{-spec}(S)$ is finite.

Next, let T be a divisor-closed subsemigroup of S and set $T' = T \cap U$. From

$$T' \subset [[T']]_S \cap U \subset T \cap U = T'$$

we obtain $[[T']]_S \cap U = T \cap U$ and hence $T = [[T']]_S$. Therefore $T^{-1}S = T'^{-1}S$. By Lemma 3.3 we have a surjective homomorphism $\mathscr{C}(U, S) \to \mathscr{C}(T'^{-1}U, T'^{-1}S)$. Hence $\mathscr{C}(T'^{-1}U, T'^{-1}S)$ is finite. By Lemma 3.1.1 we have an injective homomorphism

$$(T'^{-1}S)^\times/(T'^{-1}U)^\times \to \mathscr{C}(T'^{-1}U, T'^{-1}S).$$

Hence the group $(T'^{-1}S)^\times/(T'^{-1}U)^\times$ is finite. By assumption $(T'^{-1}U)^\times$ is finite, hence $(T'^{-1}S)^\times = (T^{-1}S)^\times = (T^{-1}T)^\times = T^{-1}T$ is finite, too.

Next we show, that any subset X of S is periodic. Intersecting X with the fibers of $[\cdot]_U^S$ and using Lemma 4.1, we may assume, that X is contained in one fiber of $[\cdot]_U^S$. Let e be the exponent of the finite group $\mathscr{C}(U, S)$. Then by Lemma 3.1.3 we have $X^e \subset U$. Hence X^e is periodic by Lemma 4.4.2. Therefore X is periodic, too.

Now let T be any divisor-closed subsemigroup of S. Since $N(T)$ is periodic we have $N(T)^{n_1+1} = N(T)^{n_1}$ for some $n_1 \in \mathbb{N}$.

Since $\mathrm{Div}(S) \to \mathrm{Div}(U)$ is bijective, also $\mathrm{Div}(T) \to \mathrm{Div}(T \cap U)$ and therefore $s\text{-spec}(T) \to s\text{-spec}(T \cap U)$ is bijective. Hence $N(T) \cap U = N(T \cap U)$.

Let d be the Davenport constant of the finite group $\mathscr{C}(U, S)$. This is the smallest integer n, such that any finite sequence of length $\geq n$ in $\mathscr{C}(U, S)$ has a subsequence, whose product is 1 [6, 5.1.4]. Then using Lemma 3.1.3 we obtain

$$\cdot N(T)^d \subset (N(T) \cap U)T = N(T \cap U)T.$$

Let $E \subset U \cap T$ be finite and $n_2 \in \mathbb{N}$ be such that $N(U \cap T)^{n_2} \subset E(U \cap T) \subset N(T \cap U)$. Then

$$N(T)^{dn_2} \subset (N(T \cap U)T)^{n_2} = N(T \cap U)^{n_2}T \subset ET \subset N(T).$$

Hence $n = \max\{n_1, dn_2\}$ and E have the properties requested in Theorem 4.6.3.

Corollary 4.8 *Let $U \subset S$ be semigroups, $\Gamma \subset S^\times$ a subgroup, such that S/Γ is quasi-finite. We suppose that s-spec$(S) \to$ s-spec(U) is injective and that U is zero complete.*

Then there exists some $m \in \mathbb{N}^+$ such that for any $s \in S$, there is some $\gamma \in \Gamma$ such that $\gamma s^m \in U$.

Proof Since S/Γ is quasi-finite we find by Lemma 4.4.1 some $m \in \mathbb{N}^+$ such that, for all $s \in S$ we have $s^{2m} = \gamma s^m$ for some $\gamma \in \Gamma$.

Let $s \in S$ and suppose $n \in \mathbb{N}^+$ and $\gamma \in \Gamma$ are such that $\gamma s^n \in U$. Now $s^{nm} = \gamma_0 s^m$ for some $\gamma_0 \in \Gamma$. Hence $\gamma^m \gamma_0 s^m = (\gamma s^n)^m \in U$. So we only have to find some $n \in \mathbb{N}^+$ and $\gamma \in \Gamma$ such that $\gamma s^n \in U$.

Set $\bar{T} = [[s]]_S$, $T = U \cap \bar{T}$. Since U is zero complete, $[U]_U^S$ is so, too. Lemma 3.5.2 implies now

$$T = U \cap [[T]]_S \subset U \cap \bar{T} = T$$

and therefore $T = U \cap \bar{T} = U \cap [[T]]_S$. Since s-spec$(S) \to$ s-spec(U) (and therefore Div$(S) \to$ Div(U)) is injective we obtain $\bar{T} = [[T]]_S$.

Let 0 be a zero of T and set $\widehat{T} = \{s' \in S \mid s'0 = 0\}$. Then $\widehat{T}$ is a subsemigroup of S such that $T \subset \widehat{T} \subset \bar{T}$. Let $v, w \in \widehat{T}$ and $s' \in \bar{T}$ be such that $s'v = w$. Then $s'0 = s'v0 = w0 = 0$, which implies $s' \in \widehat{T}$. Therefore $\widehat{T}$ is saturated in $\bar{T}$. Since $T \subset \widehat{T} \subset \bar{T} = [[T]]_S$ $\widehat{T}$ is also cofinal in $\bar{T}$. Hence $\mathscr{C}(\widehat{T}, \bar{T})$ is a group. Now the group $\mathscr{C}(\widehat{T}, \bar{T})/[\Gamma]_{\widehat{T}}^{\bar{T}}$ is a homomorphic image of $\bar{T}/\Gamma$. Since $\bar{T} \in$ Div(S) we have $\bar{T}/\Gamma \subset S/\Gamma$. If follows that $\mathscr{C}(\widehat{T}, \bar{T})/[\Gamma]_{\widehat{T}}^{\bar{T}}$ is quasi-finite, and therefore finite (being a group). Therefore, for all $s' \in \bar{T}$ we have $\gamma' s'^k \in \widehat{T}$ for some $\gamma' \in \Gamma$ and some $k \in \mathbb{N}^+$ (using again Lemma 3.1.3). In particular, we may choose $\gamma \in \Gamma$ and $n \in \mathbb{N}^+$ such that $\gamma s^n \in \widehat{T}$.

Replacing now s by γs^n (which does not change $\bar{T}$, T and $\widehat{T}$) we can assume $s \in \widehat{T}$. We will show $s^a \in U$ for some $a \in \mathbb{N}^+$, which will finish the proof.

Since $\mathscr{C}_r(\widehat{T}, \bar{T})$ is a homomorphic image of $\mathscr{C}(\widehat{T}, \bar{T})/[\Gamma]_{\widehat{T}}^{\bar{T}}$ it is finite. Hence Div$(\bar{T}) \to$ Div$(\widehat{T})$ is bijective by Lemma 3.5.3. Therefore $N(\widehat{T}) = \widehat{T} \cap N(\bar{T})$. Since $\bar{T} = [[s]]_S$ we have $s \in N(\bar{T}) \cap \widehat{T} = N(\widehat{T})$.

Finally we show $N(\widehat{T})^a = \{0\}$ for some $a \in \mathbb{N}^+$. Then $s^a = 0 \in U$. Since $\widehat{T} \subset \bar{T} \in \mathrm{Div}(S)$ we have

$$\widehat{T}/(\widehat{T}^\times \cap \Gamma) \subset \bar{T}/\Gamma \subset S/\Gamma.$$

Hence $\widehat{T}/(\widehat{T}^\times \cap \Gamma)$ is quasi-finite. Since ideal theory in $\widehat{T}$ is essentially the same as in $\widehat{T}/(\Gamma \cap \widehat{T})$ we see that $\widehat{T}$ has the property 3 of Theorem 4.6. Since 0 is zero of $\widehat{T}$ and therefore of the group $\widehat{T}^{-1}\widehat{T}$, this group is trivial. From Lemma 4.5 we obtain now some $a \in \mathbb{N}^+$ such that $N(\widehat{T})^a$ is a singleton. But $0 = s0 \in N(\widehat{T})$. Hence $N(\widehat{T})^a = \{0\}$.

We continue by giving an example of a quasi-finite semigroup, that we will be important in Sect. 7. Let R be ring. We call R quasi artinian, if R is semilocal and its Jacobson radical is nilpotent. It is well known that a Noetherian ring is quasi artinian if and only if it is artinian.

Proposition 4.9 *If R is a quasi-artinian ring, then R_{red} is quasi-finite.*

Proof Let $m_1, \ldots, m_r$ be the maximal ideals of R and choose $n \in \mathbb{N}$ such that $(m_1 \cap \cdots \cap m_r)^n = m_1^n \cap \cdots \cap m_r^n = 0$ (here of course the nth powers are ideal powers). Set $R_i = R/m_i^n$ for $i = 1, \ldots, n$. By the Chinese Remainder theorem we have an isomorphism

$$R_{\mathrm{red}} \cong \prod_{i=1}^{r} (R_i)_{\mathrm{red}}.$$

Using Remark 4.3.3 we may suppose that R is local, so that $r = 1$. Let $\pi : R \to R_{\mathrm{red}}$ be the canonical homomorphism. Then $R_{\mathrm{red}} = \pi(R^\times) \cup \pi(m_1) = \{1\} \cup \pi(m_1)$. Since $\pi(m_1)^n \subset \pi(m_1^n) = \pi(0)$ this a n-cover of R_{red}.

We close this section with a result on certain subrings of quasi-artinian rings, which is needed in the proof of Lemma 7.1.

Recall the following prime avoidance result: if an ideal in a ring is contained in a finite union of prime ideals, then it is already contained in one of those prime ideals.

Lemma 4.10 *Let D be a quasi-artinian ring and C a subring of D such that the semigroup $[C]_C^D$ is zero complete. Then C is quasi-artinian, too.*

Proof The natural map s-spec$(D) \to s$-spec(C) is surjective by Lemma 3.5.2. Now let $p \in \mathrm{spec}(C)$. Then we have seen $p = q \cap C$ for some $q \in s$-spec(D). But D is quasi-artinian, and has therefore only finitely many prime ideals. Since q is a union of prime ideals, we obtain $p = C \cap q'$ for some prime ideal q' of D by prime avoidance. Hence $\mathrm{spec}(D) \to \mathrm{spec}(C)$ is surjective. It follows that C has only finitely many prime ideals. In particular, C is semilocal.

It is now enough to show $\dim(C) = 0$. Indeed, suppose this. Since $\mathrm{spec}(D) \to \mathrm{spec}(C)$ is surjective, we have $J(C) = C \cap J(D)$. Therefore $J(C)$ is nilpotent.

We come to the proof of $\dim(C) = 0$. For that we choose some minimal prime p of C and we show that p is also maximal. We set

$$X = \bigcup_{c \in C \setminus p} (C :_D c)$$

and claim $sX \subset C$ for some $s \in C \setminus p$. Indeed, since $[C]_C^D$ is zero complete we can choose $s \in C \setminus p$ such that

$$[st]_C^D = [s]_C^D, \tag{2}$$

for all $t \in C \setminus p$. Now let $x \in X$. Then $tx \in C$ for some $t \in C \setminus p$. From (2) it follows $sx \in C$.

We proceed by setting

$$I = \bigcap_{\substack{m \in \mathrm{spec}(D) \\ m \cap C = p}} m, \quad J = \bigcap_{\substack{m \in \mathrm{spec}(D) \\ m \cap C \neq p}} m.$$

Since D is quasi-artinian there exists some $n \in \mathbb{N}^+$ such that $I^n \cap J^n = 0$ (here of course the powers are ideal powers). Since $\mathrm{spec}(D) \to \mathrm{spec}(C)$ is surjective we have $I^n \cap C \subset p$. We claim $J^n \cap C \not\subset p$. Indeed, suppose $J^n \cap C \subset p$. Then $J \cap C \subset p$. Hence $m \cap C \subset p$ for some $m \in \mathrm{spec}(D)$ with $m \cap C \neq p$. But since p is minimal this is a contradiction.

We set $\bar{D} = D/I^n$, let $g \colon D \to \bar{D}$ be the canonical homomorphism and set $\bar{C} = g(C) \cong C/(C \cap I^n)$. Since $C \cap I^n \subset p$, $\bar{p} = g(p)$ is a minimal prime of $\bar{C}$ and it is enough to show, that $\bar{p}$ is maximal.

If $x \in \bar{C} \setminus \bar{p}$ then $x \in \bar{D}^\times$ by construction. Hence we may suppose $\bar{C} \subset \bar{C}_{\bar{p}} \subset \bar{D}$. We show now $\bar{C}_{\bar{p}} = g(X)$. The inclusion $\supset$ follows from the definition of X. Conversely, let $d \in D$ be such that $g(d) \in \bar{C}_{\bar{p}}$. Then we can choose $t \in C \setminus p$ such that $g(td) \in \bar{C}$, hence $td \in C + I^n$. Choose some $t_1 \in (C \cap J^n) \setminus p$. Then

$$(tt_1)d \in t_1 C + t_1 I^n \subset C + I^n \cap J^n = C.$$

Since $tt_1 \in C \setminus p$ this implies $g(d) \in g(X)$.

Finally, choose $s \in C \setminus p$ such that $sX \subset C$. Then we obtain

$$\bar{C}_{\bar{p}} = g(s)\bar{C}_{\bar{p}} = g(sX) \subset \bar{C}.$$

Hence $\bar{C} = \bar{C}_{\bar{p}}$ and $\bar{p}$ is maximal.

5 Weakly C-Monoids

We recall the definition of weakly C-monoids [7, Definition 4.1].

Definition 5.1 The monoid H is called a weakly C-monoid if it is a submonoid of a factorial monoid F, with representative set of primes P, such that the following two conditions are satisfied:

(C1) H is v-Noetherian, $(H : \widehat{H}) \neq \emptyset$ and $\widehat{H} \subset F$ is saturated and cofinal.
(C2) There exists an equivalence relation $\sim$ on P and $\lambda \in \mathbb{N}^+$ such that

- $P/\sim$ is finite, and
- for all $p_1, p_1', \ldots, p_\lambda, p_\lambda' \in P$ with $p_1 \sim p_1' \sim \ldots p_\lambda \sim p_\lambda'$ there exists $\varepsilon \in F^\times$ such that $[p_1 \cdot \ldots \cdot p_\lambda]_H^F = [\varepsilon p_1' \cdot \ldots \cdot p_\lambda']_H^F$.

If these conditions hold we say that H is a weakly C-monoid defined in F.

Note that **(C2)** holds if and only if we may write $\{P\}_H^F$ as a finite union of sets P_i such that P_i^λ is a singleton.

Let H be a v-Noetherian monoid. We consider the following property of H:

(C) $(H : \widehat{H}) \neq \emptyset$ and there exists a factorial monoid F, containing $\widehat{H}$ as a cofinal and saturated monoid, such that $\mathscr{C}_r(H, F)$ is quasi-finite.

Proposition 5.2 *Let H be a v-Noetherian monoid satisfying* **(C)** *and let F be a factorial monoid as in* **(C)**. *Then H is a weakly C-monoid defined in F such that $\mathscr{C}_r(\widehat{H}, F)$ is finite.*

Moreover for any $T \in \mathrm{Div}(H)$ we have: $\widehat{T} \subset [[T]]_{\widehat{H}}$ is saturated and cofinal and its class group $\mathscr{C}_r(\widehat{T}, [[T]]_{\widehat{H}})$ is finite. In particular, H satisfies the condition **(C3)** *of [7, Theorem 5.3].*

Proof That H is a weakly C-monoid defined in F is clear.

Now let $T \in \mathrm{Div}(H)$. By [7, Lemma 3.7] $\widehat{T} \subset [[T]]_F =: F_T$ is cofinal and saturated and we have $T = H \cap F_T$. By Lemma 3.1.4 we obtain an isomorphism $[F_T]_H^F \to \mathscr{C}(T, F_T)$. Since $F_T^\times = F^\times$ this induces an isomorphism $\{F_T\}_H^F \to \mathscr{C}_r(T, F_T)$. Now $\{F_T\}_H^F$ is quasi-finite as a subsemigroup of $\mathscr{C}_r(H, F)$. Therefore $\mathscr{C}_r(T, F_T)$ is quasi-finite, too. Applying Lemma 3.1.5 to $T \subset \widehat{T} \subset F_T$ we obtain an epimorphism $\mathscr{C}(T, F_T) \to \mathscr{C}(\widehat{T}, F_T)$ which induces an epimorphism $\mathscr{C}_r(T, F_T) \to \mathscr{C}_r(\widehat{T}, F_T)$. We conclude that $\mathscr{C}_r(\widehat{T}, F_T)$ is quasi-finite. Being a group it is finite. Applying this to $T = H$, we obtain that $\mathscr{C}_r(\widehat{H}, F)$ is finite.

We set $\widehat{H}_T = [[T]]_{\widehat{H}}$. Note that $\widehat{T} \subset \widehat{H}$. Since $T \subset \widehat{T}$ is cofinal we have $\widehat{T} \subset \widehat{H}_T$ and this inclusion is cofinal. As already noted $\widehat{T} \subset F_T$ is saturated. Hence $\widehat{T} \subset \widehat{H}_T$ is cofinal and saturated, too. We show that $\widehat{H}_T \subset F_T$ is saturated. So let $x, y \in \widehat{H}_T$ and $z \in F_T$ such that $xz = y$. Since $\widehat{H} \subset F$ is saturated, we obtain $z \in \widehat{H}$, which implies $z \in \widehat{H}_T$.

By Corollary 3.2 we obtain isomorphism

$$\mathscr{C}_r(\widehat{T}, \widehat{H}_T) \cong \frac{\mathsf{q}(\widehat{H}_T)}{\mathsf{q}(\widehat{T})\widehat{H}_T^\times} \overset{\widehat{H}_T \subset F_T \text{ is saturated}}{=} \frac{\mathsf{q}(\widehat{H}_T)}{\mathsf{q}(\widehat{T})\left(\mathsf{q}(\widehat{H}_T) \cap F_T^\times\right)}$$
$$\overset{\mathsf{q}(\widehat{T}) \subseteq \mathsf{q}(\widehat{H}_T)}{=} \frac{\mathsf{q}(\widehat{H}_T)}{\mathsf{q}(\widehat{H}_T) \cap \left(\mathsf{q}(\widehat{T})F_T^\times\right)}.$$

Therefore, we obtain a monomorphism

$$\mathcal{C}_r(\widehat{T}, \widehat{H}_T) \hookrightarrow \frac{\mathsf{q}(F_T)}{\mathsf{q}(\widehat{T})F_T^\times} \cong \mathcal{C}_r(\widehat{T}, F_T).$$

Hence $\mathcal{C}_r(\widehat{T}, \widehat{H}_T)$ is finite.

To verify property **(C)** the following result is helpful.

Proposition 5.3 *Let H be a v-Noetherian monoid such that $(H : \widehat{H}) \neq \emptyset$ (then $\widehat{H}$ is a Krull monoid by [6, Theorem 2.3.5.3]). Then the following are equivalent:*

1. *H satisfies* **(C)***.*
2. *$\mathcal{C}_r(H, \widehat{H})$ is quasi-finite and $\mathcal{C}_v(\widehat{H})$ is finite.*

Proof Suppose first that 1 holds and let F be a factorial monoid as in **(C)**. By Proposition 5.2 $\mathcal{C}_r(\widehat{H}, F)$ is finite. It follows from [6, Theorem 2.4.8.3] that $\mathcal{C}_v(\widehat{H})$ is finite. By Proposition 3.4 we have a monomorphism $\mathcal{C}_r(H, \widehat{H}) \hookrightarrow \mathcal{C}_r(H, F)$, which implies that $\mathcal{C}_r(H, \widehat{H})$ is quasi-finite.

Conversely, assume that 2 holds. It follows from [6, Proposition 2.4.5] that we can find a reduced factorial mononoid F_0 containing $\widehat{H}_{\mathrm{red}}$ as a cofinal and saturated submonoid, such that $\mathcal{C}_v(\widehat{H}_{\mathrm{red}})$ is isomorphic to the cokernel of the inclusion $\mathsf{q}(\widehat{H}_{\mathrm{red}}) \hookrightarrow \mathsf{q}(F_0)$. By Corollary 3.2 we obtain an isomorphism $\mathcal{C}_v(\widehat{H}_{\mathrm{red}}) \cong \mathcal{C}(\widehat{H}_{\mathrm{red}}, F_0)$. By [6, Theorem 2.4.8.2] we have an isomorphism $\widehat{H} \cong \widehat{H}^\times \times \widehat{H}_{\mathrm{red}}$. It follows that we may embed $\widehat{H}$ as a cofinal and saturated submonoid in the factorial monoid $F = \widehat{H}^\times \times F_0$ such that

$$\mathcal{C}_r(\widehat{H}, F) \cong \mathcal{C}(\widehat{H}_{\mathrm{red}}, F_0) \cong \mathcal{C}_v(\widehat{H}_{\mathrm{red}}) \cong \mathcal{C}_v(\widehat{H})$$

is finite.

We set $U = \{\widehat{H}\}_H^F$ and $S = \mathcal{C}_r(H, F)$. By Proposition 3.4 we know that U is cofinal and saturated in S and that there are isomorphisms $U \cong \mathcal{C}_r(H, \widehat{H})$ and $\mathcal{C}(U, S) \cong \mathcal{C}_r(\widehat{H}, F)$. Using Corollary 4.7.2 we deduce that $S = \mathcal{C}_r(H, F)$ is quasi-finite.

We use this result to generalize [7, Theorem 6.7.1] to not necessarily semilocal Mori domains. For that we need the following result.

Lemma 5.4 *Let $A \subset B$ be integral domains and I a nonzero ideal of B such that $I \subset A$ (so that I is an ideal of A and $A/I \subset B/I$), and $S \subset A^\bullet$ a submonoid, consisting only of nonzero divisors of the A-module B/A. Let $p\colon S^{-1}B^\bullet \to S^{-1}B/S^{-1}I$ be the restriction of the canonical homomorphism. Then there exist isomorphisms*

$$f\colon \mathcal{C}(A^\bullet, B^\bullet) \to \mathcal{C}(S^{-1}A^\bullet, S^{-1}B^\bullet),$$
$$g\colon \mathcal{C}(S^{-1}A^\bullet, S^{-1}B^\bullet) \to \mathcal{C}(S^{-1}A/S^{-1}I, S^{-1}B/S^{-1}I)$$

such that

$$f([b]_{A^\bullet}^{B^\bullet}) = [b]_{S^{-1}A^\bullet}^{S^{-1}B^\bullet}, \quad g\left([b']_{S^{-1}A^\bullet}^{S^{-1}B^\bullet}\right) = [p(b')]_{S^{-1}A/S^{-1}I}^{S^{-1}B/S^{-1}I}$$

for all $b \in B^\bullet$ *and all* $b' \in S^{-1}B^\bullet$.

Proof By Lemma 3.3 there exists a surjective homomorphism

$$f\colon \mathscr{C}(A^\bullet, B^\bullet) \to \mathscr{C}(S^{-1}A^\bullet, S^{-1}B^\bullet)$$

such that

$$f([b]_{A^\bullet}^{B^\bullet}) = [b]_{S^{-1}A^\bullet}^{S^{-1}B^\bullet},$$

for all $b \in B^\bullet$. Since S consists of nonzero divisors of B/A we have

$$(A^\bullet :_{B^\bullet} b) = (S^{-1}A^\bullet :_{S^{-1}B^\bullet} b) \cap B^\bullet,$$

for all $b \in B^\bullet$. Therefore, f is also injective.

For the construction of g we may replace A by $S^{-1}A$, B by $S^{-1}B$ and I by $S^{-1}I$. Since $p^{-1}(A/I) = A^\bullet$ we have

$$(A^\bullet :_{B^\bullet} b) = p^{-1}(A/I :_{B/I} p(b)),$$

for all $b \in B^\bullet$. Therefore, there exists an injective homomorphism

$$g\colon \mathscr{C}(A^\bullet, B^\bullet) \to \mathscr{C}(A/I, B/I)$$

such that

$$g([b]_{A^\bullet}^{B^\bullet}) = [p(b)]_{A/I}^{B/I},$$

for all $b \in B$.

Since I is nonzero, p, and therefore also g, is surjective.

Let R be a Mori domain with complete integral closure $\widehat{R}$ such that $(R : \widehat{R}) \neq 0$. Let S be the set of those $r \in R^\bullet$ that are nonzero divisors of the R-module $\widehat{R}/R$. Then S ist just the set of all regular elements of $R^\bullet$ [6, Definition 2.3.1.3]. It is obviously a divisor-closed submonoid of $R^\bullet$.

Theorem 5.5 *Let* R *be Mori domain with complete integral closure* $\widehat{R}$ *and* S *its monoid of regular elements. We suppose that* $(R : \widehat{R}) \neq 0$, *that the groups* $\mathscr{C}_v(R)$ *and* $\mathscr{C}_v(\widehat{R})$ *are finite, and that the ring* $S^{-1}\widehat{R}/S^{-1}(R : \widehat{R})$ *is quasi-artinian. Then* $R^\bullet$ *satisfies* **(C)**.

Proof By Proposition 5.3 we need only show, that $\mathscr{C}_r(R^\bullet, \widehat{R}^\bullet)$ is quasi-finite. We use the isomorphism

$$f: \mathscr{C}(R^\bullet, \widehat{R}^\bullet) \to \mathscr{C}(S^{-1}R^\bullet, S^{-1}\widehat{R}^\bullet)$$

from Lemma 5.4. Then f induces an ismomorphism

$$\mathscr{C}_r(R^\bullet, \widehat{R}^\bullet) \cong \mathscr{C}(S^{-1}R^\bullet, S^{-1}\widehat{R}^\bullet)/[\widehat{R}^\times]^{S^{-1}\widehat{R}^\bullet}_{S^{-1}R^\bullet}.$$

From Lemma 3.1.1 we obtain an isomorphism

$$[S^{-1}\widehat{R}^\times]^{S^{-1}\widehat{R}^\bullet}_{S^{-1}R^\bullet}/[\widehat{R}^\times]^{S^{-1}\widehat{R}^\bullet}_{S^{-1}R^\bullet} \cong \frac{S^{-1}\widehat{R}^\times}{\widehat{R}^\times S^{-1}R^\times}.$$

Now by [6, Theorem 2.10.9.7] we have a monomorphism

$$\frac{S^{-1}\widehat{R}^\times}{\widehat{R}^\times S^{-1}R^\times} \hookrightarrow \mathscr{C}_v(R).$$

Hence the group

$$[S^{-1}\widehat{R}^\times]^{S^{-1}\widehat{R}^\bullet}_{S^{-1}R^\bullet}/[\widehat{R}^\times]^{S^{-1}\widehat{R}^\bullet}_{S^{-1}R^\bullet}$$

is finite. Using Corollary 4.7.1 it is now enough to show, that

$$\frac{\mathscr{C}(S^{-1}R^\bullet, S^{-1}\widehat{R}^\bullet)/[\widehat{R}^\times]^{S^{-1}\widehat{R}^\bullet}_{S^{-1}R^\bullet}}{[S^{-1}\widehat{R}^\times]^{S^{-1}\widehat{R}^\bullet}_{S^{-1}R^\bullet}/[\widehat{R}^\times]^{S^{-1}\widehat{R}^\bullet}_{S^{-1}R^\bullet}} \cong \mathscr{C}_r(S^{-1}R^\bullet, S^{-1}\widehat{R}^\bullet)$$

is quasi-finite.

For that we show first that $S^{-1}\widehat{R}$ is semilocal. By [6, Theorem 2.3.5.3, Proposition 2.3.10.2] S is a finite union of prime ideals. Hence $S^{-1}R$ is semilocal. It follows now from [7, Lemma 6.6], that $S^{-1}\widehat{R}$ contains only finitely many maximal ideals, that do not contain $S^{-1}(R:\widehat{R})$. But since $S^{-1}\widehat{R}/S^{-1}(R:\widehat{R})$ is quasi-artinian, in particular semilocal, $S^{-1}\widehat{R}$ contains also only finitely many maximal ideals, that contain $S^{-1}(R:\widehat{R})$. Therefore $S^{-1}\widehat{R}$ is semilocal.

For easier notation we set $A = S^{-1}R/S^{-1}(R:\widehat{R})$ and $B = S^{-1}\widehat{R}/S^{-1}(R:\widehat{R})$. We now use the isomorphism

$$g: \mathscr{C}(S^{-1}R, S^{-1}\widehat{R}) \cong \mathscr{C}(A, B)$$

from Lemma 5.4. Since $S^{-1}\widehat{R}$ is semilocal, the canonical homomorphism $S^{-1}\widehat{R} \to B$ induces an epimorphism $S^{-1}\widehat{R}^\times \to B^\times$. Hence g defines an isomorphism

$$\mathscr{C}_r(S^{-1}R, S^{-1}\widehat{R}) \to \mathscr{C}_r(A, B)$$

which implies that $\mathcal{C}_r(S^{-1}R, S^{-1}\widehat{R})$ is a homomorphic image of B_{red}, which is quasi-finite by Proposition 4.9. Hence $\mathcal{C}_r(S^{-1}R, S^{-1}\widehat{R})$ is quasi-finite.

We close this section with two examples. In the first one we construct a Noetherian domain R, that satisfies **(C)**, but such that (in the notation of Theorem 5.5) the ring $S^{-1}\widehat{R}/S^{-1}(R:\widehat{R})$ is not quasi-artinian. In the second one we give an example of a Noetherian domain, that satisfies the assumption of Theorem 5.5, and hence $R^\bullet$ satisfies **(C)**, but for which the semigroup $\mathcal{C}(R^\bullet, \widehat{R^\bullet})$ is infinite (but of course quasi-finite).

Example 5.6 We choose integral domains $A_0 \subset A_1 \subset A_2$, such that A_0 is a Noetherian, semilocal, and one-dimensional domain with quotient field K, A_1 is its integral closure (implying $A_1 = \widehat{A_0}$) and A_2 is the integral closure of A_1 (and hence of A_0) in a finite and proper extension L of K. We suppose further that A_2 is a finitely generated A_0-module. Then our assumptions imply that A_1 and A_2 are semilocal principal ideal domains. We suppose further that the canonical map $\text{spec}(A_2) \to \text{spec}(A_1)$ is bijective, so that no prime of A_1 splits in A_2. Finally we assume $A_0 \neq A_1$.

We set

$$R = A_0 + XA_2[X] \subset A_2[X].$$

Then $A_0[X] \subset R \subset A_2[X]$ and $XA_2[X] \subset R$. Hence R is Noetherian and $A_2[X]$ is its integral (and therefore complete integral) closure. An easy calculation shows $XA_2[X] = (R : A_2[X])$ and that

$$S = A_0^\times + XA_2[X]$$

is the set of regular elements of R. Therefore $S^{-1}\widehat{R}/S^{-1}(R:\widehat{R}) \cong A_2$, which is not quasi-artinian. We show that nevertheless R has the property **(C)**. Since $\widehat{R} = A_2[X]$ is factorial, we need only show that $\mathcal{C}_r(R^\bullet, A_2[X]^\bullet)$ is quasi-finite. By Lemma 5.4 we have an isomorphism

$$\mathcal{C}(R^\bullet, A_2[X]^\bullet) \cong \mathcal{C}(A_0, A_2).$$

Since $A_2[X]^\times = A_2^\times$ it induces an isomorphism

$$\mathcal{C}_r(R^\bullet, A_2[X]^\bullet) \cong \mathcal{C}_r(A_0, A_2).$$

We are left to show that $\mathcal{C}_r(A_0, A_2)$ is quasi-finite. Obviously, we have

$$\mathcal{C}_r(A_0, A_2) = \{\{0\}_{A_0}^{A_2}\} \cup \{A_2^\bullet\}_{A_0}^{A_2} \quad \text{and} \quad \{A_2^\bullet\}_{A_0}^{A_2} \cong \mathcal{C}_r(A_0^\bullet, A_2^\bullet).$$

If

$$\{A_2^\bullet\}_{A_0}^{A_2} = X_1 \cup \cdots \cup X_r$$

is a n-cover of $\{A_2^\bullet\}_{A_0}^{A_2}$ then

$$\mathscr{C}_r(A_0, A_2)) = \{\{0\}_{A_0}^{A_2}\} \cup X_1 \cup \cdots \cup X_r$$

is one of $\mathscr{C}_r(A_0, A_2)$. Therefore, it is enough to show that $\mathscr{C}_r(A_0^\bullet, A_2^\bullet)$ is quasi-finite.

Since $A_0^\bullet \subset A_2^\bullet$ is cofinal and $A_1^\bullet \subset A_2^\bullet$ is saturated, we know by Proposition 3.4 that

$$\{A_1^\bullet\}_{A_0^\bullet}^{A_2^\bullet} \subset \mathscr{C}_r(A_0^\bullet, A_2^\bullet)$$

is cofinal and saturated and that we have isomorphisms

$$\{A_1^\bullet\}_{A_0^\bullet}^{A_2^\bullet} \cong \mathscr{C}_r(A_0^\bullet, A_1^\bullet), \quad \mathscr{C}(\{A_1^\bullet\}_{A_0^\bullet}^{A_2^\bullet}, \mathscr{C}_r(A_0^\bullet, A_2^\bullet) \cong \mathscr{C}_r(A_1^\bullet, A_2^\bullet).$$

Now since A_0 is Noetherian, semilocal, $\dim(A_0) = 1$ and $A_1 = \widehat{A}_0$ is a finitely generated A_0-module, the ring A_0 satisfies the assumptions of Theorem 5.5. Hence $\mathscr{C}_r(A_0^\bullet, A_1^\bullet)$ is quasi-finite by Proposition 5.3. Using Corollary 4.7.2 we are left to show that the group $\mathscr{C}_r(A_1^\bullet, A_2^\bullet)$ is finite. By Corollary 3.2 this group is isomorphic to the cokernel of the natural homomorphism $K^\times/A_1^\times \to L^\times/A_2^\times$. Note that this homomorphism is injective. Since A_1 is a semilocal, principal ideal domain $K^\times/A_1^\times$ is a free abelian group, whose rank equals the number of maximal ideals of A_1. The same holds for $L^\times/A_2^\times$. By our assumption A_1 and A_2 have the same number of maximal ideals, which implies that the cokernel of $K^\times/A_1^\times \to L^\times/A_2^\times$, and hence $\mathscr{C}_r(A_1^\bullet, A_2^\bullet)$, is finite.

Example 5.7 Let k be an infinite field and let $D = k[[X, Y]]$ be the power series ring in two variables. Let I be the ideal of D generated by X^2 and Y^2 and set $R = k + I$. Then $D = R + RX + RY + RXY$. By Eakins theorem R is Noetherian. Clearly $(R : D) = I$ which implies that D is the (complete) integral closure of R. R is local and I is its maximal ideal. We show that the v-class groups of R and $\widehat{R} = D$ are trivial. Then R satisfies all assumptions of Theorem 5.5. For D this is clear, since D is factorial.

Since $I = (R : \widehat{R})$ is the maximal ideal of R, the monoid S of regular elements of R equals $R^\times$. Hence $S^{-1}\widehat{R}^\times = \widehat{R}^\times$ and therefore the group

$$\frac{S^{-1}\widehat{R}^\times}{S^{-1}R^\times\widehat{R}^\times}$$

is trivial. By [6, Theorem 2.10.9.7] we have an exact sequence

$$1 = \frac{S^{-1}\widehat{R}^\times}{S^{-1}R^\times\widehat{R}^\times} \to \mathscr{C}_v(R) \to \mathscr{C}_v(\widehat{R}) = 1,$$

which implies that $\mathscr{C}_v(R)$ is trivial.

We show now that $\mathscr{C}_r(R^\bullet, D^\bullet)$ is infinite. Identifying k with its image in $\bar{D} = D/I$ we have an isomorphism

$$\mathscr{C}(R^\bullet, D^\bullet) \cong \mathscr{C}(k, \bar{D})$$

by Lemma 5.4. Since $D^\times \to \bar{D}^\times$ is surjective, we obtain an isomorphism

$$\mathscr{C}_r(R^\bullet, D^\bullet) \cong \mathscr{C}_r(k, \bar{D}).$$

We show now that $\mathscr{C}_r(k, \bar{D})$ is infinite. Note that D is a regular local ring and (X^2, Y^2) is a regular sequence. Therefore $\bar{D}$ is Gorenstein [5, Proposition 3.1.19]. Since $\dim(\bar{D}) = 0$ we have

$$(0 : (0 : J)) = J$$

for any ideal J of $\bar{D}$ [12, Satz 1.44]. Let m be the maximal ideal of D and for $x \in m$ let $\bar{x}$ be its image in m/m^2. Note that

$$m/m^2 \cong \langle X, Y\rangle/\langle X, Y\rangle^2$$

so that m/m^2 is a two-dimensional k-vector space. Since k is infinite, it has infinitely many one-dimensional subspaces. Let now $x, y \in m$ be such that $\{x\}_k^{\bar{D}} = \{y\}_k^{\bar{D}}$. We will show $k\bar{x} = k\bar{y}$, which implies that $\mathscr{C}_r(k, \bar{D})$ is infinite.

Since $x \in m$ we have $ax \notin k^\times$, for all $a \in \bar{D}$. Hence $(k : x) = (0 : x) = (0 : \bar{D}x)$. Since $[x]_k^{\bar{D}} = [\varepsilon y]_k^{\bar{D}}$ for some $\varepsilon \in \bar{D}^\times$ we obtain

$$\bar{D}x = (0 : (0 : \bar{D}x)) = (0 : (k : x)) = (0 : (k : \varepsilon y)) = (0 : (0 : \varepsilon y)) = (0 : (0 : y)) = \bar{D}y,$$

which implies $k\bar{x} = k\bar{y}$.

6 Arithmetic of Weakly C-Monoid

We first quickly recall the definitions of the arithmetical invariants, which we will study here. For a more thorough discussion of these invariants we refer to [6]. In the following H is an atomic monoid.

Let A be the set of atoms of H_{red} and $\mathsf{Z}(H)$ the free abelian monoid with basis A. Then there exists a unique homomorphism $\pi : \mathsf{Z}(H) \to H_{\text{red}}$ extending the identity of A. If $z \in \mathsf{Z}(H)$, then z has a unique (up to order) representation $z = u_1 \cdot \ldots \cdot u_r$ with $u_1, \ldots, u_r \in A$. We set $|z| = r$. For $z, z' \in \mathsf{Z}(H)$ we define

$$\mathsf{d}(z, z') = \max\left\{|z \gcd(z, z')^{-1}|, |z' \gcd(z, z')^{-1}|\right\}.$$

Let $h \in H$ and let $\bar{h}$ be its image in H_{red}. Then $\mathsf{Z}(h) = \pi^{-1}(\bar{h})$ is called the set of factorizations of h, and

$$\mathsf{L}_H(h) = \mathsf{L}(h) = \{|z| \mid z \in \mathsf{Z}(h)\}$$

is called the set of lengths of h. H is called a BF-monoid, if $\mathsf{L}(h)$ is finite, for all $h \in H$. For $h \in H \setminus H^\times$

$$\rho(h) = \frac{\sup \mathsf{L}(h)}{\min \mathsf{L}(h)} \in \mathbb{Q}_{>0} \cup \{\infty\}$$

is called the elasticity of h. If $h \in H^\times$ we set $\rho(h) = 1$. The number

$$\rho(H) = \sup \{\rho(h) \mid h \in H\} \in \mathbb{R}_{>0} \cup \{\infty\}$$

is called the elasticity of H. We will need also the finer invariants $\rho_k(H)$ which are defined for $k \in \mathbb{N}^+$ by

$$\rho_k(H) = \sup \{\sup \mathsf{L}(h) \mid h \in H, k \in \mathsf{L}(h)\},$$

if $H \neq H^\times$ and by $\rho_k(H) = k$ if $H = H^\times$. Then

$$\rho(H) = \sup \left\{\rho_k(H) \mid k \in \mathbb{N}^+\right\} = \lim_{k \to \infty} \frac{\rho_k(H)}{k}.$$

Next we define the tame degrees. For $h \in H$ and $x \in \mathsf{Z}(H)$ let $\mathsf{t}(h, x)$ be the smallest $N \in \mathbb{N}^+ \cup \{\infty\}$ having the following property:

If $\mathsf{Z}(h) \cap x\mathsf{Z}(H) \neq \emptyset$ and $z \in \mathsf{Z}(h)$, there exists some $z' \in \mathsf{Z}(h) \cap x\mathsf{Z}(H)$ such that $\mathsf{d}(z, z') \leq N$.

H is called locally tame, if for each $u \in A$

$$\mathsf{t}(H, u) := \sup \{\mathsf{t}(x, u) \mid x \in \mathsf{Z}(H)\} < \infty.$$

H is called tame, if

$$\mathsf{t}(H) := \sup \{\mathsf{t}(x, u) \mid u \in A, x \in \mathsf{Z}(H)\} < \infty.$$

It follows from [7, Theorem 5.3] and Proposition 5.2 that any v-Noetherian monoid satisfying **(C)** is locally tame.

For $h \in H$ let $\omega(H, h)$ be the smallest $N \in \mathbb{N} \cup \{\infty\}$ such that the following holds:

Whenever $a_1, \ldots, a_m \in H$ are such that $h \mid a_1 \cdot \ldots \cdot a_m$, then there exists $I \subset \{1, \ldots, m\}$ such that $|I| \leq N$ and $h \mid \prod_{i \in I} a_i$.

We set

$$\omega(H) = \sup \{\omega(H, u) \mid u \text{ is an atom of } H\}.$$

If $H \neq H^\times$ is factorial then $\omega(H) = 1$ and $\mathsf{t}(H) = 0$. If H is not factorial, then $\omega(H) \leq \mathsf{t}(H) \leq \omega(H)^2$ (see the explanations after Definition 3.1 and Proposition 3.5 in [9]). In particular, H is tame if and only if $\omega(H) < \infty$.

The last arithmetical invariant we will study is the catenary degree $\mathsf{c}(H)$. It is defined as the smallest $N \in \mathbb{N} \cup \{\infty\}$, such that for all $h \in H$ and all $z, z' \in \mathsf{Z}(h)$ there exists a finite sequence $(z_i)_{0 \leq i \leq m}$ in $\mathsf{Z}(h)$ such that $z = z_0, z' = z_m$ and $\mathsf{d}(z_i, z_{i+1}) \leq N$, for all $i = 0, \dots, m-1$.

Let now H be a weakly C-monoid and choose a factorial monoid F such that H is defined in F. We let P be a representative set of primes of F. If $Q \subset P$ and $x \in F$ we set

$$v_Q(x) = \sum_{p \in Q} v_p(x), \quad \mathrm{supp}(x) = \{p \in P \mid v_p(x) > 0\}.$$

A subset $Q \subset P$ is called H-essential, if $Q = \mathrm{supp}(h)$ for some $h \in H$. Some $p \in P$ is called H-essential, if $\{p\}$ is H-essential. Let E be the set of all $p \in P$ that are H-essential. Finally we call H simple in F, if each minimal nonempty H-essential subset of H is a singleton, and we say that H is nicely embedded in F, if any $p \in P$ is contained in some minimal H-essential subset of P. Then we have the following results. Recall that A is the set of atoms of H_{red}. For an atom u of H, let $\bar{u} \in A$ be its image in H_{red}.

Lemma 6.1 *In the notation just introduced, assume that H is locally tame and $\mathscr{C}_r(\widehat{H}, F)$ is finite. Then we have*

1. $\sup\{v_E(u) \mid u$ *is an atom of* $H\} < \infty$.
2. *There exists some $K \in \mathbb{N}$ such that* $\min \mathsf{L}(h) \leq v_E(h) + K$, *for all $h \in H$.*
3. *There exists some $C, D \in \mathbb{N}$ such that* $\mathsf{t}(H, \bar{u}) \leq C v_P(u) + D$, *for all atoms u of H.*
4. *There exists some $a \in \mathbb{Q}_{>0}$ such that* $\omega(H, h) \geq a v_P(h)$, *for all $h \in H$.*

Proof 1–3 are contained in [7]: 1 is Lemma 5.4.2, 2 follows from Theorem 5.3 and Proposition 5.8 and 3 is part of Theorem 5.3.

4. Let m be the exponent of $\mathscr{C}_r(\widehat{H}, F)$, choose $z \in (H : \widehat{H})$ and set $a = 1/(v_P(z) + m)$. Let $h \in H$, say $h = \varepsilon p_1 \cdot \ldots \cdot p_n$, where $\varepsilon \in F^\times$ and $p_1, \dots, p_n \in P$. By Definition of m there exists for $i = 1, \dots, n$ some $\delta_i \in F^\times$ such that $\delta_i p_i^m \in \widehat{H}$. Then

$$h^m = \varepsilon^m (\delta_1 \cdot \ldots \cdot \delta_n)^{-1} (\delta_1 p_1^m) \cdot \ldots \cdot (\delta_n p_n^m).$$

We set $x_1 = \varepsilon^m (\delta_1 \cdot \ldots \cdot \delta_n)^{-1} \delta_1 p_1^m$ and $x_i = \delta_i p_i^m$ for $i = 2, \dots, n$. Then $x_2, \dots, x_n \in \widehat{H}$ and $h^m = x_1 \cdot \ldots \cdot x_n$. Since $\widehat{H}$ is saturated in F, we obtain $x_1 \in \widehat{H}$, too. Then $z x_i \in H, i = 1, \dots, n$ and $h^m \mid (z x_1) \cdot \ldots \cdot (z x_n)$. By the Definition of ω we may now assume that $h^m \mid (z x_1) \cdot \ldots \cdot (z x_r)$ for some $r \leq \min\{n, \omega(H, h^m)\}$. Therefore

$$m v_P(h) = v_P(h^m) \leq v_P(z x_1) + \cdots + v_P(z x_r) = r v_P(z) + r m = \frac{r}{a}.$$

Since $\omega(H, h^m) \leq m\omega(H, h)$ [8, Lemma 3.3.1] we obtain

$$\omega(H, h) \geq \frac{1}{m}\omega(H, h^m) \geq \frac{r}{m} \geq av_P(h).$$

Theorem 6.2 *Let H be a weakly C-monoid and let F be a factorial monoid, with representative set of primes P, such that H is defined in F. We assume that H is locally tame and that $\mathscr{C}_r(\widehat{H}, F)$ is finite. Then the following hold:*

1. *H has finite catenary degree.*
2. *The following are equivalent:*
   a. *$\rho(H) < \infty$.*
   b. *$\rho_k(H) < \infty$, for all $k \in \mathbb{N}^+$.*
   c. *H is simple in F.*
   d. *For each $h \in H \setminus H^\times$ the set $\{\min \mathsf{L}(h^n) \mid n \in \mathbb{N}^+\}$ is not bounded.*
3. *The following are equivalent:*
   a. *$\mathsf{t}(H) < \infty$.*
   b. *$\sup\{\max \mathsf{L}_{\widehat{H}}(u) \mid u$ is an atom of $H\} < \infty$.*
   c. *$\sup\{v_P(u) \mid u$ is an atom of $H\} < \infty$.*
4. *Suppose that H is nicely embedded in F. Then the following are equivalent:*
   a. *$\mathsf{t}(H) < \infty$.*
   b. *$\rho(H) < \infty$.*
   c. *Each $p \in P$ is H-essential.*
   d. *There exists some $m \in \mathbb{N}^+$ such that, for all $a \in \widehat{H}$ we have $\varepsilon a^m \in H$ for some $\varepsilon \in \widehat{H}^\times$.*

Proof We choose a representative set of primes P of F and use all notations introduced earlier in this section.

1. Has already been proven in [7, Theorem 6.3].

2. The implication a $\Rightarrow$ b is trivial. To prove the implications b $\Rightarrow$ c and d $\Rightarrow$ c assume by contradiction that H is not simple in F. Then there exists some $h \in H \setminus H^\times$ such that $\mathrm{supp}(h)$ is a minimal nonempty, H-essential subset, that contains at least two elements. Then $\mathrm{supp}(h) \cap E = \emptyset$. We choose now $K \in \mathbb{N}^+$ as in Lemma 6.1.2. Then $\min \mathsf{L}(h^n) \leq K$, for all $n \in \mathbb{N}^+$. Hence d does not hold. Since $\min \mathsf{L}(h^n) \leq K$, for all $n \in \mathbb{N}^+$ there exists an infinite set $T \subset \mathbb{N}^+$ and an integer $K_1 \leq K$ such that $\min \mathsf{L}(h^n)) = K_1$, for all $n \in T$. Now it follows that $\rho_{K_1}(H) \geq \max \mathsf{L}(h^n) \geq n$, for all $n \in T$ and thus $\rho_{K_1}(H) = \infty$.

c $\Rightarrow$ a. So suppose H is simple in F. Then for any $h \in H \setminus H^\times$ we have $E \cap \mathrm{supp}(h) \neq \emptyset$ and therefore $v_E(h) \geq 1$. By Lemma 6.1.1 we may choose $d \in \mathbb{N}^+$ such that $v_E(u) \leq d$, for all atoms u of H.

Let now $h \in H \setminus H^\times$. If $u_1, \ldots, u_r$ are atoms of H such that $h = u_1 \cdot \ldots \cdot u_r$ we obtain

$$v_E(h) = v_E(u_1) + \cdots + v_E(u_r) \geq r.$$

Hence $\max \mathsf{L}(h) \le v_E(h)$. On the other hand we have

$$v_E(h) = v_E(u_1) + \cdots + v_E(u_r) \le rd$$

implying $\min \mathsf{L}(h) \ge v_E(h)/d$. Putting together we obtain

$$\rho(h) \le \frac{v_E(h)}{v_E(h)/d} = d.$$

Hence $\rho(H) < \infty$. Finally, the implication a $\Rightarrow$ d is trivial.

3. Since $\mathcal{C}_r(\widehat{H}, F)$ is finite, it follows

$$s := \sup\{v_P(u) \mid u \text{ is an atom of } \widehat{H}\} < \infty$$

(see for example [7, Lemma 2.1.3]). Now let $h \in \widehat{H} \setminus \widehat{H}^\times$ and choose atoms $v_1, \ldots v_r$ of $\widehat{H}$ such that $h = v_1 \ldots v_r$ and $r = \max \mathsf{L}_{\widehat{H}}(h)$. Then

$$r \le v_P(v_1) + \cdots + v_P(v_r) = v_P(h) \le sr.$$

We obtain

$$\frac{v_P(h)}{s} \le \max \mathsf{L}_{\widehat{H}}(h) \le v_P(h)$$

for any $h \in \widehat{H} \setminus \widehat{H}^\times$. Hence b and c are equivalent.

a $\Rightarrow$ c. If $\mathsf{t}(H) < \infty$, then also $\omega(H) < \infty$. Hence the claim follows from Lemma 6.1.4. c $\Rightarrow$ a follows from Lemma 6.1.3.

4. The implication a $\Rightarrow$ b holds for any atomic monoid ([6, Theorem 1.6.6.2]).

b $\Rightarrow$ c. If $\rho(H) < \infty$ then H is simple in F by 2. Since H is nicely embedded in F this implies c.

c $\Rightarrow$ a follows from 1 and 3 in Lemma 6.1.

c $\Rightarrow$ d. Suppose that any $p \in P$ is H-essential. Let $P = P_1 \cup \cdots P_r$ be a decomposition of P and $\lambda \in \mathbb{N}^+$ be such that $\{P_i^\lambda\}_H^F$ is a singleton for $i = 1, \ldots, r$. For each $i = 1, \ldots, r$ choose $q_i \in P_i$. Since each q_i is H-essential, there exists some $n \in \mathbb{N}^+$ and $\varepsilon_1, \ldots, \varepsilon_r \in F^\times$ such that $\varepsilon_i q_i^n \in H$ for $i = 1, \ldots, r$. We set $m = n\lambda$.

Let $p \in P$. Then $p \in P_i$ for some $1 \le i \le r$. Hence $\{p^\lambda\}_H^F = \{q_i^\lambda\}_H^F$. From that and from $\varepsilon_i q_i^n \in H$ we obtain $\varepsilon p^m \in H$ for some $\varepsilon \in F^\times$.

Now let $a \in \widehat{H}$ be arbitrary. Let $\delta \in F^\times$ and $p_1, \ldots p_k \in P$ be such that $a = \delta p_1 \ldots p_k$. For $i = 1, \ldots, k$ choose $\varepsilon_i \in F^\times$ such that $\varepsilon_i p_i^m \in H$. Then

$$\delta^{-m} \varepsilon_1 \ldots \varepsilon_k a^m \in H$$

and since $\widehat{H}$ is saturated in F we have $\delta^{-m} \varepsilon_1 \ldots v e_k \in \widehat{H} \cap F^\times = \widehat{H}^\times$.

d $\Rightarrow$ c. Let $p \in P$. Since $\mathcal{C}_r(\widehat{H}, F)$ is finite we may choose $\varepsilon \in F^\times$ and $n \in \mathbb{N}^+$ such that $\varepsilon p^n \in \widehat{H}$. By assumption we have $\delta \varepsilon^m p^{nm} \in H$ for some $\delta \in \widehat{H}^\times$. Hence p is H-essential.

We close this section with an example of a weakly C-monoid (even a C-monoid) defined in the factorial monoid F, such that H is not nicely embedded in F. For examples for nicely embedded $H \subset F$ we refer to Lemma 7.1 and the proof of Theorem 7.2.

Example 6.3 Let k be a field, $k[X, Y]$ the polynomial ring in two variables and set $R = k + Xk[X, Y]$. This domain has been studied intensively in the literature (see for example the references for Example 2 in [13]).

Since $Xk[X, Y] \subset R$, we have $\widehat{R} = k[X, Y]$. We show that $R^\bullet$ is a weakly C-monoid defined in $k[X, Y]^\bullet$, that is not nicely embedded in $k[X, Y]^\bullet$. Further we will show $\rho(R) = 1$ but $\mathsf{t}(R) = \infty$. First by [13] R is a Mori domain.

Let $Q \in k[X, Y]^\bullet$. Then we have

$$(R^\bullet : Q) = \begin{cases} k[X, Y]^\bullet & \text{if } Q \in Xk[X, Y] \\ R^\bullet & \text{if } Q \in R^\bullet \setminus Xk[X, Y] \\ Xk[X, Y]^\bullet & \text{if } Q \in k[X, Y]^\bullet \setminus R \end{cases}.$$

Hence $\mathscr{C}(R^\bullet, k[X, Y]^\bullet)$ is finite. Therefore $R^\bullet \subset k[X, Y]^\bullet$ satifies **(C)** and $R^\bullet$ is a weakly C-monoid defined in $k[X, Y]^\bullet$.

We choose now the representative set of primes $P \subset k[X, Y]$ such that $X, Y \in P$. If $a \in R$ is such that $Y \mid a$, then also $X \mid a$. Hence any $R^\bullet$-essential subset of P, that contains Y also contains X. Since X is $R^\bullet$-essential, $R^\bullet$ is not nicely embedded in $k[X, Y]^\bullet$.

By [13] we have $\rho(R) = 1$. Obviously XY^n is for any $n \in \mathbb{N}^+$ an atom of $R^\bullet$. Hence condition (c) of Theorem 6.2.3 is violated. It follows $\mathsf{t}(R) = \infty$.

7 Integral Domains

In this section we want to apply Theorem 6.2 to certain Mori domains. For that we need the following preparatory result.

Lemma 7.1 *Let R be a Mori domain such that $(R : \widehat{R}) \neq 0$ and let F be a factorial monoid containing $\widehat{R}^\bullet$ as a cofinal and saturated submonoid, such that $\mathscr{C}_r(R^\bullet, F)$ is quasi-finite (so that $R^\bullet$ satisfies* **(C)***). Let $A \subset \operatorname{spec}(\widehat{R}) \setminus \{0\}$ be a finite set of nonzero primes of $\widehat{R}$ having the following properties:*

1. *For any divisor-closed subsemigroup T of $\mathscr{C}_r(R^\bullet, \widehat{R}^\bullet)$ there exists a subset $B \subset A$ such that*
$$\left\{a \in \widehat{R}^\bullet \mid \{a\}_{R^\bullet}^{\widehat{R}^\bullet} \in T\right\} = \widehat{R}^\bullet \setminus \bigcup_{q \in B} q.$$
2. *Going Up: if $q_1, q_2 \in A$ are such that $q_1 \cap R \subset q_2 \cap R$, then $q_2 \cap R = q_3 \cap R$ for some $q_3 \in A$ such that $q_1 \subset q_3$.*

3. *Incomparability: if* $q_1, q_2 \in A$ *are such that* $q_1 \subset q_2$ *and* $q_1 \cap R = q_2 \cap R$ *then* $q_1 = q_2$.

Then $R^\bullet$ *is nicely embedded in* F. *Moreover, if the map* $\operatorname{spec}(\widehat{R}) \to \operatorname{spec}(R)$ *is injective, then the map* $s\text{-}spec(\mathscr{C}(R^\bullet, \widehat{R}^\bullet)) \to s\text{-}spec([R^\bullet]_{R^\bullet}^{\widehat{R}^\bullet})$ *is injective, too.*

Proof Choose any representative set of primes P of F. We have to show, that any $p \in P$ is contained in a minimal nonempty, $R^\bullet$-essential subset of P.

Sind $\widehat{R}^\bullet$ is cofinal and saturated in F and the group $\mathscr{C}_r(\widehat{R}^\bullet, F)$ is finite by Proposition 5.2 we can choose some $e \in \mathbb{N}^+$ and for each $p \in P$ some $\varepsilon_p \in F^\times$ such that $x_p := \varepsilon_p p^e \in \widehat{R}^\bullet$.

We set $X = \{x_p \mid p \in P\} \subset \widehat{R}^\bullet$. Since P consists of primes the map $f : P \to X$, $p \mapsto x_p$ is bijective. We call a finite, nonempty subset $X' \subset X$ essential, if there exists some $\delta \in \widehat{R}^\times$ and for any $x \in X'$ some $a_x \in \mathbb{N}^+$ such that

$$\delta \prod_{x \in X'} x^{a_x} \in R^\bullet.$$

Let $Q \subset P$ be finite and nonempty. We show that Q is $R^\bullet$-essential if and only if $f(Q)$ is essential. If $f(Q)$ is essential then clearly Q is $R^\bullet$-essential. Conversely, assume this. Then there are $\varepsilon \in F^\times$ and for $p \in Q$ some $a_p \in \mathbb{N}^+$ such that

$$r = \varepsilon \prod_{p \in Q} p^{a_p} \in R^\bullet.$$

But then for some $\delta \in F^\times$ we have

$$r^e = \delta \prod_{p \in Q} x_p^{a_p} \in R^\bullet.$$

Since $\widehat{R}^\bullet$ is saturated in F, have $\delta \in \widehat{R}^\times$ and $f(Q)$ is essential.

It is now enough to show, that any $x \in X$ is contained in a minimal nonempty, essential subset of X. We do this in several steps.

Step 1. For $a \in \widehat{R}^\bullet$ we set $A(a) = \{q \in A \mid a \in q\}$. Let $a, b \in \widehat{R}^\bullet$ such that $A(a) = A(b)$. We claim that there exists some $n \in \mathbb{N}^+$ such that

$$\{a^n\}_{R^\bullet}^{\widehat{R}^\bullet} = \{b^n\}_{R^\bullet}^{\widehat{R}^\bullet}.$$

To prove this let T be any divisor-closed subsemigroup of $\mathscr{C}_r(R^\bullet, \widehat{R}^\bullet)$. Then from our assumption 1 and $A(a) = A(b)$ we obtain

$$\{a\}_{R^\bullet}^{\widehat{R}^\bullet} \in T \iff \{b\}_{R^\bullet}^{\widehat{R}^\bullet} \in T.$$

We conclude

$$[[\{a\}_{R^\bullet}^{\widehat{R}^\bullet}]] = [[\{b\}_{R^\bullet}^{\widehat{R}^\bullet}]].$$

Let T be this divisor-closed subsemigroup of $\mathscr{C}_r(R^\bullet, \widehat{R}^\bullet)$. Then we have $\{a\}_{R^\bullet}^{\widehat{R}^\bullet}, \{b\}_{R^\bullet}^{\widehat{R}^\bullet} \in N(T)$. Since $\mathscr{C}_r(R^\bullet, \widehat{R}^\bullet)$ is quasi-finite by Proposition 5.3 the group $T^{-1}T$ is finite by Theorem 4.6. Hence

$$\frac{\{a^m\}_{R^\bullet}^{\widehat{R}^\bullet}}{1} = \frac{\{b^m\}_{R^\bullet}^{\widehat{R}^\bullet}}{1} \quad \text{in} \quad T^{-1}T$$

for some $m \in \mathbb{N}^+$. Using Lemma 4.5 and Theorem 4.6 we get some $n \in \mathbb{N}^+$ such that

$$\{a^{nm}\}_{R^\bullet}^{\widehat{R}^\bullet} = \{b^{nm}\}_{R^\bullet}^{\widehat{R}^\bullet}.$$

Step 2. We call $B \subset A$ stable, if $q \in B$, $q' \in A$, $q \cap R \subset q' \cap R$ imply $q' \in B$. Let now $X' \subset X$ be finite. We show, that X' is essential if and only if

$$A\left(\prod_{x \in X'} x\right) = \bigcup_{x \in X'} A(x)$$

is stable.

Assume first that X' is essential. Then we have

$$r = \varepsilon \prod_{x \in X'} x^{a_x} \in R^\bullet$$

for some $\varepsilon \in \widehat{R}^\times$ and some $a_x \in \mathbb{N}^+$ $(x \in X')$. Then

$$A\left(\prod_{x \in X'} x\right) = A(r)$$

is clearly stable.

Conversely, assume that $A_0 := A(\prod_{x \in X'} x)$ is stable. We first construct some $r \in R^\bullet$ such that $A(r) = A_0$. If $A_0 = A$ we may take any nonzero

$$r \in \bigcap_{q \in A} (q \cap R)$$

(recall that all $q \in A$ are nonzero).

Now assume $A_0 \neq A$. Since A_0 is stable, we may choose by prime avoidance some $r \in R$ such that

$$r \in \bigcap_{q \in A_0} (q \cap R) \quad \text{but} \quad r \notin \bigcup_{q \in A \setminus A_0} (q \cap R).$$

Then $r \neq 0$ since $A_0 \neq A$ and clearly $A(r) = A_0$.

Using Step 1 we obtain some $n \in \mathbb{N}^+$ such that

$$\{r^n\}_{R^\bullet}^{\widehat{R}} = \{\prod_{x \in X'} x^n\}_{R^\bullet}^{\widehat{R}}.$$

Since $r^n \in R^\bullet$ we obtain $\varepsilon \prod_{x \in X'} x^n \in R^\bullet$ for some $\varepsilon \in \widehat{R}^\times$ and X' is essential.

Step 3. For $q \in A$ set $A_q = \{q' \in A \mid q \subset q'\}$. We claim, that for any $q \in A$ there is some $x \in X$ such that $A(x) = A_q$.

By prime avoidance there is some $a \in \widehat{R}^\bullet$ such that $A(a) = A_q$. Now let $\varepsilon \in F^\times$ and $p_1, \ldots p_n \in P$ be such that $a = \varepsilon p_1 \cdot \ldots \cdot p_n$. Then for some $\delta \in F^\times$ we have $a^e = \delta x_{p_1} \cdot \ldots \cdot x_{p_n}$. Since $\widehat{R}^\times$ is saturated in F, this implies $\delta \in \widehat{R}^\times$. Hence

$$A_q = A(a) = A(a^e) = A(x_{p_1}) \cup \cdots \cup A(x_{p_n}).$$

Let $1 \le i \le n$ be such that $q \in A(x_{p_i})$. Then $q \in A(x_{p_i}) \subset A_q$ and hence $A_q = A(x_{p_i})$.

Step 4. Let $x_0 \in X$. We construct some minimal essential $X' \subset X$ containing x_0. Set

$$A_0 = \{q \in A \mid q' \cap R \subset q \cap R \quad \text{for some} \quad q' \in A(x_0)\}.$$

Then A_0 is the smallest stable subset of A containing $A(x_0)$. If $A(x_0) = A_0$, then $A(x_0)$ is already stable, and therefore $\{x_0\}$ is essential by Step 2. So, we may assume that $A(x_0) \neq A_0$. Let M be the set of minimal (with respect to inclusion) elements of $A_0 \backslash A(x_0)$. Then we have

$$A_0 = A(x_0) \cup \bigcup_{q \in M} A_q.$$

Let $\emptyset \neq M' \subset M$ and A_1 the union of all A_q where $q \in M'$. We show that A_1 is not stable. To do this, choose $q \in M'$ such that $q \cap R$ becomes minimal. Since $q \in A_0$, there is some $q_1 \in A(x_0)$ such that $q_1 \cap R \subset q \cap R$. By Going Up there exists some $q_2 \in A$ such that $q_1 \subset q_2$ and $q_2 \cap R = q \cap R$. We show $q_2 \notin A_1$, which shows that A_1 is not stable. Assume by condratiction $q_2 \in A_1$. Since $q_1 \subset q_2$ we have $q_2 \in A(x_0)$ and therefore $q_2 \notin M'$. Since $q_2 \in A_1$, we have $q_3 \subset q_2$ for some $q_3 \in M'$. Then $q_3 \cap R \subset q_2 \cap R = q \cap R$. The minimal choice of q implies $q_3 \cap R = q_2 \cap R$. Since $q_3 \subset q_2$ Incomparability implies $q_2 = q_3 \in M'$, contradiction.

Now choose for any $q \in M$ some $x_q \in X$ such that $A(x_q) = A_q$ (Step 3) and set

$$X' = \{x_0\} \cup \{x_q \mid q \in M\}.$$

We show that X' is a minimal essential subset of X. First we have

$$A\left(\prod_{x \in X'} x\right) = A_0$$

and hence X' is essential by Step 2.

We show now that X' is minimal. For that let $\emptyset \neq X'' \subset X'$ be an essential subset. We show $X'' = X'$. To do this set $A'' = A(\prod_{x\in X''} x)$. Then A'' is stable by Step 2. Let $M' \subset M$ be such that $X''\backslash\{x_0\} = \{x_q \mid q \in M'\}$. Then $M' \neq \emptyset$ since we assumed that $A(x_0)$ is not stable. Suppose $x_0 \notin X''$. Then

$$A'' = \bigcup_{q\in M'} A_q$$

is not stable, as we have seen above. Hence we must have $x_0 \in X''$. Then

$$A(x_0) \subset A'' \subset A_0.$$

Since A_0 is the smallest stable subset of A containing $A(x_0)$ we obtain $A'' = A_0$. Hence

$$A(x_0) \cup \bigcup_{q\in M'} A_q = A\left(\prod_{x\in X''} x\right) = A'' = A_0 = A(x_0) \cup \bigcup_{q\in M} A_q.$$

Since $M \cap A(x_0) = \emptyset$ and there are no nontrivial inclusion relation between elements of M this implies $M' = M$. Therefore $X'' = X'$.

We now prove the last statement of the Lemma. So suppose $\operatorname{spec}(\widehat{R}) \to \operatorname{spec}(R)$ is injective. Let p_1, p_2 be two prime s-ideals of $\mathscr{C}(R^\bullet, \widehat{R}^\bullet)$ that have the same intersection with $[R^\bullet]^{\widehat{R}^\bullet}_{R^\bullet}$. By 1 we may choose subsets A_1, A_2 of A such that

$$P_i := \{b \in \widehat{R}^\bullet \mid [b]^{\widehat{R}^\bullet}_{R^\bullet} \in p_i\} = \bigcup_{q\in A_i} q\backslash\{0\}$$

for $i = 1, 2$. We show $P_1 = P_2$, which implies $p_1 = p_2$. By symmetry it is enough to prove $P_1 \subset P_2$. So let $q_1 \in A_1$. Then we have to find some $q_2 \in A_2$ such that $q_1 \subset q_2$. Now $p_1 \cap [R^\bullet]^{\widehat{R}^\bullet}_{R^\bullet} = p_2 \cap [R^\bullet]^{\widehat{R}^\bullet}_{R^\bullet}$ implies $P_1 \cap R^\bullet = P_2 \cap R^\bullet$. Hence

$$\bigcup_{q\in A_1} (q \cap R) = \bigcup_{q\in A_2} (q \cap R).$$

By prime avoidance there exists some $q_2 \in A_2$ such that $q_1 \cap R \subset q_2 \cap R$. Now Going Up and the fact that $\operatorname{spec}(\widehat{R}) \to \operatorname{spec}(R)$ is injective imply $q_1 \subset q_2$.

Theorem 7.2 *Let R be a Mori domain such that $R^\bullet$ satisfies condition* **(C)**, *and assume further that R is Noetherian or satisfies the assumptions of Theorem 5.5. Then R is locally tame, has finite catenary degree and the following are equivalent:*

1. $\mathsf{t}(R) < \infty$.
2. $\rho(R) < \infty$.

3. *There exists some* $m \in \mathbb{N}^+$ *such that for all* $a \in \widehat{R}$ *we have* $\varepsilon a^m \in R$ *for some* $\varepsilon \in \widehat{R}^\times$.
4. *The map* $\operatorname{spec}(\widehat{R}) \to \operatorname{spec}(R)$ *is injective.*

Proof Since $R^\bullet$ satisfies condition **(C)** we may choose a factorial monoid F, containing $\widehat{R}^\bullet$ as a cofinal and saturated submonoid, such that $\mathscr{C}_r(R^\bullet, F)$ is quasi-finite. Using Proposition 5.2 we see that $\mathscr{C}_r(\widehat{R}^\bullet, F)$ is finite and $R^\bullet$ is a weakly C-monoid defined in F, that satisfies condition **(C3)** of [7, Theorem 5.3]. Hence R is locally tame by loc. cit.

We will construct now a finite set A of nonzero primes of $\operatorname{spec}(\widehat{R})$, that satisfies the conditions 1–3 of Lemma 7.1. For doing so we distinguish if R is Noetherian or R satisfies the assumptions of Theorem 5.5.

R is Noetherian: let A be the set of those primes q of $\widehat{R}$ such that $q \cap R$ is associated to the R-module $\widehat{R}/R$. Since R is Noetherian and $(R : \widehat{R}) \neq 0, \widehat{R}$, and hence $\widehat{R}/R$, is a finitely generated R-module. It follows that the R-module $\widehat{R}/R$ possesses only finitely many associated prime ideals. Since $\widehat{R}$ is a finitely generated R-module, the set A is finite, too. Clearly any $q \in A$ is nonzero.

We show now that A satifies 1–3 of Lemma 7.1. Since R is Noetherian $\widehat{R}$ is the integral closure of R and therefore A satisifes Going Up and Incomparability by the Cohen–Seidenberg theorem. It remains to show, that A has property 1.

So let T be a divisor-closed subsemigroup of $\mathscr{C}_r(R^\bullet, \widehat{R}^\bullet)$. By Proposition 5.3 the semigroup $\mathscr{C}_r(R^\bullet, \widehat{R}^\bullet)$ is quasi-finite. From Theorem 4.6 it follows $N(T) \neq \emptyset$. Hence we can choose $s \in N(T)$. By Lemma 4.4.1 there exists some $n \in \mathbb{N}^+$ such that $s^n = s^{2n}$. Replacing s by s^n we may suppose $s = s^2$. Then

$$T = \left\{x \in \mathscr{C}_r(R^\bullet, \widehat{R}^\bullet) \mid (xs)(ys) = s \text{ for some } y \in \mathscr{C}_r(R^\bullet, \widehat{R}^\bullet)\right\}. \tag{3}$$

The inclusion $\supset$ is clear. Conversely, if $x \in T = [[s]]$ then $xy = s^n = s$ for some $y \in \mathscr{C}_r(R^\bullet, \widehat{R}^\bullet)$. Hence $(xs)(ys) = s^3 = s$.

Now choose $a \in \widehat{R}^\bullet$ such that $s = \{a\}_{R^\bullet}^{\widehat{R}^\bullet}$ and set $M = (R :_{\widehat{R}} a), M^\bullet = M \setminus \{0\}$. Then M is an R-submodule of $\widehat{R}$. For any $b \in \widehat{R}^\bullet$ we have

$$(M^\bullet : b) = ((R^\bullet : a) : b)) = (R^\bullet : ab).$$

Hence there exists an injective map

$$f\colon \mathscr{C}_r(M^\bullet, \widehat{R}^\bullet) \to \mathscr{C}_r(R^\bullet, \widehat{R}^\bullet)$$

such that

$$f(\{b\}_{M^\bullet}^{\widehat{R}^\bullet}) = s\{b\}_{R^\bullet}^{\widehat{R}^\bullet}$$

for all $b \in \widehat{R}^\bullet$. Since $s^2 = s$ this map f is multiplicative. Moreover $f(1) = s$. Using (3) we obtain

$$\left\{b \in \widehat{R}^\bullet \mid \{b\}_{R^\bullet}^{\widehat{R}^\bullet} \in T\right\} = \left\{b \in \widehat{R}^\bullet \mid \{b\}_{M^\bullet}^{\widehat{R}^\bullet} \in \mathscr{C}_r(M^\bullet, \widehat{R}^\bullet)^\times\right\}$$
$$= \left\{b \in \widehat{R}^\bullet \mid [b]_{M^\bullet}^{\widehat{R}^\bullet} \in \mathscr{C}(M^\bullet, \widehat{R}^\bullet)^\times\right\}.$$

Let V be this divisor-closed subsemigroup of $\widehat{R}^\bullet$, so that we have to show, that V is the complement of a union of primes in A. For that we set

$$V_0 = \left\{b \in \widehat{R}^\bullet \mid [b]_{M^\bullet}^{\widehat{R}^\bullet} = 1\right\} \quad \text{and} \quad R_0 = (M :_{\widehat{R}} M).$$

Then R_0 is a subring of $\widehat{R}$ such that $R \subset R_0$ and M is an R_0-submodule of $\widehat{R}$. From $(R : \widehat{R}) \neq 0$ we see that $\widehat{R}$, and therefore also R_0, are finitely generated R-modules. In particular, R_0 is Noetherian. We claim that V_0 is the set of those $b \in R_0^\bullet$, that are nonzero divisors of the R_0-module $\widehat{R}/M$. First, if $b \in V_0$, then from $[b]_{M^\bullet}^{\widehat{R}^\bullet} = [1]_{M^\bullet}^{\widehat{R}^\bullet}$ we obtain $bM \subset M$, so that $b \in R_0$. Now let $b \in R_0^\bullet$. Then we have $b \in V_0$ if and only if

$$(M^\bullet : b) = (M^\bullet : 1) = M^\bullet,$$

which is equivalent to b being a nonzero divisor of $\widehat{R}/M$. Since R_0 is Noetherian, we have

$$V_0 = R_0^\bullet \setminus \bigcup_{q \in B_0} q,$$

where B_0 is the set of all primes of R_0, that are associated to $\widehat{R}/M$. Let B be the set of all primes of $\widehat{R}$ lying over some prime in B_0. Since $V = [[V_0]]_{\widehat{R}^\bullet}$ we obtain

$$V = \widehat{R}^\bullet \setminus \bigcup_{q \in B} q.$$

It remains to show $B \subset A$. Let $q \in B$. Then $q \cap R_0$ is associated to the R_0-module $\widehat{R}/M$. Hence $q \cap R = (q \cap R_0) \cap R$ is associated to the R-module $\widehat{R}/M$. By definition of M multiplication by a defines a monomorphism $\widehat{R}/M \to \widehat{R}/R$. Hence $q \cap R$ is associated to $\widehat{R}/R$.

R satisfies the assumptions of Theorem 5.5: let S be the submonoid of regular elements of $R^\bullet$. Let A be the set of those primes of $\operatorname{spec}(\widehat{R})$ that contain $(R : \widehat{R})$ and have an empty intersection with S. Then we have an inclusion preserving bijection

$$A \to \operatorname{spec}\left(S^{-1}\widehat{R}/S^{-1}(R : \widehat{R})\right).$$

Since $S^{-1}\widehat{R}/S^{-1}(R : \widehat{R})$ is quasi-artinian, it follows that there are no nontrivial inclusion relations in A. Hence A satisfies the Incomparability property trivially.

We set $C = S^{-1}R/(R : \widehat{R})$ and $D = S^{-1}\widehat{R}/(R : \widehat{R})$. The isomorphism

$$\mathscr{C}(R^\bullet, \widehat{R}^\bullet) \cong \mathscr{C}(C, D)$$

from Lemma 5.4 induces an isomorphism

$$[R^\bullet]_{R^\bullet}^{\widehat{R}^\bullet} \cong [C]_C^D.$$

Using Lemma 3.6 we see that $[C]_C^D$ is zero complete. From Lemma 4.10 we obtain, that C is quasi-artinian, too. It follows analogously as before, that there are no non-trivial inclusion relations in the set $\{q \cap R \mid q \in A\}$. Hence A has trivially the Going Up property.

It remains to show that A has also property 1 of Lemma 7.1. By Lemma 5.4 $\widehat{R}^\bullet \to \mathscr{C}_r(R^\bullet, \widehat{R}^\bullet)$ factors through the canonical by morphism $\widehat{R}^\bullet \to S^{-1}\widehat{R}/S^{-1}(R : \widehat{R})$. Since any divisor-closed subsemigroup of $S^{-1}\widehat{R}/S^{-1}(R : \widehat{R})$ is the complement of a union of primes, A has property 1 of Lemma 7.1.

Using Lemma 7.1 we see now that $R^\bullet$ is nicely embedded in F. Therefore all assumptions of Theorem 6.2.4 are satisfied. We conclude that 1–3 are equivalent.

The implication 3 $\Rightarrow$ 4 follows from Lemma 3.5. Suppose conversely 4. Then $s\text{-spec}(\mathscr{C}(R^\bullet, \widehat{R}^\bullet)) \to s\text{-spec}([R^\bullet]_{R^\bullet}^{\widehat{R}^\bullet}$ is injective by the last statement of Lemma 7.1. Therefore we may apply Corollary 4.8 to $U = [R^\bullet]_{R^\bullet}^{\widehat{R}^\bullet}$, which is zero complete by Lemma 3.6, $S = \mathscr{C}(R^\bullet, \widehat{R}^\bullet)$ and $\Gamma = [\widehat{R}^\times]_{R^\bullet}^{\widehat{R}^\bullet}$. We choose $m \in \mathbb{N}^+$ as in this Corollary. If now $a \in \widehat{R}^\bullet$ then $[\varepsilon a^m]_{R^\bullet}^{\widehat{R}^\bullet} \in [R^\bullet]_{R^\bullet}^{\widehat{R}^\bullet}$ for some $\varepsilon \in \widehat{R}^\times$. Hence $\varepsilon a^m \in R^\bullet$.

References

1. V. Barucci, Mori Domains, in *Non-Noetherian Commutative Ring Theory, Mathematics and Its Applications*, vol. 520 (Kluwer Academic Publishers, Dordrecht, 2000), pp. 57–73
2. V. Barucci, E. Houston, On the prime spectrum of a Mori domain. Commun. Algebr. **24**, 3599–3622 (1996)
3. V. Barucci, S. Gabelli, M. Roitman, The class group of a strongly Mori domain. Commun. Algebr. **22**, 173–211 (1994)
4. V. Barucci, S. Gabelli, M. Roitman, Complete integral closure and strongly divisorial ideals. Commun. Algebr. **31**, 5447–5465 (2003)
5. W. Bruns, J. Herzog, *Cohen-Macauly Rings* (Cambridge University Press, Cambridge, 1993)
6. A. Geroldinger, F. Halter-Koch, in *Non-Unique Factorizations. Algebraic, Combinatorial and Analytic Theory*, Pure and Applied Mathematics, vol. 278 (Chapman & Hall/CRC, Boca Raton, 2006)
7. A. Geroldinger, W. Hassler, Arithmetic of Mori domains and monoids. J. Algebr. **319**, 3419–3463 (2008)
8. A. Geroldinger, W. Hassler, Local tameness of v-noetherian monoids. J. Pure Appl. Algebr. **212**, 1509–1524 (2008)
9. A. Geroldinger, F. Kainrath, On the arithmetic of tame monoids with applications to Krull monoids and Mori domains. J. Pure Appl. Algebr. **214**, 2199–2218 (2010)
10. R. Gilmer, *Commutative Semigroup Rings* (The University of Chicago Press, Chicago, 1984)
11. P.A. Grillet, *Commutative Semigroups* (Kluwer Academic Publishers, Dordrecht, 2001)

12. J. Herzog, E. Kunz, *Der kanonische Modul eines Cohen-Macauly-Rings*, SLN 238 (Springer, Berlin, 1971)
13. T. Lucas, Examples Built with $D+M$, $A+XB[X]$ and Other Pullback Constructions, in *Non Noetherian Commutative Ring Theory, Mathematics and Its Applications*, vol. 520 (Kluwer Academic Publishers, Dordrecht, 2000), pp. 341–368
14. N. Raillard, Sur les anneaux de Mori. C. R. Acad. Sci. Paris **280**, 1571–1573 (1975)

Prüfer Domains of Integer-Valued Polynomials

K. Alan Loper and Mark Syvuk

Abstract Let D be an integral domain with quotient field K. The ring $\text{Int}(D) = \{f(x) \parallel f(D) \subseteq D\}$ has been studied as a ring for more than forty years. A major topic of interest during that time has been the question of when the construction yields a Prüfer domain. The principal question has been resolved, but interesting generalizations are still being worked on. This is a survey paper that traces the history of study of integer-valued polynomial rings with a focus on when they are Prüfer domains.

1 Introduction

Throughout this paper, D is an integral domain, K is its field of fractions, and E is a nonempty subset of K. A polynomial $f(X)$ with coefficients in K is *integer-valued* if every $d \in D$ satisfies $f(d) \in D$; i.e., $f(D) \subseteq D$. The collection of such polynomials is designated $\text{Int}(D)$. One could also consider polynomials that are *integer-valued on a subset*; more precisely, the polynomial $f(X)$ is *integer-valued on the subset* $E \subseteq D$ if every $d \in E$ satisfies $f(d) \in D$; i.e., $f(E) \subseteq D$. The collection of these polynomials is designated $\text{Int}(E, D)$.

The first studies of $\text{Int}(D)$ were by Polya [18] and Ostrowski [16] both in 1919. Although it is easy to see that $\text{Int}(D)$ is a ring, both of these papers dealt purely with the additive structure. In particular, they focused on the D-module structure of $\text{Int}(D)$ where D is a ring of algebraic integers. For the next half century $\text{Int}(D)$ was studied periodically, always still with focus on the additive/module structure. Study of the ring theoretic structure of $\text{Int}(D)$ began almost simultaneously, and independently in three

K.A. Loper (✉)
Department of Mathematics, The Ohio State University–Newark,
Newark, OH 43055, USA
e-mail: lopera@math.ohio-state.edu

M. Syvuk
Department of Mathematics, The Ohio State University,
Columbus, OH 43210, USA
e-mail: syvuk.2@osu.edu

S. Chapman et al. (eds.), *Multiplicative Ideal Theory and Factorization Theory*,
Springer Proceedings in Mathematics & Statistics 170,
DOI 10.1007/978-3-319-38855-7_9

different places. Graduate students Paul-Jean Cahen and Jean-Luc Chabert at the University of Paris, professors Hiroshi Gunji and Donald McQuillan at the University of Wisconsin, and graduate student Demetrious Brizolis at UCLA all began study of Int(D) as a ring in the early 1970s. Integer-valued polynomials have been well studied now; there are deep results in many different directions. One of the main topics from very close to the beginning has been the question of when the integer-valued construction yields a Prüfer domain. This article will trace, chronologically, the study of this specific question.

2 Int(D)

2.1 Noetherian Domains

The consideration of Int(D) being a Prüfer domain began with the work of Brizolis [1]. This actually does not appear to have been his goal. The fact that Int(D) can be a Prüfer domain proved to be useful to him in his study of problems involving generating ideals. He proved that Int(D) is a Prüfer domain for a class of Dedekind domains which includes the rings of algebraic integers, and then used this result to generalize work of Skolem from the 1940s. He did find this "intermediate" result interesting though, and questioned what necessary and sufficient conditions on a domain D would be for Int(D) to be a Prüfer domain.

Jean-Luc Chabert [4] and Donald McQuillan [13] pursued this aggressively in the succeeding years and each, independently, settled the characterization problem in the case where the ring D is Noetherian. In particular, they each essentially proved the following theorem.

Theorem 2.1 *If D is Noetherian, then* Int(D) *is a Prüfer domain if and only if D is a Dedekind domain with all residue fields finite.*

In each case the method was to solve the problem locally and then globalize the solution. In particular, they each proved that Int(V) is a Prüfer domain if V is a DVR with a finite residue field. The general case follows from this because when D is Noetherian Int(D) behaves well with respect to localization. More precisely, let V be a DVR with maximal ideal M and residue field F. Let V^* be the M-adic completion of V. Then Int(V) is a Prüfer domain. Moreover, the maximal ideals of Int(V) lying over M all have the following form.

$$M_\alpha = \{f(x) \in \text{Int}(D) \parallel f(\alpha) \in MV^*\}$$

And these maximal ideals are all distinct. So the maximal ideals are indexed in a natural way by the M-adic completion of V. What McQuillan and Chabert were able to show is that this property can be globalized. Namely, if M is a maximal ideal of a Noetherian domain D, and $S = D - M$ then $\text{Int}(D_M) = S^{-1}\,\text{Int}(D)$. So, if D is a Dedekind domain with all residue fields finite, then the maximal ideals of Int(D) lying over a maximal ideal M are naturally indexed by the elements of the M-adic completion of D exactly as in the DVR case, and this led to a proof of the above theorem.

2.2 Non-Noetherian Domains

While the Noetherian case was being settled, there remained the general case where D is not assumed to be Noetherian. The first step in this direction was a result of Chabert [4] in 1987.

Theorem 2.2 *Let D be an integral domain. If* Int(D) *is a Prüfer domain, then D is an almost Dedekind domain with all residue fields finite.*

A domain D is said to be almost Dedekind provided the localization at any maximal ideal is a DVR. Noetherian almost Dedekind domains are then exactly the Dedekind domains. So the theorem seems to be a natural extension of the Noetherian necessary and sufficient condition. However, while Chabert's result gives a necessary condition for Int(D) to be a Prüfer domain, there was no indication that the condition was sufficient. In fact, at the time it seemed that the condition might be vacuous; it seemed possible that the only almost Dedekind domains with all residue fields finite were actually Dedekind.

There were a few examples of non-Notherian almost Dedekind domains in the literature. The first example is due to Nakano [15]: the ring of integers A_K of the infinite algebraic extension $K = \mathbb{Q}(\zeta_2, \zeta_3, \zeta_5, \zeta_7, \ldots)$ of $\mathbb{Q}$, where ζ_p is a primitive pth root of unity. Subsequently, there were several constructions of such domains, all due to Gilmer along with several co-authors. (A good summary of these constructions is contained in [6].) These constructions include, for example, those obtained as a Kronecker function ring or as a monoid ring. However, all of these non-Noetherian almost Dedekind domain examples contain at least one maximal ideal with infinite residue field, and hence fail Chabert's necessary condition.

In 1990 Gilmer [6] filled this gap by providing examples of non-Noetherian almost Dedekind domains which have all finite residue fields. The construction involves infinite degree algebraic extensions of algebraic number rings (or more general Dedekind domains). In the standard setting of algebraic number theory one takes a finite degree algebraic extension of a number field. In the corresponding rings of integers a prime in the smaller ring either extends to a prime in the upper ring (inertia), or extends to a power of a prime (ramification), or to a product of primes (splitting/decomposition), or to a combination of the three.

To see what is needed in such a construction consider three different cases. In each case, let V be a valuation domain with maximal ideal M generated by d, finite residue field of order q, and quotient field K. Let L be an algebraic extension of K of degree n.

1. (Ramification) Suppose that V extends to a valuation domain W in L, but that $MW = Q^n$ where Q is the maximal ideal of W. Then W will still have a principal maximal ideal, but it will not be generated by d. Rather, d generates the nth power of Q.
2. (Inertia) Suppose that V extends to a valuation domain W and that MW is the maximal ideal of W. Then d will generate the maximal ideal of W, but the residue field in W will have order q^n.
3. (Splitting/Decomposition) Suppose that V extends to a domain W which has n maximal ideals. Then each maximal ideal is locally generated by d and each residue field has order q.

If we start then with a Dedekind domain with all residue fields finite it is intuitively clear that the way to obtain an almost Dedekind domain with all residue fields finite from an infinite degree algebraic extension is to sharply curtail both inertia and ramification in the finite algebraic extensions. The "ideal" type of extension would be one where a prime in the extension field has the same residue field and is locally generated by the same element as the prime it lies over in the lower field. This is called an immediate extension. An infinite degree extension of a one-dimensional Prüfer domain is still a one-dimensional Prüfer domain. Begin with a DVR V with maximal ideal P and with a finite residue field and then consider an infinite degree extension. Each maximal ideal of the extension corresponds to a branch of a tree following the primes at successive stages, lying over P. But if one branch involves infinitely many stages with nontrivial ramification then localization at a maximal ideal will yield a non-discrete valuation domain rather than a DVR. And if there are an infinite number of stages in a single branch that involve inertia then the resulting domain will have infinite residue fields. It is not generally possible to control the behavior of an infinite number of primes in a finite extension. Gilmer's method however, employed a deep result of Krull [8], to start with a single valuation domain and then to build a tower of finite degree extensions such that at each stage the collection of all primes (necessarily finite) is completely controlled. In particular, if we start with the unique prime P in V then follow a single line of primes lying over it then we can arrange things so that on that single branch we have only immediate extensions from some finite stage onward. This will yield an almost Dedekind domain with finite residue fields.

However, once the desired domains had been constructed, it was apparent that their behavior was not necessarily like that of Dedekind domains. Note that a finite residue field must have order a power of some prime p. In a Dedekind domain there can only be finitely many maximal ideals with residue fields of characteristic p. But in an almost Dedekind domain there can be a prime number p such that there are infinitely many maximal ideals M_i with residue fields having order a power of p. And Gilmer was able to build such a domain in which the sizes of these residue fields of characteristic p are unbounded. The idea is that on each branch the extensions are immediate from some point on, but looking from one branch to another we can have inertial behavior happening at arbitrarily high levels. In such an almost

Dedekind domain D Gilmer was able to find a distinguished maximal ideal M such that $\mathrm{Int}(D) \subseteq D_M[x]$. This demonstrates that $\mathrm{Int}(D)$ is not a Prüfer domain since $D_M[x]$ is not a Prüfer domain and all overrings of a Prüfer domain are again Prüfer domains. On the other hand, Gilmer also constructed some non-Noetherian almost Dedekind domains for which the orders of the residue fields of characteristic p is a bounded set, and in such cases he proved that $\mathrm{Int}(D)$ is a Prüfer domain. Accordingly, he posed the following question (slightly paraphrased here)?

Question 2.3 *If D is an almost Dedekind domain such that all residue fields of characteristic p are of bounded size, is* $\mathrm{Int}(D)$ *a Prüfer domain?*

Note that the question only deals with the question of sufficiency. Within the setting of construction by means of infinite degree algebraic field extension, Gilmer had proven necessity of the boundedness condition.

Chabert [5] approached Gilmer's question and answered it negatively. Chabert made use of Hasse's existence theorem [7], which, along the same lines as Gilmer's use of Krull's theorem, allowed him to find an algebraic extension in which the behavior of a finite number of primes can be completely controlled. To understand Chabert's method, suppose first that we are working in characteristic zero. Now if D is an almost Dedekind domain with finite residue field then each maximal ideal must contain a rational prime number. Start with a DVR with finite residue field such that 2 is in the maximal ideal. Since D is a DVR then 2 generates some power M^n of the maximal ideal M. In an almost Dedekind extension the exponent n such that $(2)D_M = M^n D_M$ varies from one maximal ideal M to another. Chabert's method in this example however, was to shut inertia down completely in the algebraic extensions so that the residue fields stayed small, but to include enough ramification that the exponents n satisfying $(2)D_M = M^n D_M$ were unbounded as M ranged across the maximal ideals containing 2. As with Gilmer's negative examples, in Chabert's examples that had unbounded ramification he was able to prove that $\mathrm{Int}(D)$ was not a Prüfer domain by finding a distinguished maximal ideal M such that $\mathrm{Int}(D) \subseteq D_M[x]$. Following we explain Chabert's proposed modification of Gilmer's conjecture (somewhat paraphrased here).

First, consider the following two conditions on an almost Dedekind domain D with all residue fields finite.

1. Choose a prime integer p. We say that D satisfies the first boundedness condition if there is a bound on the cardinalities of the residue fields of order a power of p for each prime p.
2. The second condition is not as simply stated. We give it in two parts.

   - If D has characteristic 0 then each maximal ideal must contain exactly one prime number. If D has characteristic p then D must contain a finite field F. Choose F to have maximal order—note that D cannot contain an infinite field, because then the residue fields would not be finite. In the characteristic p case there must also be an element $t \in D$ such that t is transcendental over F. Hence, the polynomial ring $F[t] \subseteq D$. Then each maximal ideal of D must contain exactly one irreducible polynomial from $F[t]$. These irreducible polynomials play the same role as the prime numbers do in the characteristic 0 case.

- For ease of exposition assume that D has characteristic 0. Choose a prime number p. For each maximal ideal M containing p consider the integer n such that $pD_M = (M^n)D_M$. Call n a ramification index. We say that D satisfies the second boundedness condition if the collection of ramification indices is bounded for each prime p.

An almost Dedekind domain which satisfies the above conditions is said to be doubly-bounded. This then led Chabert to the following question.

Question 2.4 *Suppose D is an almost Dedekind domain with all residue fields that is doubly-bounded. Is* Int(D) *Prüfer?*

Chabert's question turned out eventually to precisely give the necessary and sufficient conditions for Int(D) to be a Prüfer domain. As with Gilmer's question, Chabert's questions dealt only with sufficiency. The reason for this is that both were able to prove the necessity of the boundedness conditions in the special setting of the constructions they employed. In particular, they began with a Dedekind domain, took a countably generated algebraic extension of the quotient field, and produced the desired almost Dedekind domain in the field extension. So the sufficiency question was still outstanding, and the necessity question would be still outstanding if it could be shown that non-Noetherian almost Dedekind domains with finite residue fields could be constructed that were built without utilizing a countably generated algebraic field extension.

At the same time as he analyzed a two-part condition which he knew to be necessary under certain conditions, Chabert also considered a condition which he could prove was sufficient

- For M a maximal ideal of D let $S = D - M$. Then Int(D) is said to *behave well under localization* if S^{-1} Int$(D) =$ Int(D_M) for each maximal ideal M of D. Chabert proved:

Theorem 2.5 *Let D be an almost Dedekind domain with finite residue fields. If* Int(D) *behaves well under localization, it is a Prüfer domain.*

This clearly leads to a question about necessity:

Question 2.6 *If* Int(D) *is a Prüfer domain, does it necessarily behave well under localization?*

In some sense, the property of good behavior under localization would not be a satisfactory resolution of the characterization question because it attempts to equate two properties of $\text{Int}(D)$ rather than equating the Prüfer property of $\text{Int}(D)$ with a property of D. However, in the particular case of almost Dedekind domains defined by countably infinite degree algebraic field extensions, Chabert was able to show that good behavior under localization was equivalent to a property of D which he called the *immediate subextension property*. This property imposed a strong finiteness condition on the manner in which properties of valuation domains could be modified as one went up and down the ladder of an infinite degree field extension. We explain more precisely below.

- Let K_0 be a field and let K be a countably generated algebraic extension of K_0. Let D_0 be a Dedekind domain with quotient field K_0 and let D be an almost Dedekind domain with quotient field K such that every maximal ideal of D lies over a maximal ideal of D_0.
- Choose a maximal ideal M of D. Then we can associate other maximal ideals M_i of D with M by
  - Choose an intermediate field K^* between K_0 and K.
  - Contract the valuation domain D_M to a valuation domain V^* contained in K^*.
  - Consider all the valuation overrings of D which are extensions of V^*. Consider these valuation domains to be associated with D_M.
- Then we say that D has the immediate subextension property if for every D_M we can find a field K^* which is finitely generated over K_0 such that when we restrict D_M to a valuation domain V^* of K^* and then pull back up to all the valuation overrings of D which are extensions of V^*, then for all D_M and all the valuation domains thus associated with it the extensions are immediate.

A modified form of Theorem 2.5 is then

Theorem 2.7 *Let D be an almost Dedekind domain with finite residue fields. If D is constructed using a countably infinite algebraic field extension and satisfies the immediate subextension property then* $\text{Int}(D)$ *is a Prüfer domain.*

So we pose a modification of Question 2.6.

Question 2.8 *If* $\text{Int}(D)$ *is a Prüfer domain, does D have the immediate subextension property?*

This question focuses on a property of D, but it is restricted to just those domains built using countably infinite algebraic field extensions. In any case, the properties of behaving well under localization and immediate subextension turned out not to be the properties that characterize when $\text{Int}(D)$ is a Prüfer domain. Nonetheless, they are important because they illustrate the topological nature of resolving the classification problem in the general case. In particular, it seems reasonable that for a Dedekind domain, since any nonzero element is contained in only finitely many prime ideals then perhaps the only convergence that could happen with prime ideals is convergence to the zero ideal. But if an almost Dedekind domain were not

Noetherian then a nonzero element could be contained in infinitely many prime ideals and nontrivial convergence of some sort could happen. A model for this idea is the behavior of Int(Z). The maximal ideals containing a given prime p are naturally indexed by the p-adic numbers. Hence, one might expect these maximal ideals to have topological properties relative to each other matching the topology of the p-adic integers. In the negative examples of Gilmer and Chabert the proof that Int(D) was not a Prüfer domain was accomplished by finding a maximal ideal M of D such that Int$(D) \subseteq D_M[x]$. Also, in both cases there were infinite collections of maximal ideals for which a particular index was unbounded on the collection. So it seems plausible to try to locate the distinguished maximal ideal as a limit of a sequence of maximal ideals which has the relevant index going to infinity. With this intuitive idea in mind, Loper defined what seemed to be perhaps the simplest possible class of non-Noetherian almost Dedekind domains.

A *sequence domain* is a non-Noetherian almost Dedekind domain D with finite residue fields and field of fractions K such that the following conditions hold:

1. There exists a collection of maximal ideals $S = \{P_i\}_{i=1}^{\infty}$ of D such that
   a. $D = \cap_{i=1}^{\infty} D_{P_i}$,
   b. each residue field D/P_i has the same characteristic p,
   c. the collection $\{P_i\}_{i=1}^{\infty}$ does not constitute all of the maximal ideals of D.
2. There exists a collection $\{v_i\}_{i=1}^{\infty}$ of valuations on K such that
   a. $v_i^{(N)}$ is the normed valuation on K corresponding to P_i for each i,
   b. for all $d \in D\backslash\{0\}$, the sequence $\{v_i(d)\}_{i=1}^{\infty}$ is eventually constant,
   c. for all $d \in D\backslash\{0\}$, $v^*(d) = \lim_{i\to\infty} v_i(d) \in \mathbb{Z}^+ \cup \{0\}$,
   d. there is $\pi \in D$ such that for all $i \in \mathbb{Z}^+$, $v_i(\pi) = 1$.

Set $P^* = \{d \in D \parallel v^*(d) > 0\} \cup \{0\}$. It turns out that if the residue field of each P_i is finite, then the set $\{P^*, P_1, P_2, \cdots\}$ comprises all of the maximal ideals of the sequence domain D. Moreover, the primes P_i are all principal while P^* is not finitely generated. The idea here is to view P^* as the limit of the sequence $\{P_i\}$. Then the maximal ideals of Int(D) lying over P^* inherit their properties from sequences of maximal ideals lying over the P_i's rather than from the structure of Int(D_{P^*}). In particular

Theorem 2.9 *If D is a sequence domain, then* Int(D) *is Prüfer if and only if D is doubly-bounded.*

For sequence domains, double-boundedness translates to the set $\{|D/P_i|\}_{i\in\mathbb{Z}^+\cup\infty}$ being bounded and, for each $d \in D\backslash\{0\}$, the set $\{v_i^{(N)}(d)\}_{i\in\mathbb{Z}^+\cup\infty}$ being bounded. Hence in the setting of sequence domains, the classification question has a complete answer.

This setting also allows insight into whether Int(D) behaving well under localization is necessary for it to be Prüfer. When D is a sequence domain, the following result characterizes when Int(D) behaves well under localization:

Theorem 2.10 *If D is a sequence domain, then* Int(D) *behaves well under localization if and only if both of the following conditions hold:*

1. $q_i = |D/P_i| = |D/P^*|$ *for all but finitely many* $i \in \mathbb{Z}^+$.
2. $v_i = v_i^{(N)}$ *for all but finitely many* $i \in \mathbb{Z}^+$.

The key to both the Prüfer characterization and the good behavior under localization for sequence domains is the same. The behavior of maximal ideals of Int(D) that lie over P^* is determined by sequences of maximal ideals lying over the P_i's. Consider just the residue field part of this. If the sizes of the residue fields of the P_i's are unbounded then, even though the residue field of P^* is finite, we have Int$(D) \subseteq D_{P^*}[x]$. which proves that Int(D) is not Prüfer. So since P^* is a limit of primes with residue fields of cardinalities going to infinity then Int(D) behaves as if the residue field of P^* was infinite even though it is actually finite. Similarly, examples can be built such that the residue field of each P_i has order p^2 but P^* has residue field of order p and then Int(D) will have maximal ideals lying over P^* with residue field of order p^2. The integer-valued polynomial ring for such a domain is a Prüfer domain but does not behave well under localization. Thus Question 2.6 has a negative answer. The key again is that Int(D) respects the limiting process of the maximal ideals of D even when D does not.

The complete classification of all domains D such that Int(D) is a Prüfer domain came not long after the paper on sequence domains. Loper's proof that double-boundedness is sufficient in [10] was expanded by Cahen and Chabert in [2]. While not presented as such, their proof actually demonstrates sufficiency for the general case. Chabert proved necessity in the case where D is built using a countably infinite algebraic field extension. So what was left was to prove necessity in a general setting. This was done in [9] using the topological ideas in [10]. In particular, ultrafilters were used to find limit primes of unbounded sequences, yielding a maximal ideal M of D such that Int$(D) \subseteq D_M[x]$.

If D has characteristic zero the final theorem is as follows.

Theorem 2.11 *Let D be an almost Dedekind domain with finite residue fields. Then the following conditions are equivalent.*

1. Int(D) *is a Prüfer domain.*
2. *For each prime number p which is a nonunit in D, the two sets*

$$F_p = \{|D/P| \parallel p \in P\}$$

and

$$E_p = \{v_P^{(N)}(p) \parallel p \in P\}$$

are bounded sets.

The theorem remains true for fields with nonzero characteristic, provided a suitable irreducible polynomial replaces the prime number p.

3 Int(E, D)

Recall that Int(E, D) is the set of polynomials with coefficients in K that map a subset E of D into D; that is, $\text{Int}(E, D) = \{f \in K[X] \mid f(E) \subseteq D\}$. As with Int$(D)$, the question of when Int(E, D) is a Prüfer domain has been studied; however, it is far from being resolved. Recall that $E \subseteq K$ is a *fractional subset* of D if there is some nonzero element d of D such that dE is a subset of D. In almost all cases, $\text{Int}(E, D) = D$ if $E \subseteq K$ is not a fractional subset of D. In the few cases where Int(E, D) is different from D when E is not a fractional subset, D is not integrally closed. It is easy to see that Int(E, D) is not a Prüfer domain in this case. Moreover, if E is a fractional subset of D and $dE \subseteq D$ with $d \neq 0$ then Int(E, D) is naturally isomorphic to Int(dE, D). We will then assume henceforth that $E \subseteq D$.

There is then a very easy necessary condition. Choose an element $d \in E$. Then the set $\{f(x) \in \text{Int}(E, D) \mid f(d) = 0\}$ is easily seen to be a prime ideal of Int(D). It is also easy to see that the quotient of Int(D) by this prime ideal is D. Hence our necessary condition is

- If E is a fractional subset of D and Int(E, D) is a Prüfer domain, then D is a Prüfer domain.

McQuillan [14] completely settled the case when E is finite. He has shown:

Theorem 3.1 *If E is finite, then* Int(E, D) *is a Prüfer domain if and only if D is a Prüfer domain.*

Since a necessary condition is that D be a Prüfer domain, and it is reasonable to approach the problem locally, the next results consider Int(E, V) where V is a valuation domain. If V is a DVR with finite residue field and $E \subseteq V$ then Int(E, V) is an overring of the Prüfer domain Int(V) and hence a Prüfer domain.

Along this line, Cahen, Chabert, and Loper [3] considered the case of Int(E, V), where E is an infinite subset of a valuation domain V with quotient field K with particular focus on the cases where Int(V) is not a Prüfer domain. Let $I \subset V$ be an ideal of V such that $\cap(I^n) = (0)$, and consider the I-adic completions $\widehat{E}, \widehat{K}, \widehat{V}$ of E, K, V, respectively. We say that E is precompact if $\widehat{E}$ is compact in $\widehat{K}$. The main result of the paper connected to the Prüfer property is a sufficient condition.

Theorem 3.2 *If E is a precompact subset of V, then* Int(E, V) *is Prüfer.*

The key to this theorem is that if E is precompact then E hits only finitely many cosets modulo any nonzero ideal. In this regard, E has many properties in common with the collection of all elements of a DVR with finite residue field. There was no proof of the necessity of this condition.

There is also a curious example in the paper. Let T be the ring of entire functions. It is well known that T is a Bezout domain and that it has many maximal ideals of infinite height. In fact, the height of such a maximal ideal is large enough that the intersection of a chain of prime ideals contained in it cannot be the zero ideal. One consequence of this is that if we localize T at a maximal ideal of infinite height, we obtain a valuation domain V such that $\text{Int}(E, V)$ is a Prüfer domain if and only if E is finite.

Recently, Loper and Werner proved that precompactness is not a necessary condition.

To understand this result let V be a one-dimensional valuation domain that is not discrete. Let $\{d_i\}$ be a sequence of elements of V such that $v(d_i - d_{i+1})$ is an increasing sequence, but does not increase to infinity. We say then that the sequence is pseudo- convergent. If $\{d_i\}$ is a pseudo-convergent sequence and $\alpha \in V$ is such that $v(\alpha - d_i)$ is an increasing sequence then we say that α is a pseudo-limit of the sequence.

It can happen in such a valuation domain V that pseudo-convergent sequences that have pseudo-limits or that do not have pseudo-limits can both exist, with the sequences in both cases not converging in the classical sense. Consider the following examples.

1. Let k be a field. Consider the ring $k[\{x^\alpha\}]$ where α runs over the positive real numbers. We can either think of this as a polynomial ring in powers of x or as a semigroup ring over k. In any case, localize the ring at the maximal ideal generated by the powers of x. The result is a one-dimensional valuation ring V with value group the field of real numbers under addition. For a given power of x, the value is simply the exponent.
2. Consider the sequence $\{x^{\beta_i}\}$ where $\{\beta_i\}$ is an increasing sequence of real numbers converging to 2. Then the sequence $\{x^{\beta_i}\}$ is a pseudo-convergent sequence with x^2 as a pseudo-limit.
3. Let $\{\beta_i\}$ be as above. Then define $y_1 = x^{\beta_1}$ and for $n > 1$ define $y_n = x^{\beta_1} + x^{\beta_2} + \cdots + x^{\beta_n}$. The sequence $\{y_i\}$ is then pseudo-convergent, but does not have a pseudo-limit in V.

Using this type of setup Loper and Werner [12] proved:

Theorem 3.3 *There exists a nondiscrete one-dimensional valuation domain V with a subset E consisting of a pseudo-convergent sequence which does not have a pseudo-limit in V such that* $\text{Int}(E, V)$ *is a Prüfer domain, even though E is not precompact.*

Hence the question of when $\text{Int}(E, D)$ is a Prüfer domain is very far from settled. There is no complete classification for when $\text{Int}(E, D)$ is a Prüfer domain even in the special case where D is a valuation domain. And in the case where D is a valuation domain it is clear that the solution will not mirror the characterization for $\text{Int}(D)$.

4 Generalizations

Let D be domain with quotient field K and let $\overline{K}$ be an algebraic closure of K. If we let $f(x)$ be a polynomial in $K[x]$ and let $\alpha \in \overline{K}$ be integral over D then it is reasonable to ask whether $f(\alpha)$ is still integral over D. Along this line of thought we can define a generalized form of integer-valued polynomial ring.

1. Let A_α be the ring of algebraic integers in the finite degree extension $Q[\alpha]$ of the rational numbers.
2. Let A_∞ be the ring of all algebraic integers.
3. Let A_n be the set of all algebraic integers in A_∞ of degree $\leq n$ over Q.
4. Let $\text{Int}_Q[A_\alpha] = \{f(x) \in Q[x] \parallel f(A_\alpha) \subseteq A_\alpha\} = \text{Int}(A_\alpha) \cap Q[x]$
5. Let $\text{Int}_Q(A_n) = \{f(x) \in Q[x] \parallel f(A_n) \subseteq A_n\} = \bigcap_{[Q[\alpha]:Q] \leq n} \text{Int}(A_\alpha) \cap Q[x]$

Using the above constructions Loper and Werner [11] proved the following theorem.

Theorem 4.1 *Let A_α and A_n be as above. Then* $\text{Int}_Q[A_\alpha]$ *and* $\text{Int}_Q(A_n)$ *are Prüfer domains.*

Moreover, they give a strong answer to a question posed by Brizolis in the paper where Prüfer rings of integer-valued polynomials were introduced. Brizolis wondered whether a proper subring of $\text{Int}(Z)$ existed which had $Q(x)$ as quotient field and was a Prüfer domain. Chabert answered this question in [4] by demonstrating that if we let $E = \frac{1}{2}Z$ be the fractional ideal of Z generated by $1/2$ then $\text{Int}(E, Z)$ is a proper subring of $\text{Int}(Z)$ and is isomorphic to $\text{Int}(Z)$. Note however, that $2x$ lies in the ring $\text{Int}(E, Z)$ but x does not. A stronger question is whether there exists a Prüfer domain which lies properly between $Z[x]$ and $Q[x]$. The theorem above demonstrates that such domains do exist.

The paper [11] also generalizes a little farther. Let I be the $n \times n$ identity matrix, let α be a rational number and identify α with the diagonal matrix αI. With this identification we can choose a polynomial $f(x)$ over the rational numbers and evaluate at an $n \times n$ matrix M with integer entries. It is then reasonable to ask which polynomials with rational coefficients map integral $n \times n$ matrices to integral $n \times n$ matrices. Let $M_n(Z)$ be the ring of $n \times n$ matrices over the integers. We then define $\text{Int}_Q(M_n(Z))$ to be the ring of all polynomials over the rational numbers which map $M_n(Z)$ to $M_n(Z)$. Since each such matrix satisfies a monic polynomial over the integers it seems natural to identify this ring with $\text{Int}_Q(A_n)$. However, let M be a nonzero matrix such that $M^2 = 0$. Then $f(x) = x^2/n^2$ will map M to 0 for any positive integer n, but for all but finitely many integers $g(x) = x/n$ will map M to a matrix with entries not lying in the integers. This suggests that $\text{Int}_Q(M_n(Z))$ is not integrally closed. Accordingly, Loper and Werner proved the following theorem.

Theorem 4.2 $\text{Int}_Q(M_n(Z))$ *is not integrally closed but has integral closure equal to* $\text{Int}_Q(A_n)$*, which is a Prüfer domain.*

Along the same lines as the above results, Peruginelli [17] has very recently extended McQuillan's results concerning integer-valued polynomials over finite sets.

Theorem 4.3 *Let D be an integrally closed domain with quotient field K, and let A be a torsion-free, finitely generated D-algebra. Let $E \subseteq A$ be a finite set of elements and consider the ring* $\text{Int}_K(E, A)$ *of polynomials with coefficients in K which map E into A. Then the integral closure of* $\text{Int}_K(E, A)$ *is a Prüfer domain if and only if D is a Prüfer domain.*

References

1. D. Brizolis, A theorem on ideals in Prüfer rings of integral-valued polynomials. Comm. Alg. **7**(10), 1065–1077 (1979)
2. P.-J. Cahen, J.-L. Chabert, Integer-valued polynomials. Amer. Math. Soc. Surveys and Monographs, Providence (1997)
3. P.-J. Cahen, J-L. Chabert, K.A. Loper, High dimension Prüfer domains of integer-valued polynomials. J. Korean Math. Soc. **38**(5), 915-935 (2001)
4. J.-L. Chabert, Un anneau de Prüfer. J. Algebra. **107**(1), 1–16 (1987)
5. J.-L. Chabert, Integer-valued polynomials, Prüfer domains, and localization. Proc. Amer. Math. Soc. **116**(4), 1061–1073 (1993)
6. R. Gilmer, Prüfer domains and rings of integer-valued polynomials. J. Algebra. **129**(2), 502–517 (1990)
7. H. Hasse, Zwei Existenztheoreme über algebraische Zahlkörper. Math. Ann. **95**(1), 229–238 (1926)
8. W. Krull, Über einen Existenzsatz der Bewertungstheorie. Abh. Math. Sem. Univ. Hamburg. **23**, 29–35 (1959)
9. K.A. Loper, A classification of all D such that Int(D) is a Prüfer domain. Proc. Amer. Math. Soc. **126**(3), 657–660 (1998)
10. K.A. Loper, Sequence domains and integer-valued polynomials. J. Pure Appl. Algebra. **119**(2), 185–210 (1997)
11. K.A. Loper, N. Werner, Generalized rings of integer-valued polynomials. J. Number Theory. **132**(11), 2481–2490 (2012)
12. Loper, K. A., Werner, N. Pseudo-convergent sequences and Prüfer domains of integer-valued polynomials. J. of Comm. Algebra. (to appear)
13. D. McQuillan, On Prüfer domains of polynomials. J. reine Angew. Math. **358**, 162–178 (1985)
14. D. McQuillan, Rings of integer-valued polynomials determined by finite sets. Proc. Roy. Irish Acad. Sect. A. **85**(2), 177–184 (1985)
15. N. Nakano, Idealtheorie in einem speziellen unendlichen algebraischen Zahlkörper. J. Sci. Hiroshima Univ. Ser. A. **16**, 425–439 (1953)
16. A. Ostrowski, Über ganzwertige Polynome in algebraischen Zahlkörpern. J. reine Angew. Math. **149**, 117–124 (1919)
17. Peruginelli, G. The ring of polynomials integral-valued over a finite set of integral elements. J. Comm. Algebra (to appear)
18. G. Polya, Über ganzwertige Polynome in algebraischen Zahlkörpern. J. reine Angew. Math. **149**, 97–116 (1919)

Lobal Properties of Integral Domains

Thomas G. Lucas

Abstract The fundamental quest of this article is to attempt to characterize a given global property of certain integral domains in terms of containment relations among the ideals and elements contained in a single maximal ideal. We say that a global property **G** is lobal if there is a property **P** (implied by **G**) satisfied by the ideals and elements of a single maximal ideal such that a domain R satisfies **G** if (and only if) at least one maximal ideal satisfies **P**. For example, a domain is Laskerian if each ideal is a finite intersection of primary ideals. This turns out to be a lobal property: a domain R is Laskerian if and only if there is a maximal ideal M with the property that each ideal contained in M is a finite intersection of MP-primary ideals. An ideal $P \subseteq M$ is an M-prime if for $a, b \in M$ with $ab \in P$, at least one of a and b is in P; and an ideal $Q \subseteq M$ with radical the M-prime P is MP-primary if for $a, b \in M$ with $ab \in Q$ and $a \in M \backslash P$, we have $b \in Q$. Other lobal properties include Prüfer domains, coherent domains, h-local domains, UFDs, Krull domains, atomic domains, and HFDs.

Keywords Coherent domain · Krull domain · Atomic domain · HFD · Pseudovaluation domain

Subject Classifications [MSC 2010] Primary 13A15, 13G05 · Secondary 13F05, 13F15

1 Introduction

This article takes an alternate viewpoint with regard to discovering properties of a given integral domain. Essentially the goal is to determine global properties of an integral domain by "looking" locally without actually localizing. For example, an

T.G. Lucas (✉)
Department of Mathematics and Statistics, University of North Carolina, Charlotte, NC 28223, USA
e-mail: tglucas@uncc.edu

S. Chapman et al. (eds.), *Multiplicative Ideal Theory and Factorization Theory*, Springer Proceedings in Mathematics & Statistics 170,
DOI 10.1007/978-3-319-38855-7_10

integral domain R (that is not a field) is quasilocal if and only if there is a maximal ideal M such that for each nonzero element $b \in M$, each principal ideal aR that is properly contained in bR can be factored as $aR = bRcR$ for some $c \in M$.

Throughout the article R represents an integral domain that is properly contained in its quotient field and M denotes a maximal ideal of R. We impose the following four restrictions on what properties we are allowed to consider about the elements and ideals of R that are contained in a given maximal ideal M.

(A1) If $t \in M$ and $I \subseteq M$ is an ideal of R, then we can determine when t is in I and when t is not in I. In addition, we can "pool" such knowledge for a nonempty set $X \subseteq M$, either $X \subseteq I$ or $X \nsubseteq I$.
(A2) If I and J are ideals contained in M, then we can determine when I is contained in J and when I is not contained in J.
(A3) For nonzero elements $t, a \in M$ such that $tR \subseteq aR$, we can determine either that there is a $b \in M$ such that $tR = aRbR$ (and that $t = ac$ for some $c \in M$) or that no such b exists.
(A4) For a collection of elements or of ideals which satisfy some "knowable" property, we can determine whether the collection is finite or infinite. Also for a pair of finite lists (perhaps with repetitions in one or both), we can determine whether there is a one-to-one correspondence between the lists.

Note that one can combine (A1) and (A3) to have the following: for a pair of nonzero elements $b, c \in M$ and nonzero ideal $I \subseteq M$, we can determine whether or not bI is a contained in cI by considering the products by for $y \in I$ individually: we can first see if $byR \subseteq cR$, and then if it is, we look to see if $byR = cxR$ for some $x \in I$.

For a given global property **G** of (certain) integral domains, we say that **G** is *lobal* if there is a property **P** that we can derive from (A1)–(A4) with respect to a single maximal ideal that is equivalent to **G**. In the case there is a property **Q** that each maximal ideal satisfies that collectively is equivalent to **G**, we say that **G** is *weakly lobal*. By default, a lobal property is also a weakly lobal one. The implications for lobal characterization are these: **P** for at least one maximal ideal $\Rightarrow$ **G** for $R \Rightarrow$ **P** for every maximal ideal. On the other hand for a weakly lobal characterization all we are sure of is **G** $\Leftrightarrow$ **Q** for each maximal ideal.

Using only (A1) and (A2) (for a single maximal ideal) it is possible to determine when a particular set is a generating set for an ideal I contained in M (without actually using the set to generate I): if $X \subseteq M$ is such that $X \subseteq I$ and each ideal $J \subseteq M$ that contains X also contains I, then X generates I. Hence, we can determine when R is a PID based solely on the ideals that are contained in a single maximal ideal: R is a PID if and only if there is a maximal ideal M such that for each ideal $I \subseteq M$, there is an element $x \in I$ such that each ideal $J \subseteq M$ that contains x also contains I ([5, Theorem 2.2]). Also with the additional help of (A4), we can determine when a particular ideal $B \subseteq M$ is finitely generated: B is finitely generated if and only if there is a finite set $Y \subseteq B$ such that each ideal $A \subseteq M$ that contains Y also contains B. Also for a finite collection of ideals $\{A_1, A_2, \ldots, A_n\}$ where each A_i is contained in M, the intersection of the A_is is the ideal I that is contained in each A_i and contains each ideal that is contained in each A_i.

When considering a particular maximal ideal M, we do not consider properties of elements and/or ideals that lie outside of M. For example to try to determine whether a particular finitely generated ideal I is invertible or not, we do not consider II^{-1} specifically as the product of I with its inverse as I^{-1} is not contained in M. However, there is a way to determine when II^{-1} is not contained in M without looking at II^{-1} at all. In the special case that M is the only maximal ideal that contains I, this allows us to determine when I is invertible. The special case for a two-generated ideal (contained in unique maximal ideal) was considered in [5, Corollary 2.7]. In Theorem 3.1 we show that for a nonzero finitely generated ideal I contained in a (specified) maximal ideal M, it is possible to determine when I is invertible using (A1)–(A4). Using this we give a new characterization of a Prüfer domain as a lobal property.

For a nonzero nonunit $b \in R$, we let $\mathscr{P}(b) = \{aR \mid a \in bR\}$. For each maximal ideal M containing b, $\mathscr{P}(b)$ can be partitioned into two sets: $\mathscr{F}_M(b) = \{aR \in \mathscr{P}(b) \mid aR = bcR$ for some $c \in M\}$ and $\mathscr{N}_M(b) = \mathscr{P}(b) \backslash \mathscr{F}_M(b)$. Note that $bR \in \mathscr{N}_M(b)$ and $b^2R \in \mathscr{F}_M(b)$. By (A3) it is possible to determine which of the sets $\mathscr{N}_M(b)$ and $\mathscr{F}_M(b)$ contains a given principal ideal $aR \in \mathscr{P}(b)$. One use of the sets $\mathscr{N}_M(y)$ is in describing the ideal $IR_M \cap R$ (for $I \subseteq M$) without localizing and contracting: for a nonzero ideal $I \subseteq M$, $IR_M \cap R = I_{(M)} := \{x \in M \mid yR \in \mathscr{N}_M(x)$ for some $y \in I\} \cup \{0\}$ [5, Theorem 3.7].

Recall that a domain R is said to be *Laskerian* if each ideal has a (finite) primary decomposition. In [5], we introduced the notion of an ideal $P \subseteq M$ being an M-prime, meaning that if $a, b \in M$ are such that $ab \in P$, then at least one of a and b is contained in P. We also introduced MP-primary ideals (where P is an M-prime) as an ideal Q such that $t^n \in Q$ for each $t \in P$ and for $c, d \in M$ with $c \in M \backslash P$, $cd \in Q$ implies $d \in Q$. Clearly, each prime ideal that is contained in M is an M-prime and each primary ideal (with prime radical P), contained in M is MP-primary. By [5, Theorem 3.8] an M-prime P contained in M is prime if and only if $P_{(M)} = P$. Similarly, an MP-primary ideal Q contained in M is primary if and only if $Q_{(M)} = Q$ [5, Theorem 3.9]. We start by characterizing the M-primes that are not prime and the MP-primary ideals that are not primary (Theorems 2.2 and 2.3 respectively). We then show that the Laskerian property is lobal (Theorem 2.4).

For a given nonzero ideal I of R and a maximal ideal M that contains I, there is a way to determine whether I is contained in infinitely many maximal ideals or only finitely many by examining a related set of ideals that are contained in M [5, Theorem 3.4]. Using this it is possible to determine when R has finite character (each nonzero ideal/element is contained in only finitely many maximal ideals) by looking only at the nonzero ideals that are contained in a single maximal ideal [5, Theorem 3.6]. As a special case, it is possible to determine when a particular nonzero prime contained in M is contained in no other maximal ideal [5, Corollary 3.5]. By making use of what we refer to as an M-maximal ideal (defined below), we will give a new (simpler) lobal characterization of finite character. In addition, in Theorem 3.9 we show that the property that each nonzero prime ideal is contained in a unique maximal ideal is lobal. Thus we obtain a lobal characterization of h-local domains. Note that [5, Theorem 3.10] shows that being h-local is a weakly lobal property.

Other lobal properties (new to this article) include being completely integrally closed, being a Krull domain, being a UFD, being atomic, being an HFD, and being a PVD (pseudovaluation domain).

2 Prime and Primary Ideals

We start by recalling the characterization of the prime and primary ideals that are contained in a given maximal ideal. In Theorems 2.2 and 2.3, we finish the characterization of M-primes and MP-primary ideals.

Theorem 2.1 (cf. [5, Theorems 3.8 and 3.9]) *Let M be a maximal ideal of a domain R and let J be a nonzero ideal contained in M.*

1. *J is a prime ideal of R if and only if $J = J_{(M)}$ is an M-prime of R.*
2. *J is a primary ideal if and only if J is MP-primary where $P = \sqrt{J}$ is an M-prime and $J_{(M)} = J$ (or $P_{(M)} = P$).*

Statements (2) and (3) in the next theorem provide a lobal type characterizations of when an M-prime is not a prime ideal of R.

Theorem 2.2 *Let M be a maximal ideal of a domain R. The following are equivalent for an ideal $P \subsetneq M$.*

1. *P is an M-prime that is not a prime ideal of R.*
2. *P is an M-prime and $P_{(M)} = M$.*
3. *P is an M-prime such that $P \subsetneq P_{(M)}$.*
4. *$P = M \cap Q$ for some prime ideal Q that is comaximal with M.*
5. *$P = M \cap (P :_R M)$ and $(P :_R M)$ is a prime ideal of R that is comaximal with M.*
6. *$(P :_R M)$ is a prime ideal of R that is comaximal with M.*
7. *P is an M-prime and $(P :_R M)$ is an ideal of R that is comaximal with M.*

Proof [(1) $\Rightarrow$ (2)] Suppose P is an M-prime that is not a prime ideal of R. Then by [5, Theorem 3.8], $P \subsetneq P_{(M)}$.

We first show that $P_{(M)}$ is a prime ideal of R. It is clear that $(P_{(M)})_{(M)} = P_{(M)}$. So by [5, Theorem 3.8] it suffices to show that $P_{(M)}$ is an M-prime. Suppose $a, b \in M$ are such that $ab \in P_{(M)}$. If $ab \in P$, then at least one of a and b is in P, so without loss of generality we may assume $a, b, ab \notin P$. Then there is an $x \in R\backslash M$ (not a unit) such that $xab \in P$. As $xa, a, xb, b \in M$ with $a, b \notin P$ and P is an M-prime, both xa and xb are in P so both $a, b \in P_{(M)}$. Thus $P_{(M)}$ is a prime ideal of R. Next let $n \in M\backslash P$. Then $an \in P_{(M)}$ and so for some $y \in R\backslash M$, $yan \in P$. Since $a \in M\backslash P$ and $yn \in M$, $yn \in P$ and therefore $n \in P_{(M)}$. It follows that $P_{(M)} = M$ when P is an M-prime that is not prime.

It is clear that (2) implies (3). Also if $P \subsetneq P_{(M)} = PR_M \cap R$, then P is not a prime ideal of R. So (3) implies (1).

[(4) $\Rightarrow$ (1)] It is clear that P is not a prime if it is the intersection M and a prime Q that is comaximal with M (as $P = QM$ in this case). To see that P is M-prime, let $a, b \in M$ be such that $ab \in P$. Then $ab \in Q$ and so at least one of a and b is in Q and thus in $M \cap Q = P$. Hence, P is an M-prime.

[(2) $\Rightarrow$ (5)] Suppose P is an M-prime such that $P_{(M)} = M(\supsetneq P)$. Let $m \in M\backslash P$. Since $PR_M = MR_M$, there is an element $r \in R\backslash M$ such that $rm \in P$. Choose any other $n \in M\backslash M$. Then we have $n(rm) = (nr)m \in P$ which implies $nr \in P$ as P is an M-prime and $m \in M\backslash P$. It follows that $rM \subseteq P$.

To simplify notation, let $N = (P :_R M)$. This is a proper ideal of R that contains r and so is comaximal with M. We have $P \subseteq N \cap M = NM \subseteq P$.

Suppose $x, y \in R$ are such that $xy \in N$ with $x \notin N$. Then $xym \in P$ for each $m \in M$, in particular, for each $m \in M\backslash P$. On the other hand, there is an element $n \in M$ such that $xn \notin P$. For an arbitrary $k \in M$, $xnyk \in P$ with $xn \in M\backslash P$ and $yk \in M$. Hence $yM \subseteq P$ and we have $y \in N$. Therefore, $N = (P :_R M)$ is a prime ideal of R such that $P = M \cap (P :_R M)$.

It is clear that (5) implies both (4) and (6). Also if $(P :_R M) + M = R$, then $P \subseteq M \cap (P :_R M) = M(P :_R M) \subseteq P$. Thus (6) implies (5).

Finally with regard to (7), if $(P :_R M)$ is comaximal with M, then $PR_M = MR_M$. Thus (7) implies (2). Conversely, (7) easily follows from the combination of the equivalent conditions (2) and (5).

Note that in statement (4), since the prime Q is comaximal with M, we have $P = M \cap Q = MQ$ and thus Q is unique and equal to $(P :_R M)$.

Next we wish to characterize the MP-primary ideals in M. Note that by Theorems 2.1 and 2.2, if I is an ideal contained in M and P is a prime ideal that contains I, then $M \cap P$ is an M-prime that contains I. Hence $\sqrt{I}$ is the intersection of the M-primes that contain I. Also Theorem 2.1 contains a characterization of the MP-primary ideals that are also primary ideals of R.

In the next theorem we restrict to looking at MP-primary ideals where P is an M-prime that is not a prime ideal of R. In this case, an MP-primary ideal is not a primary ideal. Statement (2) together with having $P \neq P_{(M)}$ gives a lobal type characterization of when an MP-primary ideal is not a primary ideal of R.

Theorem 2.3 *Let M be a maximal ideal and let P be an M-prime. Also let $N = (P :_R M)$. The following are equivalent for an ideal $Q \subseteq P$.*

1. *Q is MP-primary but it is not a primary ideal of R.*
2. *Q is MP-primary and $Q \neq Q_{(M)}$.*
3. *$Q = M \cap Q'$ for some N-primary ideal Q' and $P \subsetneq N$.*
4. *$Q = M \cap QR_N$ and $\sqrt{Q} = P \neq P_{(M)}$.*
5. *$\sqrt{Q} = P$ and $Q = M \cap J$ for some primary ideal J of R that is comaximal with M.*

Proof We start by showing (3) implies (1), (4) and (5). Suppose $Q = M \cap Q'$ where Q' is an N-primary ideal and $P \neq N$. Since $P \neq N$, P is not a prime ideal of R and thus $N + M = R$ (Theorem 2.2). It follows that M and N are the only minimal primes of Q. Hence, $\sqrt{Q} = P$ and Q is not primary. To see that Q is MP-primary, suppose $a, b \in M$ are such that $a \notin P$ and $ab \in Q$. Then $a \notin N$ and from this we have $b \in Q' \cap M = Q$. Hence, Q is MP-primary. In addition, $QR_N \cap R$ is an N-primary ideal and so by minimality $Q' = QR_N \cap R$. Thus (3) implies (1), (4) and (5).

If $Q = M \cap QR_N$ and $\sqrt{Q} = P \neq P_{(M)}$, then N is minimal over Q and thus $QR_N \cap R = Q'$ is an N-primary ideal. We then have $Q = M \cap Q'$. Also P is not a prime ideal of R. Hence, (3) and (4) are equivalent.

Next suppose $P = \sqrt{Q}$ and $Q = M \cap J$ for some primary ideal J that is comaximal with M. Since $P = M \cap N$, N is minimal over Q and so must contain J (and be minimal over it). Also P is not a prime. Hence, (3) and (5) are equivalent.

To finish the proof we show that (2) implies (3). Assume Q is MP-primary and $Q \neq Q_{(M)}$. Then Q is not a primary ideal by Theorem 2.1. Also by Theorem 2.2, we may assume Q is properly contained in P. By definition $P = \sqrt{Q}$. Also we have that M and N are the only minimal primes of Q. Let $Q' = QR_N \cap R$. Then Q' is an N-primary ideal. Clearly $P = \sqrt{M \cap Q'}$ so that $M \cap Q'$ is an MP-primary ideal. In addition, $M \cap Q' = MQ'$ since M and Q' are comaximal and it is clear that $(MQ')_{(M)} = M$.

Since M is minimal over Q, $Q_{(M)}$ is M-primary. Let $a \in Q_{(M)} \backslash N$ (such an element exists since $Q_{(M)}$ and N are comaximal). Then there is an element $y \in R \backslash M$ such that $ay \in Q \subseteq N$. Thus $y \in N$ and hence $y \in Q'$. Next, let $b \in Q_{(M)} \backslash Q$. The product ayb is in Q and both a and by are in M with $a \notin P$. It follows that $yb \in Q$. Next let $d \in Q' \backslash M$. Then there is an element $z \in R \backslash N$ such that $dz \in Q$. It follows that $z \in M \backslash P$. For the element b, we have $bdz \in Q$ and thus $bd \in Q$. Moreover, for each $f \in M$, $fdz \in Q$ with $z \in M \backslash P$ implies $fd \in Q$. Hence, $Q_{(M)} = M$ and $MQ' \subseteq Q \subseteq M \cap Q' = MQ'$.

Recall from above that a ring R is said to be *Laskerian* if each ideal has a (finite) primary decomposition. We next show that for domains, being Laskerian is a lobal property.

We say that a maximal ideal M satisfies **Lask** if for each nonzero ideal $I \subseteq M$ there are finitely many MP_i-primary ideals $Q_1, Q_2, \ldots, Q_n$ with each Q_i corresponding to an M-prime P_i such that $I = Q_1 \cap Q_2 \cap \cdots \cap Q_n$. Note that if each P_i is a prime ideal (equivalently $(P_i)_{(M)} = P_i$ for each i), then each Q_i is a primary ideal of R.

Theorem 2.4 *The following are equivalent for a domain R.*

1. *R is Laskerian.*
2. *Each maximal ideal satisfies **Lask**.*
3. *At least one maximal ideal satisfies **Lask**.*

Proof First suppose R is Laskerian and let M be a maximal ideal of R. Then for each nonzero ideal $I \subseteq M$, there are finitely many primary ideals $Q'_1, Q'_2, \ldots, Q'_n$ such that $I = Q'_1 \cap Q'_2 \cap \cdots \cap Q'_n$. For each i, let $P_i = \sqrt{M \cap Q'_i}$. Since $N_i = \sqrt{Q'_i}$ is a prime ideal, each P_i is an M-prime. In the case $Q_i' \subseteq M$, $P_i = N_i$ and we set $Q_i = Q_i'$. Otherwise, $P_i = M \cap N_i$ is an M-prime that is not a prime ideal and $Q_i = Q_i' \cap M$ is an MP_i-primary ideal (that is not a primary ideal of R). Since $I \subseteq M$, we have $I = Q_1 \cap Q_2 \cap \cdots \cap Q_n$. Hence, each maximal ideal satisfies **Lask**.

To complete the proof it suffices to prove (3) implies (1). Let M be a maximal ideal of R that satisfies **Lask**. We start by showing each nonzero ideal $I \subseteq M$ has a primary decomposition. For a given I, we have $I = Q_1 \cap Q_2 \cap \cdots \cap Q_n$ where each Q_i is an MP_i-primary ideal for some corresponding M-prime P_i. If each Q_i is a primary ideal of R, we have found a primary decomposition for I. Hence, we may assume at least one Q_j is not a primary ideal of R. Then the corresponding P_j is an M-prime that is not a prime ideal.

Let $P \subseteq M$ be a minimal prime of I. Then P is minimal over at least one Q_i. By Theorem 2.3, if P is properly contained in M, then each $Q_i \subseteq P$ is P-primary. On the other hand, we could have $P = M$. Then $I(M)$ is M-primary and we can add this ideal to the intersection to guarantee at least one Q_i is a primary ideal of R. Thus we may assume there is an integer $1 \leq k < n$ such that Q_i is a primary ideal of R for each $1 \leq i \leq k$ and Q_j is an MP_j-primary ideal that is not a primary ideal of R for each $k+1 \leq j \leq n$.

For $k+1 \leq j \neq n$, $P_j = M \cap N_j$ for some prime ideal N_j that is comaximal with M and there is an N_j-primary ideal Q_j' such that $Q_j = M \cap Q_j'$.

It follows that $I = Q_1 \cap \cdots \cap Q_k \cap Q_{k+1}' \cap \cdots \cap Q_n'$ is a primary decomposition for I.

Next suppose J is a nonzero ideal that is comaximal with M. Then $J \cap M = JM$ is a nonzero ideal contained in M. As such it has a primary decomposition (of distinct primary ideals) $J \cap M = Q_1 \cap Q_2 \cap \cdots \cap Q_n$, necessarily with $Q_i = M$ for some i. Without loss of generality we may assume $Q_1 = M$. Then checking locally we have $J = Q_2 \cap Q_3 \cap \cdots \cap Q_n$. Therefore, R is Laskerian.

3 Invertible Ideals, Coherent Domains, H-Local Domains and PVDs

For a nonzero two-generated ideal $I = aR + bR$ and maximal ideal M that contains I, M does not contain II^{-1} if and only if at least one of $\mathcal{N}_M(a) \cap \mathcal{P}(b)$ and $\mathcal{N}_M(b) \cap \mathcal{P}(a)$ is nonempty ([5, Theorem 2.6]). It follows that such an I is invertible if M is the only maximal ideal that contains it. We wish to give a lobal-like characterization that does not need the assumption that M is the only maximal ideal that contains I. With this we are able to give a lobal characterization of Prüfer domains in terms of invertible ideals.

First we introduce the notion of M-maximal ideals. For a given maximal ideal M, we say that an M-prime N is *M-maximal* if $N \neq M = N_{(M)}$ and there are no M-primes properly between M and N (equivalently, $M = N_{(M)}$ is the only M-prime that properly contains N). Using Theorem 2.2 it is easy to see that an M-prime P is M-maximal if and only if $P = P' \cap M$ for some maximal ideal $P' \neq M$.

Theorem 3.1 *Let M be a maximal ideal of a domain R and let I be a nonzero finitely generated ideal that is contained in M. Then I is invertible if and only if (1) there is an element $b \in I$ such that $I_{(M)} = bR_{(M)}$ and (2) for each M-maximal ideal N', there is an element $s \in M\backslash N'$ and an element $a \in I$ such that $sI \subseteq aR$.*

Proof Let Q' be an M-maximal ideal. Then $Q' = Q \cap M$ for some maximal ideal $Q \neq M$. If I is invertible, then $IR_M = bR_M$ for some $b \in I$ and thus $I_{(M)} = bR_{(M)}$. Also for Q, there is an element $t \in I$ such that $IR_Q = tR_Q$. Since I is finitely generated, there is an element $y \in R\backslash Q$ such that $yI \subseteq tR$. Choose an element $x \in M\backslash Q$. Then $xy \in M\backslash Q'$ is such that $xyI \subseteq tR$.

For the converse, assume both (1) and (2) hold for I. From (1) and the fact that I is finitely generated, we have that II^{-1} is not contained M: if $I_{(M)} = bR_{(M)}$ for some $b \in I$, then $IR_M = bR_M$ so that $b^{-1}IR_M \subseteq R_M$, by finite generation, there is an element $z \in R\backslash M$ such that $zb^{-1} \in I^{-1}$ which puts $z \in II^{-1}$. Next let $N' = M \cap N$ where N is a maximal ideal other than M. Then let $a \in I$ and $s \in M\backslash N'$ be such that $sI \subseteq aR$. As $s \notin N$, we have $IR_N \subseteq aR_N \subseteq IR_N$. Thus I is locally principal and hence invertible.

Corollary 3.2 *The following are equivalent for an integral domain R.*

1. *R is a Prüfer domain.*
2. *For each maximal ideal N, each finitely generated nonzero ideal $B \subseteq N$ is such that $B_{(N)} = (bR)_{(N)}$ for some $b \in B$ and for each N-maximal ideal Q, there is an element $s \in N\backslash Q$ and an element $a \in B$ such that $sB \subseteq aR$.*
3. *There is a maximal ideal M such that for each finitely generated nonzero ideal $I \subseteq M$ there is an element $b \in I$ such that $I_{(M)} = bR_{(M)}$ and for each M-maximal ideal P, there is an element $s \in M\backslash P$ and an element $a \in I$ such that $sI \subseteq aR$.*

It is also possible to give a lobal characterization of Prüfer domains via valuation domains. First we need a lobal-like characterization of when R_Q is a valuation domain for a given prime ideal Q (no matter whether Q is contained in the given maximal ideal M or not).

Theorem 3.3 *Let M be a maximal ideal of a domain R and let Q' be a nonzero M-prime. For the corresponding prime Q such that $Q' = Q \cap M$, R_Q is a valuation domain if and only if for all pairs of nonzero elements $a, b \in Q'$ either (i) $Q' = M$ and $aR_{(M)}$ and $bR_{(M)}$ are comparable, or (ii) $Q' \neq M$ and there are elements $r, s \in M$ with at least one not in Q' such that $ra = sb$.*

Proof Note that $Q' = M$ if and only if $Q = M$. Assume R_Q is a valuation domain. In the case $Q' = M$, we have aR_M and bR_M are comparable. That $aR_{(M)}$ and $bR_{(M)}$ are comparable follows from the fact that $I_{(M)} = IR_M \cap R$ for all ideals $I \subseteq M$. Next consider the case $Q' \neq M$. In this case, aR_Q and bR_Q are comparable. Without loss of generality we may assume $aR_Q \subseteq bR_Q$. Then there are elements $x, y \in R$ with

$y \in R\backslash Q$ such that $ya = xb$. Choose an element $m \in M\backslash Q$. Then we have $mya = mxb$ with $my \in M\backslash Q'$ as desired.

For the converse, first assume $Q' = M$ and it is always the case that for all pairs of nonzero $a, b \in M$, $aR_{(M)}$ and $bR_{(M)}$ are comparable. Then we have aR_M and bR_M comparable and from this we may conclude that $R_M = R_Q$ is a valuation domain. Next consider the case that $Q' \neq M$. Let $m \in M\backslash Q'$. For an arbitrary pair of nonzero elements $c, d \in Q$, mc and md are in Q'. Without loss of generality we may assume there is an element $p \in M\backslash Q'$ and an element $k \in R$ such that $pmc = kmd$. As the element pm is in $M\backslash Q$, $cR_Q \subseteq dR_Q$. So as in the case $Q' = M$, R_Q is a valuation domain.

Using the previous theorem we can give another lobal characterization of Prüfer domains.

Corollary 3.4 *The following are equivalent for a domain R.*

1. *R is a Prüfer domain.*
2. *For each maximal ideal N and each nonzero N-prime (N-maximal) $Q \neq N$, (i) $aR_{(N)}$ and $bR_{(N)}$ are comparable for all nonzero $a, b \in N$, and (ii) for all nonzero $c, d \in Q$, there are elements $r, s \in N$ with at least one of r and s not in Q such that $rc = sd$.*
3. *There is a maximal ideal M such that (i) $aR_{(M)}$ and $bR_{(M)}$ are comparable for all nonzero $a, b \in M$, and (ii) for each M-prime (M-maximal) $Q \neq M$ and all nonzero $c, d \in Q$, there are elements $r, s \in M$ with at least one of r and s not in Q such that $rc = sd$.*

A domain R is a *finite conductor domain* if the intersection of a pair of principal ideals is always finitely generated. If the intersection of a finite number of principal ideals is always finitely generated, then R is *quasi-coherent*. Finally R is *coherent* if the intersection of each pair of finitely generated ideals is finitely generated. All three of these are easy to classify as lobal properties. For a maximal ideal M of R we say that M satisfies **FC** if $aR \cap bR$ is finitely generated for each pair of elements $a, b \in M$. Also M satisfies **QC** if $a_1R \cap a_2R \cap \cdots a_nR$ is finitely generated for each finite subset $\{a_1, a_2, \ldots, a_n\} \subseteq M$. Finally, M satisfies **Coh** if $I \cap J$ is finitely generated for each pair of finitely generated ideals I and J contained in M.

Theorem 3.5 *The following are equivalent for an integral domain R (that is not a field).*

1. *R is a coherent domain.*
2. *Each maximal ideal of R satisfies* **Coh**.
3. *There is a maximal ideal M of R that satisfies* **Coh**.

Proof It is clear that (1) implies (2), and (2) implies (3). To complete the proof suppose M is a maximal ideal of R that satisfies **Coh** and let I and J be finitely generated ideals of R. Choose a nonzero element $x \in M$. Then xI and xJ are finitely generated ideals that are contained in M. Hence, $xI \cap xJ = x(I \cap J)$ is finitely generated. It follows that $I \cap J$ is finitely generated. Therefore R is coherent.

The proof above is easily adaptable to quasi-coherent domains and finite conductor domains.

Theorem 3.6 *The following are equivalent for an integral domain R (that is not a field).*

1. *R is a quasi-coherent conductor domain.*
2. *Each maximal ideal of R satisfies* **QC**.
3. *There is a maximal ideal M of R that satisfies* **QC**.

Theorem 3.7 *The following are equivalent for an integral domain R (that is not a field).*

1. *R is a finite conductor domain.*
2. *Each maximal ideal of R satisfies* **FC**.
3. *There is a maximal ideal M of R that satisfies* **FC**.

An alternate characterization of quasi-coherent is that $(R : I)$ is a finitely generated fractional ideal for each nonzero finitely generated ideal I of R. Essentially this follows from the fact that $(R : I) = \bigcap a_i^{-1}R$ when $I = a_1R + a_2R + \cdots + a_nR$ (simply multiply the intersection by the product of the a_is to obtain a finite intersection of integral principal ideals). Like invertible ideals there is a lobal-like characterization of when a given I in a particular maximal ideal M is such that $(R : I)$ is finitely generated. For example: $(R : I)$ is finitely generated if and only if there is a nonzero element $a \in I$ and a (corresponding) finitely generated ideal $J \subseteq aR$ such that for $d \in aR$, $dI \subseteq a^2R$ if and only if $d \in J$. We leave the proof to the interested reader.

In [3], Jaffard introduced the notion of a ring being of *Dedekind type* if for each nonzero ideal I there is a finite set of pairwise comaximal ideals $I_1, I_2, \ldots, I_n$ such that $I = I_1I_2 \ldots I_n$ with each I_j in a unique maximal ideal. He showed that in the terminology introduced by Matlis [6] (much later), a domain has Dedekind type if and only if it is h-local [3, Théorème 6]. We wish to provide lobal characterizations using both Jaffard's factoring definition and Matlis' finite character type definition.

First we give a simple lobal characterization of finite character. After that we give a lobal way to determine that each nonzero prime ideal is contained in a unique maximal ideal.

Theorem 3.8 *The following are equivalent for a domain R.*

1. *R has finite character.*
2. *For each maximal ideal N, each nonzero element of N is contained in at most finitely many N-maximal ideals.*
3. *There is a maximal ideal M such that each nonzero element of M is contained in at most finitely many M-maximal ideals.*

Proof From Theorem 2.2, an ideal Q is N-maximal for some maximal ideal N if and only if $Q = Q' \cap N$ for some maximal ideal $Q' \neq N$. Thus it is clear that (1) implies (2). Only somewhat more complicated is to show that (3) implies R has finite character. If (3) holds we clearly have that each nonzero element of M is contained in only finitely many maximal ideals. For a nonzero nonunit $n \in R\backslash M$, simply multiply by a nonzero $m \in M$ to obtain $nm \in M$. Clearly each maximal ideal that contains n contains nm. Hence n is contained in only finitely many maximal ideals.

Theorem 3.9 *The following are equivalent for a domain R.*

1. *Each nonzero prime ideal of R is contained in a unique maximal ideal.*
2. *For each maximal ideal N, if P is a nonzero N-prime, then either no N-maximal ideal contains P, or exactly one N-maximal ideal contains P with* $P_{(N)} = N$.
3. *There is a maximal ideal M such that for each nonzero M-prime P, either no M-maximal ideal contains P or exactly one M-maximal ideal contains P with* $P_{(M)} = M$.

Proof If P is a nonzero prime that is contained in a unique maximal ideal N, then $P = P_{(N)}$ and for all other maximal ideals M, $Q = P \cap M$ is an M-prime such that $Q_{(M)} = M$ and $N \cap M$ is the only M-maximal ideal that contains Q. Thus (1) implies (2) follows from the characterization of M-primes given in Theorems 2.1 and 2.2.

It is clear that (2) implies (3). So to complete the proof it suffices to show (3) implies (1). Assume there is a maximal ideal M such that for each nonzero M-prime Q, either no M-maximal ideal contains Q or exactly one M-maximal ideal contains Q with $Q_{(M)} = M$. Let P be a nonzero prime ideal of R. Then $P' = P \cap M$ is an M-prime. If no M-maximal ideal contains P', then M is the only maximal ideal that contains P and $P = P' = P'_{(M)}$. On the other hand if N is the only M-maximal ideal that contains P' and $P'_{(M)} = M$, then M does not contain P and $N = N' \cap M$ for some maximal ideal N'. We have $N' \supseteq P$ with N' the only maximal ideal that contains P.

For a nonzero ideal I of R, we say that I has a *Jaffard factorization* if there is a finite set of pairwise comaximal ideals $I_1, I_2, \ldots, I_n$ such that $I = I_1 I_2 \ldots I_n$ and each I_j is contained in a unique maximal ideal.

Let M be a maximal ideal of R. We say that M satisfies **Jaf** if for each nonzero ideal $I \subseteq M$ there are ideals $C_0, C_1, C_2, \ldots, C_n$ (all contained in M) that satisfy the following conditions:

(i) $I = C_0 \cap C_1 \cap \cdots \cap C_n$,
(ii) $C_0 = I_{(M)}$ and no M-maximal ideal contains C_0,
(iii) for $1 \leq i \leq n$, $(C_i)_{(M)} = M$ and exactly one M-maximal ideal contains C_i,
(iv) for $1 \leq i \leq n$, M is the only ideal that contains both C_0 and C_i.

For a Matlis type characterization, we say a maximal ideal M satisfies **Mat** if both of the following hold.

(i) Each nonzero ideal contained in M is contained in at most finitely many M-maximal ideals.
(ii) For each nonzero M-prime P, either no M-maximal ideal contains P or exactly one M-maximal ideal contains P with $P_{(M)} = M$.

Theorem 3.10 *The following are equivalent for a domain R.*

1. *R is h-local.*
2. *Each maximal ideal satisfies* **Mat**.
3. *At least one maximal ideal satisfies* **Mat**.
4. *Each maximal ideal satisfies* **Jaf**.
5. *At least one maximal ideal satisfies* **Jaf**.

Proof The equivalence of (1), (2) and (3) follows from Theorems 3.8 and 3.9. Also the implication (4) ⇒ (5) is trivial. Next we show that (1) implies (4).

Suppose R is h-local and let M be a maximal ideal of R. Then for each nonzero ideal $I \subseteq M$, there are finitely many pairwise comaximal ideals $I_0, I_1, \ldots, I_n$ such that $I = I_0 I_1 \cdots I_n$ and each I_j is contained in a unique maximal ideal. Without loss of generality, we may assume $I_0 \subseteq M$. Since the I_js are pairwise comaximal, the product is the same as the intersection $I_0 \cap I_1 \cap \cdots \cap I_n$. For each i, let $C_i = I_i \cap M$. Trivially, we have $I_0 = C_0$ and $I = C_0 \cap C_1 \cap \cdots \cap C_n$.

For $1 \leq i \leq n$, $I_i + M = R$. Thus $C_i = I_i M$ and $C_i R_M = M R_M$. It follows that $(C_i)_{(M)} = M$. We also have that C_i is contained in exactly two maximal ideals, M and the unique maximal ideal N_i that contains I_i. Since M is the only maximal ideal that contains C_0, it certainly is the only maximal ideal that contains $C_0 + C_i$. Since $C_i R_M = M R_M$ and $(C_i + C_0)R_N = R_N = M R_N$ for all maximal ideals $N \neq M$, we have $M = C_0 + C_i$. Therefore M satisfies **Jaf**.

For the converse, suppose M satisfies **Jaf**. We first show that each nonzero prime ideal P that is contained in M is contained in no other maximal ideal M-maximal ideal.

Let $P \subseteq M$ be a nonzero prime ideal and let $Q_0, Q_1 \ldots, Q_n$ be ideals that satisfy conditions (i)–(iv) for P: $P = Q_0 \cap Q_1 \cap \cdots \cap Q_n$ with $Q_0 = P_{(M)}$ contained in no M-maximal ideal. As $P_{(M)} = P$, P is contained in no M-maximal ideal and it follows that M is the only maximal ideal that contains P.

Next let $I \subseteq M$ be a nonzero ideal and let $C_0, C_1, \ldots, C_n$ be ideals that satisfy (i)–(iv) for I. Since $M \supseteq C_0$ and no M-maximal ideal contains C_0, we are done if $I = C_0$ as will be the Jaffard factorization of I. Thus we may assume $n \geq 1$.

For each $1 \leq i \leq n$, as there is a unique M-maximal ideal that contains C_i, there is exactly one other maximal ideal N_i that contains C_i (other than M), the ideal $N_i' = N_i \cap M$ is the unique M-maximal ideal that contains C_i. Also $N_i \neq N_j$ for $i \neq j$.

Let $J_i = C_i R_{N_i} \cap R$. By [5, Lemma 3.11], each minimal prime of J_i is contained in N_i. None of these is contained in M since each nonzero prime contained in M is contained in no other maximal ideal. It follows that J_i is not contained in M. Since C_i is contained in exactly two maximal ideals, J_i is contained in N_i and no other maximal ideal. Checking locally we see that $J_i \cap M = C_i$. Since $I \subseteq C_0 \subseteq M$, $C_0 \cap J_1 \cap J_2 \cap \cdots \cap J_n = C_0 \cap C_1 \cdots \cap C_n = I$. Since C_0 and the J_is are pairwise comaximal, $I = C_0 J_1 J_2 \cdots J_n$ is a Jaffard factorization of I.

To finish the proof we need to show that each ideal that is comaximal with M has a Jaffard factorization. For this let J be a nonzero ideal that is not contained in M. Then $J \cap M$ is contained in M and has a Jaffard factorization, necessarily with M as the factor that is contained in M. Hence, we have $J \cap M = M \cap B_1 \cap B_2 \cap \cdots \cap B_m = MB_1B_2 \cdots B_m$ where the B_is are pairwise comaximal and each is comaximal with M. In addition, each B_i is contained in a unique maximal ideal N_i. Clearly, the N_i are the only maximal ideals that contain J. In addition, $JR_{N_i} = B_iR_{N_i}$ and therefore $J = B_1B_2 \cdots B_m$ is a Jaffard factorization of J. Hence R is h-local.

Note that in place of one single maximal ideal satisfying **Jaf**, it is enough to have a maximal ideal M such that each nonzero ideal $I \subseteq M$ is contained in at most finitely many M-maximal ideals, and have another maximal ideal N, such that for each nonzero N-prime P either no N-maximal ideal contains P or exactly one N-maximal ideal contains P with $P_{(N)} = N$.

Recall that a domain R is a pseudovaluation domain, PVD for short, if it is quasilocal and its maximal ideal is also the maximal ideal of a valuation domain (necessarily with the same quotient field). By [2, Theorem 1.4], a quasilocal domain R with maximal ideal M is a PVD if and only if for all ideals I and J contained in M, either $I \subseteq J$ or $J \subseteq IM$. A lobal characterization of being quasilocal is given in [5, Theorem 2.9]. Thus the PVD property is lobal. The next result gives an alternate lobal characterization for PVDs.

Theorem 3.11 *The following are equivalent for a domain R.*

1. *R is a PVD.*
2. *For each maximal ideal N of R, each pair of nonzero elements $r, s \in N$ satisfies exactly one of the following:*

   a. $rR = rR$,
   b. $rR \subsetneq sR$,
   c. $sR \subsetneq rR$, *or*
   d. $rN = sN = rR \cap sR$.

3. *There is a maximal ideal M of R such that for each pair of nonzero elements $x, y \in M$ exactly one of the following holds:*

   a. $xR = yR$,
   b. $xR \subsetneq yR$,
   c. $yR \subsetneq xR$, *or*
   d. $xM = yM = xR \cap yR$.

Proof If R is a PVD, then it has a unique maximal ideal M. Moreover there is a valuation domain V with maximal ideal M (necessarily with $R \subseteq V \subseteq K$). For nonzero $x, y \in M$, exactly one of the following holds: $xV = yV$, $xV \subsetneq yV$, or $yV \subsetneq xV$. In the case $xV \subsetneq yV$, $x = ym$ for some $m \in M$ and hence $xR \subsetneq yR$. Conversely, if $xR \subsetneq yR$, then $x = yf$ for some $f \in M$, in which case $xV \subsetneq yV$. Similarly $yR \subsetneq xR$ if and only if $yV \subsetneq xV$. If $xV = yV$, then x/y is a unit of V. If $x/y \in R$, then x/y is also a unit of R and we have $xR = yR$ in this case. Otherwise $x/y \notin R$ and xR and yR are incomparable ideals of R. Since $MV = M$, we at least have $xM = yM$. It is clear that in this case $xM = yM \subseteq xR \cap yR$. For the reverse containment, suppose $w \in xR \cap yR$ and write $w = ax = by$ for some $a, b \in R$. If a is a unit of R, then we have $x \in yR$ which then implies $xR \subseteq yR$, a contradiction. Hence, $a \in M$ (and we also have $b \in M$). Thus $xM = yM = xR \cap yR$.

For the converse, assume M is a maximal ideal that satisfies (a)–(d) (in statement (3)). We first show that M is the only maximal ideal of R. By way of contradiction suppose there is a nonunit $f \in R\backslash M$ and let $x \in M\backslash\{0\}$ be such that $fR + xR = R$. Then neither x/f nor f/x is in R. Thus there is no containment relation between the principal ideals x^2R and xfR. Localizing at M we have $x^2R_M \subsetneq xR_M = xfR_M$ and so $x^2MR_M \subsetneq xfMR_M$, consequently $x^2M \neq xfM$. Therefore, it must be that M is the only maximal ideal of R.

Note that if there are no $x, y \in M$ that satisfy (d), then R is a valuation domain as all principal ideals are comparable. So for the remainder of the proof we assume there are pairs that satisfy (d).

Let $T = R[\mathscr{X}]$ where $\mathscr{X} = \{x/y \in K \mid xM = yM, x, y \in M\backslash\{0\}\}$. It is clear that $x/y \in \mathscr{X}$ if and only if $y/x \in \mathscr{X}$. It is also the case that M is an ideal of T and $\mathscr{X}$ is closed to finite products. Both follow easily from the fact that $xM = yM$ if and only if $(x/y)M = M$.

Let $v \in T\backslash R$. Then $v = r_0 + r_1(x_1/y_1) + \cdots + r_n(x_n/y_n)$ where each $r_j \in R$ and each $x_i/y_i \in \mathscr{X}$. We can rewrite the sum as $v = (r_0y + r_1x_1' + r_2x_2' + \cdots + r_nx_n')/y$ where $y = y_1y_2\cdots y_n$ and $x_i' = x_iy/y_i$. Let $z = r_0y + r_1x_1' + \cdots + r_nx_n'$. Since $x_i, y_i \in M$ for each i, $z \in M$. As v is not in R, we do not have $zR \subseteq yR$. Also, if $yR \subsetneq zR$, then $v = 1/g$ for some $g \in M$ which is impossible since M is an ideal of T. Thus we must have $zM = yM = zR \cap yR$ and therefore $v \in \mathscr{X}$ is a unit of T. Hence T is quasilocal with maximal ideal M.

To complete the proof it suffices to show that T is a valuation domain.

Let $g = c/d \in K\backslash T$ with $c, d \in R$. Then d is not a unit of R and $cR \neq dR$. If c is a unit of R, then $g^{-1} = d/c \in M$. Similarly, if $dR \subsetneq cR \subseteq M$, then $g^{-1} = d/c \in M$. In the case $cR \subsetneq dR$, we get $g \in R$, a contradiction. Finally, $cM = dM$ puts both g and g^{-1} in $\mathscr{X}$, another contradiction. Thus $g^{-1} \in M \subseteq T$ and T is a valuation domain. It follows that R is a PVD.

4 UFDs, Krull Domains, Atomic Domains, and HFDs

In this section, we take a different lobal approach to characterizing Krull domains. Here we will develop a lobal way to show that R_P is a rank one discrete valuation domain for each height one prime P and $R = \bigcap\{R_P \mid P \in \text{Spec}(R) \text{ has height one}\}$ is a finite character intersection. In addition we give lobal characterizations of UFDs, HFDs, and atomic domains.

Given a maximal ideal M, we say that an M-prime $P \neq M$ has *M-height one* if no nonzero M-prime is properly contained in P. In addition, we say that M has M-height one if $Q_{(M)} = M$ for each nonzero M-prime Q.

Lemma 4.1 *Let M be a maximal ideal and let P be a nonzero M-prime. Then P has M-height one if and only if there is a height one prime P' such that $P = M \cap P'$. Moreover, if Q is a height one prime, then $Q \cap M$ is an M-height one M-prime.*

Proof If M is a height one prime of R and $Q \neq M$ is a nonzero M-prime, then $Q = Q' \cap M$ for some prime ideal Q' that is comaximal with M. It follows that $Q_{(M)} = M$. Since $P_{(M)} = P$ for each prime $P \subsetneq M$, we have that M has M-height one if and only if it has height one.

Next suppose $P \neq M$ is a nonzero M-prime. Then there is a unique prime P' such that $P = P' \cap M$ (obviously with $P = P'$ if and only if $P' \subseteq M$ if and only if P is a prime ideal). We split the proof into two cases, the first where $P = P'$ and the second where $P \subsetneq P'$ (necessarily with $P' + M = R$).

Assume $P = P'$. If Q is a nonzero M-prime that is not a prime ideal, then $Q = Q' \cap M = Q'M$ for some prime ideal Q' that is comaximal with M. Since P is a prime ideal that is properly contained in M, it cannot contain Q. Thus in this case P has M-height one if and only if it has height one.

For the second case, suppose $Q \subseteq P$ is a nonzero M-prime with $P \neq P'$. There is a unique prime ideal Q' such that $Q = Q' \cap M$. Since $P' \neq M$, P' contains Q'. It follows that P is M-height one if and only if P' is height one.

For each maximal ideal M, we let $\mathscr{H}_M(1)$ denote the set of M-height one primes. We split this set into the set $\mathscr{I}_M(1)$ consisting of M-height one primes P such that $P = P_{(M)}$ (so the same as the set of height one primes that are contained in M) and the set $\mathscr{O}_M(1)$ consisting of the M-height one primes Q such that $Q \neq Q_{(M)}$. Of course, if R admits no height one primes, then both $\mathscr{I}_M(1)$ and $\mathscr{O}_M(1)$ are empty.

Next we give an almost lobal characterization for each height one prime to be principal. The "almost" refers to the fact that in the case R contains at least one height one prime, we restrict the choice of the single maximal ideal M to one that contains a height one prime. For Krull domains and UFDs, each maximal ideal contains a height one prime.

Theorem 4.2 *The following are equivalent for a domain R that contains at least one height one prime.*

1. *Each height one prime is principal.*
2. *For each maximal ideal N that contains a height one prime,*
   a. *$P \in \mathscr{I}_N(1)$ implies there is an element $r \in N$ such that $P = rR$, and*
   b. *each $Q \in \mathscr{O}_N(1)$ is such that there is a $P \in \mathscr{I}_N(1)$ and an element $c \in N$ where $Q \cap P = cR$.*
3. *There is a maximal ideal M that contains a height one prime such that*
   a. *for each $P \in \mathscr{I}_M(1)$, there is an element $r \in M$ such that $P = rR$, and*
   b. *for each $Q \in \mathscr{O}_M(1)$, there is a $P \in \mathscr{I}_M(1)$ and an element $c \in M$ such that $Q \cap P = cR$.*

Proof Suppose each height one prime is principal and let M be a maximal ideal that contains at least one height one prime. Let $P \in \mathscr{I}_M(1)$ and let $Q \in \mathscr{O}_M(1)$. Then $Q = Q' \cap M$ for some height one prime Q' that is not contained in M. We have $Q' = sR$ and $P = rR$ for some elements $r, s \in R$ and so $Q \cap P = Q' \cap P = rsR$ with $rs \in M$ (since both r and s are prime elements of R).

To complete the proof it suffices to show that (3) implies (1). Let M be a maximal ideal that contains a height one prime and satisfies both (3a) and (3b). Thus each height one prime that is contained in M is principal. Let Q be a height one prime that is not contained in M. Then $Q \cap M$ is an M-height one prime in the set $\mathscr{O}_M(1)$. By assumption, there is a height one prime $P \subseteq M$ and an element $c \in M$ such that $Q \cap P = Q \cap M \cap P = cR$. We have $P = rR$ for (prime) element $r \in P$ (so in M). Thus $c = br$ for some element $b \in R$, necessarily with $b \in Q$. We will show $Q = bR$. Let $t \in Q$. Then $tr \in Q \cap P$ so $tr = qc = qbr$ for some $q \in R$ and thus $t = qb$. It follows that $Q = bR$ and therefore each height one prime of R is principal.

Next we recall a rather simple characterization of UFDs involving nonzero principal primes.

Theorem 4.3 [4, Theorem 5] *An integral domain is a UFD if and only if each nonzero prime ideal contains a nonzero principal prime ideal.*

Theorem 4.4 *The following are equivalent for a domain R.*

1. *R is a UFD.*
2. *For each maximal ideal N,*
   a. *each nonzero N-prime contains an N-height one N-prime,*
   b. *each $P \in \mathscr{I}_N(1)$ is principal, and*
   c. *for each $Q \in \mathscr{O}_N(1)$, there is an N-prime $P \in \mathscr{I}_N(1)$ and an element $c \in N$ such that $Q \cap P = cR$.*
3. *There is a maximal ideal M such that*
   a. *each nonzero M-prime contains an M-height one M-prime,*
   b. *each $P \in \mathscr{I}_M(1)$ is principal, and*
   c. *for each $Q \in \mathscr{O}_M(1)$, there is a $P \in \mathscr{I}_M(1)$ and an element $c \in M$ such that $Q \cap P = cR$.*

Proof Lemma 4.1 together with Theorems 4.2 and 4.3 show that (1) implies (2). To see that (3) implies (1), note that if P is a nonzero prime ideal that is properly contained in an M that satisfies the conditions in statement (3), then P contains an M-height one prime Q. As we have seen before, having P properly contained in M implies Q must be a prime ideal and hence a height one prime ideal. Also note that by Theorem 4.2, each height one prime is principal.

For a prime N that is not contained in M, $N \cap M$ is a nonzero M-prime and so contains an M-height one prime Q'. By Lemma 4.1, $Q' = Q'' \cap M$ for some height one prime Q'', necessarily with $Q'' \subseteq N$. Thus each nonzero prime ideal contains a principal prime and therefore R is a UFD (Theorem 4.3).

Recall that an element t in the quotient field of a domain R is almost integral over R if there is a nonzero element $r \in R$ such that $rt^n \in R$ for each positive integer n. The ideal I generated by the set $\{rt^n \mid n \geq 0\}$ is such that $tI \subseteq I$. Conversely, if there is a nonzero ideal J of R such that $tJ \subseteq J$, then t is almost integral over R.

There is no way to determine whether $t \in K \backslash R$ is almost integral over R directly (by considering rt^n and/or tI) from the above characterization using the "lobal" approach since t is not even in R, but there is a way to do it indirectly as a lobal property.

Theorem 4.5 *Let M be a maximal ideal of a domain R and let b and c be a pair of nonzero elements of M such that $b \notin cR$. Then the following are equivalent for the element $t = b/c \in K \backslash R$.*

1. *t is almost integral over R.*
2. *There is a nonzero ideal $I \subseteq M$ such that for each nonzero element $y \in I$, there is a nonzero element $x \in I$ such that $yb = xc$.*

Proof If t is almost integral over R, then there is a nonzero ideal J such that $tJ \subseteq J$. Choose a nonzero $m \in M$. The ideal $I = mJ$ is such that $tI \subseteq I$. It follows that $bI \subseteq cI$ (equivalently, for each $y \in I$ there is an $x \in I$ such that $yb = xc$).

For the converse, suppose $I \subseteq M$ is a nonzero ideal such that for each $y \in I$, there is an $x \in I$ where $yb = xc$. It follows easily that $tI \subseteq I$ and thus t is almost integral over R. For a nonzero ideal J of R, it is clear that $tJ \subseteq J$ if and only if $bJ \subseteq cJ$ (if and only if for each $y \in J$, there is an $x \in J$ such that $by = cx$). If there is such an ideal J, then $b(bJ) \subseteq c(bJ)$ where $bJ = I$ is an ideal contained in M. Hence (1) and (3) are equivalent.

Theorem 4.5 provides a lobal way to determine when R is not completely integrally closed (and when it is). In the following corollaries we use $bI \subseteq cI$ in place of considering the products by for $y \in I$ individually. Also note that for $t \in K$, if $t = f/g$ for nonzero $f, g \in R$, then choose any nonzero $m \in M$ to have $t = fm/gm$ with both $fm, gm \in M$.

Corollary 4.6 *The following are equivalent for an integral domain R.*

1. *R is not completely integrally closed.*
2. *For each maximal ideal M, there is a pair of nonzero elements $b, c \in M$ and a corresponding nonzero ideal $I \subseteq M$ such that $bI \subseteq cI$ but $bR \nsubseteq cR$.*
3. *There is a maximal ideal M which contains a pair of nonzero elements b and c and a corresponding nonzero ideal $I \subseteq M$ such that $bI \subseteq cI$ but $bR \nsubseteq cR$.*

Corollary 4.7 *The following are equivalent for an integral domain R.*

1. *R is completely integrally closed.*
2. *For each maximal ideal M, if $b, c \in M \backslash \{0\}$ are such that there is a nonzero ideal $I \subseteq M$ where $bI \subseteq cI$, then $bR \subseteq cR$.*
3. *There is a maximal ideal M with the property that if $b, c \in M \backslash \{0\}$ are such that there is a nonzero ideal $I \subseteq M$ where $bI \subseteq cI$, then $bR \subseteq cR$.*

One of the many characterizations of Krull domains is that they are precisely the completely integrally closed domains that satisfy the ascending chain conditions on divisorial ideals. The notion of an M-divisorial ideal was introduced in [5] and was used to show that the ascending chain conditions on divisorial ideals is a lobal property [5, Theorems 3.16]. Combined with Corollary 4.7 we have that Krull is a lobal propery. Using M-height one M-primes we give a very different lobal characterization for Krull domains.

We say that a maximal ideal M satisfies **Kr1** if for each M-height one prime $P \neq M$, there is a nonzero $b \in P$ such that for each $a \in P$, there are elements $r, s \in M$ with $r \notin P$ such that $ra = sb$; and if M is M-height one, then $M = cR_{(M)}$ for some $c \in M$. Also we say that M satisfies **Kr2** if for nonzero elements $a, b \in M$ with $a \notin bR$, either (i) M is M-height one and $\mathcal{N}_M(a) \cap \mathcal{P}(b) = \emptyset$, or (ii) there is an M-height one prime $P \neq M$ such that for $q \in M$, $qa \in bR$ implies $q \in P$.

Theorem 4.8 *The following are equivalent for a domain R.*

1. *R_P is a rank one discrete valuation domain for each height one prime P.*
2. *Each maximal ideal satisfies* **Kr1**.
3. *There is a maximal ideal that satisfies* **Kr1**.

Proof Suppose R_P is a rank one discrete valuation domain for each height one prime P and let M be a maximal ideal. By Lemma 4.1, $P \cap M$ is always an M-height one M-prime and each M-height one prime has this form.

First we consider the case that M has height one. In this case $MR_M = cR_M$ for some $c \in M$ and it follows that $M = cR_{(M)}$.

Next suppose $Q' \neq M$ is an M-height one M-prime with corresponding height one prime Q such that $Q' = Q \cap M$. There is an element $q \in Q$ such that $QR_Q = qR_Q$. Thus for each element $f \in Q'$, there are elements $t, v \in R$ with $v \in R \backslash Q$ such that $vf = tq$. Since $Q' \neq M$, there is an element $m \in M \backslash Q$ and we have $(mv)f = (mt)q$ with $mv \in M \backslash Q'$. Thus M satisfies **Kr1**.

It is clear that (2) implies (3). So assume M is a maximal ideal that satisfies **Kr1**. If M has M-height one (so height one), we have $M = cR_{(M)}$ and so $MR_M = cR_M$ with MR_M height one. Hence R_M is a discrete rank one valuation domain in this case. Next, let $P \neq M$ be a height one prime of R. Then $P' = P \cap M$ is an M-height one prime. Since $P \neq M$, there is a nonzero $b \in P'$ such that for each $a \in P'$, there are elements $r, s \in M$ with $r \notin P'$ where $ra = sb$. It follows that $a \in bR_P$. For each $d \in P \backslash M$, $rd \in P'$ and we have elements $r', s' \in M$ with $r' \notin P'$ such that $r'rd = s'b$ which puts $d \in bR_P$. Thus $PR_P = bR_P$. Since P has height one, R_P is a discrete rank one valuation domain.

Therefore (3) implies (1).

Theorem 4.9 *The following are equivalent for a domain R.*

1. *$R = \bigcap\{R_P \mid P \in \mathrm{Spec}(R)$ is height one$\}$.*
2. *Each maximal ideal satisfies* **Kr2**.
3. *There is a maximal ideal that satisfies* **Kr2**.

Proof Suppose $R = \bigcap\{R_P \mid P \in \text{Spec}(R)$ is height one$\}$ and let M be a maximal ideal of R. Also let $a, b \in M$ be such that $a \notin bR$. Then $a/b \notin R$. We may assume that for each M-height one M-prime $P' \neq M$, there is an element $q \in M\backslash P'$ such that $qa \in bR$. For such a P', there is a (unique) height one prime $P(\neq M)$ such that $P' = P \cap M$. Then $q \notin inP$ implies $a/b \in R_P$. Thus it must be that $a/b \notin R_M$ and M is height one. In this case, $M \supseteq (bR :_R a)$ and no principal ideal is contained in both $\mathscr{N}_M(a)$ and $\mathscr{P}(b)$.

Assume M is a maximal ideal that satisfies **Kr2** and let $t \in \bigcap\{R_P \mid P \in \text{Spec}(R)$ has height one$\}\backslash\{0\}$. We may assume $t = a/b$ for some $a, b \in M$. By way of contradiction we assume $t \notin R$, equivalently $a \notin bR$. If M has height one, then $t \in R_M$ implies there are elements $r, s \in R$ with $r \in R\backslash M$ such that $ra = sb$. The ideal $raR \in \mathscr{N}_M(a) \cap \mathscr{P}(b)$. Hence there must be at least one M-height one M-prime $P' \neq M$ that satisfies condition (ii) of **Kr2**. Let P be the corresponding height one prime such that $P' = P \cap M$. We have $t \in R_P$ so there are elements $x, y \in R$ with $x \in R\backslash P$ such that $xa = yb$. Also there is an element $m \in M\backslash P$. We have $mx, my \in M$ with $mx \notin P'$ such that $mxa = myb$, contradicting that P' satisfies (ii). Hence it must be that $t \in R$.

Theorem 4.10 *The following are equivalent for a domain R.*

1. *R is a Krull domain.*
2. *For each maximal ideal N, N satisfies both* **Kr1** *and* **Kr2** *and each nonzero* $r \in N$ *is contained in only finitely many N-height one N-primes.*
3. *There is a maximal ideal M that satisfies both* **Krl** *and* **Kr2** *and each nonzero* $r \in M$ *is contained in only finitely many M-height one M-primes.*

Proof If R is a Krull domain, then each nonzero nonunit is contained in only finitely many height one primes. Thus a nonzero element in a particular maximal ideal N is contained in only finitely many N-height one N-primes. Also by Theorems 4.8 and 4.9, each maximal ideal satisfies **Kr1** and **Kr2**.

To complete the proof suppose M is a maximal ideal that satisfies both **Kr1**and **Kr2** where each nonzero element in M is contained in only finitely many M-height one M-primes. By Theorem 4.9, $R = \bigcap\{R_P \mid P \in \text{Spec}(R)$ has height one$\}$. It follows that height one primes exist and R_P is a (rank one) discrete valuation domain for each such prime P.

Let t be a nonzero element of the quotient field of R. Then there are nonzero elements $a, b \in M$ such that $t = a/b$. By assumption, at most finitely many M-height one M-primes contain a and at most finitely many contain b. It follows that the set of height one primes P where t is not in unit of R_P is finite. Hence R is a Krull domain.

For a nonzero element r in a maximal ideal M, r is an irreducible element of R if and only if there is no $s \in M$ such that $rR \subsetneq sR$. We let $\text{Irr}(R)$ denote the set of irreducible elements of R and let $\text{Irr}(M) = M \cap \text{Irr}(R)$. To handle the irreducible elements that are not in M, we define an element $t \in M$ to be <u>M-irreducible</u> if there is an element $r \in \text{Irr}(M)$ such that (i) $tR \subsetneq rR$, (ii) $tR \in \mathscr{N}_M(r)$, and (iii) there is no $d \in M$ such that $tR \subsetneq dR \subsetneq rR$. We let M-Irr denote the set of M-irreducible elements of M. Also for a given $t \in M$-Irr, we let M-Irr(t) denote the set of elements $r \in \mathrm{Irr}(M)$ such that (i) $tR \subsetneq rR$, (ii) $tR \in \mathcal{N}_M(r)$ and (iii) there is no $d \in M$ such that $tR \subsetneq dR \subsetneq rR$.

Lemma 4.11 *For a given maximal ideal M of a domain R, M-Irr $= \{t \in R \mid t = wr$ for some $r \in \mathrm{Irr}(M)$ and irreducible $w \in R\backslash M\}$.*

Proof Let $t = wr \in M$ be such that $w \in \mathrm{Irr}(R)\backslash M$ and $r \in \mathrm{Irr}(M)$. Clearly, $tR \subsetneq rR$ and $tR \in \mathcal{N}_M(r)$. Suppose $d \in M$ is such that $tR \subseteq dR \subsetneq rR$. Then $d = rs$ and $t = dv$ for some elements $s, v \in R$ with s not a unit. We have $rw = t = dv = rsv$ and thus $w = sv$. Since w is irreducible and s is not a unit it must be that v is a unit and we have $tR = dR$. It follows that $t \in M$-Irr.

Next suppose $f \in M$-Irr and let $g \in \mathrm{Irr}(M)$ be such that (i) $fR \subsetneq gR$, (ii) $fR \in \mathcal{N}_M(g)$, and (iii) there is no element $h \in M$ such that $fR \subsetneq hR \subsetneq gR$. We have $f = gk$ for some nonunit $k \in R\backslash M$. If k is not irreducible, then there are nonunits $b, c \in R\backslash M$ such that $k = bc$. But in such a case $fR = gkR \subsetneq gb \subsetneq gR$, a contradiction. Hence k is irreducible.

In general, it is possible to have an element $f \in M$-Irr that is also a product of irreducibles inside M. For example, let $R = F[X^2, XY, Y^2]$ where F is a field. The maximal ideal $M = (X, Y + 1)F[X, Y] \cap R$ contains both X^2 and XY but not Y^2. Also all three of X^2, XY and Y^2 are irreducible in R. The element $f = X^2Y^2$ is in M-Irr but it also factors as $f = (XY)^2$ with $XY \in \mathrm{Irr}(M)$. For a maximal ideal M of R we say that M satisfies **Atom** if for each nonzero $f \in M$, there are elements $m_1, m_2, \ldots, m_r \in \mathrm{Irr}(M)$ such that $fR \in \mathcal{N}_M(m)$ for $m = m_1m_2 \cdots m_r$, and for each such list of irreducibles in M, either $fR = mR$ or there are elements $a_1, a_2, \ldots, a_s \in M$-Irr with corresponding $q_j \in M$-Irr(a_j) such that $fq_1q_2 \cdots q_s = m_1m_2 \cdots m_ra_1a_2 \cdots a_s$. In the case $fR = mR$, there is a unit $x \in R$ such that $f = xm = (xm_1)m_2 \cdots m_r$ with $xm_1 \in \mathrm{Irr}(M)$. Next we show that being an atomic domain is lobal.

Theorem 4.12 *The following are equivalent for a domain R.*

1. *R is atomic.*
2. *Each maximal ideal satisfies* **Atom**.
3. *There is a maximal ideal that satisfies* **Atom**.

Proof Suppose that R is atomic and let f be a nonzero element in the maximal ideal M. Then there are atoms $n_1, n_2, \ldots, n_k$ such that $f = n_1n_2 \cdots n_k$. At least one of the n_is is in M. If all are, then $fR = nR \in \mathcal{N}_M(n)$ where $n = n_1n_2 \cdots n_k$. Otherwise, we may assume $n_1, n_2, \ldots, n_t \in M$ for some $1 \le t < k$ and $n_{t+1}, \ldots, n_k \in R\backslash M$. In this case $fR \in \mathcal{N}_M(n')$ with $fR \subsetneq n'R$ where $n' = n_1n_2 \cdots n_t$. Next choose a $p \in \mathrm{Irr}(M)$. By Lemma 4.11 (and its proof), $a_j = pn_j \in M$-Irr with $p \in M$-Irr(a_j) for each $t < j \le k$. In addition $fp^{k-t} = n_1n_2 \cdots n_ta_{t+1}a_{t+2} \cdots a_k$.

Finally, for $m_1, m_2, \ldots, m_r \in \mathrm{Irr}(M)$ with $fR \in \mathcal{N}_M(m)$ where $m = m_1m_2 \ldots m_r$, we have $f = mx$ where $x \in R\backslash M$. In the case $fR = mR$, x is a unit and thus $xm_1 \in \mathrm{Irr}(M)$. For the case, $fR \subsetneq mR$, x is a nonunit and thus factors as a finite product of atoms in $R\backslash M$. Continue as in the case some $n_i \in R\backslash M$ to get a factorization that shows M satisfies **Atom**.

For the reverse implication, suppose M satisfies **Atom**. There is nothing to prove for a nonzero element $g \in M$ that is a finite product of irreducibles in $\mathrm{Irr}(M)$. Next, let f be a nonzero element of M with a corresponding factorization $fq_1q_2\cdots q_s = m_1m_2\ldots m_ra_1a_2\cdots a_s$ where each $m_i \in \mathrm{Irr}(M)$ with $fR \in \mathcal{N}_M(m_1\cdots m_r)$ and each $a_j \in M$-Irr with corresponding $q_j \in \mathrm{Irr}(M)$ such that $q_j = M\text{-}\mathrm{Irr}(a_j)$. By Lemma 4.11, there is an irreducible element $h_j \in \mathrm{Irr}(R)\backslash M$ such that $a_j = h_jq_j$. By cancellation we have $f = m_1\cdots m_rh_1\cdots h_s$ is factorization of f into atoms.

Finally suppose $b \in R\backslash M$ is a nonunit. The set $\mathrm{Irr}(M)$ is nonempty, so choose an arbitrary $m \in \mathrm{Irr}(M)$ and consider the element $c = mb$. We have $cR \in \mathcal{N}_M(m)$ with $cR \subsetneq mR$ and $m \in M\text{-}\mathrm{Irr}(c)$. By **Atom**, there are elements $d_1, d_2, \ldots, d_k \in M$-Irr and corresponding $p_j \in M\text{-}\mathrm{Irr}(d_j)$ such that $cp = mbp = md_1d_2\cdots d_k$ where $p = p_1p_2\cdots p_k$. By Lemma 4.11, there are irreducibles $w_1, w_2, \ldots, w_k$ such that $d_j = p_jw_j$ for each j. It follows that $b = w_1w_2\cdots w_k$. Hence R is atomic.

In our final result we show that being an HFD is lobal. For a maximal ideal M, we say that M satisfies **HFD** if for each nonzero $f \in M$ whenever there are irreducibles $m_1, m_2, \ldots, m_r \in \mathrm{Irr}(M)$ with $f \in \mathcal{N}_M(m)$ for $m = m_1m_2\cdots m_r$ and either $fR = mR$ or $fq = ma$ for $a_1, a_2, \ldots, a_s \in M$-Irr with corresponding $q_j \in M\text{-}\mathrm{Irr}(a_j)$ where $a = a_1a_2\cdots a_s$ and $q = q_1q_2\cdots q_s$, and there are irreducibles $n_1, n_2, \ldots, n_t \in \mathrm{Irr}(M)$ with $f \in \mathcal{N}_M(n)$ for $n = n_1n_2\cdots n_t$ and either $fR = nR$ or $fp = nb$ for $b_1, b_2, \ldots b_u \in M$-Irr with corresponding $p_k \in M\text{-}\mathrm{Irr}(b_k)$ where $b = b_1b_2\cdots b_u$ and $p = p_1p_2\ldots p_u$, then there is a one-to-one correspondence between the lists $m_1, m_2, \ldots, m_r, a_1, a_2, \ldots, a_s$ and $n_1, n_2, \ldots, n_t, b_1, b_2, \ldots b_u$ (perhaps with no a_js and/or no b_ks).

Theorem 4.13 *The following are equivalent for a domain R.*

1. *R is an HFD.*
2. *Each maximal ideal satisfies* **Atom** *and* **HFD**.
3. *There is a maximal ideal that satisfies* **Atom** *and* **HFD**.

Proof If R is an HFD, it is atomic and thus each maximal ideal satisfies **Atom** (Theorem 4.12). Let $f \in M\backslash\{0\}$ where M is a maximal ideal. Then for a pair of factorizations $f = g_1g_2\cdots g_n$ and $f = h_1h_2\cdots h_m$ where each g_i and each h_j is irreducible, we have $n = m$. If each g_i and h_j is in M, we are done for these two factorizations. So next we consider the case that at least one g_i is not in M. Since $f \in M$, some g_i is in M, so we may assume the indexing is such that $g_1, \ldots, g_r \in M$ and $g_{r+1}, \ldots, g_n \notin M$. For $r+1 \le k \le n$, choose an irreducible $z_k \in M$ (for example $z_k = g_1$). Then $c_k = z_kg_k \in M$-Irr for $r+1 \le k \le n$ with $z_k \in M\text{-}\mathrm{Irr}(c_k)$. Let $z = z_{r+1}z_{r+2}\cdots z_n$, $c = c_{r+1}c_{r+2}\cdots c_n$ and $g' = g_1g_2\cdots g_r$. Then $fz = g'c$ with a total of n factors from $\mathrm{Irr}(M) \cup M$-Irr on the right hand side: $g_1, g_2, \ldots, g_r, c_{r+1}, \ldots, c_n$. A similar factorization holds for f with respect to the h_ks if some h_ks are not in N. To conclude, suppose there are irreducibles $p_1, p_2, \ldots, p_s \in \mathrm{Irr}(M)$ and M-irreducibles $a_1, a_2, \ldots, a_t \in M$-Irr with corresponding $q_j \in M\text{-}Irr(a_j)$ for each j such that $fq = pa$ where $p = p_1p_2\cdots p_s$, $a = a_1a_2\cdots a_t$ and $q = q_1q_2\cdots q_t$. Then corresponding to each pair a_j, q_j there is an irreducible element $d_j \in \text{Irr}(M)$ such that $a_j = d_j q_j$ (Lemma 4.11). Cancelling the common factors of q_j, we have that $f = p_1 p_2 \cdots p_s d_1 d_2 \cdots d_t$ is a factorization of f into irreducibles. Since R is an HFD, $s + t = n$. Therefore M satisfies **HFD**.

To complete the proof it suffices to show (3) implies (1). Suppose M satisfies both **Atom** and **HFD**. Then R is atomic and thus $\text{Irr}(M)$ is nonempty. Let f be a nonzero nonunit of R. We first consider the case that $f \in M$. Then, as above, there are irreducible elements $g_1, g_2, \ldots, g_n$ such that $f = g_1 g_2 \ldots g_n$. Suppose there are also irreducible elements $h_1, h_2, \ldots, h_m$ such that $f = h_1 h_2 \cdots h_m$. If each g_i and h_j is in M, then the assumption that M satisfies **HFD** establishes that $n = m$. Suppose some g_i is not in M. Since $f \in M$, we may assume that $g_1, g_2, \ldots, g_r \in M$ and $g_{r+1}, \ldots, g_n \in R \backslash M$. For $r + 1 \leq k \leq n$, choose a irreducible $z_k \in M$. Then $c_k = g_k z_k \in M$-Irr by Lemma 4.11. We have $fz = g'c$ where $g' = g_1 g_2 \cdots g_r$, $c = c_{r+1} c_{r+2} \cdots c_n$ and $z = z_{r+1} z_{r+2} \cdots z_n$. If each $h_j \in M$, we have $n = r + (n - r) = m$. Or in the case some h_j is not in M, we may assume $h_1, h_2, \ldots, h_s \in M$ and $h_{s+1}, \ldots, h_m \in R \backslash M$. As with the g_is that are not in M, we have irreducible elements $w_{s+1}, \ldots, w_m \in M$ with $d_j h_j w_j \in M$-Irr for $s + q \leq j \leq m$ and $fw = h'd$ where $h' = h_1 h_2 \cdots h_s$, $w = w_{s+1} w_{s+2} \cdots w_m$ and $d = d_{s+1} d_{s+2} \cdots d_m$. Since M satisfies **HFD**, we have $m = s + (m - s) = r + (n - r) = n$.

For $f \in R \backslash M$, choose any irreducible $x \in M$. Then $xf \in M$. Since R is atomic, $f = y_1 y_2 \ldots y_t$ where each $y_i \in Irr(R)$ and so $xf = xy_1 \ldots y_t$. By the argument above, if $f = u_1 u_2 \ldots u_s$ is another factorization of f into irreducible, then $xy_1 y_2 \ldots y_t = xf = xu_1 u_2 \ldots u_s \in M$ implies $1 + t = 1 + s$ and therefore $t = s$ and R is an HFD.

References

1. R. Gilmer, *Multiplicative Ideal Theory, Queen's Papers in Pure and Applied Mathematics*, vol. 90 (Queen's University Press, Kingston, 1992)
2. J. Hedstrom, E. Houston, Pseudo-valuation domains. Pac. J. Math. **75**, 137–147 (1978)
3. P. Jaffard, Théorie arithmétique des anneau du type de Dedekind. Bull. Soc. Math. France **80**, 61–100 (1952)
4. I. Kaplansky, Commutative Rings, (rev. ed.) (Polygonal Publishing House, Washington D.C., 1994)
5. T.G. Lucas, Localizing global properties to individual maximal ideals, in *Recent Advances in Commutative Rings, Integer-valued Polynomials, and Polynomial Functions (Graz, 2012)* (Springer, New York, 2014). pp. 239–254
6. E. Matlis, Cotorsion modules, in *Memoirs of the American Mathematical Society* vol. 49 (Providence RI, 1964)

Noetherian Semigroup Algebras and Beyond

Jan Okniński

Abstract A selection of results on Noetherian semigroup algebras is presented. They are of structural, arithmetical, and combinatorial nature. Starting with the case of Noetherian group algebras, where several deep results are known, a lot of attention is later given to the case of algebras of submonoids of groups. The role of algebras of this type in the general theory of Noetherian semigroup algebras is explained and sample structural results on arbitrary Noetherian semigroup algebras, based on this approach, are presented. A special emphasis is on various classes of algebras with good arithmetical properties, such as maximal orders and principal ideal rings. In this context, several results indicating the nature and applications of the structure of prime ideals are presented. Recent results on the prime spectrum and arithmetics of a class of non-Noetherian orders are also given.

1 Introduction

The aim of this paper is to present selected representative results on Noetherian semigroup algebras $K[S]$, where S is a monoid and K is a field. The results are both of structural, arithmetical, and combinatorial nature. In particular, we present an approach exploiting in this context linear semigroups over division rings, and indicating the role of cancellative subsemigroups of S. An emphasis is therefore made on the case of Noetherian algebras $K[S]$ of submonoids S of polycyclic-by-finite groups. Certain concrete classes of such algebras that arise independently in other contexts and that motivate the general theory are presented. Hence, we first summarize some of the relevant results on Noetherian group algebras. Results on the structure of an arbitrary monoid S that yield certain necessary and sufficient conditions for $K[S]$ to be Noetherian are then presented. Some advantages of this

Research supported by the National Science Centre grant 2013/09/B/ST1/04408.

J. Okniński (✉)
Institute of Mathematics, Warsaw University, Banacha 2, 02-097 Warsaw, Poland
e-mail: okninski@mimuw.edu.pl

S. Chapman et al. (eds.), *Multiplicative Ideal Theory and Factorization Theory*,
Springer Proceedings in Mathematics & Statistics 170,
DOI 10.1007/978-3-319-38855-7_11

approach are illustrated, in particular in the context of polynomial identities and the Gelfand–Kirillov dimension. A special attention is given to prime Noetherian orders in simple artinian rings, and especially maximal orders. We conclude with recent developments that indicate that certain non-Noetherian orders are of interest and importance, but at the same time they seem to be very difficult to study.

We start in Sect. 2 with important results on Noetherian group algebras that are relevant for the rest of the paper. In Sect. 3 an approach to the structure of general Noetherian semigroup algebras is presented. This is via semigroups of matrices over a field, or a division ring, and this is based on finite ideal chains of some specific type that arise in a natural way. In particular, this explains the role of cancellative subsemigroups of the given semigroup S in the general theory. And this also explains our focus on the case where S is a submonoid of a polycyclic-by-finite group. An intriguing class of examples arising from a different context is presented in Sect. 4. In Sect. 5 we discuss certain classical arithmetical properties, especially in the context of orders in simple artinian algebras. In particular, these include maximal orders and principal ideal rings. The special role that is played by the prime ideals is explained. In the final Sect. 6 a recent example of a non-Noetherian order, arising from considerations in noncommutative geometry, is presented. It motivates a new area of study, namely non-Noetherian orders coming from submonoids of nilpotent groups. We conclude with some recent results in this direction.

Throughout the paper, K will denote a field and S a monoid (a semigroup with a unity element) with operation written multiplicatively. The corresponding semigroup algebra is denoted by $K[S]$. If S has a zero element θ then $K\theta$ is a 1-dimensional ideal of $K[S]$ and $K_0[S] = K[S]/K\theta$ is called the contracted semigroup algebra of S over K. In other words, we identify the zero of S with the zero of the algebra.

Our basic references for the results and methods of the theory of group algebras and semigroup algebras are [38, 51, 54, 57, 58], while we refer to [27, 50] for an extensive background on noncommutative Noetherian rings.

2 Group Algebras—Introductory Results

The class of Noetherian group algebras is one of our starting points. Recall that a polycyclic-by-finite group is a group with a finite subnormal series whose every factor is either finite or cyclic, [61]. We have the following classical result.

Theorem 2.1 *Let G be a polycyclic-by-finite group. Then $K[G]$ is Noetherian.*

The idea of the proof is easy. It is based on an induction on the length of a subnormal chain of G with finite and cyclic factors. Let $H \subseteq F$ be two consecutive factors of such a chain. Assume that $K[H]$ is Noetherian. If $[F : H] < \infty$, then we have a finite module extension $K[H] \subseteq K[F]$. So $K[F]$ is Noetherian. If F/H is infinite cyclic, then an argument similar to that in the proof of Hilbert basis theorem is used to show that $K[F]$ is also Noetherian.

We note that it is not known whether there exist classes of Noetherian group algebras other than those described in Theorem 2.1.

The second point of departure is the class of commutative semigroup rings. The following result is attributed to Budach, see [23], Theorem 5.10.

Theorem 2.2 *Assume that S is a commutative monoid. Then $K[S]$ is Noetherian if and only if S is finitely generated.*

The proof of the nontrivial implication (the necessity) is based on a decomposition theory for congruences of a commutative monoid with acc on congruences, on properties of irreducible congruences and of cancellative congruences.

In many other important cases one can also show that 'noetherian' implies 'finitely generated'. Recall that the Gelfand–Kirillov dimension of a finitely generated algebra R over K is finite if the growth function $d_V(n)$ is bounded by a polynomial in n. Here V is a finite dimensional generating subspace of R and $d_V(n) = \dim_K(V + V^2 + \cdots + V^n)$. Then $\limsup(\log d_V(n)/\log(n))$ is called the Gelfand–Kirillov dimension of R and is denoted by $\mathrm{GKdim}(R)$, see [47]. In general, it is not an integer. On the other hand, in the class of commutative algebras, this dimension coincides with the classical Krull dimension.

Theorem 2.3 ([33, 38]) *Assume $K[S]$ is right Noetherian. Then S is finitely generated in each of the following cases:*

1. *S satisfies acc on left ideals (this holds in particular if $K[S]$ is also left Noetherian),*
2. *$K[S]$ satisfies a polynomial identity,*
3. *the Gelfand–Kirillov dimension of $K[S]$ is finite.*

It is not known whether the assertion of the above theorem is true for an arbitrary right Noetherian algebra $K[S]$. This is not known even in the case where S is cancellative (in this case, S has group of classical right quotients G, because of the acc on right ideals; whence $K[G]$ is a classical localization of $K[S]$ and it is also Noetherian).

There are several deep results on the prime spectrum of Noetherian group algebras. We mention some highlights, that will be also used in Sect. 6. The first is due to Zalesskii, see [57], Corollary 11.4.6. Recall that a prime ideal P of $K[G]$ is faithful if the normal subgroup $\{g \in G \mid g - 1 \in P\}$ of G is trivial. By $\mathrm{Z}(G)$ we denote the center of G.

Theorem 2.4 *Assume that G is a finitely generated torsion free nilpotent group. There is a bijection between the set of faithful primes in $K[G]$ and faithful primes of $K[\mathrm{Z}(G)]$, given by:*

$$Q \longrightarrow Q \cap K[\mathrm{Z}(G)], \quad P \longrightarrow P \cdot K[G].$$

The above, together with a reduction to a torsion free subgroup of finite index, is one of the steps of the following result of Smith, [57], Theorem 11.4.9. Recall that the Hirsch length $\mathrm{h}(G)$ of a polycyclic-by-finite group G is defined as the number of infinite cyclic factors in a subnormal chain in G with cyclic or finite factors (which is independent of the chosen chain). By clKdim(R) we denote the classical Krull dimension of an algebra R.

Theorem 2.5 *Assume that G is a finitely generated nilpotent group. Then*

$$\mathrm{clKdim}(K[G]) = \mathrm{h}(G).$$

Let us note that in the more general polycyclic-by-finite case, a more complicated invariant, called the plinth length of G (in general, not exceeding $\mathrm{h}(G)$), see [58], page 192, plays the role of $\mathrm{h}(G)$, by a result of Roseblade [60].

The known Noetherian group algebras share a very important property of finitely generated commutative algebras, called catenarity. Recall that the latter means that every two saturated chains of primes between any two given prime ideals $P \subset P'$ have equal lengths.

Theorem 2.6 ([49]) *The group algebra $K[G]$ of a polycyclic-by-finite group G is catenary.*

The following is an immediate consequence of the fact that polycyclic-by-finite groups are finitely presented, see [61], Theorem 8.4.

Theorem 2.7 *If G is a polycyclic-by-finite group, then the algebra $K[G]$ is finitely presented.*

As a consequence of the structural characterization obtained in Theorem 2.11, one can prove the following corollary, which settles a general framework for the results presented in Sect. 4.

Corollary 2.8 ([34]) *Let S be a submonoid of a polycyclic-by-finite group. If S satisfies the ascending chain condition on right ideals, then S is a finitely presented monoid. In particular, the semigroup algebra $K[S]$ is finitely presented.*

From the point of view of the theory of orders in division rings, or more generally in simple artinian rings, the following classical results of Connell on prime rings, see [57], Theorem 4.2.10, and of Farkas and Snider, [57], Theorem 13.4.18, and Cliff [11] (domains of zero and positive characteristic, respectively) are of basic interest.

Theorem 2.9 *Let G be a group. Then*

1. *$K[G]$ is prime if and only if G has no nontrivial finite normal subgroups.*
2. *If G is polycyclic-by-finite, then $K[G]$ is a domain if and only if G is torsion free.*

Orders of the form $K[G]$ are interesting also from the point of view of the associated division rings, as they supply a rich class of examples.

Theorem 2.10 ([15]) *Let G, H be non-isomorphic finitely generated nilpotent torsion free groups. Then the classical division rings of quotients $Q_{cl}(K[G])$ and $Q_{cl}(K[H])$ are not isomorphic.*

As said above, if $K[S]$ is right Noetherian for a submonoid S of a group G then S has a group of right quotients isomorphic to $SS^{-1} \subseteq G$ and $K[G]$ is Noetherian. The case where S is a submonoid of a polycyclic-by-finite group is therefore of special interest; first because of Theorem 2.1, second, because of some important examples discussed in Sect. 4, third because of a general structural approach explained in Sect. 3.

The following complete result comes from [37], while some partial steps were earlier made in [33, 34].

Theorem 2.11 ([37]) *Let S be a submonoid of a polycyclic-by-finite group. Then the following conditions are equivalent:*

1. *$K[S]$ is right Noetherian,*
2. *S satisfies acc on right ideals,*
3. *S has a group of quotients G and there exists a normal subgroup H of G such that: $[G : H] < \infty$, $S \cap H$ is finitely generated and the derived subgroup $[H, H] \subseteq S$,*
4. *$K[S]$ is left Noetherian.*

In the above notation, let $F = [H, H]$. So, in some sense, such $K[S]$ can be approached in two steps: from the perspective of the Noetherian group algebra $K[F] \subseteq K[S]$ and of the Noetherian PI-algebra $K[S/F] \subseteq K[G/F]$. Recall that the general theory provides additional strong tools in the class of Noetherian PI-algebras, [50]. In particular, finitely generated PI-algebras are catenary, see [50], Corollary 13.10.13.

3 A General Structural Approach

In this section we present a structural approach to arbitrary Noetherian semigroup algebras $K[S]$. It is based on finite ideal chains of S of a very special type. Such chains arise naturally in the study of linear semigroups and for this reason they seem unavoidable in the context of Noetherian algebras $K[S]$. On the one hand, they allow to prove certain necessary and sufficient conditions for S in order that $K[S]$ is Noetherian. On the other hand, they allow to reduce several problems to submonoids of groups, and hence to group algebras. They also are very useful in the case of certain families of algebras arising from other contexts, which will be reflected in Sect. 4.

Let X, Y be arbitrary nonempty sets and let $P = (p_{yx})$ be a $Y \times X$-matrix with entries in $T^0 = T \cup \{0\}$, for a monoid T. So, strictly speaking, P is a mapping

$Y \times X \longrightarrow T \cup \{0\}$. Let $\mathcal{M}(T, X, Y, P)$ be the set of all $X \times Y$-matrices with entries in $T \cup \{0\}$ but with at most one nonzero entry. Such a nonzero matrix can be denoted by (g, x, y) (with $g \in T$ in position (x, y)). Multiplication, called sandwich multiplication, is defined as follows:

$$a \circ b = aPb$$

where in the right hand side one uses the standard matrix products.

Assume also that the 'sandwich matrix' P has no nonzero rows or columns and $T = G$ is a group. Then $M = \mathcal{M}(G, X, Y, P)$ is called a completely 0-simple semigroup over the group G with sandwich matrix P. It has no ideals other than M and $\{0\}$ and it can bee considered as a semigroup analogue of a simple artinian ring. Such semigroups play a prominent role in semigroup theory, see [12], §2.7 and §3.2. The nonzero maximal subgroups of $\mathcal{M}(G, X, Y, P)$ are all isomorphic to G, they are of the form $G_{xy} = \{(g, x, y) \mid g \in G\}$, where $p_{yx} \neq 0$.

A subsemigroup S of $\mathcal{M}(G, X, Y, P)$ such that S intersects nontrivially every set $M_{xy} = \{(g, x, y) \mid g \in G\}$, $x \in X, y \in Y$, is called a uniform (sub)semigroup.

The case when $X = Y$ and $P = \Delta$, the identity matrix, is of special interest. If $|X| = r < \infty$ then we write $\mathcal{M}(G, r, r, \Delta)$. In this case, the contracted semigroup algebra $K_0[\mathcal{M}(G, r, r, \Delta)]$ is isomorphic to the matrix algebra $M_r(K[G])$. A uniform subsemigroup S of $\mathcal{M}(G, r, r, \Delta)$ is called a semigroup of generalized matrix type. So $K_0[S] \subseteq M_r(K[G])$.

Let $S \subseteq M = \mathcal{M}(G, X, Y, P)$ be a uniform subsemigroup. One can show that there exists a unique subgroup H of G and a sandwich matrix Q over H^0 so that $S \subseteq M \cong \mathcal{M}(H, X, Y, Q)$ and (if S is identified with a subsemigroup of $\mathcal{M}(H, X, Y, Q)$) every maximal subgroup of $\mathcal{M}(H, X, Y, Q)$ is generated as a group by its intersection with S. So, intuitively, one is tempted to think of $\mathcal{M}(G, X, Y, P)$ as a 'semigroup of quotients of S'. If one prefers, one can consider S as an order in $\mathcal{M}(G, X, Y, P)$. This can be given a very precise meaning if additionally H is a group of quotients of $S \cap H$, see [17]. The latter holds for example if $K[S]$ satisfies a polynomial identity or if S has acc on right ideals.

If I is an ideal of a semigroup S then the Rees factor S/I is defined as the set $(S \setminus I) \cup \{0\}$ with the operation $s \cdot t = st$ if $st \in S \setminus I$ and $s \cdot t = 0$ otherwise. A structure theorem, obtained in [52], see also [54], Theorem 3.5, reads as follows.

Theorem 3.1 *If S is a subsemigroup of the multiplicative monoid $M_n(F)$ of all $n \times n$-matrices over a field F, then S has a finite ideal chain $I_1 \subseteq I_2 \subseteq \cdots \subseteq I_k = S$ with I_1 and every factor I_j/I_{j-1} nilpotent or a uniform semigroup. The same applies if F is a division ring and S satisfies the ascending chain condition on right ideals.*

In particular, the second part applies to the case where $K[S]$ is right Noetherian and embeds into $M_n(D)$ for a division ring D.

Clearly, the simplest example is $S = M_n(F)$, for a field F. Then the chain $M_1 \subseteq M_2 \subseteq \cdots \subseteq M_n = M_n(F)$ defined by $M_j = \{a \in M_n(F) \mid rank(a) \leq j\}$ has all factors completely 0-simple. The maximal subgroups of S are of the form $H_e = \{a \in eM_n(F)e \mid \text{rank}(a) = \text{rank}(e)\}$, where $e = e^2 \in M_n(F)$ and they are isomorphic to the corresponding full linear groups $Gl_j(F), j = rank(e)$.

The following important theorem is an extension of the classical result of Malcev saying that a finitely generated commutative algebra is embeddable in a matrix ring over a field.

Theorem 3.2 ([2]) *Let R be a finitely generated right Noetherian PI-algebra. Then R embeds into the matrix ring $M_n(F)$ over a field extension F of the base field K.*

So, in view of Theorem 2.3, Theorem 3.1 can be applied if $R = K[S]$ is right Noetherian and satisfies a polynomial identity. Moreover, since every semiprime Noetherian algebra has a semisimple artinian classical quotient ring, it follows that Theorem 3.1 applies also to $K[S]/B(K[S])$ ($B(K[S])$ denoting the prime radical of $K[S]$) as well as to every prime homomorphic image $K[S]/P$ of $K[S]$. So, such an S has a finite ideal chain with all factors nilpotent or uniform.

One can show that even more is true in certain other cases.

Theorem 3.3 ([38, 55]) *Let S be a monoid such that $K[S]$ is left and right Noetherian and* $\text{GKdim}(K[S]) < \infty$. *If for every $a, b \in S$ one has*

$$a\langle a, b\rangle \cap b\langle a, b\rangle \neq \emptyset \neq \langle a, b\rangle a \cap \langle a, b\rangle b,$$

then S has an ideal chain $S_1 \subseteq S_2 \subseteq \cdots \subseteq S_n = S$ such that S_1 and every factor S_i/S_{i-1} is either nilpotent or a semigroup of generalized matrix type.

More importantly, the following partial converse of this theorem holds.

Theorem 3.4 ([55]) *Let S be a finitely generated monoid with an ideal chain $S_1 \subseteq S_2 \subseteq \cdots \subseteq S_n = S$ such that S_1 and every factor S_i/S_{i-1} is either nilpotent or a semigroup of generalized matrix type. If* $\text{GKdim}(K[S]) < \infty$ *and S satisfies the ascending chain condition on right ideals, then $K[S]$ is right Noetherian.*

The assumptions in the theorem imply that cancellative subsemigroups of uniform factors S_i/S_{i-1} and S_1 have groups of quotients that are finitely generated and nilpotent-by-finite (so polycyclic-by-finite, in particular).

As an example of an application of this strategy to some problems of a combinatorial nature, we state the following result. Recall that the prime radical $B(K[S])$ of $K[S]$ is nilpotent if $K[S]$ is a right Noetherian algebra.

Theorem 3.5 ([53]) *Assume that $K[S]$ a right Noetherian algebra. Then*

1. *the Gelfand–Kirillov dimension of $K[S]$ is finite if and only if for every cancellative subsemigroup T of S we have* $\text{GKdim}(K[T]) < \infty$.
2. *Moreover, in this case* $\text{GKdim}(K[S]/B(K[S])) = \text{GKdim}(K[T])$ *(and it is an integer) for a cancellative subsemigroup T of the image $\overline{S}$ of S in $K[S]/B(K[S])$ and*

$\mathrm{GKdim}(K[S]) \leq r \cdot \mathrm{GKdim}(K[T])$, where r is the nilpotency index of $B(K[S])$. Moreover, T has a finitely generated nilpotent-by-finite quotient group.

Notice, that by a celebrated result of Gromov, the Gelfand–Kirillov dimension of a finitely generated group algebra $K[G]$ is finite if and only if G is nilpotent-by-finite, and in this case due to the formula of Bass it is an integer expressible in terms of the ranks of the (torsion free) factors of the upper central series of a nilpotent subgroup of finite index in G, see [47], Chap. 11. Moreover, $\mathrm{GKdim}(K[T]) = \mathrm{GKdim}(K[G])$ if G is the group of quotients of its submonoid T, by a result of Grigorchuk, see [51], Chap. 8.

4 Important Motivating Examples—Algebras with Homogeneous Quadratic Relations

Important classes of examples of Noetherian semigroup algebras include algebras corresponding to the set theoretic solutions of the Yang-Baxter equation. Recall that by a set theoretic solution of the Yang-Baxter equation we mean a map $r : X \times X \to X \times X$, where X is a nonempty set, such that

$$r_{12}r_{13}r_{23} = r_{23}r_{13}r_{12},$$

where r_{ij} denotes the map $X \times X \times X \to X \times X \times X$ acting as r on the (i, j) factor and as the identity on the remaining factor. We will focus on the case where $X = \{x_1, \dots, x_n\}$ is finite.

The problem of finding all such solutions was posed in [13], and turned out to be very difficult. In particular, one considers solutions that are involutive ($r^2 = id$) and non-degenerate (this condition will be defined later). This area leads to a fascinating class of Noetherian algebras, referred to as Yang-Baxter algebras. Namely, one associates to r an algebra defined by the presentation $K\langle x_1, \dots, x_n\rangle/J$ where J consists of relations of the form $xy = x'y'$ if $r(x, y) = (x', y')$. This implies that J consists of $\binom{n}{2}$ relations and it follows also that every monomial xy, $x, y \in X$, appears in at most one relation.

These algebras arose independently in several other contexts, including homological methods developed for an important class of algebras, called Sklyanin algebras, [62]. The following theorem summarizes their main properties.

Theorem 4.1 ([21]) *These algebras are isomorphic to $K[S]$, where S is a submonoid of a finitely generated torsion free abelian-by-finite group. They are Noetherian PI domains of finite homological dimension and they are maximal orders.*

Actually, S has a group of quotients that is solvable [14], see also [31], and embeds into the semidirect product $F_n \rtimes S_n$, where S_n is the symmetric group acting on the free commutative group F_n of rank n by the natural permutation of the basis, [14, 38]. Simplest examples include commutative polynomial rings $K[x_1, \dots, x_n]$, arising from the free commutative monoids S, and the algebra of the monoid defined by the presentation $S = \langle x, y \mid x^2 = y^2 \rangle$.

These algebras have several other properties similar to the properties of commutative polynomial rings, including nice homological properties. Certain families of such algebras are known, but new examples are very difficult to construct.

Height one prime ideals P of these algebras have been described [31, 36]. In particular, if $P \cap S \neq \emptyset$ then $P = aK[S] = K[S]a$ for some $a \in S$, and there are finitely many such height one primes. While prime ideals of $K[S]$ not intersecting S come from primes of the group algebra $K[SS^{-1}]$ (see Sect. 5). In particular, this can be used to prove that $K[S]$ is a maximal order.

There exist important more general classes of semigroup algebras which fit in this context. We say that an algebra A is defined by homogeneous semigroup relations if it is defined by a presentation $A = K\langle x_1, \ldots, x_n \mid R\rangle$, with every relation of the set R of defining relations of the form $v = w$, where v, w are words in the free monoid on $x_1, \ldots, x_n$ and v, w have equal lengths. Exploiting the approach presented in the previous section, one can prove the following result.

Theorem 4.2 ([20]) *Assume that an algebra $A = K[S]$ is right Noetherian and* $\mathrm{GKdim}(K[S]) < \infty$. *If A is defined by homogeneous semigroup relations, then A satisfies a polynomial identity.*

In particular, consider the following class of quadratic algebras, that generalizes the Yang-Baxter algebras. These are semigroup algebras of monoids with generators $x_1, x_2, \ldots, x_n$ subject to $\binom{n}{2}$ quadratic relations of the form $x_i x_j = x_k x_l$ with $(i, j) \neq (k, l)$ and, moreover, every monomial $x_i x_j$ appears at most once in one of the defining relations. One of the origins of these algebras comes from [18]. Recently, further combinatorial aspects of such algebras have been studied in [19].

For every $x \in X = \{x_1, \ldots, x_n\}$, let

$$f_x : X \to X$$

and

$$g_x : X \to X$$

be the maps such that

$$r(x, y) = (f_x(y), g_y(x)).$$

One says that S is a non-degenerate quadratic monoid if each f_x and each g_x is bijective, with $x \in X$.

The following result was obtained in [20] in the special case of square-free defining relations, and in full generality in [41].

Theorem 4.3 *Let S be a non-degenerate quadratic monoid. Then $K[S]$ is right and left Noetherian, it satisfies a polynomial identity and embeds into a matrix algebra over a field extension of K.*

The proof uses the structural approach explained before and several other results. First, a finite ideal chain in S is constructed from the combinatorial data. Every factor of this chain is either of generalized matrix type or it is nilpotent. Independently, one shows that $K[S]$ has finite Gelfand–Kirillov dimension and that S satisfies acc on one-sided ideals. This allows us to prove that $K[S]$ is Noetherian, by applying Theorem 3.4. Then, using Theorem 4.2, one shows that the algebra satisfies a polynomial identity. Finally, using the embedding theorem of Anan'in, Theorem 3.2, we get the last assertion.

5 Prime Ideals and Arithmetical Properties of $K[S]$

There are several classical important arithmetical properties that have been extensively studied in the class of commutative semigroup rings. These include in particular Krull domains, integrally closed domains, principal ideal rings. For the main results and general techniques of this theory we refer to Gilmer's book [23]. And for general results on commutative orders, and integrally closed domains in particular, to [16]. Several methods and results of the multiplicative ideal theory are valid for both commutative rings and monoids, as they depend only on the multiplicative structure of the ring. The philosophy that such results should be derived as far as possible without making reference to the additive structure of the ring, is presented in particular in [28].

There has been also an extensive work done on noncommutative orders, that we will discuss in this section. Some of this is based on earlier general work on noncommutative orders (in particular, see [1, 8] and its bibliography), some has been developed for special classes of noncommutative semigroups in [64], and more recently in [22].

Recall that a monoid S which has a left and right group of quotients G is called an order. Then S is called a maximal order if there does not exist a submonoid S' of G properly containing S and such that $aS'b \subseteq S$ for some $a, b \in G$.

For subsets $A, B \subseteq G$ we define $(A :_l B) = \{g \in G \mid gB \subseteq A\}$, $(A :_r B) = \{g \in G \mid Bg \subseteq A\}$. Then S is a maximal order if and only $(I :_l I) = (I :_r I) = S$ for every fractional ideal I of S. A nonempty subset I of G is called a fractional ideal of S if $SIS \subseteq I$ and $cI, Id \subseteq S$ for some $c, d \in S$.

Assume now that S is a maximal order. Then $(S :_r I) = (S :_l I)$ for any fractional ideal I. One denotes this set as $(S : I)$. Define $I^* = (S : (S : I))$, the divisorial closure of I. If $I = I^*$ then I is said to be divisorial. Then S is said to be a Krull order (or a Krull monoid in the terminology of [22]) if S satisfies also the ascending chain condition on divisorial ideals contained in S. In this case the divisor group $D(S)$ (also defined as in ring theory) is a free abelian group with basis the set of prime divisorial ideals. The latter are minimal prime ideals of S.

The following result is our starting point. Notice in particular that the obtained description is expressed in terms of the underlying semigroup only. Here $U(S)$ denotes the unit group of the monoid S.

Theorem 5.1 ([10]) *A commutative monoid algebra $K[S]$ is a Krull domain if and only if S is a submonoid of a torsion free abelian group which satisfies the ascending chain condition on cyclic subgroups and S is a Krull order in its group of quotients.*

Furthermore, S is a Krull order if and only if $S = U(S) \times S_1$, where S_1 is a submonoid of a free abelian group F such that S_1 is the intersection of the quotient group of S_1 with the positive cone of F. Moreover, in this situation the class group of $K[S]$ coincides with the class group of S.

This result extended an earlier work of Anderson, [3, 4], on commutative Noetherian maximal orders. Notice that the class of commutative Noetherian maximal orders $K[S]$ coincides with the class of finitely generated integrally closed domains.

The last property mentioned in the theorem allows one to simplify the calculation of the class group in several concrete classes of algebras, and it also shows that the height one primes of $K[S]$ determined by the minimal primes of S are crucial. In particular, for certain concrete finitely presented commutative algebras this invariant was calculated in [26].

Theorem 5.2 *Let A be a finitely generated commutative algebra over a field K with a presentation $A = K[X_1, \ldots, X_n|R]$, where R is a set of monomial relations in the generators $X_1, \ldots, X_n$. So $A = K[S]$, the semigroup algebra of the monoid $S = \langle X_1, \ldots, X_n|R\rangle$. A characterization, purely in terms of the defining relations, is given of when A is an integrally closed domain, provided R contains at most two relations. Also the class group of such algebras A is calculated.*

Also within the noncommutative ring theory, Noetherian orders in simple algebras form an important class of rings. Maximal orders have been studied in this context, in particular for the class of group algebras of polycyclic-by-finite groups. Recall that the infinite dihedral group $\langle a, b \mid ba = a^{-1}b,\ b^2 = 1\rangle$ is denoted by D_∞. A group G is said to be dihedral-free if the normalizer of any subgroup H isomorphic with D_∞ is of infinite index in G, (equivalently, H has infinitely many conjugates in G).

Theorem 5.3 ([5]) *Let G be a polycyclic-by-finite group. The group algebra $K[G]$ is a prime maximal order if and only if*

1. *G has no nontrivial finite normal subgroups,*
2. *G is dihedral-free.*

The first condition in the theorem is equivalent with the group algebra being prime, see Theorem 2.9. Brown also determined when the height one prime ideals are principal. By $\Delta(G)$ we denote the finite conjugacy subgroup of G.

Theorem 5.4 ([5]) *Let G be a polycyclic-by-finite group. If $K[G]$ is a prime maximal order, then the following conditions are equivalent for a height one prime ideal P of $K[G]$:*

1. *P is right principal,*
2. *P is invertible, that is, $Q_{cl}(K[G])$ contains a $K[G]$- bimodule J with $IJ = JI = K[G]$,*
3. *P is right projective;*
4. *$P = K[G]n = nK[G]$ for some $n \in K[\Delta(G)]$,*
5. *P contains a nonzero central element,*
6. *P contains a nonzero normal element,*
7. *P contains an invertible ideal.*

If these conditions hold for all height one primes of $K[G]$, then $K[G]$ is a UFR (unique factorization ring) in the sense of Chatters and Jordan, [9]. Some earlier partial results on this topic can be found in [42, 43, 63]. The following consequence for the case of PI-algebras that are domains is of special interest.

Theorem 5.5 ([5]) *Let G be a finitely generated torsion free abelian-by-finite group. Then the group algebra $K[G]$ is a Noetherian maximal order. Moreover, all height one primes of $K[G]$ are principally generated by a normal element.*

Only for very few classes of noncommutative semigroups S it has been determined when the semigroup algebra is a Noetherian maximal order. Apart from the Yang-Baxter algebras, see Sect. 4, Wauters in [64] dealt with cancellative semigroups S consisting of normal elements (so $aS = Sa$ for every $a \in S$) and with the cancellative semigroups of the regular elements of a prime Goldie ring. Various aspects of arithmetical properties of noncommutative monoids were recently studied in [22]. We will summarize results obtained on algebras of submonoids of a polycyclic-by-finite group G, obtained in [24, 25, 32, 36].

Recall that G has a normal subgroup of finite index H that is torsion free. Then $K[G]$ can be considered as a ring graded by the finite group G/H in a natural way. Therefore, known deep results on the correspondence of prime ideals for rings graded by finite groups [58], Theorem 17.9, allow to establish a strong link between the primes in $K[G]$ and in $K[H]$ (incomparability, going up, going down). Hence, the information on prime ideals in the torsion free case is essential, [24]. Crucial results on prime ideals in case $K[S]$ is Noetherian and $G = SS^{-1}$, based also on Theorem 2.11, were proved in [24, 34].

Proposition 5.6 ([24]) *Let S be a submonoid of a torsion free polycyclic-by-finite group. Assume that $K[S]$ is right Noetherian. Then*

1. *$K[S \cap P]$ is a prime ideal in $K[S]$ for any prime ideal P in $K[S]$ with $P \cap S \neq \emptyset$.*
2. *$K[Q]$ is a prime ideal in $K[S]$ for any prime ideal Q in S.*
3. *the set of height one prime ideals of $K[S]$ intersecting S nontrivially coincides with the set of the ideals of the form $K[Q]$, where Q is a minimal prime ideal of S.*

In view of the above theorem, the study of prime ideals of $K[S]$ splits into two cases, one leading to primes of the group algebra $K[SS^{-1}]$ and one leading to the primes of the monoid S. Recall that if C is a right Ore subset consisting of regular elements in a ring R, we denote by R_C the classical localization of R with respect to C. If either R is right Noetherian or R satisfies a polynomial identity and R_C is right Noetherian, then the maps $P \mapsto PR_C, J \mapsto J \cap R$ are inverse bijections between the sets of prime ideals in R not intersecting C and the set of primes in R_C, see [27], Theorems 10.18 and 10.20 and its proof. This also holds in the following case.

Lemma 5.7 ([29]) *Let S be a submonoid of a nilpotent group and let G be the group of quotients of S. Assume that P is a prime ideal of $K[S]$. If $P \cap S = \emptyset$, then*

1. *$PK[G]$ is a two-sided ideal of $K[G]$,*
2. *$Q = PK[G]$ is a prime ideal of $K[G]$, $Q \cap K[S] = P$ and $K[G]/Q$ is a localization (with respect to an Ore set) of $K[S]/P$.*

We say that a prime Goldie ring R is a Krull order if R is a maximal order that satisfies the ascending chain condition on divisorial integral ideals. In the next theorem we collect some of the essential properties of these orders in the case of algebras satisfying a polynomial identity. In this case, our definition coincides with that of Chamarie. For details we refer the reader to his work [7, 8]. The prime spectrum of R is denoted by $Spec(R)$, and the set of height one prime ideals of R by $X^1(R)$.

In view of the structural result on Noetherian algebras $K[S]$, Theorem 2.11, it is natural to consider first the case where G is a finitely generated abelian-by-finite group. Recall that the group algebra of a finitely generated group G satisfies a polynomial identity if and only if G is abelian-by-finite, see [57], Theorems 5.3.7 and 5.3.9.

Theorem 5.8 ([24]) *Let R be a prime Krull order satisfying a polynomial identity. Then the following properties hold:*

1. *The divisorial ideals form a free abelian group with basis $X^1(R)$, the height one primes of R.*
2. *If $P \in X^1(R)$ then $P \cap Z(R) \in X^1(Z(R))$, and furthermore, for any ideal I of R, $I \subseteq P$ if and only if $I \cap Z(R) \subseteq P \cap Z(R)$.*
3. *$R = \bigcap R_{Z(R)\setminus P}$, where the intersection is taken over all height one primes of R, and every regular element $r \in R$ is invertible in almost all (that is, except possibly finitely many) localizations $R_{Z(R)\setminus P}$. Furthermore, each $R_{Z(R)\setminus P}$ is a left and right principal ideal ring with a unique nonzero prime ideal.*
4. *For a multiplicatively closed set of ideals M of R, the (localized) ring $R_M = \{q \in Q_{cl}(R) \mid Iq \subseteq R$, for some $I \in M\}$ is a Krull order, and $R_M = \bigcap R_{Z(R)\setminus P}$, where the intersection is taken over those height one primes P for which $R_M \subseteq R_{Z(R)\setminus P}$.*

If S is a monoid with a torsion free abelian-by-finite group of quotients G (so $K[S]$ is a PI-domain), the maximal order property of $K[S]$ is determined by the structure of S and can be reduced to some 'local' monoids S_P, with P a minimal prime ideal of S. Here

$$S_P = \{g \in G \mid Cg \subseteq S \text{ for some } G\text{-conjugacy class } C \text{ of } G \text{ contained in } S \text{ with } C \not\subseteq P\}.$$

The next theorem comes from [35], see also [38], Theorems 7.2.5 and 7.2.7.

Theorem 5.9 *Let S be a submonoid of a finitely generated torsion free abelian-by-finite group. Then the monoid algebra $K[S]$ is a Noetherian maximal order if and only if the following conditions are satisfied:*

1. *S satisfies the ascending chain condition on one-sided ideals,*
2. *S is a maximal order in its group of quotients,*
3. *for every minimal prime ideal P of S the monoid S_P has only one minimal prime ideal.*

Furthermore, in this case, each S_P is a maximal order satisfying the ascending chain condition on one-sided ideals.

This result was extended in [24] to the case of a submonoid of an arbitrary finitely generated abelian-by-finite group. The final step was made in [25], where a further extension was obtained.

Theorem 5.10 ([25]) *Let S be a submonoid of a polycyclic-by-finite group such that the semigroup algebra $K[S]$ is Noetherian, i.e., there exist normal subgroups F and N of $G = SS^{-1}$ such that $F \subseteq S \cap N$, N/F is abelian, G/N is finite and $S \cap N$ is finitely generated. Suppose that for every minimal prime P of S the intersection $P \cap N$ is G-invariant. Then, the semigroup algebra $K[S]$ is a prime maximal order if and only if the monoid S is a maximal order in its group of quotients G, the group G is dihedral-free and has no nontrivial finite normal subgroups.*

Suppose that in the previous theorem one also assumes that the group G is abelian-by-finite. Then, in [24], it is shown that the condition 'for every minimal prime P of S the intersection $P \cap N$ is G-invariant' is necessary for $K[S]$ to be a maximal order. However, no example of a maximal order S in a polycyclic-by-finite group $G = SS^{-1}$ (with G dihedral-free and $K[G]$ prime) is known so that $K[S]$ is Noetherian but not a maximal order. We note that for a submonoid S of a torsion free polycyclic-by-finite group certain necessary and certain sufficient conditions for a Noetherian $K[S]$ to be a unique factorization ring in the sense of Chatters and Jordan were studied in [44, 45].

The following result allows to construct several concrete examples of maximal orders in the PI-case. As we shall see, this is in contrast to the situation described in Sect. 6, where no such a general construction is known.

Proposition 5.11 ([24, 38]) *Let A be an abelian normal subgroup of finite index in a group G. Suppose that B is a submonoid of A so that $A = BB^{-1}$ and B is a finitely generated maximal order. Let S be a submonoid of G such that $G = SS^{-1}$ and $S \cap A = B$. Then S is a maximal order that satisfies the ascending chain condition on right ideals if and only if S is maximal among all submonoids T of G with $T \cap A = B$.*

Substantial results have been also obtained on semigroup algebras that are principal ideal rings. This story begins with the case of group algebras, settled by Passman in [56], and concludes with the results obtained in [30]. References to several partial intermediate results can be found in [23, 38]. The rest of this section is devoted to a presentation of these results. We will always assume that a principal ideal ring contains an identity element, though in this section S is not necessarily a monoid. First we state Passman's result on the group algebra case. We follow [46], where this result is stated in the slightly more general context of matrices over group algebras, that will be needed later.

Proposition 5.12 ([56]) *Let G be a group and $R = M_n(K)$, a matrix ring over K. The following conditions are equivalent:*

1. *$R[G] = M_n(K[G])$ is a principal right ideal ring,*
2. *$R[G]$ is right Noetherian and the augmentation ideal $\omega(R[G])$ is a principal right ideal,*
3. *if char $K = 0$, then G is finite or finite-by-infinite cyclic,*
   if char $K = p > 0$, then G is finite p'-by-cyclic p or G is finite p'-by-infinite cyclic.

This result was then extended to semigroup algebras of cancellative monoids as follows.

Proposition 5.13 ([46]) *Let T be a cancellative monoid and K a field of characteristic p (possibly zero). The following conditions are equivalent:*

1. *$K[T]$ is a principal right ideal ring,*
2. *T is a semigroup satisfying one of the following conditions:*
   a. *T is a group satisfying the conditions of Proposition 5.12,*
   b. *T contains a finite p'-subgroup H and a nonperiodic element x such that $xH = Hx$, $T = \bigcup_{i\in \mathbb{N}} Hx^i$ and the central idempotents of $K[H]$ are central in $K[T]$.*

As explained in Sect. 3, the structure theorem for linear semigroups provides a link between a linear semigroup and some of its cancellative subsemigroups. In order to apply this approach to semigroup algebras of arbitrary semigroups that are principal ideal rings one first has to reduce the problem to linear semigroups. This is guaranteed by Theorem 3.2 together with the following result.

Theorem 5.14 ([30]) *Let $K[S]$ be a principal right ideal ring. Then $K[S]$ satisfies a polynomial identity.*

Using the structure theorem of linear semigroups, explained in Sect. 3, one now can prove the following results.

Proposition 5.15 ([30]) *If $K[S]$ is a principal right ideal ring, then the Gelfand–Kirillov dimension of $K[S]$ is equal to its classical Krull dimension and* *it is 0 or 1. In the former case S is finite. Moreover, every prime artinian homomorphic image of $K[S]$ is finite dimensional over K.*

Theorem 5.16 ([30]) *Let S be a semigroup and K a field of characteristic p (possibly zero). The following conditions are equivalent:*

1. *$K_0[S]$ is a principal (left and right) ideal ring;*
2. *there exists an ideal chain*
$$I_1 \subseteq \cdots \subseteq I_t = S$$
such that I_1 and every factor I_j/I_{j-1} is of the form $\mathcal{M}(T, n, n, P)$ for an invertible over $K_0[T]$ sandwich matrix P, and one of the following conditions holds:
   a. *T is a group of the type described in Proposition 5.12;*
   b. *T is a monoid with a finite group of units H such that $T = \bigcup_{i\geq 0} Hx^i$ for some $x \in T$, and either this union is disjoint or $x^n = \theta$ for some $n \geq 1$. Also $Hx = xH$, the central idempotents of $K[H]$ commute with x, and $p = 0$ or $p \nmid |H|$.*

In case the equivalent conditions are satisfied it follows that

$$K_0[S] \cong K_0[I_1] \oplus K_0[I_2/I_1] \oplus \cdots \oplus K_0[I_t/I_{t-1}].$$

Moreover, $K_0[S]$ is a finite module over its center, which is finitely generated.

It is not known whether the left-right symmetric hypothesis in Theorem 5.16 is essential.

The above theorem applies to finite dimensional algebras $K[S]$, since a finite dimensional algebra is a principal right ideal ring if and only if it is a principal left ideal ring. One can also show that semiprime principal right ideal semigroup algebras are necessarily principal left ideal rings as well.

Theorem 5.17 ([30]) *Let S be a semigroup and K a field of characteristic p (possibly zero). Then $K_0[S]$ is a semiprime principal right ideal ring if and only if there exists an ideal chain*

$$I_1 \subseteq \cdots \subseteq I_t = S$$

such that I_1 and every factor I_j/I_{j-1} is of the form $\mathcal{M}(T, n, n, P)$ for an invertible over $K_0[T]$ sandwich matrix P and a monoid T such that

1. *either T is a group as in Proposition 5.12 so that $K[T]$ is semiprime,*
2. *or T is a monoid with finite group of units H such that $T = \bigcup_i Hx^i$ is a disjoint union, for some $x \in T$. Also $Hx = xH$, the central idempotents of $K[H]$ commute with x, and $p = 0$ or $p \nmid |H|$. Furthermore, for every primitive central idempotent $e \in K[H]$, either $K[H]ex = 0$ or $K[H]ex^i \neq 0$ for all $i \geq 1$.*

Moreover, if the equivalent conditions are satisfied, then $K_0[S]$ is a principal left ideal ring.

Corollary 5.18 *$K_0[S]$ is a prime principal right ideal ring if and only if*

$$S \cong \mathcal{M}(\{1\}, n, n, Q), \quad S \cong \mathcal{M}(\langle x \rangle, n, n, Q) \quad or \quad S \cong \mathcal{M}(\langle x, x^{-1} \rangle, n, n, Q)$$

where Q is invertible in $M_n(K)$, $M_n(K[x])$ or $M_n(K[x, x^{-1}])$ respectively. Hence, $K_0[S] \cong M_n(K), M_n(K[x])$, or $M_n(K[x, x^{-1}])$.

6 Why Should We Look at the Non-Noetherian Case? Motivation and First Results

We start with an interesting example of a finitely presented algebra, denoted by $R(\mathbb{1})$, that has recently played an important role in certain aspects of noncommutative geometry. This algebra is not Noetherian, but it leads to a family of deformations that consists of Noetherian algebras [59]. It turns out that it is based on a relatively simple construction of a semigroup algebra of a submonoid of the Heisenberg group (a nilpotent group of class 2):

$$G = gr(a, b, c \mid ac = ca, ab = ba, bc = acb).$$

On one hand, this example shows that computations in such algebras may be quite difficult. On the other hand, it seems to be a good motivation for studying non-Noetherian orders coming from finitely generated nilpotent groups. After explaining the nature of this example, we present some recent general results on this class of algebras.

Let

$$M = \langle x, y, z, t \mid xy = yx, zt = tz, yz = xt = zx, zy = tx = yt \rangle,$$

a finitely presented monoid, defined by homogeneous relations. So $K[M]$ carries some similarity to Yang-Baxter algebras, considered in Sect. 4. Namely, it has the 'correct' number of quadratic relations ($\binom{n}{2}$ relations), however some monomials appear in two different relations.

It can be shown that: $\phi : M \longrightarrow G$ defined by

$$x \mapsto c, y \mapsto ac, z \mapsto bc, t \mapsto abc$$

is a homomorphism which also is an embedding. Hence

$$M \cong \phi(M) \subseteq G.$$

Note that $K[M]$ is an Ore domain, but it is not Noetherian (use Theorem 2.11: G is not abelian-by-finite while M has trivial units; but this is also easy to check directly).

$K[M]$ is the algebra used by Yekutieli and Zhang [65] (as a counterexample in the context of Artin-Schelter regular rings), and recently by Rogalski and Sierra, where it plays a key role in the classification of 4-dimensional non-commutative projective surfaces, [59]. Namely, a family of deformations of $K[M]$ is considered. They are of the form:

$$R(\rho, \theta) = K\langle x_1, x_2, x_3, x_4 \mid f_i = 0, i = 1, 2, 3, 4, 5, 6\rangle$$

where

$$\begin{aligned} f_1 &= x_1(cx_1 - x_3) + x_3(x_1 - cx_3) \\ f_2 &= x_1(cx_2 - x_4) + x_3(x_2 - cx_4) \\ f_3 &= x_2(cx_1 - x_3) + x_4(x_1 - cx_3) \\ f_4 &= x_2(cx_2 - x_4) + x_4(x_2 - cx_4) \\ f_5 &= x_1(dx_1 - x_2) + x_4(x_1 - dx_2) \\ f_6 &= x_1(dx_3 - x_4) + x_4(x_3 - dx_4) \end{aligned}$$

for $c = (\theta - 1)/(\theta + 1)$ and $d = (\rho - 1)(\rho + 1)$.

Notice that $R(1, 1) \cong K[M]$ and it is embeddable in the skew polynomial ring $K(u, v)[t, \sigma]$ over the rational function field $K(u, v)$, where $\sigma(v) = v, \sigma(u) = uv$.

Theorem 6.1 ([59]) *Assume that K is algebraically closed and uncountable. If ρ, θ are algebraically independent over the prime subfield of K, then $R(\rho, \theta)$ is a Noetherian domain of global dimension* 4 *and Gelfand–Kirillov dimension* 4. *And it is birational to $\mathbb{P}^2$.*

Here, for a Noetherian domain R such that $R = \bigoplus_{i\geq 0} R_i$ is connected $\mathbb{N}$-graded (meaning that the zero component $R_0 = K$ and $\dim(R_i) < \infty$ for every i), it is known that the graded ring of quotients $Q_{gr}(R) \cong D[t, t^{-1}, \sigma]$, for a division ring D. So, $Q_{gr}(R)$ is obtained by localizing with respect to the set of nonzero homogeneous elements in R. If the division ring D is a field (then $D = K(X)$ for a projective variety X), then R is said to be birational to X.

Hence, this provides a new motivation to study algebras of submonoids of nilpotent groups that are not necessarily Noetherian. The starting case is where the quotient group is nilpotent of class 2. Then we have the following surprising and very useful result.

Lemma 6.2 ([29]) *A prime ideal P of a submonoid S of a nilpotent group of class two is completely prime; that is, $st \in P$ implies $s \in P$ or $t \in P$, for $s, t \in S$. In particular, if S is finitely generated, then S has only finitely many prime ideals.*

Using also Lemma 5.7 and Proposition 5.6, one can then get a partial extension of the classical result on the classical Krull dimension of a group algebra of a nilpotent group, stated in Theorem 2.5.

Theorem 6.3 ([40]) *Let S be a submonoid of a nilpotent group of class two. If the group of quotients $G = SS^{-1}$ of S is finitely generated then* $\mathrm{clKdim}(K[S]) = \mathrm{h}(G)$. *Moreover, if P is a prime ideal of $K[S]$, then $K[S]/P$ is a Goldie ring.*

In order to indicate a striking contrast with the case of higher nilpotency classes, we will construct some prime ideals in the algebra $K[S]$ of the submonoid $S = \langle b, c\rangle$ of the free nilpotent group $\mathrm{F}_3(b, c)$ of class 3. In other words, $\mathrm{F}_3(b, c)$ is defined by the following relations:

$$bc = acb, \quad ab = dba, \quad ac = eca,$$
$$db = bd, \quad dc = cd, \quad eb = be, \quad ec = ce.$$

Lemma 6.4 *For positive integers k and n, the word $(bc^k)^n$ cannot be rewritten in $S = \langle b, c\rangle \subseteq \mathrm{F}_3(b, c)$.*

Recall that a doubly infinite word in b and c is a sequence $x = (x_i)_{i\in\mathbb{Z}}$ with $x_i \in \{b, c\}$. One says that x is recurrent if every (finite) subword of x appears in x at least twice (thus, it appears infinitely many times). For example, the cyclic word $(bc^k)^\infty$ is of this type. Then,

$$J = \{s \in S \ :\ s \neq t \text{ in } S \text{ for every subword } t \text{ of } x\}$$

is an ideal of S and it is easy to check that J is a prime ideal of S. Since $\mathrm{F}_3(b, c)$ is torsion free, this, together with Proposition 5.6, is used to derive the following consequence.

Theorem 6.5 ([40]) *The submonoid $S = \langle b, c\rangle$ of the group $\mathrm{F}_3(b, c)$ has infinitely many prime ideals P that are not completely prime. Furthermore, each $K[P]$ is a prime ideal of $K[S]$ such that $K[S]/K[P]$ is an algebra satisfying a polynomial identity and* $\mathrm{clKdim}(K[S]/K[P]) = \mathrm{GKdim}(K[S]/K[P]) = 1$.

This result shows that the situation is quite different than the one in the case of nilpotency class 2, where all primes are completely prime.

A natural open question that arises is whether there exist other, more exotic, primes in $K[S]$ for a submonoid S of a finitely generated nilpotent group G of nilpotency class exceeding 2. In particular, do there exist prime homomorphic images of $K[S]$ that are not Goldie? Can $K[S]$ have infinite classical Krull dimension?

As mentioned in Sect. 5, prime ideals provide one of the main tools in dealing with maximal orders, and with related classes of algebras with nice arithmetical properties. We state some results in this direction.

Theorem 6.6 ([32]) *Let S be a submonoid of a finitely generated torsion free nilpotent group. Then the following properties hold.*

1. *S is a maximal order if and only if $K[S]$ is a maximal order.*

2. *If S satisfies the ascending chain condition on right ideals and is a maximal order, then all elements of S are normal (meaning that $aS = Sa$ for every $a \in S$).*

So, in the latter case, the theorem below applies.

Theorem 6.7 ([32]) *Let S be a submonoid of a torsion free polycyclic-by-finite group. Assume that all elements of S are normal. Then the following conditions are equivalent:*

1. *$K[S]$ is a Krull domain,*
2. *S is a Krull order,*
3. *$S/U(S)$ is an abelian Krull order.*

Using the special features of groups of nilpotency class 2, and applying Theorem 6.7, one can prove the following result. Here we define $N(S) = \{a \in S \mid aS = Sa\}$, the submonoid of normal elements of S.

Theorem 6.8 ([39]) *Assume that S is a submonoid of a torsion free nilpotent group of class two. Assume that S is a Krull order. Then*

(i) *the derived subgroup G' of the quotient group G of S is contained in S,*
(ii) *$S = N(S)$,*
(iii) *S/G' is a commutative Krull order,*
(iv) *if G is finitely generated, then $K[S]$ is a Krull domain for every field K; moreover S is finitely generated and $K[S]$ is right and left Noetherian.*

On the other hand, if $G' \subseteq S$ and S/G' is a Krull order then S is a Krull order.

So, in some sense, the class of such orders is quite restricted and carries a lot of commutative flavor. It is an open problem whether there exist maximal orders that do not satisfy the property $N = N(S)$ and, in higher nilpotency classes whether there exist Krull orders of this type.

References

1. E. Akalan, H. Marubayashi, Multiplicative ideal theory in non-commutative rings, in *Multiplicative Ideal Theory and Factorization Theory*, ed. by S.T. Chapman, M. Fontana, A. Geroldinger, B. Olberding (Springer, Heidelberg, 2016)
2. A.Z. Anan'in, An intriguing story about representable algebras, in *Ring Theory 1989, Israel Mathematical Conference Proceedings* (Weizmann, Jerusalem, 1989), pp. 31–38
3. D.F. Anderson, Graded Krull domains. Commun. Algebra **7**, 79–106 (1979)
4. D.F. Anderson, The divisor class group of a semigroup ring. Commun. Algebra **8**, 467–476 (1980)
5. K.A. Brown, Height one primes of polycyclic group rings. J. Lond. Math. Soc. **32**, 426–438 (1985)
6. K.A. Brown, Corrigendum and addendum to 'Height one primes of polycyclic group rings'. J. Lond. Math. Soc. **38**, 421–422 (1988)

7. M. Chamarie, *Anneaux de Krull non commutatifs* (Université Claude-Bernard - Lyon I, Thèse, 1981)
8. M. Chamarie, Anneaux de Krull non commutatifs. J. Algebra **72**, 210–222 (1981)
9. A.W. Chatters, D.A. Jordan, Non-commutative unique factorisation rings. J. Lond. Math. Soc. **33**(2), 22–32 (1986)
10. L.G. Chouinard, Krull semigroups and divisor class groups. Can. J. Math. **23**, 1459–1468 (1981)
11. G.H. Cliff, Zero divisors and idempotents in group rings. Can. J. Math. **32**, 596–602 (1980)
12. A.H. Clifford, G.B. Preston, *The Algebraic Theory of Semigroups*, vol. I (American Mathematical Society, Providence, 1961)
13. V.G. Drinfeld, On some unsolved problems in quantum group theory, in *Quantum Groups*, ed by P.P. Kulish, Lecture Notes in Mathematics, vol. 1510 (Springer, Heidelberg, 1992), pp. 1–8
14. P. Etingof, T. Schedler, A. Soloviev, Set-theoretical solutions to the quantum Yang-Baxter equation. Duke Math. J. **100**, 169–209 (1999)
15. D.R. Farkas, A.H. Schofield, R.L. Snider, J.T. Stafford, The isomorphism question for division rings of group rings. Proc. AMS **85**, 327–330 (1982)
16. R. Fossum, *The Divisor Class Group of a Krull Domain* (Springer, New York, 1973)
17. J. Fountain, M. Petrich, Completely 0-simple semigroups of quotients. Math. Proc. Camb. Philos. Soc. **105**, 263–275 (1989)
18. T. Gateva-Ivanova, Skew polynomial rings with binomial relations. J. Algebra **185**, 710–753 (1996)
19. T. Gateva-Ivanova, Quadratic algebras, Yang-Baxter equation, and Artin-Schelter regularity. Adv. Math. **230**, 2152–2175 (2012)
20. T. Gateva-Ivanova, E. Jespers, J. Okniński, Quadratic algebras of skew type and the underlying semigroups. J. Algebra **270**, 635–659 (2003)
21. T. Gateva-Ivanova, M. Van den Bergh, Semigroups of I-type. J. Algebra **206**, 97–112 (1998)
22. A. Geroldinger, Non-commutative Krull monoids: a divisor theoretic approach and their arithmetic. Osaka J. Math. **50**, 503–539 (2013)
23. R. Gilmer, *Commutative Semigroup Rings* (University Chicago Press, Chicago, 1984)
24. I. Goffa, E. Jespers, J. Okniński, Primes of height one and a class of Noetherian finitely presented algebras. Int. J. Algebra Comput. **17**, 1465–1491 (2007)
25. I. Goffa, E. Jespers, J. Okniński, Semigroup algebras of submonoids of polycyclic-by-finite groups and maximal orders. Algebras Represent. Theory **12**, 357–363 (2009)
26. I. Goffa, E. Jespers, J. Okniński, Normal domains with monomial presentations. Int. J. Algebra Comput. **19**, 287–303 (2009)
27. K.R. Goodearl, R.B. Warfield, *An Introduction to Noncommutative Noetherian Rings*, 2nd edn. London Mathematical Society Student Texts 61 (Cambridge University Press, Cambridge, 2004)
28. F. Halter-Koch, Ideal systems, an introduction to multiplicative ideal theory, in *Monographs and Textbooks in Pure and Applied Mathematics, 211* (Marcel Dekker Inc, New York, 1998)
29. E. Jespers, J. Okniński, Nilpotent semigroups and semigroup algebras. J. Algebra **169**, 984–1011 (1994)
30. E. Jespers, J. Okniński, Semigroup algebras that are principal ideal rings. J. Algebra **183**, 837–863 (1996)
31. E. Jespers, J. Okniński, Binomial semigroups. J. Algebra **202**, 250–275 (1998)
32. E. Jespers, J. Okniński, Semigroup algebras and maximal orders. Can. Math. Bull. **42**, 298–306 (1999)
33. E. Jespers, J. Okniński, Noetherian semigroup algebras. J. Algebra **218**, 543–562 (1999)
34. E. Jespers, J. Okniński, Submonoids of polycyclic-by-finite groups and their algebras. Algebras Represent. Theory **4**, 133–153 (2001)
35. E. Jespers, J. Okniński, Semigroup algebras and noetherian maximal orders. J. Algebra **238**, 590–622 (2001)
36. E. Jespers, J. Okniński, Monoids and groups of I-type. Algebras Represent. Theory **8**, 709–729 (2005)

37. E. Jespers, J. Okniński, Noetherian semigroup algebras. Bull. Lond. Math. Soc. **38**, 421–428 (2006)
38. E. Jespers, J. Okniński, Noetherian semigroup algebras, in *Algebra and Applications*, vol. 7 (Springer, Dordrecht, 2007)
39. E. Jespers, J. Okniński, Krull orders in nilpotent groups. Arch. Math. **103**, 27–37 (2014)
40. E. Jespers, J. Okniński, Prime ideals in algebras determined by submonoids of nilpotent groups, Algebras and Representation Theory, to appear
41. E. Jespers, J. Okniński, M. Van Campenhout, Finitely generated algebras defined by homogeneous quadratic monomial relations and their underlying monoids. J. Algebra **440**, 72–99 (2015)
42. E. Jespers, P.F. Smith, Group rings and maximal orders, in *Methods in Ring Theory* (Antwerp, 1983), pp. 185–195. (Nato Adv. Sci. Inst. Ser. C: Math. Phys. Sci. 129, Reidel, Dordrecht-Boston, 1984)
43. E. Jespers, P.F. Smith, Integral group rings of torsion-free polycyclic-by-finite groups are maximal orders. Commun. Algebra **13**, 669–680 (1985)
44. E. Jespers, Q. Wang, Noetherian unique factorization semigroup algebras. Commun. Algebra **29**, 5701–5715 (2001)
45. E. Jespers, Q. Wang, Height-one prime ideals in semigroup algebras satisfying a polynomial identity. J. Algebra **248**, 118–131 (2002)
46. E. Jespers, P. Wauters, Principal ideal semigroup rings. Commun. Algebra **23**, 5057 5076 (1995)
47. G.R. Krause, T.H. Lenagan, *Growth of Algebras and Gelfand-Kirillov Dimension, Graduate Studies in Mathematics 22* (American Mathematical Society, Providence, Rhode Island, 2000)
48. L. le Bruyn, M. Van den Bergh, F. Van Oystaeyen, *Graded Orders* (Birkhauser, Boston, 1988)
49. E.S. Letzter, M. Lorenz, Polycyclic-by-finite group algebras are catenary. Math. Res. Lett. **6**, 183–194 (1999)
50. J.C. McConnell, J.C. Robson, *Noncommutative Noetherian Rings* (Wiley Interscience, New York, 1987)
51. J. Okniński, Semigroup algebras, in *Pure and Applied Mathematics*, vol. 138 (Marcel Dekker, New York, 1991)
52. J. Okniński, Linear representations of semigroups, in *Proceedings of the Berkeley Workshop on Monoids and Semigroups with Applications* (World Scientific, Singapore, 1991), pp. 257–277
53. J. Okniński, Gelfand-Kirillov dimension of noetherian semigroup algebras. J. Algebra **162**, 302–316 (1993)
54. J. Okniński, *Semigroups of Matrices* (World Scientific, Singapore, 1998)
55. J. Okniński, In search for noetherian algebras, in *Algebra—Representation Theory, NATO ASI* (Kluwer, 2001), pp. 235–247
56. D.S. Passman, Observations on group rings. Commun. Algebra **5**, 1119–1162 (1977)
57. D.S. Passman, *The Algebraic Structure of Group Rings* (Wiley-Interscience, New York, 1977)
58. D.S. Passman, *Infinite Crossed Products* (Academic Press Inc, San Diego, 1989)
59. D. Rogalski, S.J. Sierra, Some noncommutative projective surfaces of GK-dimension 4. Compos. Math. **148**, 1195–1237 (2012)
60. J.E. Roseblade, Prime ideals in group rings of polycyclic groups. Proc. Lond. Math. Soc. **36**(3), 385–447 (1978). (Corrigenda, ibid. 38 (1979), 216–218)
61. D. Segal, *Polycyclic Groups* (Cambridge University Press, Cambridge, 1983)
62. J. Tate, M. Van den Bergh, Homological properties of Sklyanin algebras. Invent. Math. **124**, 619–647 (1996)
63. P.F. Smith, Some examples of maximal orders. Math. Proc. Camb. Philos. Soc. **98**, 19–32 (1985)
64. P. Wauters, On some subsemigroups of noncommutative Krull rings. Commun. Algebra **12**, 1751–1765 (1984)
65. A. Yekutieli, J.J. Zhang, Homological transcendence degree. Proc. Lond. Math. Soc. **93**, 105–137 (2006)

Topological Aspects of Irredundant Intersections of Ideals and Valuation Rings

Bruce Olberding

Abstract An intersection of sets $A = \bigcap_{i \in I} B_i$ is irredundant if no B_i can be omitted from this intersection. We develop a topological approach to irredundance by introducing a notion of a spectral representation, a spectral space whose members are sets that intersect to a given set A and whose topology encodes set membership. We define a notion of a minimal representation and show that for such representations, irredundance is a topological property. We apply this approach to intersections of valuation rings and ideals. In the former case, we focus on Krull-like domains and Prüfer v-multiplication domains, and in the latter on irreducible ideals in arithmetical rings. Some of our main applications are to those rings or ideals that can be represented with a Noetherian subspace of a spectral representation.

Keywords Zariski–Riemann space · Valuation ring · Krull domain · Prüfer domain · Prüfer v-multiplication ring · Spectral space

Mathematics Subject Classification (2010) 13F30 · 13F05 · 13A15

1 Introduction

The goal of this article is to develop a topological framework for recognizing and dealing with an irredundant infinite intersection of ideals, subrings, submodules, even sets. While our main interest here is in the intersection of valuation rings, we include one application to the intersection of irreducible ideals in arithmetical rings to illustrate how the framework applies in a different setting. A key requirement for our point of view is that the objects from which the intersection is formed be drawn from a spectral space whose topology encodes set membership. The Zariski topology on the set of irreducible ideals of an arithmetical ring provides one such

B. Olberding (✉)
Department of Mathematical Sciences,
New Mexico State University, Las Cruces, NM 88003-8001, USA
e-mail: olberdin@nmsu.edu

S. Chapman et al. (eds.), *Multiplicative Ideal Theory and Factorization Theory*,
Springer Proceedings in Mathematics & Statistics 170,
DOI 10.1007/978-3-319-38855-7_12

context, while the inverse topology on the Zariski–Riemann space of valuation rings of a field is another. Several other contexts to which our approach applies, and which we do not pursue, are given in Example 2.2.

Irredundance of intersections of valuation rings is often a consequential and special phenomenon. For example, if F/k is a finitely generated field extension of transcendence degree one with k algebraically closed in F, and $\mathfrak{X}$ is the set of all valuation rings containing k and having quotient field F, then $k = \bigcap_{V \in \mathfrak{X}} V$ and this intersection is irredundant. Thus, $\mathfrak{X}$ is the unique representation of k as an (irredundant) intersection of valuation rings in $\mathfrak{X}$. This is a consequence of Riemann's Theorem for projective curves and is closely related to the strong approximation theorem for such curves [25, Theorem 2.2.13]. If, however, F/k has transcendence degree >1, then k can still be represented by an irredundant intersection of valuation rings (albeit by very specially selected subsets of $\mathfrak{X}$), but there exist infinitely many such representations. Such examples can be constructed along the lines of [41, Example 6.2].

In general, the existence, much less uniqueness, of an irredundant representation of a ring can only be expected under circumstances where "few" valuation rings are needed to represent the ring. For example, Krull domains can all be represented by an irredundant intersection of valuation rings, but this ultimately depends on the fact that they can be represented by a finite character intersection of valuation rings; see Sect. 5. On the other hand, if F/k is a function field in more than one variable and k is existentially, but not algebraically, closed in F, and A is the intersection of all the valuation rings in F/k having residue field k, then no representation X of A as an intersection of valuation rings contains an irredundant member; i.e., any member of X can be omitted and the intersection will remain A; see [43, Theorem 4.7]. This last example is even a Prüfer domain and hence has the property that every valuation ring between A and its quotient field is a localization of A. Thus, even for classes of rings whose valuation theory is explicitly given by their prime spectra, intersections of valuation rings can behave in complicated ways.

In Sect. 3 we develop a topological approach to these issues for intersections of sets, where the sets themselves can be viewed as points in a spectral space. The prime ideals of a ring or the valuation rings of a field comprise such sets when viewed with the appropriate topologies, but also so do the irreducible ideals in an arithmetical ring. Throughout this article we are particularly interested in Noetherian spectral spaces, and in Sect. 2 we work out some of the properties of these spaces when viewed under the inverse or patch topologies. (These topologies are reviewed in Sect. 2.) Krull domains, and generalizations of these rings of classical interest, can be represented by intersections of valuation rings drawn from a Noetherian subspace of a spectral space, and we apply the results from Sects. 2 and 3 in Sects. 5 and 6 to intersections of valuation rings from a Noetherian subspace of the Zariski–Riemann space of a field.

Sections 4–7 contain the main applications of the article. Section 4 applies the abstract setting of spectral representations to the Zariski–Riemann space of a field. This section recasts the abstract approach in Sect. 3 into a topological framework for working with irredundance in intersections of valuation rings. Section 5 specializes the discussion to the Krull-like rings of classical interest and recaptures the representation theorems for these rings. A feature throughout Sects. 4 and 5 that is afforded by the abstract approach of spectral representations is that intersections of valuation rings can be considered relative to a subset of the ambient field. The motivation for this comes from the articles [1, 30, 38, 42, 46]. In these studies, one considers integrally closed domains A between a given domain and overring, e.g., between $\mathbb{Z}[T]$ and $\mathbb{Q}[T]$. In such cases A is an intersection of $\mathbb{Q}[T]$ and valuation rings not containing $\mathbb{Q}[T]$. Since $\mathbb{Q}[T]$ can be viewed as always present in these representations, it is helpful then to consider representations of a ring A of the form $A = (\bigcap_{V \in X} V) \cap C$, where C is a fixed ring. The approach provided by Sect. 3 makes it easy to incorporate a fixed member C of the representation into such a picture, regardless of whether C is a ring or simply a set.

The already well-understood theory of irredundance for Prüfer domains also can be recovered from our framework, and this is done in Sect. 6 in the more general setting of Prüfer v-multiplication domains. We consider existence and uniqueness for irredundant representations of such domains, with special emphasis on the case in which the space of t-maximal ideals is Noetherian. When restricted to a Prüfer domain A, these results specialize to a topological characterization of the property that every overring of A is an irredundant intersection of the valuation rings that are minimal over it.

In order to help justify the generality of the approach Sect. 3, we show in Sect. 7 how the topological framework can be applied to the study of irredundant intersections of irreducible ideals in arithmetical rings. We show in particular how intersection decomposition results involving such ideals can be recovered from our point of view. This section is independent of the valuation-theoretic Sects. 4–6.

I thank the referee for helpful comments that improved the clarity of some of the arguments.

2 Spectral Spaces

A *spectral space* is a T_0 topological space having (a) a basis of quasicompact open sets closed under finite intersections, and (b) the property that every irreducible closed subset has a unique generic point, i.e., a point whose closure is the irreducible closed set. By a theorem of Hochster [33, Corollary, p. 45], a topological space X is spectral if and only if X is homeomorphic to the prime spectrum of a ring. In the setting of this paper, it is mostly the topological features of spectral spaces that are needed rather than the connection with prime spectra of rings.

A spectral space X admits two other well-studied topologies that are useful in our context. The *inverse topology* on X has as a basis of closed sets the subsets of X that are quasicompact and open in the spectral topology. By an *inverse closed* subset of X we mean a subset that is closed in the inverse topology. The *patch topology* has as a basis of open sets the sets of the form $U \cup V$, where U is open and quasicompact in the spectral topology and V is the complement of a quasicompact open set. These basic open sets are also closed, so that the patch topology is zero-dimensional and Hausdorff. A *patch* in X is a set that is closed in the patch topology. In this section, we denote the closure of a subset Y of X in the spectral topology as $\overline{Y}$, and the closure of Y in the patch topology as $\widetilde{Y}$.

The patch topology refines both the spectral and inverse topologies. This can be made more precise using the *specialization order* of the spectral topology: If $x, y \in X$, then $x \leq y$ if and only if $y \in \overline{\{x\}}$ in the spectral topology. With this order in mind, we define for $Y \subseteq X$,

$$\begin{aligned}&\uparrow Y = \{x \in X : x \geq y \text{ some } y \in Y\} \text{ and } \downarrow Y = \{x \in X : x \leq y \text{ some } y \in Y\},\\&\text{Min } Y = \{y \in Y : y \text{ is minimal in } Y \text{ with respect to } \leq\},\\&\text{Max } Y = \{y \in Y : y \text{ is maximal in } Y \text{ with respect to } \leq\}.\end{aligned}$$

Proposition 2.1 *Let X be a spectral space with specialization order $\leq$. Then*

(1) *X with the inverse topology is a spectral space whose specialization order is the reverse of that of $(X, \leq)$.*
(2) *X with the patch topology is a spectral space, and in particular a compact Hausdorff zero-dimensional space.*
(3) *For each $Y \subseteq X$, $\overline{Y} = \uparrow(\widetilde{Y})$ and the closure of Y in the inverse topology is $\downarrow(\widetilde{Y})$.*
(4) *If Y is a patch in X, then the following statements hold.*
    (a) *Y is spectral in the subspace topology.*
    (b) *For each $y \in Y$ there exists $m \in \text{Min } Y$ with $m \leq y$.*
    (c) *The patch and spectral topologies agree on* Min Y.

Proof Statement (1) can be found in [33, Proposition 8]; statement (2) can be deduced from [33, Sect. 2]. Statement (3) is a consequence of [33, Corollary, p. 45] and (1). Statement (4)(a) follows from [33, Proposition 9]. Statement (4)(b) now follows from (a), since a spectral space has minimal elements. Finally, the spectral and patch topologies agree on the set of minimal elements of a spectral space [49, Corollary 2.6], so (4)(c) follows from (4)(a). □

We give now a list of examples of some spectral spaces in our context. We only use a few of these examples in what follows, but the intersection representation theory developed in the next section applies to all of them. As we indicate, several of these examples have appeared in the literature before, but with different proofs than what we give here. Our approach is inspired by a theorem of Hochster [33, Proposition 9] that a topological space is spectral if and only if it is homeomorphic to a patch closed subset of a power set endowed with the hull-kernel topology. Interestingly, inspection of Zariski and Samuel's proof in [51] that the Zariski–Riemann space $\mathfrak{X}$ of a field is quasicompact shows that although their work predated the notion of spectral spaces, what is proved there is that $\mathfrak{X}$ is a patch closed subset of a certain spectral space, and hence from their argument can be deduced the fact that $\mathfrak{X}$ is spectral.

To formalize the setting of the example, let S be a set. We denote by $\mathbf{2}^S$ the power set of S endowed with the hull-kernel topology having as an open basis the sets of the form $\mathscr{U}(F) := \{B \subseteq S : F \not\subseteq B\}$, where F is a finite subset of S. The complement of $\mathscr{U}(F)$ is denoted $\mathscr{V}(F)$; i.e., $\mathscr{V}(F) = \{V \subseteq S : F \subseteq B\}$. Then the sets $\mathscr{U}(F)$ are quasicompact and $\mathbf{2}^S$ is a spectral space; cf. [33, Theorem 8 and Proposition 9]. Thus by Proposition 2.1, to show that a collection X of subsets of S is a spectral space in the subspace topology, it is enough to show that X is patch closed in 2^S. Specifically, what must be shown is that X is an intersection of sets of the form $\mathscr{V}(F_1) \cup \cdots \cup \mathscr{V}(F_n) \cup \mathscr{U}(G)$, where $F_1, \ldots, F_n, G$ are finite subsets of S. This is done in each case by encoding the question of whether a given subset of S satisfies a first-order property in the relevant language into an assertion about membership in a set of the form $\mathscr{V}(F_1) \cup \cdots \cup \mathscr{V}(F_n) \cup \mathscr{U}(G)$. This amounts in most cases to rewriting a statement of the form "$p \to q$" as "(not p) or q." Because the goal is to produce patch closed subsets, statements involving universal quantifiers (which translate into intersections) are more amenable to this approach than statements involving existential quantifiers (which translate into infinite unions).

To clarify terminology, when R is a ring, the *Zariski topology* on a collection X of ideals of R is the hull-kernel topology defined above; i.e., it is simply the subspace topology on X inherited from $\mathbf{2}^R$. This agrees with the usual notion of the Zariski topology on Spec R. However, when S is a ring and and X is a collection of subrings of S, then the *Zariski topology* on X is the inverse of the hull-kernel topology; i.e., it has an open basis consisting of sets of the form $\mathscr{V}(G)$, where G is a finite subset of S. Despite the discrepancy, it is natural to maintain it in light of the fact that when R is a subring of a field F, then with these definitions, Spec R with the Zariski topology is homeomorphic to the space $\{R_P : P \in \operatorname{Spec} R\}$ of subrings of F with the Zariski topology. This discrepancy, which is due to Zariski, also allows for an identification between projective models and projective schemes; cf. [51] for the notion of a projective model.

Example 2.2 (1) *The set of all proper ideals of a ring R is a spectral space in the Zariski topology.* The set of proper ideals in R is precisely the patch closed subset of $\mathbf{2}^R$ given by

$$X_1 = \mathscr{U}(1) \cap \left(\bigcap_{a,b \in R} \mathscr{U}(a, b) \cup \mathscr{V}(a + b) \right) \cap \left(\bigcap_{a,r \in R} \mathscr{U}(a) \cup \mathscr{V}(ra) \right).$$

(2) *If R is a ring, the set of all submodules of an R-module is a spectral space in the Zariski topology.* An easy modification of (1) shows this to be the case.

(3) *The set of all radical ideals of a ring R is a spectral space in the Zariski topology.* The set of radical ideals is precisely the patch closed subset of $\mathbf{2}^R$ given by

$$X_3 = X_1 \cap \left(\bigcap_{a \in R, n > 0} (\mathscr{U}(a^n) \cup \mathscr{V}(a)) \right).$$

(4) *If R is a ring such that $aR \cap bR$ is a finitely generated ideal of R for all $a, b \in R$, then the set of all proper strongly irreducible ideals is a spectral space in the Zariski*

topology. Recall that an ideal I of R is strongly irreducible if whenever $J \cap K \subseteq I$, then $J \subseteq I$ or $K \subseteq I$; equivalently, I is strongly irreducible if and only if whenever $a, b \in R$ and $aR \cap bR \subseteq I$, it must be that $a \in I$ or $b \in I$. Thus, the set of strongly irreducible proper ideals in R is given by

$$X_4 = X_1 \cap \left(\bigcap_{a,b \in R} (\mathscr{U}(aR \cap bR) \cup \mathscr{V}(aR) \cup \mathscr{V}(bR)) \right).$$

By assumption, for each $a, b \in R$, $aR \cap bR$ is a finitely generated ideal of R, so the set $\mathscr{U}(aR \cap bR)$ is quasicompact and open. Therefore, X_4 is a patch closed subset of $\mathbf{2}^R$. This example will be used in Sect. 7.

(5) (Finocchiaro [8, Proposition 3.5]) *Let $R \subseteq S$ be an extension of rings. The set of rings between R and S with the Zariski topology is a spectral space.* The set of rings between R and S is given by the patch closed set

$$X_5 = \left(\bigcap_{r \in R} \mathscr{V}(r) \right) \cap \left(\bigcap_{a,b \in S} \mathscr{U}(a, b) \cup \mathscr{V}(a + b, ab) \right).$$

The Zariski topology on X_5 is the inverse topology of the subspace topology on X_5 inherited from $\mathbf{2}^S$, so by Proposition 2.1, X_5 is spectral in the Zariski topology.

(6) (Finocchiaro [8, Proposition 3.6]) *Let $R \subseteq S$ be an extension of rings. The set of all integrally closed rings between R and S with the Zariski topology is a spectral space.* Let $\mathscr{M}$ denote the set of monic polynomials in $S[T]$, and for each $f \in \mathscr{M}$, let $c(f)$ denote the set of coefficients of f. The set of integrally closed rings between R and S is given by the patch closed set

$$X_6 = X_5 \cap \left(\bigcap_{s \in S} \left(\bigcap_{f \in \mathscr{M}, f(s)=0} \mathscr{U}(c(f)) \cup \mathscr{V}(s) \right) \right).$$

As in (5), this implies that X_7 is spectral in the Zariski topology.

(7) (Finocchiaro–Fontana–Spirito [12, Corollary 2.14]) *Let R be a subring of a field F. The set of all local rings between R and F with the Zariski topology is a spectral space.* A ring A between R and F is local if whenever a, b are nonzero elements of R with $1/(a + b) \in R$, we have $1/a \in R$ or $1/b \in R$. Thus, the set of all local rings between R and F is given by the patch closed subset

$$X_7 = X_5 \cap \left(\bigcap_{0 \neq a,b \in F} \mathscr{U}(a, b, 1/(a + b)) \cup \mathscr{V}(1/a) \cup \mathscr{V}(1/b) \right).$$

As in (5), this implies that X_7 is spectral in the Zariski topology.

(8) *Let A be a subring of a field F. The set of all valuation rings containing A and having quotient field F is a spectral space in the Zariski topology.* This has been

proved by a number of authors; see [9, 45] for discussion and references regarding this result. A subring V between A and F is a valuation ring with quotient field F if and only if for all $0 \neq q \in F$, $q \in V$ or $q^{-1} \in V$. Thus with $R = A$ and $S = F$, we use the set X_5 from (5) to obtain the set of valuation rings of F containing A as the patch closed subset of $\mathbf{2}^F$ given by

$$X_8 = X_5 \cap \bigcap_{0 \neq q \in F} \mathscr{V}(q) \cup \mathscr{V}(q^{-1}).$$

As in (5), this implies that X_8 is spectral in the Zariski topology.

For the remainder of the section we focus on Noetherian spectral spaces, since these play a central role in later sections. A topological space is *Noetherian* if its open sets satisfy the ascending chain condition. Rush and Wallace [48, Proposition 1.1 and Corollary 1.3] have shown that a collection Y of prime ideals of a ring R is a Noetherian subspace of Spec R if and only if for each prime ideal P of R, there is a finitely generated ideal $I \subseteq P$ such that every prime ideal in Y containing I contains also P. Since every spectral space can be realized as Spec R for some ring R, we may restate this topologically in the following form.

Lemma 2.3 (Rush and Wallace) *Let X be a spectral space, and let Y be a subspace of X. Then Y is Noetherian if and only if for each irreducible closed subset C of X, $Y \cap C = Y \cap C'$ for some closed subset $C' \supseteq C$ such that $X \setminus C'$ is quasicompact.*

In later sections, we focus on spectral spaces X in which the set of maximal elements under the specialization order $\leq$ of X is a Noetherian space. The spectral spaces in our applications have the additional property that $(X, \leq)$ is a tree. In this case, as we show in Theorem 2.5, the Noetherian property for the maximal elements descends to subsets consisting of incomparable elements.

Lemma 2.4 *Let X be a spectral space whose specialization order $\leq$ is a tree. Suppose that* Max X *is a Noetherian space. Then, a subspace Y of X is Noetherian if and only if $(Y, \leq)$ satisfies the ascending chain condition.*

Proof If Y is Noetherian, then the closed subsets of Y satisfy the descending chain condition, so $(Y, \leq)$ satisfies the ascending chain condition. Conversely, suppose that $(Y, \leq)$ satisfies ACC. Let C be an irreducible closed subset of X. By Lemma 2.3, to prove that Y is Noetherian, it suffices to show that there exists a closed subset $C' \supseteq C$ such that $Y \cap C = Y \cap C'$ and $X \setminus C'$ is quasicompact. By Lemma 2.3, there exists a closed subset $C_1 \supseteq C$ such that $C \cap \operatorname{Max} X = C_1 \cap \operatorname{Max} X$ and $X \setminus C_1$ is quasicompact. Since C is irreducible, there is $c \in C$ such that $C = \{x \in X : c \leq x\}$. Let $D = \bigcap_{y<c, y \in Y} \overline{\{y\}}$. Since $(X, \leq)$ is a tree and $(Y, \leq)$ satisfies ACC, C is a proper subset of D. Thus since the quasicompact open subsets of X form a basis for X, there is a closed set C_2 such that $C \subseteq C_2$, $D \nsubseteq C_2$ and $X \setminus C_2$ is quasicompact. We claim that $Y \cap C = Y \cap C_1 \cap C_2$. The containment "$\subseteq$" is clear since $C \subseteq C_1 \cap C_2$. Let $y \in Y \cap C_1 \cap C_2$. Then there exists $m \in \operatorname{Max} X$ such that $y \leq m$. Thus $m \in \operatorname{Max} X \cap$

$C_1 \cap C_2 \subseteq C$, so that $c \leq m$. Since $(X, \leq)$ is a tree and $y, c \leq m$, it must be that $y < c$ or $c \leq y$. If $y < c$, then $C \subsetneq D \subseteq \overline{\{y\}}$. However, since $y \in C_2$, this forces $D \subseteq C_2$, a contradiction. Thus $c \leq y$, and hence $y \in C$. This shows that $Y \cap C = Y \cap C_1 \cap C_2$. Finally, since $X \setminus C_1$ and $X \setminus C_2$ are quasicompact, so is their union $X \setminus (C_1 \cap C_2)$. Thus with $C' = C_1 \cap C_2$ the claim is proved. □

Theorem 2.5 *Let X be a spectral space whose specialization order ≤ is a tree. Then* Max X *is a Noetherian space in the spectral topology if and only if every subset of X consisting of elements that are incomparable under ≤ is discrete in the inverse topology.*

Proof Suppose Max X is a Noetherian space. Let Y be a nonempty subset of X whose elements are incomparable under $\leq$. Then by Lemma 2.4, Y is a Noetherian space. Let $y \in Y$. Since the elements of Y are incomparable, $Y \setminus \{y\}$ is open in Y. In a Noetherian space, open sets are quasicompact, so $Y \setminus \{y\}$ is inverse closed in Y, which proves that y is isolated in the inverse topology on Y.

Conversely, suppose that every subspace of X consisting of incomparable elements is discrete in the inverse topology. To prove that Max X is Noetherian, it suffices by Lemma 2.3 to show that for each irreducible closed subset C of X there exists a closed set $C' \supseteq C$ such that $C \cap \text{Max}\, X = C' \cap \text{Max}\, X$ and $X \setminus C'$ is quasicompact. Let C be an irreducible closed subset of X, and let $c \in C$ such that $C = \{x \in X : c \leq x\}$. By assumption, $\{c\} \cup ((\text{Max}\, X) \setminus C)$ is discrete in the inverse topology since the elements in this set are incomparable. Thus, there exists a quasicompact open subset U of X such that $(\text{Max}\, X) \setminus C \subseteq U$ and $c \notin U$. Since $c \notin U$ and U, being open, has the property that $U = \downarrow U$, it must be that $C \cap U = \emptyset$. Thus $U \subseteq X \setminus C$, so that $(\text{Max}\, X) \cap U \subseteq (\text{Max}\, X) \setminus C$. Since also $(\text{Max}\, X) \setminus C \subseteq U$, we conclude $(\text{Max}\, X) \setminus C = (\text{Max}\, X) \cap U$. Thus with $C' = X \setminus U$, we have $(\text{Max}\, X) \cap C = (\text{Max}\, X) \cap C'$ and $X \setminus C'$ is quasicompact. Since also $C \subseteq X \setminus U = C'$, the claim is proved. □

Corollary 2.6 *Let X be a spectral space whose specialization order ≤ is a tree. If* Max X *is a Noetherian space, then* Min C *is finite for each nonempty closed subset C of X.*

Proof Let C be a nonempty closed subset of X. By Proposition 2.1(1), Min C consists of the elements that are maximal with respect to the specialization order in the inverse topology. The set of maximal elements of a spectral space is a quasicompact subspace, so Min C is quasicompact in the inverse topology. Since by Theorem 2.5, Min C is discrete in the inverse topology, Min C is finite. □

3 Spectral Representations

Throughout this section A, C, and D are nonempty sets with $A \subsetneq C \subseteq D$. We do not assume the presence of any algebraic structure on these sets. We work under the assumption that A can be represented as an intersection of C with sets B between A

and D such that the sets B are points in a spectral space X whose specialization order is compatible with set inclusion. The set C can be viewed as a fixed component in the intersection (e.g., the case $C = D$ is often an interesting choice). The goal then is to make "efficient" choices from X to represent A. We formalize some of this with the following definition.

Definition 3.1 Let X be a collection of subsets of D, and assume that X is a *C-representation of A*, meaning that $A = \left(\bigcap_{B\in X} B\right) \cap C$.

(1) For each $F \subseteq D$, let $\mathcal{V}(F) = \{B \in X : F \subseteq B\}$ and $\mathcal{U}(F) = \{B \in X : F \nsubseteq B\}$. We say the C-representation X is *spectral* if X is a spectral space and $\{\mathcal{U}(d) : d \in D\}$ is a subbasis for X consisting of quasicompact open sets. Note that this choice of subbasis assures that the specialization order agrees with the partial order given by set inclusion.

Now assume that X is a spectral C-representation of A.

(2) Let $Z \subseteq X$ be a C-representation of A, and let $B \in Z$. Then B is *irredundant* in Z if $Z \setminus \{B\}$ is not a C-representation of A; B is *strongly irredundant* in Z if the only closed subset Y of $\mathcal{V}(B)$ such that $(Z \setminus \{B\}) \cup Y$ is a C-representation of A is $Y = \mathcal{V}(B)$; B is *tightly irredundant* in Z if $(Z \cup \mathcal{V}(B)) \setminus \{B\}$ is not a C-representation of A.
(3) A closed subset Y of X (resp., a patch) that is minimal with respect to set inclusion among closed (resp., patch) C-representations of A in X is a *minimal closed (resp., patch) C-representation of A in X*.
(4) A subspace of X of the form Min Y for some minimal closed C-representation Y of A is a *minimal C-representation of A*.

The notions of strong and tight irredundance become much clearer in the settings of Sects. 4–7; see the discussion after Definition 4.1.

Observe that in (4), since the specialization order on X agrees with the partial order given by set inclusion among the members of X, Min Y is also the minimal elements of Y with respect to set inclusion.

Lemma 3.2 *Every spectral C-representation of A contains a minimal closed C-representation of A and a minimal patch C-representation of A.*

Proof Let X be a spectral C-representation of A, and let $\mathcal{F}$ be the set of closed C-representations of A in X. Then $\mathcal{F}$ is nonempty since $X \in \mathcal{F}$. Let $\{Y_\alpha\}$ be a chain of elements in $\mathcal{F}$, and let $Y = \bigcap_\alpha Y_\alpha$. As an intersection of closed subsets, Y is closed. We claim that Y is a C-representation of A. Clearly, $A \subseteq (\bigcap_{B\in Y} B) \cap C$. Let $d \in (\bigcap_{B\in Y} B) \cap C$. Then $\bigcap_\alpha Y_\alpha = Y \subseteq \mathcal{V}(d)$, and hence $\mathcal{U}(d) \subseteq \bigcup_\alpha Y_\alpha^c$, where $Y_\alpha^c = X \setminus Y_\alpha$. Since $\mathcal{U}(d)$ is quasicompact and each Y_α^c is open, the fact that the Y_α form a chain under inclusion implies that $\mathcal{U}(d) \subseteq Y_\alpha^c$ for some α. For this choice of α, $Y_\alpha \subseteq \mathcal{V}(d)$, which since Y_α is a C-representation of A implies that $d \in (\bigcap_{B\in Y_\alpha} B) \cap C = A$. Therefore, $A = (\bigcap_{B\in Y} B) \cap C$, which shows that $Y \in \mathcal{F}$. By Zorn's Lemma, $\mathcal{F}$ contains minimal elements. Since the patch topology is spectral by Proposition 2.1(2), the final statement follows from the first. □

Lemma 3.3 *Let X be a spectral C-representation of A, let $Z \subseteq X$ be a C-representation of A and let $B \in Z$. Then*

(1) *B is irredundant in Z if and only if B is irredundant in $\widetilde{Z}$.*
(2) *B is tightly irredundant in Z if and only if B is irredundant in $\overline{Z}$.*

Moreover, if B is irredundant in Z, then B is isolated in the spectral and patch subspace topologies on Z.

Proof (1) Suppose that B is irredundant in Z. Then there exists $d \in D$ such that $d \notin B$ but d is in every other set that is in Z. Thus $Z \subseteq \mathscr{V}(d) \cup \{B\}$, and since $\mathscr{V}(d)$ and $\{B\}$ are patches in X, we have $\widetilde{Z} \subseteq \mathscr{V}(d) \cup \{B\}$. Hence $\widetilde{Z} \setminus \{B\} \subseteq \mathscr{V}(d)$. Since $d \notin B$, this implies that B is irredundant in $\widetilde{Z}$. The converse is clear since $Z \subseteq \widetilde{Z}$.

(2) Suppose that V is tightly irredundant in Z. Then, there exists $d \in D$ such that d is not in B but d is in every any other set in $Z \cup \mathscr{V}(B)$. Thus $(Z \cup \mathscr{V}(B)) \setminus \{B\} \subseteq \mathscr{V}(d)$, which implies that $\mathscr{V}(d) \cup \mathscr{V}(B) = \mathscr{V}(d) \cup \{B\}$. Since $\mathscr{V}(d) \cup \mathscr{V}(B)$ is closed, we have $\overline{Z} \subseteq \mathscr{V}(d) \cup \mathscr{V}(B) = \mathscr{V}(d) \cup \{B\}$. Therefore, $\overline{Z} \setminus \{B\} \subseteq \mathscr{V}(d)$, and hence d is in every set in $\overline{Z} \setminus \{B\}$. Since $d \notin B$, we conclude that B is irredundant in $\overline{Z}$. Conversely, if B is irredundant in $\overline{Z}$, then since $\mathscr{V}(B) \subseteq \overline{Z}$, it follows that B is tightly irredundant in Z.

It remains to prove the last statement of the lemma. Suppose that B is irredundant in Z. Let $Z' = Z \setminus \{B\}$. Then there exists $c \in (\bigcap_{B' \in Z'} B') \cap C$ with $c \notin B$. Since the set $\mathscr{V}(c)$ is closed in the spectral and patch topologies and this set contains Z' but not B, we have that B is an isolated point in the spectral and patch subspace topologies on Z. □

Proposition 3.4 *Let X be a spectral C-representation of A. If $Z \subseteq X$ is a tightly irredundant C-representation of A, then Z is contained in a minimal C-representation of A. Thus the number of minimal C-representations of A is greater than the number of tightly irredundant C-representations of A.*

Proof By Lemma 3.3, the members of Z are irredundant in the C-representation $\overline{Z}$ of A. By Lemma 3.2, there exists a minimal C-representation Z_1 contained in $\overline{Z}$. Since the members of Z are irredundant in $\overline{Z}$, it must be that $Z \subseteq Z_1$. To prove the last claim of the proposition, it suffices to show that distinct irredundant C-representations of A are contained in distinct minimal C-representations of A. Suppose $Y \subseteq X$ is another tightly irredundant C-representation of A with $Y \neq Z$. Then the members of Y are irredundant in a minimal C-representation Y_1 of A. If $Y_1 = Z_1$, then the members of Y and Z are irredundant in Y_1, which, since Y and Z are C-representations of A, implies that $Y = Z$, a contradiction that implies $Y_1 \neq Z_1$. □

Example 4.3 shows that neither tightly irredundant nor minimal representations need be unique.

Lemma 3.5 *Let X be a spectral C-representation of A, and let Z be a minimal C-representation of A in X. Then $\overline{Z}$ is a minimal closed C-representation of A, $\widetilde{Z}$ is a minimal patch C-representation of A, $Z = \text{Min } \widetilde{Z} = \text{Min } \overline{Z}$ and $\overline{Z} = {\uparrow} Z$.*

Proof Since Z is a minimal C-representation, there exists a minimal closed C-representation Y of A such that $Z = \text{Min } Y$. Since $Z \subseteq Y$ and Y is closed, $\overline{Z} \subseteq Y$. Thus the minimality of Y forces $\overline{Z} = Y$, and hence $\overline{Z}$ is a minimal closed C-representation of A. Also, note that this implies that $\text{Min } \overline{Z} = \text{Min } Y = Z$.

Suppose that Y is a patch C-representation of A with $Y \subseteq \widetilde{Z}$. We claim that $Y = \widetilde{Z}$, and we prove this by first showing that $Z \subseteq Y$. By Proposition 2.1(3), $\uparrow Y$ is a closed C-representation of A. Also by Proposition 2.1(3), $\uparrow Y \subseteq \uparrow(\widetilde{Z}) = \overline{Z}$, so that the minimality of $\overline{Z}$ forces $\uparrow Y = \overline{Z}$. Since $\overline{Z}$ is closed, $\uparrow(\text{Min } \overline{Z}) = \overline{Z} = \uparrow Y$. We have established that $Z = \text{Min } \overline{Z}$, so $\uparrow Y = \uparrow Z$. Now $\text{Min} \uparrow Y = \text{Min } Y$, and since Z consists of pairwise incomparable elements, $\text{Min} \uparrow Z = Z$. Thus $\text{Min } Y = Z$, so that $Z \subseteq Y$. Since Y is a patch, $\widetilde{Z} \subseteq Y$. Therefore, $Y = \widetilde{Z}$. Since $\text{Min } Y = Z$, we conclude from $\widetilde{Z} = Y$ that $\text{Min } \widetilde{Z} = \text{Min } Y = Z$. All that remains is to show that $\overline{Z} = \uparrow Z$. Since $\text{Min } \overline{Z} = Z$, it follows that $\overline{Z} = \uparrow \text{Min } \overline{Z} = \uparrow Z$. □

We obtain now a topological characterization of irredundance in minimal C-representations. Example 4.6 shows that without the restriction to minimal representations, irredundance may not have a similar topological expression.

Theorem 3.6 *Let X be a spectral C-representation of A, and let Z be a minimal C-representation of A in X. Then the spectral and patch subspace topologies agree on Z, and the following are equivalent for $B \in Z$.*

(1) *B is irredundant in Z.*
(2) *B is strongly irredundant in Z.*
(3) *B is isolated in the spectral (equivalently, patch) subspace topology on Z*

Proof By assumption, there exist a minimal closed C-representation Y of A such that $Z = \text{Min } Y$. Since Y is a patch in X, we have by Proposition 2.1(4)(a) that Y is spectral in the spectral subspace topology. Also by Proposition 2.1(4)(c), the spectral topology on the minimal points of a spectral space is the same as the patch topology. Thus, the spectral and patch topologies agree on Z.

That (2) implies (1) is clear, and that (1) implies (3) follows from Lemma 3.3. It remains to prove that (3) implies (2). Suppose B is isolated in Z, so that $B \notin \overline{Z \setminus \{B\}}$. To see that B is strongly irredundant in Z, let F be a closed subset of $\mathscr{V}(B)$ such that $(Z \setminus \{B\}) \cup F$ is a C-representation of A. Now $F \subseteq \mathscr{V}(B) \subseteq \overline{Z}$, so $Z' := \overline{(Z \setminus \{B\})} \cup F$ is a closed C-representation of A contained in $\overline{Z}$. By Lemma 3.5, $\overline{Z}$ is a minimal closed C-representation of A, so $Z' = \overline{Z}$, and hence $B \in Z'$. Since $B \notin \overline{Z \setminus \{B\}}$, we conclude that $B \in F$ and hence $F = \mathscr{V}(B)$. Thus B is strongly irredundant in Z. □

Corollary 3.7 *Let X be a spectral C-representation of A, and let Z be a minimal C-representation of A in X. Then Z contains a (strongly) irredundant C-representation of A if and only if the set of isolated points in Z is dense in Z. Hence there is at most one irredundant C-representation of A in $\widetilde{Z}$.*

Proof By Theorem 3.6 the patch and spectral topologies agree on Z, so in the proof we work exclusively in the patch topology. Suppose Z is a minimal C-representation of A in X that contains an irredundant C-representation Y of A. Since Y is a C-representation of A, so is $\widetilde{Y}$. Thus since by Lemma 3.5, $\widetilde{Z}$ is a minimal patch C-representation of A, we have $\widetilde{Y} = \widetilde{Z}$. Therefore, Y is dense in Z. Moreover, by Lemma 3.3, each member of Y is irredundant in $\widetilde{Y} = \widetilde{Z}$, hence in Z. Therefore, by Theorem 3.6, each member of Y is isolated in Z.

Conversely, suppose that the set Y of isolated points in Z is dense in Z. If Y is not a C-representation of A, then there exists $B' \in Z$ and $c \in (\bigcap_{B \in Y} B) \cap C$ with $c \notin B'$. Thus $Y \subseteq \mathscr{V}(c)$ and $B' \notin \mathscr{V}(c)$, so that $\mathscr{U}(c) \cap Y = \emptyset$ while $\mathscr{U}(c) \cap Z \neq \emptyset$, a contradiction to the assumption that Y is dense in Z. Therefore, Y is a C-representation of A, and hence by Theorem 3.6, the members of Y are strongly irredundant in the C-representation Z.

To prove the last claim of the corollary, suppose Y is an irredundant C-representation of A in $\widetilde{Z}$. By Lemma 3.3, the elements of Y are isolated points in $\widetilde{Z}$ with respect to the patch topology. Thus for each $y \in Y$, $\{y\}$ is open in $\widetilde{Z}$, so that since Z is dense in $\widetilde{Z}$, we must have $y \in Z$. Therefore, $Y \subseteq Z$. Since Y is an irredundant C-representation of A, Theorem 3.6 implies that Y is the set of isolated points of Z. Hence ,there is at most one irredundant C-representation of A in $\widetilde{Z}$, namely the set of isolated points of Z. □

A topological space X is *scattered* if every nonempty subspace Y of X contains a point that is isolated in Y; equivalently, in every nonempty subset Y of X the set of isolated points in Y is dense in Y.

Corollary 3.8 *Let X be a spectral C-representation of A. If X is scattered in the spectral or patch topologies, then X contains a strongly irredundant C-representation of A.*

Proof Since the patch topology refines the spectral topology, to be scattered in the spectral topology implies the space is scattered in the patch topology. Thus, we assume that X is scattered in the patch topology. Let Z be a minimal C-representation of A. Since X is scattered, the set of isolated points in Z is dense in Z with respect to the patch topology, so by Corollary 3.7, Z contains a strongly irredundant C-representation of A. □

Corollary 3.9 *If X is a countable spectral C-representation of A, then X contains a strongly irredundant C-representation of A.*

Proof Since X is spectral, the patch topology on X is compact and Hausdorff. A countable compact Hausdorff space is homeomorphic to an ordinal space [39], and hence scattered since an ordinal space is well ordered. Thus X is scattered in the patch topology, and by Corollary 3.8, X contains a strongly irredundant C-representation of A. □

We single out next the members of X that must appear in every closed C-representation of A. These members play an important role in the applications to intersections of valuation rings in Sects. 5 and 6.

Definition 3.10 Let X be a spectral C-representation of A. Let Y be the intersection of all closed C-representations of A in X (so that Y is a closed set in X, but not necessarily a C-representation of A). An element $B \in X$ is *critical* for the C-representation X if $B \in Y$. Since Y is a closed subset of the spectral space X, Y contains minimal elements. We define $\mathscr{C}(X) = \mathrm{Min}\, Y$.

Proposition 3.11 *Let X be a spectral C-representation of A. Then B is critical in X if and only if whenever $A = A_1 \cap \cdots \cap A_n \cap C$, where each A_i is an intersection of members of X, it must be that $A_i \subseteq B$ for some i.*

Proof Suppose B is critical in X and $A = A_1 \cap \cdots \cap A_n \cap C$, where each A_i is an intersection of members of X. Then $\mathscr{V}(A_1) \cup \cdots \cup \mathscr{V}(A_n)$ is a closed C-representation of A in X. Since B is critical in X, $B \in \mathscr{V}(A_i)$ for some i, and hence $A_i \subseteq B$. Conversely, suppose that whenever $A = A_1 \cap \cdots \cap A_n \cap C$, where each A_i is an intersection of members of X, we have $A_i \subseteq B$ for some i. Let Y be a closed C-representation of A in X. Since Y is closed and X is spectral, Y is an intersection of sets of the form $\mathscr{V}(F_1) \cup \cdots \cup \mathscr{V}(F_n)$, where each F_i is a finite subset of C. Thus to show that $B \in Y$ it suffices to show B is in every set of this form that contains Y. Let $F_1, \ldots, F_n$ be finite subsets of C such that $Y \subseteq \mathscr{V}(F_1) \cup \cdots \cup \mathscr{V}(F_n)$. For each i, let $A_i = \bigcap_{E \in \mathscr{V}(F_i)} E$. Since Y is a C-representation of A, we have $A = A_1 \cap \cdots \cap A_n \cap C$. By assumption, $A_i \subseteq B$ for some i, so $F_i \subseteq B$. Thus $B \in V(F_i)$, which proves the proposition. □

The next corollary shows that for critical members of X, being irredundant in a representation is the same as being strongly irredundant.

Corollary 3.12 *Let X be a spectral C-representation of A. If $B \in X$ is critical in X and B is irredundant in some C-representation Z of A in X, then B is strongly irredundant in Z.*

Proof Suppose there is a nonempty closed subset Y of $\mathscr{V}(B)$ such that $(Z \setminus \{B\}) \cup Y$ is a C-representation of A. We claim that $B \in Y$. Since Y is closed, Y is an intersection of sets of the form $\mathscr{V}(F_1) \cup \cdots \cup \mathscr{V}(F_n)$, where $F_1, \ldots, F_n$ are finite subsets of D. Let $F_1, \ldots, F_n$ be finite subsets of D such that $Y \subseteq \mathscr{V}(F_1) \cup \cdots \cup \mathscr{V}(F_n)$. Then

$$A = \left(\bigcap_{B' \in \mathscr{V}(F_1)} B' \right) \cap \cdots \cap \left(\bigcap_{B' \in \mathscr{V}(F_n)} B' \right) \cap \left(\bigcap_{B' \in Z \setminus \{B\}} B' \right) \cap C \subseteq B.$$

Since B is irredunant in Z, we have $\bigcap_{B' \in Z \setminus \{B\}} B' \not\subseteq B$. Thus since B is critical in X, we conclude that $F_i \subseteq \bigcap_{B' \in \mathscr{V}(F_i)} B' \subseteq B$ for some i. Hence $B \in \mathscr{V}(F_1) \cup \cdots \cup \mathscr{V}(F_n)$, which shows that $B \in Y$. □

Next we prove a uniqueness theorem for strongly irredundant C-representations when $\mathscr{C}(X)$ has enough members to be itself a C-representation of A. This case is important in Sect. 6, where we work with v-domains, a class of rings that can be represented as an intersection of their critical valuation overrings.

Theorem 3.13 *Let X be a spectral C-representation of A. Then $A = (\bigcap_{B \in \mathscr{C}(X)} B) \cap C$ if and only if A has a unique minimal C-representation in X. If this is the case, then the following statements hold for the set*

$$S = \{B \in X : B \text{ is strongly irredundant in some } C\text{-representation of } A \text{ in } X\}.$$

(1) *$S \subseteq \mathscr{C}(X)$ and hence every member of S is critical in X.*
(2) *Each $B \in S$ is strongly irredundant in the C-representation $\mathscr{C}(X)$.*
(3) *If A has a strongly irredundant C-representation Z in X, then $Z = S$.*
(4) *X contains at most one strongly irredundant C-representation of A.*

Proof Observe first that $\uparrow\mathscr{C}(X)$ is the intersection of all the closed C-representations of A in X. Thus $A = (\bigcap_{B \in \mathscr{C}(X)} B) \cap C$ if and only if $\uparrow\mathscr{C}(X)$ is a C-representation of A, if and only if there is a unique minimal closed C-representation of A, if and only if there is a unique minimal C-representation of A in X.

(1) Let $B \in S$. Then, there is $Y \subseteq X$ such that $Y \cup \{B\}$ is a C-representation of A and B is strongly irredundant in $Y \cup \{B\}$. For $E \in \mathscr{C}(X)$, Proposition 3.11 implies $B \subseteq E$ or $\bigcap_{B' \in Y} B' \subseteq E$. If $\bigcap_{B' \in Y} B' \subseteq E$ for every $E \in \mathscr{C}(X)$, then since by assumption $\mathscr{C}(X)$ is a C-representation of A, this forces $A = (\bigcap_{B' \in Y} B') \cap C$, contrary to the irredundance of B in $\{B\} \cup Y$. Therefore, $Z := (\uparrow\mathscr{C}(X)) \cap \mathscr{V}(B)$ is nonempty, and for every $E \in \mathscr{C}(X)$ with $B \nsubseteq E$, it must be that $\bigcap_{B' \in Y} B' \subseteq E$. Thus, since $\mathscr{C}(X)$ is a C-representation of A, so is $Z \cup Y$. Since $\uparrow\mathscr{C}(X)$ and $\mathscr{V}(B)$ are closed subsets of X, so is Z. Now $Z \subseteq \mathscr{V}(B)$, so since B is strongly irredundant in the C-representation $\{B\} \cup Y$, it must be that $Z = \mathscr{V}(B)$. Thus $B \in \uparrow\mathscr{C}(X)$.

Next we show that $B \in \mathscr{C}(X)$. There exists $E \in \mathscr{C}(X)$ such that $E \subseteq B$. Since $A = B \cap (\bigcap_{B' \in Y} B') \cap C$ and E is critical, Proposition 3.11 implies that $B = E$ or $\bigcap_{B' \in Y} B' \subseteq E$. Since $E \subseteq B$ and B is irredundant in $\{B\} \cup Y$, the latter cannot occur, so $B = E$. Therefore, B is minimal in $\mathscr{C}(X)$.

(2) As in (1), every $E \in \mathscr{C}(X)$ with $E \neq B$ contains $\bigcap_{B' \in Y} B'$, so that if B is not irredundant in $\mathscr{C}(X)$, then since $\mathscr{C}(X)$ is a C-representation of A, we have $A = (\bigcap_{B' \in Y} B') \cap C$, a contradiction. Thus B is irredundant in $\mathscr{C}(X)$, and by Corollary 3.12, B is strongly irredundant in $\mathscr{C}(X)$.

(3) Suppose Z is a strongly irredundant C-representation in X. By (1), $Z \subseteq \mathscr{C}(X)$ and the members of Z are strongly irredundant in $\mathscr{C}(X)$. Also by (2), the members of S are strongly irredundant in $\mathscr{C}(X)$. It follows that $S = Z$.

(4) This is clear from (3). □

4 Irredundance in Intersections of Valuation Rings

In this section, we reinterpret the material of Sect. 3 for the Zariski–Riemann space of a field. We assume the following notation throughout this section.

- A is a proper integrally closed subring of a field F.
- C is a set (not necessarily a ring) such that $A \subsetneq C \subseteq F$.
- $\mathfrak{X}$ denotes the set of valuation rings of F containing A.

Zariski introduced a topology on $\mathfrak{X}$ (the *Zariski topology*) by designating as a basis of open sets the sets of the form $\{V \in \mathfrak{X} : x_1, \ldots, x_n \in V\}$, where $x_1, \ldots, x_n \in F$. With this topology, the same topology as in Example 2.2(8), $\mathfrak{X}$ is a spectral space and is termed the *Zariski–Riemann space* of F/A. For some recent articles emphasizing a topological approach to the Zariski–Riemann space, see [8–13, 42, 44, 45].

Comparison of the basic opens in the Zarsiki topology on $\mathfrak{X}$ with the topology in Sect. 3 shows that it is the inverse topology on $\mathfrak{X}$ rather than the Zariski topology that is needed in order to deal with issues of irredundance. The inverse topology is also a natural one to consider here since under this topology the specialization order on $\mathfrak{X}$ agrees with the usual order on $\mathfrak{X}$ given by set inclusion. To avoid confusing the two topologies, we denote by $\mathfrak{X}^{-1}$ the set $\mathfrak{X}$ with the inverse topology. Then $\mathfrak{X}^{-1}$ is a spectral C-representation of A, and so all the results of Sect. 3 can be translated into the context of the Zariski–Riemann space of F by working inside the spectral C-representation $\mathfrak{X}^{-1}$.

In the spirit of Sect. 3, we work throughout this section relative to the set C and consider *C-representations X of A*; that is, $A = (\bigcap_{V \in X} V) \cap C$, where X is a collection of valuation rings in $\mathfrak{X}$. That C need only be a subset in most cases is a byproduct of the approach in Sect. 2. While we do not have an application for the level of generality that working with a set rather than a ring affords, we do so anyway since it comes at no extra expense. When $C = F$, we abbreviate "C-representation" to "representation." Thus a *representation* of A is a subset X of $\mathfrak{X}$ such that $A = \bigcap_{V \in X} V$.

In this section, $\mathfrak{X}^{-1}$ will play the role that X did in Sect. 3 of an ambient spectral representation. In this section, we use "X" then for not necessarily spectral subsets of $\mathfrak{X}$. A C-representation X of A then is a subspace of $\mathfrak{X}^{-1}$. In particular:

When applying the results of Sect. 3, *the default topology on the C-representations of A is the inverse topology. Thus, the specialization order coincides with set inclusion among the valuation rings, and the operators* Min (−) *and* Max (−) *yield the minimal and maximal elements, respectively, of a collection of valuation rings with respect to set inclusion.*

Definition 4.1 Let $X \subseteq \mathfrak{X}$. We define $\mathrm{cl}(X)$, $\mathrm{inv}(X)$ and $\mathrm{patch}(X)$ to be the closure of X in the Zariski, inverse and patch topologies, respectively. We denote by $\mathrm{gen}(X)$ the set of generalizations of the valuation rings in X; that is,

$$\mathrm{gen}(X) = \{V \in \mathfrak{X} : W \subseteq V \text{ for some } W \in X\}.$$

We interpret now the results of Sect. 3 in the setting of the Zariski–Riemann space. The notions of irredundance from Definition 3.1(2) can be simplified for valuation rings. Let X be a subset of $\mathfrak{X}$ such that $A = (\bigcap_{V \in X} V) \cap C$. Then $V \in X$ is *irredundant* in the C-representation X if V cannot be omitted from this intersection; V is *strongly irredundant* if V cannot be replaced in this intersection by a valuation overring[1]; and V is *tightly irredundant* if V cannot be replaced by an intersection of valuation overrings that properly contain V.

Combining Lemmas 3.2 and 3.5, and observing that in the notation of Sect. 2, $\text{gen}(X) = \downarrow X$, we have the following existence result for minimal representations.

(4.2) *In every inverse closed subset X of $\mathfrak{X}$ there is a minimal C-representation of A; that is, there exists in X a collection Z of pairwise incomparable valuation rings such that* $\text{gen}(Z)$ *is a minimal inverse closed C-representation of A and* $\text{patch}(Z)$ *is a minimal patch C-representation of A.*

In general, there can exist infinitely many such minimal C-representations of A. This is illustrated by Example 4.3.

Example 4.3 In [41, Example 6.2], an integrally closed overring A of $K[X, Y, Z]$, with K any field and X, Y, Z indeterminates, is constructed such that A has uncountably many strongly irredundant representations. Since every valuation overring of $K[X, Y, Z]$ has finite Krull dimension, a valuation ring in a representation of A is strongly irredundant if and only if it is tightly irredundant (see the discussion after (4.4)). By Proposition 3.4, A has uncountably many minimal representations.

Applying Lemma 3.3, we have

(4.4) *If $A = (\bigcap_{V \in X} V) \cap C$, then $V \in X$ is irredundant in X if and only if V is irredundant in* $\text{patch}(X)$; *V is tightly irredundant in X if and only if V is irredundant in* $\text{inv}(X)$.

If V has finite Krull dimension, then since there are only finitely many overrings of V, V is strongly irredundant in the C-representation X if and only if V is tightly irredundant in X. More generally, if the maximal ideal of V is not the union of the prime ideals properly contained in it, then the notions of strong and tight irredundance coincide for V. In particular, if $V \in X$ has rank one, then V is irredundant in the C-representation X if and only if V is irredundant in $\text{inv}(X)$.

While an irredundant member V of a C-representation X is by Lemma 3.3 an isolated point in the inverse and patch topologies on Z, the converse need not be true, as illustrated by Example 4.6. However, by restricting to minimal C-representations we obtain from Theorem 3.6 that irredundance is topological for such representations.

(4.5) *Suppose X is a minimal C-representation of A, as in (4.2). A valuation ring $V \in X$ is irredundant in X if and only if V is strongly irredundant in X; if and only if V is isolated in X in the inverse (equivalently, patch) topology.*

Example 4.6 A valuation ring V in a C-representation X of A may be isolated in the inverse topology on X but be redundant in X. For example, let A be an integrally closed Noetherian local domain of Krull dimension >1, let $X = \{A_P : P \text{ is a height one prime ideal of } A\}$, and let V be a DVR overring of A that dominates A. Write the maximal ideal M of A as $M = (a_1, \ldots, a_n)$. Then, X is a subset of the inverse

[1] By an *overring* of a domain R we mean a ring between R and its quotient field.

closed set $Y := \{W \in \mathfrak{X} : 1/a_i \in W \text{ for some } i = 1, \ldots, n\}$, while $V \notin Y$. Thus V is an isolated point in $\{V\} \cup X$ with respect to the inverse topology. However, since $A = \bigcap_{W \in X} W$, V is redundant in the representation $\{V\} \cup X$ of A. (The notion of a minimal representation remedies this: X is a minimal representation so that V is excluded from consideration since it is not an element of X.)

By Corollary 3.7, the existence of a strongly irredundant C-representation within a minimal C-representation depends only on the topology of the minimal representation:

(4.7) *Suppose X is a minimal C-representation of A, as in (4.2). Then X contains a strongly irredundant C-representation Y of A if and only if the set of isolated points in X is dense in X with respect to the inverse topology.*

In such a case the only choice for Y is the set of isolated points of X, and hence there exists at most one such irredundant C-representation of A in X, hence also in patch(X) (Lemma 3.3). However, moving outside of patch(X), Example 4.3 shows there can exist infinitely many distinct strongly irredundant C-representations of A. This example involves an intersection of valuation overrings of a three-dimensional Noetherian domain. By contrast, strongly irredundant representations over two-dimensional Noetherian domains are much better behaved and have a number of uniqueness properties [41].

One consequence of the topological approach of Sect. 3 is an existence result for strongly irredundant C-representations of A in the countable case. This result, which follows from Corollary 3.9, is revisited in the next section in Theorem 6.8.

(4.8) *If* $A = (\bigcap_{V \in X} V) \cap C$ *for some countable patch X in* $\mathfrak{X}$, *then X contains a strongly irredundant C-representation of A.*

It is important here that we work with a countable patch rather than simply a countable subset of $\mathfrak{X}$. This is illustrated by the next example.

Example 4.9 Suppose A is a countable integrally closed local Noetherian domain with maximal ideal M and quotient field F. Suppose also that A has Krull dimension >1. Let X be the collection of all DVR overrings V of A that are are centered in A on M and such that V is a localization of the integral closure of some finitely generated A-subalgebra of F. Since A is countable and Noetherian, there are countably many such valuation rings. Moreover, A is the intersection of the valuation rings in X, since if $x \in F \setminus A$, then there exists a valuation ring V in X whose maximal ideal contains x^{-1}, so that $x \notin V$. If $V \in X$ is an irredundant representative of A, then since the value group of V is a subgroup of the group of rational numbers, V is a localization of A [29, Lemma 1.3], a contradiction to the fact that A has Krull dimension >1 and V is centered on the maximal ideal of A. Therefore, although X is countable, X contains no irredundant representatives of A. It follows from (4.8) that X is not a patch closed subspace of $\mathfrak{X}$.

Adapting the terminology from Sect. 3, we say a valuation ring $V \in \mathfrak{X}$ is *C-critical for A* if V is an element of every inverse closed C-representation of A. Thus by Proposition 3.11 and the fact that every integrally closed A-subalgebra of F is an intersection of valuation rings in $\mathfrak{X}$, we have

(4.10) *V is C-critical for A if and only if whenever $A_1, \ldots, A_n$ are integrally closed A-subalgebras of F such that $A = A_1 \cap \cdots \cap A_n \cap C$, it must be that $A_i \subseteq V$ for some i.*

Also, from Corollary 3.12 we see that if V is C-critical for A and irredundant in some C-representation X of A, then V is strongly irredundant in X. By restricting to the case where C is an A-submodule of F, we obtain an important class of C-critical valuation rings; these are the valuation rings that play an important role in the next sections. A valuation ring $V \in \mathfrak{X}$ is *essential for A* if $V = A_P$ for a prime ideal P of A.

Proposition 4.11 *Let $V \in \mathfrak{X}$ such that $C \not\subseteq V$. If C is an A-submodule of F and V is essential for A, then V is C-critical for A.*

Proof We use Proposition 3.11 to prove the claim. Let P be a prime ideal of A such that $A_P = V$, and let $A_1, \ldots, A_n$ be integrally closed A-subalgebras of F such that $A = A_1 \cap \cdots \cap A_n \cap C$. Then since localization commutes with finite intersections, we have $V = A_P = (A_1)_P \cap \cdots \cap (A_n)_P \cap C_P$. Since V is a valuation ring, the set of V-submodules between V and F forms a chain. Therefore, since $C \not\subseteq V$, there is i such that $A_i \subseteq V$, and hence by Proposition 3.11, V is C-critical. □

Example 4.12 A valuation overring that is C-critical for A need not be essential. Suppose A has quotient field F. Then A is said to be *vacant* if it has a unique Kronecker function ring [7]. (Kronecker function rings are discussed after 4.13.) As we see in (4.15), this implies that A has a unique minimal representation. Hence A is vacant if and only if every valuation overring of A is critical (see also [7]). As discussed in [7] there exist vacant domains that are not Prüfer domains, and hence such a domain has a critical valuation overring that is not essential.

Example 4.3 shows that in general A need not be an intersection of critical valuation overrings; equivalently, A need not have a unique minimal C-representation. However, for some well-studied classes of rings, such as those in the next two sections, it is possible to represent A with critical valuation rings. In this case, strong properties hold for A. For example, applying Theorem 3.13(1), we have the following fact.

(4.13) *Suppose $A = (\bigcap_{V \in \mathscr{C}(\mathfrak{X})} V) \cap C$, where $\mathscr{C}(\mathfrak{X})$ is the set of minimal C-critical valuation rings in $\mathfrak{X}$. If $V \in \mathfrak{X}$ is strongly irredundant in some C-representation of A, then V is in $\mathscr{C}(\mathfrak{X})$ and V is strongly irredundant in $\mathscr{C}(\mathfrak{X})$.*

Thus $\mathscr{C}(\mathfrak{X})$ collects all the strongly irredundant representatives of A, and so, as in Theorem 3.13, having a strongly irredundant representation is a matter of having enough strongly irredundant representatives.

(4.14) *Suppose* $A = (\bigcap_{V \in \mathscr{C}(\mathfrak{X})} V) \cap C$, *so that* $\mathscr{C}(X)$ *is a* C*-representation of* A. *Then* A *has a strongly irredundant* C*-representation if and only if* A *is an intersection of* C *with valuation rings in the set*

$$\{V \in \mathfrak{X} : V \text{ is strongly irredundant in some } C\text{-representation of } A\}.$$

Thus A *has at most one strongly irredundant* C*-representation.*

There is a long tradition of using Kronecker function rings to represent integrally closed rings in the field F with a Bézout domain in a transcendental extension $F(T)$ of F. We depart from this tradition because of our emphasis on the more general topological approach via spectral representations as in Sect. 3. However, in the present context of Zariski–Riemann spaces there is a precise connection between minimal representations and maximal Kronecker function rings. In fact, minimal representations play for us a role similar to that played by the Kronecker function ring in articles such as [2, 3, 24, 42]. We outline this connection here.

Let T be an indeterminate for F. For each valuation ring $V \in \mathfrak{X}$, let $V^* = V[T]_{\mathfrak{M}_V[T]}$, where $\mathfrak{M}_V$ is the maximal ideal of V. Then V^* is a valuation ring with quotient field $F(T)$ such that $V = V^* \cap F$. For a nonempty subset X of $\mathfrak{X}$, the *Kronecker function ring of* X is the ring

$$\mathrm{Kr}(X) = \bigcap_{V \in X} V^*.$$

Then $\mathrm{Kr}(X)$ is a Bézout domain with quotient field $F(T)$; cf. [9, Corollary 3.6], [28, Theorem 2.2] and [32, Corollary 2.2]. When X is a C-representation of A, then $A = \mathrm{Kr}(X) \cap C$, and we say that $\mathrm{Kr}(X)$ is a *Kronecker* C*-function ring of* A. Thus to every C-representation of A corresponds a Kronecker C-function ring of A.

For each $X \subseteq \mathfrak{X}$, let $X^* = \{V^* : V \in X\}$. The mapping $\mathfrak{X} \to \mathfrak{X}^*$ is a homeomorphism with respect to the Zariski topology (see [9, Corollary 3.6] or [32, Proposition 2.7]), and hence is a homeomorphism in the inverse and patch topologies also. The subset X is inverse closed in $\mathfrak{X}$ if and only if X^* is the set of localizations at prime ideals of $\mathrm{Kr}(X)$; i.e., X^* is the Zariski–Riemann space of the Bézout domain $\mathrm{Kr}(X)$ [45, Proposition 5.6]. Moreover, we have the following connection between C-representations and Kronecker C-function rings, which can be deduced from [45, Corollary 5.8 and Proposition 5.10].

(4.15) *The inverse closed* C*-representations of* A *bijectively correspond to the Kronecker* C*-function rings of* A. *The minimal* C*-representations of* A *bijectively correspond to the maximal Kronecker* C*-function rings of* A. *Moreover, a subset* X *of* $\mathfrak{X}$ *is a minimal* C*-representation of* A *if and only if* X^* *consists of the localizations at maximal ideals of a maximal Kronecker* C*-function ring of* A.

5 Generalizations of Krull Domains

In this section, we assume the same notation as Sect. 4. Thus A is an integrally closed subring of the field F, C is a set between A and F, and $\mathfrak{X}$ is the Zariski–Riemann space of F/A. Intersection representations play an important role in the theory of *Krull domains*, those integral domains that can be represented by a finite character intersection of rank one discrete valuation rings (DVRs). (A subset X of $\mathfrak{X}$ has *finite character* if each $0 \neq x \in F$ is a unit in all but at most finitely many valuation rings in X.) Finite character representations of a Krull domain A are well understood: The collection $X = \{A_P : P$ a height one prime ideal of $A\}$ is a finite character, irredundant representation of DVRs. Krull [35] proved more generally that if A has a finite character representation consisting of valuation rings whose value groups have rational rank one, then this collection can be refined to one in which every valuation ring is essential for A; see also [40, Corollary 5.2]. Examples due to Griffin [26, Sect. 4], Heinzer and Ohm [30, 2.4] and Ohm [40, Example 5.3] show that the same is not true if the value groups of the valuation rings are assumed only to have rank one rather than rational rank one.

Griffin defines the ring A to have *Krull type* if A has a finite character representation X consisting of essential valuation rings [26, 27]. In [26, Theorem 7] he gives necessary conditions for a ring A having a finite character representation of valuation rings to be a ring of Krull type. Pirtle [47, Corollary 2.5] showed that when in addition the valuations in X have rank one, X is an irredundant representation of A. More generally, Brewer and Mott [3, Theorem 14] prove that if A has a finite character representation X of valuation rings (no restriction on rank), then A has an irredundant finite character representation, and if also the valuation rings in X have rank one, then A has one, and only one, irredundant finite character representation consisting of rank one valuation rings [3]. In [2, Theorem 1.1], Brewer proves that if A has Krull type, then A has an irredundant finite character representation X consisting of essential valuation rings, and that X is unique among such representations.

In both the articles [2, 3], the authors prove their results by passing to a maximal Kronecker function ring of A and applying Gilmer and Heinzer's theory of irredundant representations of Prüfer domains to work out the problem of irredundance in a Prüfer setting. This method of passage to a Kronecker function ring, and hence to a Prüfer domain, is applied in [44] to domains A that can be represented with a collection of valuation rings from a Noetherian subspace of the Zariski–Riemann space, a class of representations that subsumes the finite character ones. In such a case, A can be represented by a strongly irredundant Noetherian space of valuation rings [44, Theorem 4.3]. The results in [44] are in fact framed in terms of C-representations, where C is a ring.[2]

[2]The results in [44] also apply to representations consisting of integrally closed rings, not just valuation rings. In light of Finocchiaro's theorem that the space of integrally closed subrings of F is a spectral space (see Example 2.2(6)), it seems likely that this level of generality might be handled with spectral C-representations also.

The introduction of finite character rank one C-representations to generalize the theory of Krull domains is due to Heinzer and Ohm [30]. This allows for considerably more flexibility in applying results to settings in which one considers, say, integrally closed rings between A and some integrally closed overring C. Even when A is a two-dimensional Noetherian domain and C is chosen a PID, the analysis of the integrally closed rings between A and C is quite subtle; see for example [1, 6, 38, 42, 46]. Regardless of the choice of A and C, Heinzer and Ohm [30, Corollary 1.4] prove that finite character rank one C-representations remain as well behaved as in the classical case of $C = F$: If C is a ring and A has a C-representation consisting of rank one valuation rings, then A has a unique irredundant finite character representation consisting of rank one valuation rings.

In this section, we recover the above results using the topological methods developed in Sect. 3 and elaborated on in Sect. 4. Whereas in the articles [2, 3, 44] the strategy is to pass to a maximal Kronecker function ring and treat irredundance there, we work instead with the topology of minimal representations to obtain our results. We also need only that C is a set. A C-representation X is *Noetherian* if X is a Noetherian subspace of $\mathfrak{X}$ with respect to the Zariski topology.

Theorem 5.1 *If* $A = (\bigcap_{V \in X} V) \cap C$, *where* X *is a Noetherian subspace of* $\mathfrak{X}$, *then* $\text{gen}(X)$ *contains a Noetherian strongly irredundant* C*-representation of* A.

Proof Since X is Noetherian, X is quasicompact, and hence $\text{inv}(X) = \text{gen}(X)$ [45, Proposition 2.2]. Thus by (4.2), there exists a minimal C-representation Y of A in $\text{gen}(X)$. By Proposition 2.1(1), $\text{gen}(X)$ is a spectral space under the Zariski topology. The elements of $\text{Min}\, X$ are the maximal elements of $\text{gen}(X)$ under the specialization order of the Zariski topology. In particular, $\text{Min}\, X \subseteq X$. Thus since $\text{Min}\, X$ is Noetherian in the Zariski topology, Theorem 2.5 implies that Y is discrete in the inverse topology, so that each valuation ring in Y is an isolated point in Y in the inverse topology. By (4.5), the valuation rings in the minimal C-representation Y are strongly irredundant. Also, by Lemma 2.4, Y is Noetherian in the Zariski topology. □

Remark 5.2 In general, the strongly irredundant C-representation in Theorem 5.1 is not unique. For example, the uncountably many strongly irredundant representations of the ring A discussed in Example 4.3 are Noetherian spaces in the Zariski topology. The ring A in this case is an overring of a three-dimensional Noetherian domain. By contrast, when C is integrally closed and A is an overring of a two-dimensional Noetherian domain, with A representable by a Noetherian space of valuation overrings, then A has a unique strongly irredundant C-representation [41, Corollary 5.7]. In the case in which A is an overring of a two-dimensional Noetherian domain, the existence of a Noetherian C-representation has strong implications for the structure of A; see [42].

From the theorem, we deduce a corollary that recovers a number of the results discussed at the beginning of the section, with the additional feature that the valuation rings in the representation are strongly irredundant rather than just irredundant. When the valuations are essential, we also obtain uniqueness across all strongly irredundant representations, not just the finite character ones.

Corollary 5.3 *If* $A = (\bigcap_{V \in X} V) \cap C$*, where* X *is a finite character subset of* $\mathfrak{X}$*, then* $\mathrm{gen}(X)$ *contains a strongly irredundant* C*-representation* Y *of* A*. If also each valuation ring in* X *is essential for* A*, then* Y *is the only strongly irredundant* C*-representation of* A *in* $\mathfrak{X}$*.*

Proof A finite character collection of valuation rings in $\mathfrak{X}$ is Noetherian in the Zariski topology [44, Proposition 3.2], so by Theorem 5.1, $\mathrm{gen}(X)$ contains a strongly irredundant C-representation Y. If also every valuation ring in X is essential for A, then every valuation ring in $\mathfrak{X}$ that does not contain C is C-critical for A (Proposition 4.11), so the assertion of uniqueness follows from (4.13). □

If A has quotient field F and $A = V \cap R$, where V is a rational valuation overring of A (i.e., V has value group isomorphic to a subgroup of the rational numbers) and R is a subring of F properly containing A, then $V = A_P$, where P is the prime ideal of A that is contracted from the maximal ideal of V [29, Lemma 1.3]. Applying this to the setting of Corollary 5.3, we recover the result of Krull discussed at the beginning of the section, but strengthened to guarantee uniqueness across all strongly irredundant representations.

Corollary 5.4 *Suppose* C *is a ring and* $A = (\bigcap_{V \in X} V) \cap C$*, where* X *is a finite character representation of* A *consisting of valuation rings of rational rank one. Then* X *contains a strongly irredundant* C*-representation* Y *of* A*, and* Y *is the only strongly irredundant* C*-representation of* A*.*

Remark 5.5 Heinzer and Ohm prove a version of Corollary 5.4 for finite character C-representations X of A when A consists of rank one valuation rings (in their terminology, *A is a C-domain of finite real character*). Thus, their approach includes rank one valuation rings with irrational value group also. Unlike rational valuation rings, such valuation rings can be strongly irredundant in X but not essential for A; see [30, Sect. 2]. They prove that if A has quotient field F, C is a ring and $A = (\bigcap_{V \in X} V) \cap C$, where X is a finite character subset of $\mathfrak{X}$ consisting of rank one valuation rings, then any valuation ring that is irredundant in some C-representation of A is a member of every finite character C-representation of A that consists of rank one valuation rings, and the collection of all such valuation rings is a C-representation of A [30, Corollary 1.4]. It seems plausible that our approach can recover this result also, but more information is needed about C-representations.

6 Prüfer *v*-Multiplication Domains

We assume throughout this section that A is an integrally closed domain with quotient field F, and that $\mathfrak{X}$ is the Zariski–Riemann space of F/A. We no longer work with an intermediate set C, or more precisely, we work with $C = F$. Hence, we drop C

from our usual terms such as "C-representation" and "C-critical" and simply write "representation" for "F-representation" and "critical" for "F-critical."

The concept of a Prüfer v-multiplication domain encompasses that of a Krull domain and a Prüfer domain, as well as polynomial rings over these domains. In this section, we apply the point of view developed in Sect. 4 to the issue of irredundance in representations of Prüfer v-multiplication domains. While the results in this section shed additional light on the domains of Krull type considered in the last section, the real impetus for the section comes from the theory of Prüfer domains. We use the topological methods of Sects. 3 and 4 to recover irredundance results for this class of rings as well as generalize them to Prüfer v-multiplication domains.

For an ideal I of A, let $I_v = (A :_F (A :_F : I))$. An integral domain A is a *v-domain* if whenever I, J, K are ideals of A such that $(IK)_v = (JK)_v$, then $I_v = J_v$. Examples of such domains include completely integrally closed domains and Prüfer v-multiplication domains; for a recent survey of this class of rings, see [17]. A v-domain A has a unique maximal Kronecker function ring [22, Theorem 28.1], so by (4.14), A has a unique minimal representation. In particular, A is an intersection of its critical valuation rings.

Theorem 6.1 *A v-domain has at most one strongly irredundant representation.*

Proof Since a v-domain A is an intersection of critical valuation overrings, we may apply (4.14) in the case where $C = F$ to obtain the theorem. □

Remark 6.2 In [24, p. 310], Gilmer and Heinzer ask whether it is the case that if A is a v-domain that is an irredundant intersection of valuation rings, then the unique maximal Kronecker function ring of A is an irredundant intersection of valuation rings. We can answer this question in the affirmative under the stipulation that A is represented as a *strongly* irredundant intersection of valuation overrings. In this case, by (4.13), each strongly irredundant representative of A is contained in the minimal representation $\mathscr{C}$ of A consisting of the minimal critical valuation rings for A. By (4.15), $\mathrm{Kr}(\mathscr{C})$ is the unique maximal Kronecker function ring of A. Since $\mathscr{C}$ is a minimal representation of A, (4.7) implies $\mathscr{C}$ contains a dense set of isolated points in the inverse topology. Thus $\mathscr{C}^*$ has a dense set of isolated points, so that by (4.7), $\mathrm{Kr}(\mathscr{C})$ has a strongly irredundant representation.

For each ideal I of A, let $I_t = \sum_J J_v$, where J ranges over the finitely generated ideals of A contained in I. An ideal I of A is a *t-ideal* if $I = I_t$. If I is maximal among t-ideals, then I is a *t-maximal ideal.* A t-ideal I is a *t-prime ideal* if I is prime. A t-maximal ideal is a prime, hence t-prime, ideal. The set of t-prime ideals is denoted t-Spec$A$, while the set of t-maximal ideals is denoted t-Max$A$. The domain A is a *Prüfer v-multiplication domain (PvMD)* if every nonzero finitely generated ideal of A is t-invertible; equivalently, A_M is a valuation domain for each $M \in t$-Max$A$ [34, Theorem 3.2]. Every PvMD A is an *essential domain*, meaning that A is an intersection of essential valuation overrings.

In a recent article, Finocchiaro and Tartarone [14, Corollary 2.6] show that an essential domain is a PvMD if and only if the set of its essential valuation overrings is patch closed. We use this characterization to interpret PvMDs in terms of critical valuation rings.

Lemma 6.3 *The domain A is a PvMD if and only if A is an essential domain for which every critical valuation ring of A is essential.*

Proof Since A is a PvMD, A is an essential domain. By [14, Corollary 2.6], the set E of essential valuation rings is patch closed in the Zariski–Riemann space of A. If $V \in E$, then so is every overring of V, so $\text{gen}(E) = E$. Thus Proposition 2.1(3) implies that E is an inverse closed representation of A, and hence every critical valuation ring of A is in E.

Conversely, suppose every critical valuation ring of A is essential. Then by Proposition 4.11, the set E of essential valuation rings is equal to the set of critical valuation overrings of A. Therefore, E is inverse closed, hence patch closed, so that by [14, Corollary 2.6], A is a PvMD. □

Theorem 6.4 *Suppose A is a PvMD. Let V be a valuation overring of A, and let P be the center of V in A. Then, the following are equivalent.*

(1) *V is strongly irredundant in some representation of A.*
(2) *$V = A_P$, $P \in t\text{-Max}A$ and P is isolated in $t\text{-Max}A$ in the inverse topology.*
(3) *$V = A_P$, $P \in t\text{-Max}A$ and P contains a finitely generated ideal that is not contained in any other t-maximal ideal.*

Proof (1) $\Rightarrow$ (2) Suppose V is strongly irredundant in some representation of A. Since A is a PvMD, A is an essential domain, and hence, A is an intersection of its critical valuation rings. Therefore, by (4.13), V is in the collection $\mathscr{C}$ of the valuation rings that are minimal among critical valuation rings for A. By Lemma 6.3, the set of critical valuation rings for A coincides with the set of essential valuation overrings of A. Thus $V = A_P$, and since $V \in \mathscr{C}, P \in t\text{-Max}A$. By (4.13), V is strongly irredundant in $\mathscr{C}$, so that by (4.5), V is isolated in $\mathscr{C}$ with respect to the inverse topology. Since $\mathscr{C} = \{A_Q : Q \in t\text{-Max}A\}$, the map that sends a valuation overring of A to its center in A restricts to a homeomorphism from $\mathscr{C}$ onto $t\text{-Max}A$. (That this map is continuous and closed follows from [51, Lemma 5, p. 119].) Thus P is a t-maximal ideal of A that is isolated in the inverse topology of $t\text{-Max}A$.

(2) $\Rightarrow$ (3) By (2), there is a Zariski quasicompact open subset of $t\text{-Max}A$ whose complement in $t\text{-Max}A$ is $\{P\}$. Statement (3) now follows.

(3) $\Rightarrow$ (2) This is clear.

(2) $\Rightarrow$ (1) As in the proof that (1) implies (2), the canonical map $\mathscr{C} \to t\text{-Max}A$ is a homeomorphism, so we conclude that V is isolated in the inverse topology on $\mathscr{C}$. Since $\mathscr{C}$ is a minimal representation of A, (1) follows from (4.5). □

Remark 6.5 If V is a valuation overring of the domain A that is irredundant in some representation of A as an intersection of valuation overrings and V is essential, then, as discussed after (4.10), V is strongly irredundant. Thus, when A is a Prüfer domain, V is irredundant in a representation of A if and only if it is strongly irredundant. Therefore, when A is a Prüfer domain, "strongly irredundant" can be replaced by "irredundant" in Theorem 6.4(1). With this replacement, the theorem recovers a characterization of irredundant representatives of Prüfer domains due to Gilmer and Heinzer; cf. [24, Proposition 1.4 and Lemma 1.6].

In light of Remark 6.5, the next corollary is a version of [24, Theorem 1.10] for PvMDs.

Corollary 6.6 *Suppose A is a PvMD. Then, A can be represented (uniquely) as a strongly irredundant intersection of valuation overrings if and only if there is a collection X of t-maximal ideals of A such that*

(a) *each nonzero element of A is contained in a member of X, and*
(b) *each $P \in X$ contains a finitely generated ideal that is not contained in any other t-maximal ideal of A.*

Proof Suppose that A has a strongly irredundant representation Z. By Theorem 6.1 this representation is unique, and by Theorem 6.4 there is a subset X of t-Max$A$ such that $Z = \{A_P : P \in X\}$. Since $A = \bigcap_{P\in X} A_P$, every nonzero element of A is contained in a member of X. Moreover, by Theorem 6.4, each t-maximal ideal in X contains a finitely generated ideal that is not contained in any other t-maximal ideal. Conversely, if $X \subseteq t$-Max$A$ such that (a) and (b) hold for X, then by (a), $Z := \{A_P : P \in X\}$ is a representation of A, and by Theorem 6.4, each $V \in Z$ is strongly irredundant in some representation of A. Therefore, by Theorem 6.1, A has a strongly irredundant representation. □

Remark 6.7 The ring of integer-valued polynomials Int($\mathbb{Z}$) is the set of all polynomials $f(X) \in \mathbb{Q}[X]$ such that $f(\mathbb{Z}) \subseteq \mathbb{Z}$. Among the many well-known properties of this interesting ring is that it is a Prüfer domain; see [5]. In the recent article [6], Chabert and Peruginelli characterize all the rings R between Int($\mathbb{Z}$) and $\mathbb{Q}[X]$ that can be represented as an irredundant intersection of valuation overrings. These are precisely the intermediate rings R such that for each prime integer p, the set $\{\alpha \in \widehat{\mathbb{Z}}_p : \mathfrak{M}_{p,\alpha}R \subsetneq R\}$ is dense with respect to the p-adic topology in the ring $\widehat{\mathbb{Z}}_p$ of p-adic integers [6, Remark 5.7]. Here $\mathfrak{M}_{p,\alpha} = \{f \in \mathrm{Int}(\mathbb{Z}) : f(\alpha) \in p\widehat{\mathbb{Z}}_p\}$, a maximal ideal of Int($\mathbb{Z}$).

Let A be a PvMD. A *subintersection* of A is an overring of A of the form $\bigcap_{V\in X} V$, where $X \subseteq \{A_P : P \in t\text{-Spec}A\}$. Equivalently, by Lemma 6.3, a subintersection of A is an intersection of critical valuation overrings of A.

Corollary 6.8 *If A is a PvMD such that t-Spec$A$ is countable, then each subintersection B of A can be represented as a strongly irredundant intersection of essential valuation overrings of A. This representation is the only strongly irredundant representation of B.*

Proof Let $X = \{A_P : P \in t\text{-Spec}A\}$, and let $Y \subseteq X$. Then patch(Y) is a representation of $B = \bigcap_{V\in Y} V$. Since by [14, Corollary 2.6], X is a patch closed representation of A, patch(Y) $\subseteq X$, and hence patch(Y) is countable. Therefore, by (4.8), B has a strongly irredundant representation in patch(Y). Since the valuation rings in X are essential for A, hence essential for B, Proposition 4.11 implies that the valuation rings in Y are critical for B,. Thus, the ring B has a strongly irredundant representation consisting of critical valuation rings. As a subintersection of A, B is a Prüfer v-multiplication domain, hence a v-domain. Therefore, by Theorem 6.1, there is only one strongly irredundant representation of B. □

When A is a PvMD for which t-Max$A$ is a Noetherian subspace of Spec$A$, then $\{A_P : P \in t\text{-Max}A\}$ is a Noetherian space of valuation overrings that represents A. Therefore, by Theorem 5.1, A has a strongly irredundant representation in $\{A_P : P \in t\text{-Max}A\}$. However, the fact that A is a PvMD allows us to assert the stronger claim that $\{A_P : P \in t\text{-Max}A\}$ itself is a strongly irredundant representation of A, and that this property is inherited by similarly constituted representations of subintersections of A. We prove this in the next theorem.

Theorem 6.9 *Suppose A is a PvMD. Then, t-*Max$A$ *(resp., t-*Spec$A$*) is Noetherian in the Zariski topology if and only if each collection X of incomparable essential valuation overrings is a strongly (resp., tightly) irredundant representation of* $\bigcap_{V\in X} V$.

Proof Suppose t-Max$A$ is a Noetherian space. By [14, Corollary 2.6], the set $E = \{A_P : P \in t\text{-Spec}A\}$ of essential valuation overrings of A is patch closed. Thus by Proposition 2.1(4)(a), E is spectral in the Zariski topology. Since Min$E = \{A_P : P \in t\text{-Max}A\}$ is by assumption Noetherian in the Zariski topology, Theorem 2.5 implies every subset X of E consisting of incomparable valuation rings is discrete in the inverse topology.

Now let X be a subset of E consisting of incomparable valuation rings. The ring $B := \bigcap_{V\in X} V$ is again a PVMD [34, Corollary 3.9]. We claim that $X = \{B_P : P \in t\text{-Max}B\}$. To this end, let P be a maximal t-ideal of B. Then since by Lemma 2.4, X is Noetherian and $B = \bigcap_{V\in X} V$, we have $B_P = \bigcap_{V\in X}(VB_P)$ [44, Theorem 3.5] and $\{VB_P : V \in X\}$ is a Noetherian subspace of $\mathfrak{X}$ in the Zariski topology [44, Theorem 3.7]. Thus $\{VB_P : V \in X\}$ satisfies DCC, and since $B_P = \bigcap_{V\in X}(VB_P)$ is a valuation ring, this forces $B_P = VB_P$ for some $V \in X$, and hence $V \subseteq B_P$. Since V is essential for A, hence for B, $V = B_Q$ for some t-prime ideal Q of B. Therefore, since P is a t-maximal ideal and $B_Q \subseteq B_P$, we have $B_P = V \in X$, which shows $\{B_P : P \in t\text{-Max}B\} \subseteq X$.

To verify the reverse inclusion, let $W \in X$. Then since W is essential, $W = B_Q$ for some prime ideal Q in B, and hence Q is a t-prime ideal of B. Let L be a maximal t-ideal of B containing Q. Then, by what we have shown, $B_L \in X$, so since $B_L \subseteq W$ and the members of X are incomparable, it must be that $B_L = W = B_Q$, proving that $Q \in t\text{-Max}B$. This proves that $X = \{B_P : P \in t\text{-Max}B\}$. Therefore, it follows from [51, Lemma 5, p. 119] that X is homeomorphic to t-Max$B$, so that t-Max$B$ is discrete in the inverse topology. By Theorem 6.4 each member of X is strongly irredundant in some representation of B. By Proposition 4.11, the valuation rings in X are critical for B, so by (4.14), X is a strongly irredundant representation of B.

Conversely, suppose each collection X of incomparable valuation rings in $\{A_P : P \in t\text{-Spec}A\}$ is a strongly irredundant representation of $\bigcap_{V\in X} V$. Then by Theorem 6.4 each such collection X is discrete in the inverse topology, and consequently, each collection of incomparable prime ideals in t-Spec$A$ is discrete in the inverse topology. Therefore, by Theorem 2.5, t-Max$A$ is Noetherian in the Zariski topology.

Now suppose that t-Spec$A$ is Noetherian. As noted above, $E = \{A_P : P \in t\text{-Spec}A\}$ is a spectral space; it is also Noetherian since t-Spec$A$ is. Therefore, for any $V \in E$, the set $\{W \in E : V \subseteq W\}$ satisfies DCC. Thus, if X is any collection of incomparable valuation rings in E, since we have established that each valuation ring in X is strongly irredundant in the intersection $\bigcap_{V \in X} V$, it follows that each valuation ring in X is tightly irredundant.

Conversely, suppose that each collection X of incomparable critical valuation overrings is a tightly irredundant representation of $\bigcap_{V \in X} V$. We have established already that this implies that t-Max$A$ is Noetherian. Thus by Lemma 2.4, to prove that t-Spec$A$ is Noetherian, we need only verify that for each $V \in E$, the set $\{W \in E : V \subsetneq W\}$ has a minimal element with respect to $\subseteq$. Let $V \in E$. Then by assumption V is tightly irredundant in the representation $\{V\}$ of the ring V, so the intersection of the valuation rings in $\{W \in E : V \subsetneq E\}$ is again in this same set. Therefore, t-Spec$A$ is Noetherian. □

Gabelli, Houston and Lucas [20] define a domain A to have property ($t\#$) if whenever X is a collection of pairwise incomparable t-prime ideals and $P \in X$, $\bigcap_{Q \neq P} A_Q \subsetneq A_P$, where Q ranges over the prime ideals in X distinct from P. Using Theorem 6.9, additional characterizations of PvMDs with Noetherian t-maximal spectrum can be deduced from the work of Gabelli, Houston and Lucas; see for example Propositions 2.4, 2.6, and 2.8 of [20]. Similarly, for a PvMD A, the characterization of Noetherian spectral spaces given in Lemma 2.3 can be used to link the property in which t-Spec$A$ is Noetherian to the work of Gabelli, Houston, and Lucas, specifically to the equivalent characterizations in Propositions 2.11 and Theorem 2.14 in [20]. For additional applications, see [4]. For example, it follows from Corollary 2.6 and [4, Theorem 3.9] that A is a generalized Krull domain if and only if A is a PvMD for which t-Max$A$ is Noetherian and $P \neq (P^2)_t$ for each t-prime ideal P of A.

The ($t\#$) property extends to non-Prüfer settings the property (#) introduced for Prüfer domains by Gilmer [21] and studied further by Gilmer and Heinzer in [23]. A Prüfer domain A is said to satisfy (#) if for each maximal ideal M of A, A_M is irredundant in the representation $\{A_N : N \in \text{Max}\,A\}$ of A. Property (#) and the stronger version (##), which requires that every overring has (#), play an important role in the local–global theory of Prüfer domains; see for example [15, 16]. Since every maximal ideal of a Prüfer domain is a t-maximal ideal, the properties (#) and ($t\#$) coincide for Prüfer domains. Thus, we have the following topological characterization of Prüfer domains satisfying (##). The corollary, which is immediate from Theorem 6.9, is implicit in [19, Theorem 5.14], where it is proved using the work of Rush and Wallace [48].

Corollary 6.10 *Suppose A is Prüfer domain. Then Max A is a Noetherian space if and only if R satisfies* (##). □

7 Irredundance in Intersections of Irreducible Ideals

An ideal A of the ring R is *irreducible* if A is not an intersection of two ideals properly containing it. Every ideal A is the intersection of the irreducible ideals containing it. Indeed, if $r \in R \setminus A$, then by Zorn's Lemma, there exists an ideal of R maximal with respect to containing A and not containing r. This ideal is necessarily irreducible, from which it follows that A is an intersection of irreducible ideals. In this section, we are interested in when A is an irredundant intersection of irreducible ideals. We show how the topological approach of Sect. 3 can be applied to intersection decompositions of irreducible ideals in *arithmetical rings*, those rings R for which the ideals of R_M form a chain for each maximal M of R. The ambient spectral space from which these intersection representations are drawn is $\operatorname{Irr} R$, the set of proper irreducible ideals of R, viewed as a topological space having a basis of closed sets of the form $\mathscr{V}(A) := \{B \in \operatorname{Irr} R : A \subseteq B\}$, where A ranges over all the ideals of R. Since $R \notin \operatorname{Irr} R$, the maximal elements in $\operatorname{Irr} R$ are the maximal ideals of R. By a *representation* of A we mean a subset of $\operatorname{Irr} R$ whose intersection is A.

Lemma 7.1 *Let R be an arithmetical ring. Then for each proper ideal A of R, $\mathscr{V}(A)$ is a spectral representation of A that does not contain any proper closed representations of A.*

Proof Every irreducible ideal in an arithmetical ring is strongly irreducible [31, Lemma 2.2(3)]. Also, the intersection of two finitely generated ideals in an arithmetical ring is finitely generated [50, Corollary 1.11], so by Example 2.2(4), $\mathscr{V}(A)$ is a spectral representation of A. Finally, suppose X is a closed subset of $\mathscr{V}(A)$ that is a representation of A. Then $X = \mathscr{V}(B)$ for some proper ideal B of R. Since every proper ideal of R is an intersection of irreducible ideals, B is the intersection of the ideals in $\mathscr{V}(B)$. Since $\mathscr{V}(B)$ is a representation of A, this forces $A = B$. Therefore, no proper closed subset of $\mathscr{V}(A)$ is a representation of A. □

Let A be a proper ideal of the arithmetical ring R. Since $\mathscr{V}(A)$ is a spectral space and the specialization order agrees with set inclusion, $\mathscr{V}(A)$ contains minimal elements with respect to set inclusion and A is an intersection of these minimal irreducible ideals. Using the ideas developed in Sect. 3, along with the results about Noetherian spectral spaces in Sect. 2, we obtain a version of a theorem proved in [19] by different methods. We recall that a few notions from [19]. A *Krull associated prime* of an ideal A of a ring R is a prime ideal that is a union of ideals of the form $A : r = \{s \in R : rs \in A\}$ with $r \in R$. If P is a prime ideal of R, we set $A_{(P)} := \{r \in R : br \in A$ for some $b \in R \setminus P\}$. A Zorn's Lemma argument shows that the set of Krull associated primes of A contains maximal elements. Let $\mathscr{X}_A$ denote the set of these maximal elements. Then (with R arithmetical) we have $\{A_{(P)} : P \in \mathscr{X}_A\} = \operatorname{Min} \mathscr{V}(A)$; see [19, Theorem 5.8].

Theorem 7.2 (cf. [19, Theorem 5.14]) *If R is an arithmetical ring for which* $\operatorname{Max} R$ *is Noetherian in the Zariski topology, then for each proper ideal A of R the set of irreducible ideals that are minimal over A is a strongly irredundant representation* *of A, and this is the unique irredundant representation of A as an intersection of irreducible ideals.*

Proof Let A be a proper ideal of R. As discussed before the theorem, $\text{Min}\,\mathscr{V}(A) = \{A_{(P)} : P \in \mathscr{X}_A\}$. Let $P \in \mathscr{X}_A$. Then by Lemma 2.3 there is a finitely generated ideal $I \subseteq P$ such that I is not contained in any maximal ideal that does not contain P. Since R is an arithmetical ring, the prime ideals of R form a tree under inclusion, so the only prime ideal in $\mathscr{X}_A$ that contains I is P. Thus since P is a union of ideals of the form $A : r$, $r \in R$, there are $r_1, \ldots, r_n \in R$ such that P is the only ideal in $\mathscr{X}_A$ containing $(A : r_1) + \cdots + (A : r_n)$. Since R is arithmetical, the latter ideal is equal to $A : (r_1 R \cap \cdots \cap r_n R)$ [37, Exercise 18(c), p. 151]. Hence $r_1 R \cap \cdots \cap r_n R \subseteq \left(\bigcap_{Q \neq P} A_{(Q)}\right) \setminus A_{(P)}$, where Q ranges over the prime ideals in $\mathscr{X}_A \setminus \{P\}$. This shows that the representation $\text{Min}\,\mathscr{V}(A) = \{A_{(P)} : P \in \mathscr{X}_A\}$ of A is irredundant.

Next, since by Lemma 7.1, $\mathscr{V}(A)$ is a minimal closed representation of A, we have by Theorem 3.6 that $\text{Min}\,\mathscr{V}(A)$ is a strongly irredundant representation of A.

Finally, since $\mathscr{V}(A)$ is a minimal closed representation of A, all the irreducible ideals in $\mathscr{V}(A)$ are critical for A in the spectral representation $\mathscr{V}(A)$ of A. Therefore, by Corollary 3.12, every irredundant representation of A is strongly irredundant, and hence by Theorem 3.13(4), there is a unique irredundant representation of A. □

Remark 7.3 By [19, Theorem 5.14], the converse of Theorem 7.2 is also true: If every ideal A of a ring R can be written uniquely as an irredundant intersection of the irreducible ideals that are minimal with respect to containing A, then R is an arithmetical ring with Noetherian maximal spectrum.

Remark 7.4 The ideas in Sect. 3 can also be applied to the intersections of prime ideals. Let A be a radical ideal of a ring R. Then, $\mathscr{V}(A) = \{P \in \text{Spec}\,R : A \subseteq P\}$ is a spectral representation of A, each member of which is critical for A. Thus A has a strongly irredundant representation if and only if the set of minimal primes of R/A contains a dense set of isolated points with respect to the Zariski topology (Corollary 3.7 and Theorem 3.13). Also, every irredundant representative of A is strongly irredundant (Corollary 3.12), and there is at most one irredundant representation of A (Theorem 3.13). Finally, if $\mathscr{V}(A)$ is countable, then A has a strongly irredundant representation (Corollary 3.9).

References

1. S. Abhyankar, P. Eakin, W. Heinzer, On the uniqueness of the coefficient ring in a polynomial ring. J. Algebra **23**, 310–342 (1972)
2. J. Brewer, Integral domains of finite character. II. J. Reine Angew. Math. **251**, 7–9 (1971)
3. J. Brewer, J. Mott, Integral domains of finite character. J. Reine Angew. Math. **241**, 34–41 (1970)
4. S. El Baghdadi, On a class of Prüfer v-multiplication domains. Comm. Algebra **30**, 3723–3742 (2002)

5. P.J. Cahen, J.L. Chabert, *Integer-valued Polynomials, Mathematical Surveys and Monographs*, vol. 48 (American Mathematical Society, Providence, RI, 1997)
6. J.L. Chabert, G. Peruginelli, Polynomial overrings of Int($\mathbb{Z}$), J. Commut. Algebra 8 (2016), 1–28
7. A. Fabbri, Integral domains having a unique Kronecker function ring. J. Pure Appl. Algebra **215**, 1069–1084 (2011)
8. C. Finocchiaro, Spectral spaces and ultrafilters. Comm. Algebra **42**, 1496–1508 (2014)
9. C. Finocchiaro, M. Fontana, K.A. Loper, The constructible topology on spaces of valuation domains. Trans. Am. Math. Soc. **365**, 6199–6216 (2013)
10. C. Finocchiaro, M. Fontana, K.A. Loper, Ultrafilter and constructible topologies on spaces of valuation domains. Comm. Algebra **41**(5), 1825–1835 (2013)
11. C. Finocchiaro, M. Fontana, K.A. Loper, Some closure operations in Zariski-Riemann spaces of valuation domains: a survey, Commutative algebra (Springer, New York, 2014), 153–173
12. C. Finocchiaro, M. Fontana, D. Spirito, New distinguished classes of spectral spaces: a survey, this volume
13. C. Finocchiaro, D. Spirito, Some topological considerations on semistar operations. J. Algebra **409**, 199–218 (2014)
14. C. Finocchiaro, F. Tartarone, On a topological characterization of Prüfer v-multiplication domains among essential domains, arXiv:1410.4037
15. M. Fontana, E. Houston, T. Lucas, *Lecture Notes of the Unione Matematica Italiana*, vol. 14, Factoring Ideals in Integral Domains (Springer, Heidelberg, 2013). UMI, Bologna
16. M. Fontana, J. Huckaba, I. Papick, *Prüfer Domains* (Marcel Dekker, New York, 1997)
17. M. Fontana, M. Zafrullah, *On v-Domains: A Survey, in Commutative Algebra-Noetherian and Non-Noetherian Perspectives* (Springer, New York, 2011), pp. 145–179
18. L. Fuchs, On primal ideals. Proc. Am. Math. Soc. **1**, 1–6 (1950)
19. L. Fuchs, W. Heinzer, B. Olberding, Commutative ideal theory without finiteness conditions: primal ideals. Trans. Am. Math. Soc. **357**(7), 2771–2798 (2005)
20. S. Gabelli, E. Houston, T. Lucas, The t#-property for integral domains. J. Pure Appl. Algebra **194**, 281–298 (2004)
21. R. Gilmer, Overrings of Prüfer domains. J. Algebra **4**, 331–340 (1966)
22. R. Gilmer, Multiplicative ideal theory. Queen's Papers in Pure and Applied Mathematics, vol. 12, Queen's University, Kingston, Ont (1968)
23. R. Gilmer, W. Heinzer, Overrings of Prüfer domains. II. J. Algebra **7**, 281–302 (1967)
24. R. Gilmer, W. Heinzer, Irredundant intersections of valuation rings. Math. Z. **103**, 306–317 (1968)
25. D. Goldschmidt, Algebraic Functions and Projective Curves, Grad. Texts in Math, vol. 215 (Springer, New York, 2003)
26. M. Griffin, Families of finite character and essential valuations. Trans. Am. Math. Soc. **130**, 75–85 (1968)
27. M. Griffin, Rings of Krull type. J. Reine Angew. Math. **229**, 1–27 (1968)
28. F. Halter-Koch, Kronecker function rings and generalized integral closures. Comm. Algebra **31**(1), 45–59 (2003)
29. W. Heinzer, J. Ohm, Noetherian intersections of integral domains. Trans. Am. Math. Soc. **167**, 291–308 (1972)
30. W. Heinzer, J. Ohm, Defining families for integral domains of real finite character. Can. J. Math. **24**, 1170–1177 (1972)
31. W. Heinzer, L. Ratliff, D. Rush, Strongly irreducible ideals of a commutative ring. J. Pure Appl. Algebra **166**(3), 267–275 (2002)
32. O. Heubo-Kwegna, Kronecker function rings of transcendental field extensions. Comm. Algebra **38**(8), 2701–2719 (2010)
33. M. Hochster, Prime ideal structure in commutative rings. Trans. Am. Math. Soc. **142**, 43–60 (1969)
34. B.G. Kang, Prüfer v-multiplication domains and the ring $R[X]_{N_v}$. J. Algebra **123**, 151–170 (1989)

35. W. Krull, Über die Zerlegung der Hauptideale in allgemeinen Ringen. Math. Ann. **105**(1), 1–14 (1931)
36. O. Kwegna-Heubo, Kronecker function rings of transcendental field extensions. Comm. Algebra **38**, 2701–2719 (2010)
37. M. Larsen, P.J. McCarthy, *Multiplicative theory of ideals, Pure and Applied Mathematics*, vol. 43 (Academic Press, New York-London, 1971)
38. K.A. Loper, F. Tartarone, A classification of the integrally closed rings of polynomials containing $\mathbb{Z}[x]$. J. Comm. Algebra **1**, 91–157 (2009)
39. S. Mazurkiewicz, W. Sierpinski, Contribution á la topologie des ensembles dénombrables. Fund. Math. **1**, 17–27 (1920)
40. J. Ohm, Some counterexamples related to integral closure in $D[[x]]$. Trans. Am. Math. Soc. **122**, 321–333 (1966)
41. B. Olberding, Irredundant intersections of valuation overrings of two-dimensional Noetherian domains. J. Algebra **318**, 834–855 (2007)
42. B. Olberding, Overrings of two-dimensional Noetherian domains representable by Noetherian spaces of valuation rings. J. Pure Appl. Algebra **212**, 1797–1821 (2008)
43. B. Olberding, Holomorphy rings in function fields, in *Multiplicative Ideal Theory in Commutative Algebra* (Springer, Berlin, 2006), pp. 331–348
44. B. Olberding, Noetherian space of integrally closed domains with an application to intersections of valuation rings. Comm. Algebra **38**(9), 3318–3332 (2010)
45. B. Olberding, Affine schemes and topological closures in the Zariski–Riemann space of valuation rings. J. Pure Appl. Algebra **219**(5), 1720–1741 (2015)
46. B. Olberding, F. Tartarone, Integrally closed rings in birational extensions of two-dimensional regular local rings. Math. Proc. Camb. Phil. Soc. **155**, 101–127 (2013)
47. E. Pirtle, Families of valuations and semigroups of fractionary ideal classes. Trans. Am. Math. Soc. **144**, 427–439 (1969)
48. D. Rush, L. Wallace, Noetherian maximal spectrum and coprimely packed localizations of polynomial rings. Houston J. Math. **28**, 437–448 (2002)
49. N. Schwartz, M. Tressl, Elementary properties of minimal and maximal points in Zariski spectra. J. Algebra **323**, 698–728 (2010)
50. T. Shores, R. Wiegand, Rings whose finitely generated modules are direct sums of cyclics. J. Algebra **32**, 152–172 (1974)
51. O. Zariski, P. Samuel, *Commutative Algebra*, vol. 29, Vol. II. Graduate Texts in Mathematics (Springer, New York-Heidelberg, 1975)

Idempotent Pairs and PRINC Domains

Giulio Peruginelli, Luigi Salce and Paolo Zanardo

Dedicated to Franz Halter-Koch on the occasion of his 70th birthday.

Abstract A pair of elements a, b in an integral domain R is an idempotent pair if either $a(1-a) \in bR$, or $b(1-b) \in aR$. R is said to be a PRINC domain if all the ideals generated by an idempotent pair are principal. We show that in an order R of a Dedekind domain every regular prime ideal can be generated by an idempotent pair; moreover, if R is PRINC, then its integral closure, which is a Dedekind domain, is PRINC, too. Hence, a Dedekind domain is PRINC if and only if it is a PID. Furthermore, we show that the only imaginary quadratic orders $\mathbb{Z}[\sqrt{-d}]$, $d > 0$ square-free, that are PRINC and not integrally closed, are for $d = 3, 7$.

Keywords Orders · Conductor · Primary decomposition · Dedekind domains · Principal ideals · Projective-free

2010 Mathematics Subject Classification: 13G05 · 13F05 · 13C10 · 11R11

G. Peruginelli · L. Salce (✉) · P. Zanardo
Department of Mathematics, University of Padova, Via Trieste 63, 35121 Padua, Italy
e-mail: salce@math.unipd.it

G. Peruginelli
e-mail: gperugin@math.unipd.it

P. Zanardo
e-mail: pzanardo@math.unipd.it

S. Chapman et al. (eds.), *Multiplicative Ideal Theory and Factorization Theory*, Springer Proceedings in Mathematics & Statistics 170,
DOI 10.1007/978-3-319-38855-7_13

1 Introduction

Let R be an integral domain, $M_n(R)$ the ring of matrices of order n with entries in R, T any singular matrix in $M_n(R)$ (i.e., $\det T = 0$). A natural question is to find conditions on R to ensure that T is always a product of idempotent matrices. This problem was much investigated in past years, see [17] for comprehensive references. The case when R is a Bézout domain (i.e., the finitely generated ideals of R are all principal) is crucial. In fact, we recall three fundamental results, valid for matrices with entries in a Bézout domain. Laffey [9] showed that every square singular matrix T with entries in an Euclidean domain is a product of idempotent matrices if and only if the same property is satisfied just for 2×2 matrices. This result was extended to Bézout domains in [17]. Arguably, the most important result in this subject is due to Ruitenburg [16], who proved that every singular matrix $T \in M_n(R)$ is a product of idempotent matrices if and only if every invertible matrix $U \in M_n(R)$ is a product of elementary matrices. From Ruitenburg's theorem and a result by O'Meara [13], we derive that every singular matrix with entries in a Bézout domain R is a product of idempotents if and only if R admits a *weak Euclidean algorithm* (see [17] for the definitions and other details). As a matter of fact, a major problem in this subject is to establish whether the property that any singular matrix T with entries in R is a product of idempotent matrices implies that R is a Bézout domain.

While investigating this problem in [17], the second and third authors were led to define the property (princ) of an integral domain R. We rephrase the property in a way more suitable to our discussion.

Two elements a, b of a commutative integral domain R are said to form an *idempotent pair* if they are the entries of a row, or of a column, of a 2×2 nonzero idempotent singular matrix. Since a nonzero matrix $T = (a_{ij}) \in M_2(R)$, different from the identity, is idempotent if and only if $\det(T) = 0$ and T has trace 1, it is easily seen that $a, b \in R$ form an idempotent pair if and only if either $a(1-a) \in bR$, or $b(1-b) \in aR$. We say that an integral domain R satisfies the (princ) property if all the ideals generated by idempotent pairs are principal; such a ring will be called *PRINC* domain. The class of PRINC domains obviously includes Bézout domains. This class of domains was also investigated by McAdam and Swan in [10] and [11] under the name UCFD (unique comaximal factorization domain) and from a different point of view (see Remark 8).

In [17] it is proved that if a PRINC domain satisfies the condition that 2×2 singular matrices are product of idempotent matrices, then it is necessarily a Bézout domain. A similar result was proved recently in [1] by Bashkara Rao, who assumed the domain R to be projective-free instead than PRINC; recall that a domain R is projective-free if finitely generated projective R-modules are free. Actually, Bashkara Rao's result is a particular case of the mentioned result of [17], since the class of PRINC domains, as proved in [17], includes, besides Bézout domains, the projective-free domains and the unique factorization domains. In [17] it was also claimed (without proof) that the ring $\mathbb{Z}[\sqrt{-3}]$ is an example of a nonintegrally closed PRINC domain.

In Sect. 1 of this paper we provide a characterization of ideals generated by idempotent pairs, related to invertible ideals in domains of finite character and PRINC domains.

In Sect. 2 we consider orders in rings of algebraic integers. We prove that a prime ideal of such an order O which is comaximal with the conductor of O is generated by an idempotent pair. This fact implies that the maximal ideals of Dedekind domains are generated by idempotent pairs; it follows that a Dedekind domain which is not a PID cannot be a PRINC domain. Another relevant consequence is that if the order O is a PRINC domain, then its integral closure is necessarily a PID. However, concerning the orders $\mathbb{Z}[\sqrt{-d}]$ in quadratic imaginary extensions of $\mathbb{Q}$ whose integral closures are PIDs, we prove that they are not PRINC domains, when $d = 11, 19, 43, 67, 163$.

On the other hand, in Sect. 3 we prove that the rings $\mathbb{Z}[\sqrt{-3}]$ and $\mathbb{Z}[\sqrt{-7}]$ are PRINC domains.Therefore, from this point of view, $\mathbb{Z}[\sqrt{-3}]$ and $\mathbb{Z}[\sqrt{-7}]$ are exceptional among the imaginary quadratic orders $\mathbb{Z}[\sqrt{-d}]$, $d > 0$ square-free. We also prove that the invertible ideals of these two rings are principal, and from this fact we deduce that they are projective-free. Let us note that a first proof that $\mathbb{Z}[\sqrt{-3}]$ is a PRINC domain was privately communicated by U. Zannier to the third author; that proof used arguments different from those used in Theorem 20 of the present paper.

2 Ideals Generated by an Idempotent Pair

In what follows, every ring R considered will be a commutative integral domain. Some results proved in this section are valid also for commutative rings. If I is an ideal of R, generated by $x_1, \ldots, x_n \in R$, we will use the notation $I = \langle x_1, \ldots, x_n \rangle$.

We recall the definitions given in the introduction: $a, b \in R$ are said to be an idempotent pair if either $a(1-a) \in bR$, or $b(1-b) \in aR$, and R satisfies property (princ) if all the ideals generated by idempotent pairs are principal. For short, we will say that R is a PRINC domain.

Lemma 1 *Let I be an invertible ideal I of a UFD R. Then I is principal.*

Proof Assume, for a contradiction, that the minimal number of generators of I is $n \geq 2$, say $I = \langle x_1, \ldots, x_n \rangle$ (in particular, I is a proper ideal). We may assume, without loss of generality, that $GCD(x_1, \ldots, x_n) = 1$. Take any element $y/z \in I^{-1}$, where $GCD(y, z) = 1$. Then, from $x_i y/z \in R$ for all $i \leq n$, it follows that z divides x_i for every $i \leq n$. We conclude that z is a unit of R, hence $I^{-1} \subseteq R$, since y/z was arbitrary. Then $R = II^{-1} \subseteq I$, impossible. □

Recall that two ideals I and J of a commutative ring R are said to be comaximal if $I + J = R$.

Lemma 2 *Let I and J be two comaximal ideals of a commutative ring R. Then there exists an element $a \in I$ such that $a - 1 \in J$, implying that $a^2 - a \in IJ$ and $I = aR + IJ$.*

Proof Since $I + J = R$, there exists an element $a \in I$ such that $a - 1 \in J$. Hence $a^2 - a = a(a-1) \in IJ$. Also, $aR + IJ = aR + I(aR + J) = aR + I = I$, since aR and J are comaximal. Thus $I = aR + IJ$. Similarly, $J = (a-1)R + IJ$. □

The next theorem gives different characterizations for ideals generated by an idempotent pair.

Theorem 3 *Let I be a nonzero ideal of an integral domain R. The following conditions are equivalent:*

1. *I is generated by an idempotent pair.*
2. *$I = \langle a, b\rangle = \langle a^2, b\rangle$, for some $a, b \in R$.*
3. *There exists an ideal J such that I and J are comaximal and such that IJ is a principal ideal.*

In particular, if one of the above equivalent conditions holds, then I is invertible.

Proof (1) ⇒ (3): Suppose that $I = \langle a, b\rangle$, where a, b is an idempotent pair, so $a^2 - a = bc$, for some $c \in R$. Let $J = \langle a-1, b\rangle$. Then I and J are comaximal and we have

$$IJ = \langle a, b\rangle\langle a-1, b\rangle = \langle bc, ab, b(a-1), b^2\rangle = b\langle c, a, a-1, b\rangle = bR,$$

(3) ⇒ (2): By assumption there exists an ideal J such that $I + J = R$ and $IJ = bR$, for some $b \in R$. By Lemma 2, there exists $a \in I$ such that $I = \langle a, b\rangle$, with $a^2 - a \in bR$. In particular, it follows that $\langle a, b\rangle = \langle a^2, b\rangle$.

(2) ⇒ (1): Since $\langle a, b\rangle = \langle a^2, b\rangle$, there exist $\lambda, \mu \in R$ such that $a = \lambda a^2 + \mu b$. This implies that $a - \lambda a^2 = a(1 - \lambda a) = \mu b$, so $\lambda a(1 - \lambda a) = \lambda\mu b$, and $\lambda a, b$ form an idempotent pair. Obviously, $\langle \lambda a, b\rangle \subseteq I$. Conversely, since $a \in \langle \lambda a, b\rangle$, also $a^2 \in \langle \lambda a, b\rangle$, consequently $I = \langle a^2, b\rangle \subseteq \langle \lambda a, b\rangle$. So $I = \langle \lambda a, b\rangle$.

The last claim follows immediately from the characterization at the point (3). □

Recall that R is said to be a Bézout domain if every finitely generated ideal of R is principal, and R is called projective-free if every finitely generated projective R-module is free.

Proposition 4 *If an integral domain R is either Bézout, or UFD, or projective-free, then R satisfies property (princ).*

Proof A Bézout domain is trivially a PRINC domain. Moreover, by Theorem 3, every ideal I of R generated by an idempotent pair is invertible. Then such I is free, hence principal, when R is projective-free. Finally, if R is a UFD, then I is principal by Lemma 1. □

Corollary 5 *Let R be an integral domain, b an element of R such that bR is a finite product of primary ideals that are pairwise comaximal. Let Q be such a primary ideal and let P denote its radical. Then there exists $a \in Q$ such that a, b form an idempotent pair, and P is the radical of $\langle a, b \rangle$.*

Proof By assumption, $bR = QJ$, where Q and J are comaximal. Hence, by Lemma 2 (or Theorem 3, (3)), there exists $a \in Q$ such that a, b is an idempotent pair and $Q = \langle a, b \rangle$. □

In the case of a domain of finite character, the last claim of Theorem 3 can be reversed. We recall that a domain is said to be of finite character if each nonzero element is contained in finitely many maximal ideals; moreover, an invertible ideal I of a domain of finite character is $1\frac{1}{2}$ generated, that is, I is generated by two elements and one of the two generators can be arbitrarily chosen among the nonzero elements of the ideals (see [5, Proposition 2.5e, p. 12]).

Corollary 6 *Let R be an integral domain of finite character and I an invertible ideal. Then I is generated by an idempotent pair.*

Proof Choose $0 \neq a \in I$. By the aforementioned Proposition 2.5 in [5], there exists $b \in I$ such that $\langle a^2, b \rangle = I$. Now, $I = \langle a^2, b \rangle \subseteq \langle a, b \rangle \subseteq I$, so each of the previous containments is indeed an equality. By Theorem 3, I is generated by an idempotent pair. □

As an application to domains of finite character, we derive the following

Corollary 7 *Let R be a domain with finite character. Then R is a PRINC domain if and only if all invertible ideals are principal. In particular, if R is also Prüfer, then R is a Bézout domain.*

Proof The sufficiency holds for any domain R, by [17, Proposition 4.2]. The necessity follows from the preceding proposition. □

Remark 8 McAdam and Swan defined the notion of an S-ideal I in [10] as in condition 2. of Theorem 3, in the context of a definition analog to that of unique factorization domain. We recall this definition here. Let R be an integral domain. A nonzero non-unit element b of R is pseudo-irreducible if it is not possible to factor b as $b = cd$ with c and d comaximal non-units. The domain R is called *comaximal factorization domain* (CFD) if any nonzero non-unit element b has a complete comaximal factorization, namely, $b = b_1 \cdot \ldots \cdot b_m$, where the b_i's are pairwise comaximal pseudo-irreducible elements of R. A CFD is a *unique comaximal factorization domain* (UCFD) if complete comaximal factorizations are unique. In [10, Theorem 1.7] they show that if R is a CFD, then R is UCFD if and only if every S-ideal is principal, that is, R is a PRINC domain, by Theorem 3. They prove also that a domain with finite character is a CFD ([10, Lemma 1.1]), from which our Corollary 7 also follows.

3 Orders in Number Fields and Idempotent Pairs

We recall the definition of order.

Definition 1 An integral domain O is an order if its integral closure D in its quotient field is a Dedekind domain which is finitely generated as an O-module.

By a well-known result of Eakin [4], it follows that O is Noetherian as well. So, an order is a one-dimensional Noetherian domain. We say that an order is proper if it is not integrally closed. We recall that a Dedekind domain is characterized by the fact that each ideal can be written uniquely as an intersection, or equivalently as a product, of powers of prime ideals. On the other hand, since an order is a one-dimensional Noetherian domain, each ideal of an order can be written uniquely as a product of primary ideals (see for example [18, Theorem 9, Chap. IV, Sect. 5, p. 213]). It follows that in a proper order there are some primary ideals of the order O which are not equal to a power of a prime ideal. We recall that the conductor of the integral closure D of an order O is defined as

$$\mathfrak{f} \doteqdot (O : D) = \{x \in O \mid xD \subseteq O\}.$$

The conductor is the largest ideal of O which is also an ideal of D. Since D is a finitely generated O-module, $\mathfrak{f}$ is nonzero. Following the terminology of [12], we call an ideal of O (or of D) comaximal with $\mathfrak{f}$ a *regular* ideal.

We will need the following easy fact.

Lemma 9 *Let O be an order and P a prime ideal. Then the set of P-primary ideals of O is linearly ordered if and only if P is regular. If this condition holds, then each P-primary ideal is equal to a power of P.*

Proof Since the local ring O_P is a local one-dimensional noetherian ring, every nonzero ideal is PO_P-primary. We recall that there is a one-to-one correspondence between P-primary ideals of O and PO_P-primary ideals of O_P. Hence, the set of P-primary ideals of O is linearly ordered if and only if the set of ideals of O_P is linearly ordered, that is, O_P is a valuation domain. By [12, Proposition 12.10] this condition is equivalent to the fact that P is a regular prime.

If one of these equivalent conditions holds, then O_P is a DVR, so its ideals are powers of the maximal ideal PO_P. Hence, the P-primary ideals of O are powers of P. □

The following result is a consequence of Corollary 5, since an order is a domain of finite character; we include a direct proof, which makes use of a different technique.

Proposition 10 *Let O be an order and P a regular prime ideal of O. Then there exists an idempotent pair that generates P. In particular, if O is integrally closed, then every prime ideal is generated by an idempotent pair.*

Proof Let

$$\mathfrak{f} = \prod_{i=1}^{k} Q_i$$

be the primary decomposition of $\mathfrak{f}$ in O, where the Q_i's are primary ideals of O with distinct radicals $P_i, i = 1, \ldots, k$. Now take $x \in P \setminus P^2$. Since P is comaximal with $\mathfrak{f}$, there exists $b \in O$ which satisfies the following conditions:

$$b \equiv x \pmod{P^2}$$
$$b \equiv 1 \pmod{P_i}, \quad \forall\, i = 1, \ldots, k.$$

In particular, bO is comaximal with $\mathfrak{f}$, hence the primary components of bO are comaximal with the conductor, so, in particular, they are regular. Hence, by Lemma 9, we get a primary decomposition of bO of the form

$$bO = P \prod_{j=1}^{n} P_j'^{e_j}$$

where the P_j''s are prime (i.e., maximal) ideals of O distinct from the P_i's.

We are in the position to apply Corollary 5, hence there exists $a \in P$ such that a, b form an idempotent pair and P is the radical of $\langle a, b \rangle$. Since $\langle a, b \rangle$ is a product of primary ideals, $b \notin P^2$ and P is regular, by Lemma 9 it follows that $P = \langle a, b \rangle$, as required.

Finally, note that O is integrally closed if and only if $\mathfrak{f} = O$. In this case, every prime ideal of O is comaximal with the conductor, so every prime ideal can be generated by an idempotent pair. □

Remark 11 The congruences in the proof of Proposition 10 are not necessary. Indeed, the crucial conditions are:

(i) the P-primary component of bO is regular.

(ii) $b \in P \setminus P^2$.

In fact, if

$$bO = P \prod_{j=1}^{n} Q_j$$

where Q_j are primary ideals of O (not necessarily regular), we may take $a \in O$ satisfying the conditions:

$$a \equiv b \pmod{P}$$
$$a \equiv b + 1 \pmod{Q_j}, \ \forall j = 1, \ldots, k$$

and the same conclusion follows: $a(1 - a) \in bO$, and P is the only primary ideal that contains both a and b, so that $\langle a, b \rangle = P$, since $b \notin P^2$ and the P-primary ideals are regular.

On the other hand, if P were not regular we may not get the same conclusion, even imposing the condition $a \equiv b \pmod{P}$, since in this case the P-primary ideals are not linearly ordered. We still get that $\langle a, b\rangle$ is P-primary, though. We will see in Sect. 5 that the conductor of the order $\mathbb{Z}[\sqrt{-3}]$ is an instance of this phenomenon.

Corollary 12 *If an order O is a PRINC domain, then every regular ideal I of O is principal.*

Proof Let $I \subset O$ be a regular ideal. In particular, the primary components of I are regular primary ideals. By Lemma 9, each of these primary components is equal to a power of its own radical. By Proposition 10, each of these radicals is generated by a suitable idempotent pair, hence they are all principal, since O is a PRINC domain. Hence, all the primary components of I are principal as well. It follows that I is equal to a product of principal ideals, so it is principal. □

The next corollary is a consequence of Proposition 10; it also follows from Corollary 7.

Corollary 13 *A Dedekind domain is PRINC if and only if it is a PID.*

Corollary 14 *If an order O is a PRINC domain then its integral closure D is a PRINC domain, or, equivalently, a PID.*

Proof We will show that each prime ideal P of D is principal. Without loss of generality, we may just consider the case of a regular prime ideal P. In fact, there are only finitely many prime ideals of D that divide the conductor, and, by Claborn [3, Corollary 1.6], a Dedekind domain which is not a PID has an infinite number of non-principal prime ideals. Since $P \cap O$ is regular, too (see the remark below), it follows by Corollary 12 that $P \cap O$ is principal. Since P is an extended ideal, $P = (P \cap O)D$, so P is principal, too. □

Remark 15 By the arguments of the proof of Corollary 12, the following conditions are equivalent:

(1) each regular ideal of O is principal.
(2) each regular prime ideal of O is principal.

If one of the two conditions holds, then D is a PID, exactly by the same argument of the proof of Corollary 14: each regular prime ideal P of D is extended, so it is equal to the extension of its contraction, which is principal. More generally, there is a 1–1 correspondence between the regular ideals of D and the regular ideals of O (see for example [14, Lemma 2.26, p. 389]). In particular, each regular ideal of O is contracted and each regular ideal of D is extended.

4 $\mathbb{Z}[\sqrt{-d}]$ Not Satisfying (Princ)

Let $\eta = \sqrt{-d}$, where $d > 0$ is a square-free integer. In this section, we want to establish when the order $O = \mathbb{Z}[\eta]$ fails to be a PRINC domain. By Corollary 14, we know that O cannot be PRINC if its integral closure D is not a PID. So, it is enough to examine the cases when D is a PID, namely when $d = 1, 2, 3, 7, 11, 19, 43, 67, 163$ (see, for instance, [15], p. 81). Of course, when $d = 1, 2$ we get $O = D$, hence O is trivially PRINC. We will focus on the remaining cases.

The next proposition is valid for any $d > 0$.

Proposition 16 *Let $O = \mathbb{Z}[\eta]$, with $\eta = \sqrt{-d}$, where $d > 0$ is any integer (not necessarily square-free). Then we have:*

(1) The element $1 + \eta$ is irreducible in O.

(2) If $a \in \mathbb{Z} \setminus \{\pm 1\}$ properly divides $1 + d$, then $\langle 1 + \eta, a \rangle$ is a proper non-principal ideal of O.

Proof (1) Assume that $1 + \eta = (x + \eta y)(z + \eta t)$ for suitable $x, y, z, t \in \mathbb{Z}$. Taking norms, we get

$$1 + d = (x^2 + dy^2)(z^2 + dt^2),$$

which implies that $y = 0$ or $t = 0$. Assuming that $y = 0$, it follows that $x \in \mathbb{Z}$ divides $1 + \eta \in \mathbb{Z}[\sqrt{-d}]$, which implies that $x = \pm 1$.

(2) Firstly, let us show that $I = \langle 1 + \eta, a \rangle$ is a proper ideal of O. Assuming that I is not proper, we obtain that

$$O = \langle 1 + \eta, a \rangle \langle 1 - \eta, a \rangle \subseteq aO$$

which is a contradiction.

Let us see that I is not principal. Otherwise, we should get $I = (1 + \eta)O$, since $1 + \eta$ is irreducible, hence, in particular, $a = (1 + \eta)(x + \eta y)$, where $x, y \in \mathbb{Z}$, $y \neq 0$. But this is impossible, since

$$a^2 = (1 + d)(x^2 + dy^2) \geq (1 + d)^2 > a^2. \qquad \square$$

It follows from Proposition 16 that, if $1 + d = a(a - 1)$ for some $a \in \mathbb{Z}, a \neq -1$, then O does not satisfy (princ). For example, $\mathbb{Z}[\sqrt{-11}]$ and $\mathbb{Z}[\sqrt{-19}]$ do not satisfy (princ). However, we will prove below a stronger result (Proposition 18).

Like Proposition 16, also the next lemma is valid for any $d > 0$, not necessarily square-free.

Lemma 17 *In the above notation, if $p \in \mathbb{Z}$ is a prime which divides $1 + d$, then the ideal $\langle p, 1 + \eta \rangle$ of $\mathbb{Z}[\eta]$ is prime.*

Proof Since $1+d = pb$, for some $b \in \mathbb{Z}$, then $X^2+d = (X+1)(X-1)+pb \in \langle p, 1+X\rangle\mathbb{Z}[X]$. Hence, if $\pi : \mathbb{Z}[X] \to \mathbb{Z}[\eta]$ is the canonical homomorphism sending X to η, then $\pi^{-1}(\langle p, 1+\eta\rangle\mathbb{Z}[\eta]) = \langle p, 1+X\rangle\mathbb{Z}[X]$, since the latter ideal contains $\langle X^2+d\rangle$, the kernel of π. Therefore,

$$\frac{\mathbb{Z}[\eta]}{\langle p, 1+\eta\rangle} \cong \frac{\mathbb{Z}[X]}{\langle p, 1+X\rangle} \cong \mathbb{Z}/p\mathbb{Z}.$$ □

Proposition 18 *Let $d \in \{11, 19, 43, 67, 163\}$. Then $O = \mathbb{Z}[\sqrt{-d}]$ does not satisfy (princ).*

Proof Note that for each of the relevant d there exists a prime $p \neq 2$ that properly divides $1+d$. Lemma 17 shows that $P = \langle p, 1+\eta\rangle$ is a prime ideal of O, which is comaximal with the conductor $\mathfrak{f}$, since $2 \in \mathfrak{f}$ and p is odd. Then P is generated by an idempotent pair, by Proposition 10. Moreover, since p properly divides $1+d$ (which is the norm of $1+\eta$), by Proposition 16 the same ideal $\langle p, 1+\eta\rangle$ is not principal. We conclude that O does not satisfy (princ). □

The argument in the proof of Proposition 18 neither applies to $\mathbb{Z}[\sqrt{-3}]$ nor to $\mathbb{Z}[\sqrt{-7}]$, since $1+3=4$ and $1+7=8$ are not divisible by an odd prime. In fact, we note that the above proof shows that, in the orders $\mathbb{Z}[\sqrt{-d}]$ with $d \in \{11, 19, 43, 67, 163\}$, there are regular prime ideals that are not principal, but become principal after extending them to their integral closures, which are PIDs. This means that the generator of such an extended ideal lies in the integral closure but not in the corresponding (proper) order. This phenomenon does not happen with the orders $\mathbb{Z}[\sqrt{-3}]$ and $\mathbb{Z}[\sqrt{-7}]$, as we will see in the next section.

5 $\mathbb{Z}[\sqrt{-3}]$ and $\mathbb{Z}[\sqrt{-7}]$ are PRINC Domains

We start recalling some well-known facts on $\mathbb{Z}[\sqrt{-d}] = \mathbb{Z}[\eta]$. Let $d \in \mathbb{Z}$ be a positive square-free integer, which is congruent to 3 modulo 4. Then the ring of integers of $\mathbb{Q}(\eta)$ is $D = \mathbb{Z}[\frac{1+\eta}{2}]$. The conductor $\mathfrak{f}$ of $\mathbb{Z}[\frac{1+\eta}{2}]$ into the order $O = \mathbb{Z}[\eta]$ is equal to:

$$\mathfrak{f} = 2D = \langle 2, 1+\eta\rangle O \subset O.$$

Clearly, the conductor is a maximal ideal of O, since the quotient ring $O/\mathfrak{f}$ is isomorphic to $\mathbb{Z}/2\mathbb{Z}$. As an ideal of O, $\mathfrak{f}$ is not principal. In fact, Proposition 16 applies to $\mathfrak{f} = \langle 2, 1+\eta\rangle$, since 2 properly divides $1+d$. More generally, it is straightforward to show that the conductor of a proper order is always not principal. Moreover, a simple computation shows that $\mathfrak{f}^2 = 2\mathfrak{f}$; it follows that $\mathfrak{f}$ is not invertible, since 2 does not generate $\mathfrak{f}$.

The next technical lemma will be a main ingredient for the proof of the following Theorem 20. We denote by D^* the multiplicative group of the units of a domain D.

Lemma 19 *Let $D = \mathbb{Z}[\frac{1+\sqrt{-3}}{2}]$ and $O = \mathbb{Z}[\sqrt{-3}]$. Then for each $z \in D$, there exists a unit $u \in D^*$ such that $zu \in O$.*

Proof We recall that $D^* = \{\pm 1, \frac{\pm 1 \pm \sqrt{-3}}{2}\}$, namely, the multiplicative group of the 6th roots of unity. If $z \in O$ the claim follows immediately. Let $z \in D \setminus O$; then we may write $z = \frac{a+b\sqrt{-3}}{2}$, for suitable integers a, b, with $a \equiv b \equiv 1 \pmod 2$. We have

$$\frac{a+b\sqrt{-3}}{2} \cdot \frac{1+\sqrt{-3}}{2} = \frac{(a-3b)+(a+b)\sqrt{-3}}{4}$$

and

$$\frac{a+b\sqrt{-3}}{2} \cdot \frac{1-\sqrt{-3}}{2} = \frac{(a+3b)-(a-b)\sqrt{-3}}{4}$$

Looking at the residue classes modulo 4, a direct check shows that either $a - 3b \equiv a + b \equiv 0 \pmod 4$ or $a + 3b \equiv a - b \equiv 0 \pmod 4$, for any possible choice of the odd integers a, b. We conclude that $zu \in O$ for some $u \in D^*$. □

We remark that $\mathbb{Q}(\sqrt{-1})$ and $\mathbb{Q}(\sqrt{-3})$ are the only imaginary quadratic number fields which contains roots of unity distinct from ± 1.

Theorem 20 *Let $\eta = \sqrt{-d}$, where $d \in \{3, 7\}$. Let $P \subset \mathbb{Z}[\eta] = O$ be a prime ideal containing an odd prime p. Then P is principal.*

Proof Let $D = \mathbb{Z}[\frac{1+\eta}{2}]$ be the integral closure of O. Let P be any prime ideal of O containing an odd prime p. In particular, P is regular and so it is a contracted ideal, namely, $PD \cap O = P$ (see Remark 15). We firstly examine the case when $p = d$. Then p is ramified in D, and the unique prime ideal of D above p is $\sqrt{-p}D$. It follows that P is principal, equal to $\sqrt{-p} \cdot O$. Suppose now that p is an odd prime different from d. If pD is a prime ideal, then P is equal to $pD \cap O = pO$, and so P is principal. Suppose that p decomposes in D, so that P is one of the two distinct prime ideals above p in O. We know that PD has norm p (and so P as well, since $p \neq 2$). Since D is an Euclidean domain, it follows that PD is principal, generated by an element $\alpha \in D$ of norm p.

We distinguish now the two cases.

(i) $d = 3$

By Lemma 19, we may multiply α by a suitable unit of $\mathbb{Z}[\frac{1+\sqrt{-3}}{2}]$, to get an element $\alpha' \in \mathbb{Z}[\sqrt{-3}]$ which is associated to α. In particular, α' is a generator of PD which lies in $\mathbb{Z}[\sqrt{-3}]$, so that $\alpha' O = \alpha' D \cap O = PD \cap O = P$.

(ii) $d = 7$

Since α is an element of $\mathbb{Z}[\frac{1+\sqrt{-7}}{2}]$, we may write $\alpha = \frac{a+b\sqrt{-7}}{2}$, for suitable integers a, b, with $a \equiv b \pmod 2$. We have $N(\alpha) = \frac{a^2+7b^2}{4} = p$, or, equivalently,

$$a^2 + 7b^2 = 4p \tag{21}$$

If $a \equiv b \equiv 1 \pmod 2$, then $a^2 \equiv b^2 \equiv 1 \pmod 8$, so, looking at (21), we get $0 \equiv 4p \pmod 8$, a contradiction. This means that, necessarily, $a \equiv b \equiv 0 \pmod 2$, i.e., the generator α of PD actually in $\mathbb{Z}[\sqrt{-7}]$, hence $P = \alpha\mathbb{Z}[\sqrt{-7}]$ follows. □

Corollary 22 *Let $d \in \{3, 7\}$. Let $I \subset \mathbb{Z}[\sqrt{-d}]$ be a regular ideal. Then I is principal.*

Proof The proof follows by Remark 15 and by Theorem 20. □

Corollary 23 *The orders $\mathbb{Z}[\sqrt{-3}]$ and $\mathbb{Z}[\sqrt{-7}]$ are PRINC domains.*

Proof Let $O = \mathbb{Z}[\sqrt{-d}]$, where $d \in \{3, 7\}$. Take any idempotent pair $a, b \in O$; assume, without loss of generality, that $a(1 - a) = bc$, for some $c \in O$. Let us show that $\langle a, b\rangle$ is a principal ideal of O. Assume, for a contradiction, that $\langle a, b\rangle$ is not principal. Then Corollary 22 shows that $\langle a, b\rangle$ is contained in the conductor $\mathfrak{f}$, because $\mathfrak{f}$ is a maximal ideal. Then, obviously, the ideal $\langle 1 - a, b\rangle$ is comaximal with $\mathfrak{f}$, hence it is principal in O. However, recall that $\langle a, b\rangle\langle 1 - a, b\rangle = bO$ (cf. the proof of Theorem 3). Since $\langle 1 - a, b\rangle$ is principal, it follows that also $\langle a, b\rangle$ is principal, impossible. □

We recall that U. Zannier privately communicated, to the third author, a first direct proof of the fact that $\mathbb{Z}[\sqrt{-3}]$ is a PRINC domain.

Since by Corollary the rings 23 $\mathbb{Z}[\sqrt{-3}]$ and $\mathbb{Z}[\sqrt{-7}]$ are PRINC domains of finite character, by Corollary 7 each projective ideal (hence, invertible) of these domains is principal. In the next proposition we give an *ad hoc* argument of this result.

Proposition 24 *Let O be either $\mathbb{Z}[\sqrt{-3}]$ or $\mathbb{Z}[\sqrt{-7}]$. Then every invertible ideal of O is principal.*

Proof Take an arbitrary invertible ideal I of O. Corollary 22 shows that every ideal of O not contained in the conductor $\mathfrak{f}$ is principal. By the unique factorization of ideals of O into primary ideals, it follows that any element $s \in O \setminus \mathfrak{f}$ is a product of prime elements of O, and any ideal I contained in $\mathfrak{f}$ has the form $I = sQ$, where $s \notin \mathfrak{f}$ and Q is $\mathfrak{f}$-primary. Therefore, to prove our statement, it suffices to consider the case when $I = Q$ is $\mathfrak{f}$-primary.

We must show that the invertible ideal Q is principal. Let us consider the localization $O_\mathfrak{f}$ of O at $\mathfrak{f}$. Then the extended ideal $Q_\mathfrak{f}$ is invertible in $O_\mathfrak{f}$, hence it is principal (since local domains are projective-free), say $Q_\mathfrak{f} = aO_\mathfrak{f}$, where we may take $a \in Q$.

By the unique factorization of ideals of O into primary ideals, we readily get $aO = sQ_1$, where Q_1 is $\mathfrak{f}$-primary, and $s \in O \setminus \mathfrak{f}$, so it is a product of prime elements of O. Say $a = sa_1$; then a_1O is $\mathfrak{f}$-primary, equal to Q_1, and we have $Q_\mathfrak{f} = a_1O_\mathfrak{f}$.

Pick now any element $b \in Q$. Then $b = a_1y/t$, for some $t \in O \setminus \mathfrak{f}$. Say $t = \prod_1^n p_i$, where the p_i are prime elements of O. Then from $tb = a_1y$ and $a_1 \notin p_iO$ we get $y/p_i \in O$, for $i = 1, \ldots, n$. It readily follows that $y/t \in O$, whence $b \in a_1O$. Since $b \in Q$ was arbitrary, we conclude that $Q = a_1O$, as required. □

We summarize the previous results in a final statement.

Theorem 25 *Let $d > 0$ be a square-free integer. Then $\mathbb{Z}[\sqrt{-d}]$ does not satisfy property (princ), except $\mathbb{Z}[\sqrt{-1}]$, $\mathbb{Z}[\sqrt{-2}]$, which are PIDs, and $\mathbb{Z}[\sqrt{-3}]$, $\mathbb{Z}[\sqrt{-7}]$.*

We remark that the next proposition can be proved using, instead of the UCS property, Corollary 2.6 in [8], quoted by Heitmann as "Serre's Theorem".

Proposition 26 *The two rings $\mathbb{Z}[\sqrt{-3}]$ and $\mathbb{Z}[\sqrt{-7}]$ are projective-free.*

Proof Let O be one of the two considered rings. O is almost local–global, since its proper quotients are zero-dimensional, hence local–global. Then, by a result by Brewer and Klingler [2] (see also Theorem 4.7 in Chap. V of [5]), O satisfies the UCS-property, that is, every finitely generated submodule M of a free module F with unit content contains a rank-one projective direct summand of F, and hence of itself. By Proposition 24, this direct summand must be cyclic. Let now assume that M is finitely generated projective, and let I be its content. Let $F = M \oplus N$ be a free module containing M as a summand. Then $M \subseteq IF = IM \oplus IN$ implies that $M = IM$, so that, by Nakayama's lemma, there exists an element of O of the form $1 - a$, with $a \in I$, such that $(1 - a)M = 0$. But M is torsion-free, hence $a = 1$ and $I = O$. By the UCS property, M contains a cyclic summand xO, that is, $M = M \oplus xO$. Now an easy induction on the rank of the projective module shows that M is free. □

The following question naturally arises: are there PRINC domains which are neither UFD, nor projective-free? An example of a domain of this kind was exhibited in [10, Sect. 4]. The authors refer also to a paper by Gilmer [6], where an example of a domain containing an n-generated invertible ideal (n an arbitrary positive integer) was given. In [10, Remark p. 189], the authors show that the domain in the example of Gilmer is a PRINC domain which is neither UFD nor projective-free.

The following question is still open.

Question: Let R be a Prüfer domain which is PRINC. Is R a Bézout domain?

We remark that a positive answer was given in [10, Corollary 1.9] under the additional hypothesis that the Prüfer domain is CFD.

Acknowledgments Research supported by "Progetti di Eccellenza 2011/12" of Fondazione CARIPARO and by the grant "Assegni Senior" of the University of Padova. We are grateful to the referee for suggesting many improvements to the paper, especially the useful characterization in Theorem 3.

References

1. K.P.S. Bhaskara Rao, Products of idempotent matrices over integral domains. Linear Algebra Appl. **430**, 2690–2695 (2009)
2. J.W. Brewer, L. Klingler, Pole assignability and the invariant factor theorem. J. Algebra **111**, 536–545 (1987)

3. L. Claborn, Dedekind domains and rings of quotients. Pacific J. Math. **15**, 59–64 (1965)
4. P.M. Eakin Jr, The converse to a well known theorem on Noetherian rings. Math. Ann. **177**, 278–282 (1968)
5. L. Fuchs, L. Salce, *Modules over Non-Noetherian Domains*, Mathematical Surveys and Monographs (American Mathematical Society, Providence, 2001)
6. R. Gilmer, A note on generating sets for invertible ideals. Proc. Amer. Math. Soc. **22**, 426–427 (1969)
7. R. Gilmer, *Multiplicative Ideal Theory* (Dekker, New York, 1972)
8. R. Heitmann, Generating non-Noetherian modules efficiently. Michigan Math. J. **31**(2), 167–180 (1984)
9. T.J. Laffey, Products of idempotent matrices. Linear Multilinear Algebra **14**(4), 309–314 (1983)
10. S. McAdam, R.G. Swan, Unique comaximal factorization. J. Algebra **276**(1), 180–192 (2004)
11. S. McAdam, R.G. Swan, Arithmetical Properties of Commutative Rings and Monoids, *A Special Type of Invertible Ideal*, Lecture Notes in Pure and Applied Mathematics (Chapman & Hall/CRC, Boca Raton, 2005), pp. 356–362
12. J. Neukirch, *Algebraic Number Theory* (Springer, Berlin, 1999)
13. O.T. O'Meara, On the finite generation of linear groups over Hasse domains. J. Reine Angew. Math. **217**, 79–128 (1965)
14. M. Post, H. Zassenhaus, *Algorithmic Algebraic Number Theory*, Encyclopedia of Mathematics and its Applications (Cambridge University Press, Cambridge, 1989)
15. P. Ribenboim, *Algebraic Numbers* (Wiley, New York, 1972)
16. W. Ruitenburg, Products of idempotents matrices over Hermite domains. Semigroup Forum **46**(3), 371–378 (1993)
17. L. Salce, P. Zanardo, Products of elementary and idempotent matrices over integral domains. Linear Algebra Appl. **452**, 130–152 (2014)
18. O. Zariski, P. Samuel, *Commut. Algebra*, vol. I (Springer, New York, 1975)

Some Recent Results and Open Problems on Sets of Lengths of Krull Monoids with Finite Class Group

W.A. Schmid

Für Herrn Professor Halter-Koch zum siebzigsten Geburtstag

Abstract Some of the fundamental notions related to sets of lengths of Krull monoids with finite class group are discussed, and a survey of recent results is given. These include the elasticity and related notions, the set of distances, and the structure theorem for sets of lengths. Several open problems are mentioned.

1 Introduction

Krull monoids are a central structure in factorization theory. On the one hand, many structures of interest such as maximal orders of algebraic number fields and more generally Dedekind domains are Krull monoids; we give some more examples in Sect. 2. On the other hand, Krull monoids are by definition the class of monoids one gets by considering the monoids whose arithmetic is given by direct restriction of the arithmetic of a 'surrounding' factorial monoid. Thus, there is also a purely intrinsic reason why they are a very natural type of monoid in this context, and this might be part of the reason why they arise in various areas.

The investigation of the lengths of factorizations, that is the number of irreducible factors in the factorizations, is a central subject in factorization theory. One reason for considering lengths is that the length is a simple and natural parameter of a factorization, while still containing interesting information. There are other, more technical reasons, that are explained later.

W.A. Schmid (✉)
Université Paris 13, Sorbonne Paris Cité, LAGA, CNRS, UMR 7539,
Université Paris 8, 99 Avenue Jean Baptiste Clément, 93430 Villetaneuse, France
e-mail: schmid@math.univ-paris13.fr

S. Chapman et al. (eds.), *Multiplicative Ideal Theory and Factorization Theory*,
Springer Proceedings in Mathematics & Statistics 170,
DOI 10.1007/978-3-319-38855-7_14

The idea of this survey article is to give some insight into current research on sets of lengths of Krull monoids, with an emphasis on the case of finite class group and each class containing a prime divisor. By 'current' we roughly mean obtained during the last decade, or put differently since the publication of Geroldinger and Halter-Koch's monograph [23], which covered this subject in detail (see especially Chaps. 6 and 7).

The scope is quite narrow and even in this narrow scope we do not attempt to be complete. Rather, the aim is to convey via discussion of selected subjects some of the main trends in recent research on this subject and to highlight some problems that might be interesting avenues for future research. In this vein, some effort is made to explain the *why* and not only the what. For the most part, this survey does not contain proofs of the results we mention. However, proofs of some basic constructions and lemmas are included, on the one hand since sometimes the details of these proofs are relevant for the discussion and on the other hand to convey the type of arguments used.

No attempt is made to faithfully recount the history of the subject. Of course, we try to attribute correctly the main results we discuss, but we also often make reference to secondary sources or even give none at all when we give a proof; this is the case especially for some basic results and constructions that are very widely known and used, and that sometimes exist in numerous slightly different versions in the literature. Except for Proposition 4.14, none of the results in this survey is new.

2 Preliminaries

We denote by $\mathbb{N}$ the set of positive integers and by $\mathbb{N}_0$ the set of nonnegative integers. Intervals are intervals of integers, that is, for real numbers a, b we have $[a, b] = \{z \in \mathbb{Z} : a \le z \le b\}$.

For subsets A, B of the integers we denote by $A + B = \{a + b : a \in A,\ b \in B\}$ the sum of the sets A and B. For k an integer we denote by $k \cdot A = \{ka : a \in A\}$ the dilation of A by k.

In general we follow the notation and conventions of [23] and [18] where more detailed information could be found; the former gives an in-depth treatment of factorization theory as a whole, the latter gives an introduction to the aspects most relevant to this survey, that is, factorizations in Krull monoids and the associated zero-sum problems.

2.1 *Monoids, Factorizations, Sets of Lengths*

In this paper, a monoid is a commutative, cancelative semigroup with identity, which we usually simply denote by 1. We typically use multiplicative notation for monoids. The multiplicative semigroup of nonzero elements of an integral domain is a good example to keep in mind. Let $(H, \cdot)$ be a monoid. We denote by $H^{\times}$ the set of invertible elements of H; we call the monoid reduced if 1 is the only invertible element. By $\mathcal{A}(H)$ we denote the set of irreducible elements of H, also called atoms, that is the elements $a \in H \setminus H^\times$ such that $a = bc$ implies that b or c is invertible. Moreover, we recall that an element a is called prime if $a \mid bc$ implies that $a \mid b$ or $a \mid c$. Every prime is irreducible; the converse is not necessarily true.

We denote by $H_{\text{red}} = H/H^\times$ the reduced monoid associated to H. We say that elements $a, b \in H$ are associated, in symbols $a \simeq b$, if $a = \varepsilon b$ with an invertible element $\varepsilon \in H^\times$.

A monoid F is called free abelian if there exists a subset P (of prime elements) such that every $a \in F$ has a unique representation of the form

$$a = \prod_{p \in P} p^{\mathsf{v}_p(a)}, \text{ where } \mathsf{v}_p(a) \in \mathbb{N}_0 \text{ with } \mathsf{v}_p(a) = 0 \text{ for all but finitely many } p \in P.$$

We use the notation $\mathcal{F}(P)$ to denote the free abelian monoid with P as set of prime elements. We call $|a| = \sum_{p \in P} \mathsf{v}_p(a)$ the length of a.

The monoid $\mathsf{Z}(H) = \mathcal{F}\big(\mathcal{A}(H_{\text{red}})\big)$ is called the factorization monoid of H, and the monoid homomorphism

$$\pi \colon \mathsf{Z}(H) \to H_{\text{red}}$$

induced by $\pi(a) = a$ for each $a \in \mathcal{A}(H_{\text{red}})$ is the called factorization homomorphism of H.

For $a \in H$,

$$\mathsf{Z}(a) = \pi^{-1}(aH^\times)$$

is the set of factorizations of a and

$$\mathsf{L}(a) = \big\{|z| \colon z \in \mathsf{Z}(a)\big\} \subset \mathbb{N}_0$$

is the set of lengths of a. The above definition of the set of factorizations of a is a formalization of what one could describe informally as the set of distinct (up to ordering and associates) factorizations of a into irreducibles.

In the present survey, we essentially exclusively deal with lengths of factorizations, and thus we are mainly interested in $\mathsf{L}(a)$. An alternate description for $\mathsf{L}(a)$, for $a \in H \setminus H^\times$, is that it is the set of all l such that there exist $u_1, \ldots, u_l \in \mathcal{A}(H)$ with $a = u_1 \ldots u_l$; and setting $\mathsf{L}(a) = \{0\}$ for $a \in H^\times$.

Moreover, we set $\mathcal{L}(H) = \{\mathsf{L}(a) \colon a \in H\}$ the system of sets of lengths of H.

2.2 Abelian Groups and Zero-Sum Sequences

We denote abelian groups additively. Mainly we deal with finite abelian groups. Let $(G, +, 0)$ be an abelian group. Let $G_0 \subset G$ be a subset. Then $[G_0] \subset G$ denotes the subsemigroup generated by G_0, and $\langle G_0\rangle \subset G$ denotes the subgroup generated by G_0. A family of nonzero elements $(e_i)_{i\in I}$ of G is said to be independent if, for $m_i \in \mathbb{Z}$,

$$\sum_{i\in I} m_i e_i = 0 \text{ implies } m_i e_i = 0 \text{ for all } i \in I.$$

The tuple $(e_i)_{i\in I}$ is called a basis if $(e_i)_{i\in I}$ is independent and the elements e_i generate G as a group.

For $n \in \mathbb{N}$, let C_n denote a cyclic group with n elements. Suppose G is finite. For $|G| > 1$, there are uniquely determined integers $1 < n_1 \mid \cdots \mid n_r$ such that

$$G \cong C_{n_1} \oplus \ldots \oplus C_{n_r}.$$

We denote by $\mathsf{r}(G) = r$ the rank of G and by $\exp(G) = n_r$ the exponent of G. If $|G| = 1$, then $\mathsf{r}(G) = 0$ and $\exp(G) = 1$. A group is called a p-group if the exponent is a prime power.

We set $\mathsf{D}^*(G) = 1 + \sum_{i=1}^{r}(n_i - 1)$; the relevance of this number is explained at the end of this subsection.

For $(G, +)$ an abelian group, and $G_0 \subset G$, we consider $\mathcal{F}(G_0)$. It is common to call an element $S \in \mathcal{F}(G_0)$ a sequence over G_0, and to use some terminology derived from it. In particular, divisors of S are often called subsequences of S and the neutral element of $\mathcal{F}(G_0)$ is sometimes called the empty sequence.

By definition

$$S = \prod_{g\in G_0} g^{\mathsf{v}_g(S)}$$

where $\mathsf{v}_g(S) \in \mathbb{N}_0$ with $\mathsf{v}_g(S) = 0$ for all but finitely many $g \in G_0$, and this representation is unique. Moreover, $S = g_1 \ldots g_{|S|}$ with $g_i \in G_0$ for each $i \in [1, |S|]$ that are uniquely determined up to ordering.

Since the set G_0 is a subset of a group, it makes sense to consider the sum of S, that is

$$\sigma(S) = \sum_{g\in G_0} \mathsf{v}_g(S) g = \sum_{i=1}^{|S|} g_i.$$

The sequence S is called a zero-sum sequence if $\sigma(S) = 0 \in G$. A zero-sum sequence is called a minimal zero-sum sequence if it is nonempty and each proper subsequence is not a zero-sum sequence.

The set of all zero-sum sequences over G_0 is denoted by $\mathcal{B}(G_0)$; it is a submonoid of $\mathcal{F}(G_0)$. The irreducible elements of $\mathcal{B}(G_0)$ are the minimal zero-sum sequences; for brevity we denote them by $\mathcal{A}(G_0)$ rather than by $\mathcal{A}(\mathcal{B}(G_0))$.

The Davenport constant of G_0, denoted by $\mathsf{D}(G_0)$, is defined as

$$\sup\{|A| \colon A \in \mathcal{A}(G_0)\}.$$

It can be shown in general that $\mathsf{D}(G_0)$ is finite if G_0 is finite (see [23, Theorem 3.4.2]); in the special case that G_0 is a subset of a finite group, or more generally contains only elements of finite order, it however follows just by noting that in a minimal zero-sum sequence no element can appear with a multiplicity larger than its order.

For G a finite abelian group, one has $\mathsf{D}(G) \geq \mathsf{D}^*(G)$. Equality is known to hold for groups of rank at most two and for p-groups. However, for groups of rank at least four it is known that the inequality is strict for infinitely many groups. We refer to [23, Chap. 5] and [18] for more information on the Davenport constant in the context of factorization theory.

2.3 Krull Monoids and Transfer Homomorphisms

We recall some basic facts on Krull monoids. For a detailed discussion on Krull monoids we refer to the relevant chapters of Halter-Koch's monograph [33] or again [23, Chap. 2].

There are several equivalent ways to define a Krull monoid; the one we use is well suited for the current context. A monoid H is called a Krull monoid if it admits a divisor homomorphism into a free abelian monoid. This means there is some free abelian monoid $\mathcal{F}(P)$ and a monoid homomorphism $\varphi : H \to \mathcal{F}(P)$ such that $a \mid b$ if and only if $\varphi(a) \mid \varphi(b)$. Thus, the arithmetic of a Krull monoid is directly induced by the one of a free abelian, and thus factorial, monoid.

There is an essentially unique 'minimal' free abelian monoid with this property, which is characterized by the property that for each $p \in P$ there exist $a_1, \dots, a_k \in H$ such that $p = \gcd(\varphi(a_1), \dots, \varphi(a_k))$.

One calls a divisor homomorphism $\varphi : H \to \mathcal{F}(P)$ with the additional property, for each $p \in P$ there exist $a_1, \dots, a_k \in H$ such that $p = \gcd(\varphi(a_1), \dots, \varphi(a_k))$, a divisor theory. The elements of P are called prime divisors.

Every Krull monoid admits a divisor theory, which is unique up to isomorphism. More specifically, a divisor theory is given by the map from H to $\mathcal{I}_v(H)$, the monoid of divisorial ideals, mapping each element to the principal ideal it generates. This is indeed a free abelian monoid in the case of Krull monoids as every divisorial ideal is in an essentially unique way, the product (in the sense of divisorial ideals) of divisorial prime ideals.

Another characterization for Krull monoids is that they are completely integrally closed and v-noetherian, that is, they satisfy the ascending chain condition on divisorial ideals.

For $\varphi : H \to \mathcal{F}(P)$ a divisor theory, the group $G = \mathsf{q}(\mathcal{F}(P))/\mathsf{q}(\varphi(H))$ is called the class group of H. We denote the class containing some element f by $[f]$; moreover, we use additive notation for the class group. The set $G_P = \{[p] \colon p \in P\} \subset G$ is called the set of classes containing prime divisors. The set G_P generates G as a semi-group; any generating subset of G can arise in this way.

Let $\tilde{\beta} : \mathcal{F}(P) \to \mathcal{F}(G_P)$ be the surjective monoid homomorphism induced by $p \mapsto [p]$ for $p \in P$.

One can see that the image of $\tilde{\beta} \circ \varphi$ is $\mathscr{B}(G_P)$, and $\beta = \tilde{\beta} \circ \varphi : H \to \mathscr{B}(G_P)$ is called the block homomorphism.

The block homomorphism is the archetypal example of a transfer homomorphism. A monoid homomorphism $\theta : H \to B$ is called a transfer homomorphism if it has the following properties:

- $B = \theta(H)B^\times$ and $\theta^{-1}(B^\times) = H^\times$.
- If $u \in H$, $b, c \in B$ and $\theta(u) = bc$, then there exist $v, w \in H$ such that $u = vw$, $\theta(v) \simeq b$ and $\theta(w) \simeq c$.

An important property of transfer homomorphism is that $\mathsf{L}(a) = \mathsf{L}(\theta(a))$ for each $a \in H$, and $\mathscr{L}(H) = \mathscr{L}(B)$. Thus, a transfer homomorphism allows to transfer questions on sets of lengths from a monoid of interest H to a simpler auxiliary monoid B. The notion transfer homomorphism was introduced by Halter-Koch [32]; an early formalization of the block homomorphism, in the context of rings of algebraic integers, was given by Narkiewicz [37].

2.4 Examples of Krull Monoids and Related Structures

We gather some of the main examples of structures of interest to which the results recalled in this survey apply, that is, structures that are Krull monoids or structures that admit a transfer homomorphism to a Krull monoid, which then usually is a monoid of zero-sum sequences.

Before we start, we recall that a domain is a Krull domain if and only if its multiplicative monoid is a Krull monoid, as shown by Krause [35]. Thus, we include Krull domains in our list of Krull monoids without further elaboration of this point. Moreover, we recall that Dedekind domains and more generally integrally closed noetherian domains are Krull domains (see, e.g., [23, Sect. 2.11]).

The following structures are Krull monoids.

- Rings of integers in algebraic number fields and more generally holomorphy rings in global fields (see, e.g., [23], in particular Sects. 2.11 and 8.9).
- Regular congruence monoids in Dedekind domains, for example the domains mentioned above (see, e.g., [22] or [23, Sect. 2.11]).
- Rings of polynomial invariants of finite groups (see, e.g., [10, Theorem 4.1].
- Diophantine monoids (see, e.g., [7]).

Moreover, the monoid of zero-sum sequences over a subset G_0 of an abelian group is itself a Krull monoid; the embedding $\mathscr{B}(G_0) \hookrightarrow \mathscr{F}(G_0)$ is a divisor homomorphism.

Moreover, semi-groups of isomorphy classes of certain modules (the operation being the direct sum) turn out to be Krull monoids in various cases. There are many contributions to this subject; we refer to the recent monograph of Leuschke and Wiegand [36] for an overview. We mention, specifically, a recent result by Baeth and Geroldinger [2, Theorem 5.5], yielding a Krull monoid with cyclic classgroup such that each class contains a prime divisor (earlier examples often had infinite class groups).

In addition to those examples of Krull monoids, there are structures that while not Krull monoids themselves, for example as they are not commutative or not integrally closed, still admit a transfer homomorphism to a Krull monoid. Hence their system of sets of lengths is that of a Krull monoid.

We recall two recent results; the first is due to Smertnig [44, Theorem 1.1], the second due to Geroldinger, Kainrath, and Reinhart [25, Theorem 5.8] (their actual result is more general).

- Let $\mathcal{O}$ be a holomorphy ring in a global field and let A be a central simple algebra over this field. For H a classical maximal $\mathcal{O}$-order of A one has that if every stably free left H-ideal is free, then there is a transfer homomorphism from $H \setminus \{0\}$ to the monoid of zero-sum sequence over a ray class group of $\mathcal{O}$, which is a finite abelian group.
- Let H be a seminormal order in a holomorphy ring of a global field with principal order $\widehat{H}$ such that the natural map $\mathfrak{X}(\widehat{H}) \to \mathfrak{X}(H)$ is bijective and there is an isomorphism between the v-class groups of H and $\widehat{H}$. Then there is a transfer homomorphism from $H \setminus \{0\}$ to the monoid of zero-sum sequence over this v-class group, which is a finite abelian group.

In general we formulate the results we recall for Krull monoids. However, in cases where it seems to cause too much notational inconvenience, we give them for monoids of zero-sum sequences only.

3 Some General Results

In this section we collect some general results, before we focus on the more specific context of Krull monoids with finite class group in the subsequent sections.

Definition 3.1 Let H be a monoid.

1. H is called atomic if $|\mathsf{Z}(a)| > 0$ for each $a \in H$.
2. H is called factorial if $|\mathsf{Z}(a)| = 1$ for each $a \in H$.
3. H is called half-factorial if $|\mathsf{L}(a)| = 1$ for each $a \in H$.
4. H is called an FF-monoid if $1 \le |\mathsf{Z}(a)| < \infty$ for each $a \in H$.
5. H is called a BF-monoid if $1 \le |\mathsf{L}(a)| < \infty$ for each $a \in H$.

The definition directly implies that all these monoids are atomic; a factorial monoid is half-factorial; an FF-monoid is a BF-monoid. It is not hard to see that a Krull monoid is an FF-monoid, and thus a BF-monoid.

Sets of lengths are subsets of the nonnegative integers. However, sets of lengths containing 0 or 1 are very special. We make this precise in the following remark.

Remark 3.2 Let H be a monoid and let $a \in H$.

1. If $0 \in \mathsf{L}(a)$, then $\mathsf{L}(a) = \{0\}$ and $a \in H^\times$.
2. If $1 \in \mathsf{L}(a)$, then $\mathsf{L}(a) = \{1\}$ and $a \in \mathscr{A}(H)$.

If H is half-factorial, then $\mathscr{L}(H) = \{\{n\} : n \in \mathbb{N}_0\}$. Going beyond half-factorial monoids, one might have the idea to relax the condition only slightly, say by imposing that each element has factorizations of at most two distinct lengths. However, this idea is infeasible, as the following lemma illustrates.

Lemma 3.3 *Let H be an atomic monoid and let $a, b \in H$. Then $\mathsf{L}(a) + \mathsf{L}(b) \subset \mathsf{L}(ab)$. In particular, if $|\mathsf{L}(a)| > 1$, then $|\mathsf{L}(a^n)| > n$ for each $n \in \mathbb{N}$.*

Proof Let $k \in \mathsf{L}(a)$ and $l \in \mathsf{L}(b)$. Let $a = u_1 \dots u_k$ and $b = v_1 \dots v_l$ with irreducible $u_i, v_j \in \mathscr{A}(H)$ for each $i \in [1, k]$ and $j \in [1, l]$. Then $ab = u_1 \dots u_k v_1 \dots v_l$ is a factorization of ab of length $k + l$, and thus $k + l \in \mathsf{L}(ab)$. The 'in particular'-statement follows by an easy inductive argument, using the fact that for $A, B \subset \mathbb{Z}$ of cardinality at least 2, one has $|A + B| > |A|$ (in fact even $|A + B| \geq |A| + |B| - 1$).

We end this section by discussing some 'extremal' cases for Krull monoids. The first result, in the context of rings of algebraic integers, goes back to Carlitz [4]; for a proof in the context of monoids of zero-sum sequences, which suffices by the transfer result recalled in Sect. 2.3 see [23, Theorem 3.4.11.5] or [18, Proposition 1.2.4].

Theorem 3.4 *Let H be a Krull monoid such that each class contains a prime divisor. Then, H is half-factorial if an only if its class group has order at most* 2.

The subsequent result is due to Kainrath [34].

Theorem 3.5 *Let H be a Krull monoid with infinite class group such that each class contains a prime divisor. Then, every finite subset of $\mathbb{N}_{\geq 2}$ is a set of lengths.*

Thus for H a Krull monoid with class group of order at most 2, we have $\mathscr{L}(H) = \{\{n\} : n \in \mathbb{N}_0\}$; for H a Krull monoid with infinite class group such that each class contains a prime divisor we have $\mathscr{L}(H) = \{\{0\}, \{1\}\} \cup \mathbb{P}_{\text{fin}}(\mathbb{N}_{\geq 2})$, where $\mathbb{P}_{\text{fin}}(\mathbb{N}_{\geq 2})$ denotes the set of all finite subsets of $\mathbb{N}_{\geq 2}$.

For this reason we often restrict to considering the case of finite class groups of order at least 3.

4 Small Sets

As discussed, an atomic monoid that is not half-factorial always has arbitrarily large sets in its system of sets of lengths. One approach to understand the system of sets of lengths is to focus on 'small' sets, that is, those sets that arise from factoring elements that are a product of only few irreducibles (their sets of lengths thus contain some small number).

As an irreducible element u has a unique factorization and $\mathsf{L}(u) = \{1\}$, the next simplest case is to consider the product of two irreducibles. Studying the factorizations of uv, for $u, v \in \mathcal{A}(H)$, turns out to yield interesting problems.

One natural question to ask is what other lengths can there be besides 2 in a set of lengths. We start by recalling two basic constructions.

Lemma 4.1 *Let G be a finite abelian group of order at least* 3.

1. *Then $\{2, 3\} \in \mathcal{L}(G)$.*
2. *If $g \in G$ is an element of order $n \geq 3$, then $\{2, n\} \in \mathcal{L}(G)$.*

Proof Let $g \in G$ be of order $n \geq 3$. Setting $B = g^2(-2g) \cdot (-g)^2(2g)$ and noting $B = ((-g)g)^2 \cdot (-2g)2g$, it follows that $\mathsf{L}(B) = \{2, 3\}$. Note that $2g = -g$ holds for $n = 3$, but this does not affect the argument. Moreover, setting $C = g^n(-g)^n$ and noting $C = ((-g)g)^n$ we see $\mathsf{L}(C) = \{2, n\}$.

It remains to show the first part in case there is no element of order at least 3. If this is the case, there exist independent elements (e_1, e_2) each of order 2. We set $D = e_1^2e_2^2(e_1 + e_2)^2$ and noting $D = (e_1e_2(e_1 + e_2))^2$, it follows that $\mathsf{L}(D) = \{2, 3\}$.

We note that in some sense the simplest non-singleton set that can be a set of length, namely $\{2, 3\}$, is always in $\mathcal{L}(G)$ for $|G| \geq 3$, but there is no absolute bound (that is one independent of G) on the size of elements in a set of lengths containing 2. One natural question is to study this maximum size, for a given monoid H. Formally, one investigates $\sup\{\max \mathsf{L}(uv) \colon u, v \in \mathcal{A}(H)\}$ or written differently $\sup \left(\bigcup_{2 \in L,\, L \in \mathcal{L}(H)} L\right)$.

Similarly, one can consider the product of 3 or more irreducibles. More generally, one considers the following quantities.

Definition 4.2 Let H be an atomic monoid. For $M \subset \mathbb{N}_0$ let

$$\mathcal{U}_M(H) = \bigcup_{M \subset L,\, L \in \mathcal{L}(H)} L.$$

Moreover, let $\lambda_M(H) = \min \mathcal{U}_M(H)$ and $\rho_M(H) = \sup \mathcal{U}_M(H)$.

The case where M is a singleton is of particular interest. For $k \in \mathbb{N}_0$, we write $\mathcal{U}_k(H)$, $\lambda_k(H)$ and $\rho_k(H)$ for $\mathcal{U}_{\{k\}}(H)$, $\lambda_{\{k\}}(H)$ and $\rho_{\{k\}}(H)$. These constants, especially $\rho_k(H)$ are those that received most interest so far. The constants $\rho_k(H)$ and $\lambda_k(H)$ are called the upper kth elasticity of H and the lower kth elasticity of H, respectively. The sets $\mathcal{U}_k(H)$ were introduced by Chapman and Smith [9], and the generalization $\mathcal{U}_M(H)$ appeared in [3].

Moreover, the quantity $\rho(H) = \sup_{k \in \mathbb{N}} \rho_k(H)/k$ is called elasticity of the monoid, and it is also a classical constant in factorization theory. The more common way to define it is as $\sup_{a \in H \setminus H^\times} (\sup \mathsf{L}(a) / \min \mathsf{L}(a))$. We refer to [1] for an overview of classical results.

We saw that $\mathcal{U}_0(H) = \{0\}$ and $\mathcal{U}_1(H) = \{1\}$. For H a Krull monoid with finite class group G such that each class contains a prime divisor, it is not difficult to determine $\rho_{\{k\}}(H)$ for even k; it is however a challenging problem for odd k. We show the former as part of the following well-known lemma, which we prove to give a general idea of the type of argument.

Lemma 4.3 *Let H be a nonfactorial Krull monoid with set of classes containing prime divisors G_P such that the Davenport constant $\mathsf{D}(G_P)$ is finite.*

1. $\rho_k(H) \le k\mathsf{D}(G_P)/2$ *for all* $k \in \mathbb{N}$.
2. *If* $G_P = -G_P$, *then* $\rho_{k+2}(H) \ge \rho_k(H) + \mathsf{D}(G_P)$. *In particular,*

$$\rho_k(H) \ge \begin{cases} \frac{k}{2}\mathsf{D}(G_P) & k \text{ even} \\ \frac{k-1}{2}\mathsf{D}(G_P) + 1 & k \text{ odd} \end{cases}$$

and $\rho_{2l}(H) = l\mathsf{D}(G_P)$ *for every* $l \in \mathbb{N}_0$.

Proof By the transfer results recalled in Sect. 2.3 we can consider the problem in $\mathcal{B}(G_P)$. We note that $\mathsf{D}(G_P) \ge 2$ as the monoid is not factorial.

1. Let $B \in \mathcal{B}(G_P)$ with $k \in \mathsf{L}(B)$, say $B = U_1 \dots U_k$ with $U_i \in \mathcal{A}(G_P)$ for each $i \in [1, k]$. Let $B = V_1 \dots V_r$ with $V_j \in \mathcal{A}(G_P)$ for each $j \in [1, r]$.

First, suppose $0 \nmid B$. Then $|V_j| \ge 2$ for all $j \in [1, r]$, while $|U_i| \le \mathsf{D}(G_P)$ for all $i \in [1, k]$, whence $2r \le |B| \le k\mathsf{D}(G_P)$. Thus $r \le k\mathsf{D}(G_P)/2$. This shows that every element of $\mathsf{L}(B)$ is bounded above by $k\mathsf{D}(G_P)/2$, showing the claim.

Now, let $B = 0^v B'$ where $v \in \mathbb{N}$ and $0 \nmid B'$. Then $\mathsf{L}(B) = v + \mathsf{L}(B')$ and $k - v \in \mathsf{L}(B')$. Thus, $\max \mathsf{L}(B') \le (k - v)\mathsf{D}(G_P)/2$ and $\max \mathsf{L}(B) \le v + (k - v)\mathsf{D}(G_P)/2 \le k\mathsf{D}(G_P)/2$.

2. Let $U = g_1 \dots g_l \in \mathcal{A}(G_P)$. Then $-U \in \mathcal{A}(G_P)$. We have the equality $(-U)U = \prod_{i=1}^{l}(-g_i)g_i$ and $(-g_i)g_i \in \mathcal{A}(G_P)$ for all $i \in [1, l]$, it follows that $l \in \mathsf{L}((-U)U)$ and $l \le \rho_2(H)$. Let us now assume U has length $|U| = \mathsf{D}(G_P)$; such a U exists by definition of $\mathsf{D}(G_P)$.

Let $B \in \mathcal{B}(G_P)$ with $\{k, \rho_k(H)\} \subset \mathsf{L}(B)$. Then, from the inclusion $\mathsf{L}((-U)U) + \mathsf{L}(B) \subset \mathsf{L}((-U)UB)$, we have $\{k + 2, \rho_k(H) + \mathsf{D}(G_P)\} \subset \mathsf{L}((-U)UB)$ and the claim follows.

To get the 'in particular'-claim it suffices to apply this bound repeatedly, starting from $\rho_0(H) = 0$ and $\rho_1(H) = 1$.

We focus on the case that every class contains a prime divisor. Since $G_P = -G_P$ is trivially true, in this case $\rho_k(H)$ is determined for even k, and we now recall some results for the case that k is odd.

From the preceding lemma one has the inequality

$$k\mathsf{D}(G) + 1 \le \rho_{2k+1}(G) \le k\mathsf{D}(G) + \left\lfloor \frac{\mathsf{D}(G)}{2} \right\rfloor. \tag{1}$$

By a result of Gao and Geroldinger [15] it is known that for cyclic groups equality always holds at the lower bound.

Theorem 4.4 *Let H be Krull monoid with finite cyclic class group G of order at least 3 such that each class contains a prime divisor. Then $\rho_{2k+1}(H) = k|G| + 1$ for all $k \in \mathbb{N}_0$.*

The proof uses results on the structure of long minimal zero-sum sequences over cyclic groups ('long' is meant in a relative sense), see [40, 45]. As can be seen from the proof of the preceding lemma, one of the factorizations that could lead to a larger value of $\rho_{2k+1}(H)$ would have to be composed of minimal zero-sum sequences of length 'close' to $\mathsf{D}(G)$. Having knowledge on the structure of such sequences, allows to analyze this situation in a more explicit way.

However, the case of cyclic groups seems to be quite exceptional, and there are various results asserting even equality at the upper bound in the inequality above.

We recall a recent result due to Geroldinger, Grynkiewicz, Yuan [19, Theorem 4.1]. Moreover, they conjectured that cyclic groups and the group C_2^2 are the only groups for which $\rho_3(G) = \mathsf{D}(G) + 1$.

Theorem 4.5 *Let H be Krull monoid with class group G such that each class contains a prime divisor. Suppose that $G \cong \oplus_{i=1}^{r} C_{n_i}^{s_i}$ with $1 < n_1 \mid \cdots \mid n_r$ and $s_i \geq 2$ for each $i \in [1, r]$. Then, for every $k \in \mathbb{N}$,*

$$\rho_{2k+1}(H) \geq (k-1)\mathsf{D}(G) + \mathsf{D}^*(G) + \left\lfloor \frac{\mathsf{D}^*(G)}{2} \right\rfloor .$$

In particular, if $\mathsf{D}^*(G) = \mathsf{D}(G)$, then $\rho_{2k+1}(G) = k\mathsf{D}(G) + \left\lfloor \frac{\mathsf{D}(G)}{2} \right\rfloor$ for every $k \in \mathbb{N}$.

The point of considering $\mathsf{D}^*(G)$ rather than $\mathsf{D}(G)$ is that the former is explicitly known and one thus has explicit examples of minimal zero-sum sequences of the relevant length that can be used to construct examples. By contrast, $\mathsf{D}(G)$ is in general not known, and thus knowledge on zero-sum sequences of this length can only be obtained by general considerations.

For other conditions that imply equality at the upper bound in (1) see for example [23, Theorem 6.3.4]. Indeed, Geroldinger, Grynkiewicz, Yuan [19, Conjecture 3.3] put forward the conjecture that for sufficiently large k this equality always holds for noncyclic groups.

Conjecture 4.6 Let H be Krull monoid with finite noncyclic class group G such that each class contains a prime divisor. Then there exists some $k^* \in \mathbb{N}$ such that for each $k \geq k^*$ one has

$$\rho_{2k+1}(H) = k\mathsf{D}(G) + \left\lfloor \frac{\mathsf{D}(G)}{2} \right\rfloor .$$

To restrict to sufficiently large k is certainly necessary, as the following result illustrates, see Geroldinger, Grynkiewicz, Yuan [19, Theorem 5.1].

Theorem 4.7 *Let H be Krull monoid with class group G such that each class contains a prime divisor. Suppose that $G \cong C_m \oplus C_{mn}$ with $m \geq 2$ and $n \geq 1$. Then*

$$\rho_3(H) = \mathsf{D}(G) + \left\lfloor \frac{\mathsf{D}(G)}{2} \right\rfloor \quad \textit{if and only if } n = 1 \textit{ or } n = m = 2.$$

The proof uses the fact that the structure of minimal zero-sum sequences of maximal length is known for groups of rank 2 (see [16, 39, 43]). To put this in context, we remark that to know the sequences of maximal lengths allows to exclude equality at the upper bound for most groups of rank 2; to get further improved upper bounds might need knowledge on the structure of long (yet not maximum length) minimal zero-sum sequences in addition, as known and used in the case of cyclic groups.

This result allows to give examples of groups where the actual value of $\rho_3(G)$ can be neither the upper nor the lower bound in (1). An example is $C_2 \oplus C_{2n}$ for $n \geq 3$; however, in line with the above-mentioned conjecture, one still has equality of $\rho_{2k+1}(G)$ with the upper bound for $k \geq 2n - 1$ (see [19, Corollary 5.3]).

Very recently Fan and Zhong [11, Theorem 1.1] made considerable progress toward the above-mentioned conjecture. In particular, they verified it under the assumption that $\mathsf{D}(G) = \mathsf{D}^*(G)$.

Theorem 4.8 *Let H be Krull monoid with finite noncyclic class group G such that each class contains a prime divisor. Then there exists some $k^* \in \mathbb{N}$ such that for each $k \geq k^*$ one has*

$$\rho_{2k+1}(H) \geq (k - k^*)\mathsf{D}(G) + k^*\mathsf{D}^*(G) + \left\lfloor \frac{\mathsf{D}^*(G)}{2} \right\rfloor .$$

In particular, if $\mathsf{D}(G) = \mathsf{D}^*(G)$, then $\rho_{2k+1}(H) = k\mathsf{D}(G) + \left\lfloor \frac{\mathsf{D}(G)}{2} \right\rfloor$ for $k \geq k^*$.

Having discussed $\rho_k(H)$ in some detail, we turn to the other constants. However, we see that in important cases the determination of $\mathcal{U}_k(H)$ and $\lambda_k(H)$ can be reduced to the problem of determining $\rho_k(H)$.

The following result is due to Freeze and Geroldinger [12, Theorem 4.2]; for another proof of this result due to Halter-Koch see [18, Theorem 3.1.3].

Theorem 4.9 *Let H be a Krull monoid with finite class group such that each class contains a prime divisor. Then $\mathcal{U}_k(H)$ is an interval for every $k \in \mathbb{N}$.*

Thus, in this case it suffices to determine $\lambda_k(H)$ and $\rho_k(H)$ to know $\mathcal{U}_k(H)$. Moreover, it is even possible (see [18, Corollary 3.1.4]) to express (in this case) the constants $\lambda_k(H)$ in terms of $\rho_k(H)$.

Theorem 4.10 *Let H be a Krull monoid with finite class group G such that each class contains a prime divisor. Then for every $k \in \mathbb{N}_0$ we have*

$$\lambda_{k\mathsf{D}(G)+j}(H) = \begin{cases} 2k & \textit{for } j = 0 \\ 2k+1 & \textit{for } j \in [1, \rho_{2k+1}(H) - k\mathsf{D}(G)] \\ 2k+2 & \textit{for } j \in [\rho_{2k+1}(H) - k\mathsf{D}(G) + 1, \mathsf{D}(G) - 1] \end{cases}$$

We turn to results on $\mathcal{U}_M(H)$ where M is not a singleton. In view of the results above, we see that if min M and max M are too far apart then for H a Krull monoid with finite class group the sets $\mathcal{U}_M(H)$ will always be empty. Specifically, when $\min M = k$, then $\mathcal{U}_M(H)$ is empty if M contains some element greater than $\rho_k(H)$.

Considering $\mathcal{U}_{\{k,\rho_k(H)\}}(H)$ is thus an interesting extremal case. This problem was investigated recently by Baginski, Geroldinger, Grynkiewicz, Philipp [3], with a focus on groups of rank two; again, it is important to know the structure of minimal zero-sum sequences of maximal length.

We start by recalling an older result for cyclic groups and elementary 2-groups (see [23, Theorem 6.6.3]).

Theorem 4.11 *Let H be a Krull monoid with finite class group such that each class contains a prime divisor. Then, $\mathcal{U}_{\{2,\rho_2(H)\}}(H) = \{2, \rho_2(H)\}$ if and only if the class group is cyclic or an elementary 2-group.*

For groups of rank 2 the set $\mathcal{U}_{\{2,\rho_2(H)\}}(H)$ is a lot larger as shown in [3, Theorem 3.5].

Theorem 4.12 *Let H be a Krull monoid with class group $G \cong C_m \oplus C_{mn}$ where $m, n \in \mathbb{N}$ and $m \ge 2$ such that each class contains a prime divisor. Then,*

$$\mathcal{U}_{\{2,\rho_2(H)\}}(H) = \begin{cases} \{2a\colon a \in [1,n]\} \cup \{\rho_2(H)\} & \text{for } m = 2 \\ [2, \rho_2(H)] & \text{for } m \in [3,4] \\ [2, \rho_2(H)] \setminus \{3\} & \text{for } m \ge 5 \end{cases}$$

If the class group is a group of rank greater than 2, one faces the following problem. While one still knows $\rho_2(H) = \mathsf{D}(G)$, one does in general not know $\mathsf{D}(G)$ explicitly. Thus, one also has only little knowledge on the form of minimal zero-sum sequences of maximal length.

However, for most groups for which $\mathsf{D}(G) = \mathsf{D}^*(G)$ holds a description of $\mathcal{U}_{\{2,\rho_2(H)\}}(H)$ can still be obtained, as more generally, $\mathcal{U}_{\{2,\mathsf{D}^*(G)\}}(H)$ can be described almost completely for most groups of rank at least 3. The following result was obtained in [3, Theorem 4.2].

Theorem 4.13 *Let H be a Krull monoid with class group $G \cong \oplus_{i=1}^r C_{n_i}$ where $1 < n_1 \mid \cdots \mid n_r$ with $r \ge 3$ and $n_{r-1} \ge 3$ such that each class contains a prime divisor. Then, $\mathcal{U}_{\{2,\mathsf{D}^*(G)\}}(H) \supset [2, \mathsf{D}^*(G)]$. In particular, if $\mathsf{D}(G) = \mathsf{D}^*(G)$, then $\mathcal{U}_{\{2,\rho_2(H)\}} = [2, \rho_2(H)]$.*

We highlight the similarity to the results on $\rho_2(H)$, where also for general groups one resorted to $\mathsf{D}^*(G)$ instead of $\mathsf{D}(G)$.

We end this section with a small complement to the preceding theorem, investigating the relevance of the condition on n_{r-1}.

Proposition 4.14 *Let H be a Krull monoid with class group $G \cong C_2^{r-1} \oplus C_{2n}$ with $r \ge 3$ and $n \in \mathbb{N}$ such that each class contains a prime divisor. Then, $3 \in \mathcal{U}_{\{2,\mathsf{D}^*(G)\}}(H)$ if and only if $n \ge 3$.*

Proof By the transfer results that we recalled in Sect. 2.3 we can assume $H = \mathcal{B}(G)$. Let $(e_1, \dots, e_{r-1}, f)$ be a basis of G with $\operatorname{ord} e_i = 2$ for $1 \le i \le r-1$ and $\operatorname{ord} f = 2n$.

First, suppose $n \ge 3$. We note that the sequence $U = f^{2n-3}(f+e_1)^3(f+e_2)(-f+e_1+\dots+e_{r-1})e_3\dots e_{r-1}$ is a minimal zero-sum sequence of length $\mathsf{D}^*(G)$: the assertion on the sum and length are direct, and to see that it is minimal we note that $f^{2n-3}(f+e_1)^3(f+e_2)$ has no nonempty subsequence with sum 0, so that a zero-sum subsequence T of U has to contain one and then each element of $(-f+e_1+\dots+e_{r-1})e_3\dots e_{r-1}$, which implies that T contains $(f+e_1)(f+e_2)$ and thus must equal U to get sufficiently many elements containing f.

We consider $(-U)U$. Of course it has factorizations of length 2 and $\mathsf{D}^*(G)$. It remains to show that it has a factorization of length 3. To see this note that $(-U)U$ is equal to $V_1V_2V_3$ with $V_1 = f^{2n-3}(f+e_1)(f+e_2)(f+e_1+\dots+e_{r-1})e_3\dots e_{r-1}$, $V_2 = (-f)^{2n-5}(-f+e_1)^3(-f+e_2)(-f+e_1+\dots+e_{r-1})e_3\dots e_{r-1}$, and $V_3 = (f+e_1)^2(-f)^2$, and V_1, V_2, V_3 are minimal zero-sum sequences.

For $n = 1$ it is established in Theorem 4.11 that $3 \notin \mathscr{U}_{\{2,\mathsf{D}^*(G)\}}(H)$. It remains to consider $n = 2$. Note that in this case the exponent of G is 4, so G is a 2-group and $\mathsf{D}(G) = \mathsf{D}^*(G)$ (see Sect. 2). Assume for a contradiction there is a zero-sum sequence B over G such that $\{2, 3, \mathsf{D}^*(G)\} \subset \mathsf{L}(B)$. As $\mathsf{D}(G) = \mathsf{D}^*(G)$ and $\{2, \mathsf{D}^*(G)\} \subset \mathsf{L}(B)$, it follows that $B = U(-U)$ where U is a minimal zero-sum sequence of length $\mathsf{D}(G) = \mathsf{D}^*(G)$ (see the proof of Lemma 4.3). Since $3 \in \mathsf{L}(U(-U))$ it follows that $U(-U) = V_1V_2V_3$ with minimal zero-sum sequences V_1, V_2, V_3 and further for each $1 \le i \le 3$ we have $V_i = S_i(-T_i)$ with $U = S_1S_2S_3 = T_1T_2T_3$. We note that none of the S_i and T_i is the empty sequence. We have $\sigma(S_i) = \sigma(T_i)$ for each $1 \le i \le 3$ and moreover $\sigma(S_1) + \sigma(S_2) + \sigma(S_3) = 0$.

We claim that at least one of the elements $\sigma(S_1), \sigma(S_2), \sigma(S_3)$ has order 2. Since U is a minimal zero-sum sequence all three elements are nonzero, as sums of proper and nonempty subsequences of U. Denoting by $G[2]$ the subgroup of G of elements of order at most 2, we have $G/G[2]$ is a group of order 2 and as the images of $\sigma(S_1), \sigma(S_2), \sigma(S_3)$ in $G/G[2]$ form a zero-sum sequence not all of them can be the nonzero element in $G/G[2]$. Consequently, at least one of the elements has order at most 2 and as it must be nonzero it has order 2, establishing the claim.

Without loss of generality, we assume that $\sigma(S_3) = e$ has order 2. If $\gcd(S_3, T_3) = 1$, then $S_3T_3 \mid U$. As $\sigma(T_3S_3) = 2e = 0$, it follows that $S_3T_3 = U$, that is $T_3 = S_1S_2$ and $S_3 = T_1T_2$. Yet then $S_3(-T_3) = (S_1(-T_1))(S_2(-T_2))$ contradicting the fact that $S_3(-T_3)$ is a minimal zero-sum sequence.

Thus, $\gcd(S_3, T_3) \ne 1$. This implies, as $S_3(-T_3)$ is a minimal zero-sum sequence, that $|S_3| = |T_3| = 1$ and $S_3 = T_3 = e$.

If $\gcd(S_1, T_1) = \gcd(S_2, T_2) = 1$, then $S_2 = T_1$ and $S_1 = T_2$. As $\sigma(S_1) = \sigma(T_1)$, it follows that $\sigma(S_1) = \sigma(S_2)$ and thus $e = -2\sigma(S_1) \in 2 \cdot G$. However, this is not possible, as a minimal zero-sum sequence of maximal length over a 2-group must not contain an element from $2 \cdot G$ (see [23, Proposition 5.5.8]). Alternatively, one can argue that the image of Ue^{-1} in $G/\langle e \rangle \cong C_2^r$ has to be a minimal zero-sum sequence, which is not possible as its length exceeds the Davenport constant of C_2^r.

Thus, we get that $\gcd(S_2, T_2) \neq 1$. As above we get that $|S_2| = |T_2| = 1$. Yet then $|S_1(-T_1)| = 2|U| - 4 = 2\mathsf{D}(G) - 4 > \mathsf{D}(G)$, a contradiction.

The preceding results yield the following corollary.

Corollary 4.15 *Let H be a Krull monoid with class group G of rank $r \geq 3$ such that each class contains a prime divisor. The following conditions are equivalent:*

- *G is neither an elementary 2-group nor of the form $C_2^{r-1} \oplus C_4$.*
- $3 \in \mathcal{U}_{\{2,\mathsf{D}^*(G)\}}(H)$.

We mention that this corollary allows to fill what we believe to be a minor gap in the proof of [3, Theorem 5.6]; it can be invoked there instead of [3, Theorem 4.2] (that is the result we recalled as Theorem 4.13).

5 Distances

In the preceding section we discussed how spread out sets of lengths can be, in the sense of comparing their extremal values. We now turn to the question how large distances there can be between adjacent elements of the sets of lengths. Moreover, considering distances also gives another measure for the complexity of a set of lengths; highly structured sets, such as arithmetic progressions, have few distinct distances even when the set itself might be large.

Definition 5.1

- Let $A \subset \mathbb{Z}$. Then the set of distances of A, denoted by $\Delta(A)$, is the set of all differences between consecutive elements of A, formally, it is the set of all $d \in \mathbb{N}$ for which there exists $l \in A$ such that $A \cap [l, l+d] = \{l, l+d\}$.
- For an atomic monoid H, we denote by

$$\Delta(H) = \bigcup_{a \in H} \Delta\big(\mathsf{L}(a)\big) \subset \mathbb{N}$$

the set of distances of H.

It is sometimes common to denote, for $a \in H$, the set $\Delta(\mathsf{L}(a))$ by $\Delta(a)$. Since we only use it rarely, we do not use this abbreviation here.

If H is a Krull monoid with finite class group, then $\Delta(H)$ is finite. More specifically and more generally, one has the following general bound (see, e.g., [23, Theorems 3.4.11 and 1.6.3].

Lemma 5.2 *Let H be a Krull monoid and let G_P denote the set of classes containing prime divisors. Then* $\sup \Delta(H) \leq \mathsf{D}(G_P) - 2$.

In case the class group is infinite, $\Delta(H)$ can be infinite, too. In fact, if each class contains a prime divisor then $\Delta(H) = \mathbb{N}$. (This is a direct consequence of Theorem 3.5, yet it is a much simpler result; indeed, we give a partial proof below.)

The example recalled in Lemma 4.1 shows that for a Krull monoid where each class contains a prime divisor we always have $1 \in \Delta(H)$. Moreover, Geroldinger and Yuan [28] showed that for these monoids $\Delta(H)$ is an interval.

Theorem 5.3 *Let H be a Krull monoid with finite class group such that each class contains a prime divisor. Then $\Delta(H) = [1, \max \Delta(H)]$.*

Thus, in this important case the problem of determining $\Delta(H)$ is reduced to the problem of determining the maximum of this set. Before we discuss results toward this goal, we recall some well-known constructions to get some rough insight into which size of $\max \Delta(H)$ one might expect (for further details see, e.g., [23, Lemma 6.4.1]).

Lemma 5.4 *Let $G = C_{n_1} \oplus \cdots \oplus C_{n_r}$ with $|G| \ge 3$ and $1 < n_1 \mid \cdots \mid n_r$. Then*

$$[1, n_r - 2] \cup [1, -1 + \sum_{i=1}^{r} \left\lfloor \frac{n_i}{2} \right\rfloor] \subset \Delta(G)$$

Proof Let $e_1, \ldots, e_r \in G$ be independent with $\operatorname{ord} e_i = n_i$ for each $i \in [1, r]$. Let $e_0 = k_1 e_1 + \cdots + k_r e_r$, where $k_i \in \mathbb{N}_0$ and $2k_i \le \operatorname{ord} e_i$ for all $i \in [1, r]$. For

$$U = (-e_0) \prod_{i=1}^{r} e_i^{k_i},$$

we have $\mathsf{L}((-U)U) = \{2, k_1 + \cdots + k_r + 1\}$. This yields a distance of $-1 + k_1 + \cdots + k_r$ (except if $k_1 + \cdots + k_r = 1$). Since $k_1 + \cdots + k_r$ can attain any value in $[1, \sum_{i=1}^{r} \lfloor \frac{n_i}{2} \rfloor]$, we get $[1, -1 + \sum_{i=1}^{r} \lfloor \frac{n_i}{2} \rfloor] \subset \Delta(G)$.

Let $e \in G$ be nonzero. Then $\mathsf{L}(e^n((a-1)e)(-e)^{a-1}) = \{2, a\}$ for $a \in [2, \operatorname{ord}(e)]$. This yields a distance of $a - 1$. As there is an element of order n_r, we get $[1, n_r - 2]$.

Remark 5.5 Since an infinite abelian torsion group contains elements of arbitrarily large order or an infinite independent set, the above constructions show $\Delta(G) = \mathbb{N}$ for infinite torsion groups.

No element in $\Delta(G)$ larger than the ones given above is known. The bound $\max \Delta(G) \le \mathsf{D}(G) - 2$ shows that for G cyclic or an elementary 2-group, indeed, there can be no larger element. Thus, one has the following result (see [23, Theorem 6.4.7]).

Theorem 5.6 *For $r \ge 2$ and $n \ge 3$ one has $\Delta(C_2^r) = [1, r-1]$ and $\Delta(C_n) = [1, n-2]$.*

These groups are in fact the only ones for which $\max \Delta(G) = \mathsf{D}(G) - 2$. A characterization of groups for which $\max \Delta(G) = \mathsf{D}(G) - 3$ was recently given by Geroldinger and Zhong [31].

However, in general the following problem is wide open.

Problem 5.7 Let $G \cong C_{n_1} \oplus \cdots \oplus C_{n_r}$ with $|G| \geq 3$ and $1 < n_1 \mid \cdots \mid n_r$. Is

$$\max \Delta(G) = \max \left\{ n_r - 2, -1 + \sum_{i=1}^{r} \left\lfloor \frac{n_i}{2} \right\rfloor \right\} ?$$

We recall results that give upper bounds on $\max \Delta(H)$. It turned out that the following quantity is a useful tool to this end. It was introduced in [21]. The problem of determining $\max \Delta(H)$ and problems of distances more generally are often studied in combination or even via a notion called catenary degree. The catenary degree is a notion of factorization theory that does not only take the length of factorizations into account, which is why we do not discuss it here.

Definition 5.8 Let H be an atomic monoid. Let

$$\daleth(H) = \sup \left\{ \min \big(\mathsf{L}(uv) \setminus \{2\} \big) : u, v \in \mathscr{A}(H) \right\},$$

with the convention that $\min \emptyset = \sup \emptyset = 0$.

We point out that we again study sets of lengths of a product of two irreducible elements; other aspects of this problem were discussed in the preceding section. The following lemma is essentially a direct consequence of the definition.

Lemma 5.9 *Let H be an atomic monoid. Then $\daleth(H) \leq 2 + \sup \Delta(H)$.*

While equality does not always hold (for an example see below), it can be shown to hold for Krull monoids under certain assumptions on the class group. Informally, this then means that the largest possible distance is already attained in the sets of lengths of the product of two irreducible elements, which simplifies the task of actually determining this distance.

The following result is a special case of [21, Corollary 4.1].

Theorem 5.10 *Let H be a Krull monoid with class group $G \cong C_{n_1} \oplus \cdots \oplus C_{n_r}$ where $1 < n_1 | \ldots | n_r$ and $|G| \geq 3$ such that each class contains a prime divisors. If*

$$\left\lfloor \frac{1}{2} \mathsf{D}(G) + 1 \right\rfloor \leq \max \left\{ n_r, \ 1 + \sum_{i=1}^{r} \left\lfloor \frac{n_i}{2} \right\rfloor \right\} .$$

Then $\daleth(H) = 2 + \max \Delta(H)$.

We discuss the technical condition. Since

$$1+\sum_{i=1}^{r}\left\lfloor\frac{n_i}{2}\right\rfloor=\frac{1+\mathsf{r}_2(G)+\mathsf{D}^*(G)}{2},$$

where $\mathsf{r}_2(G)$ denotes the number of even n_is, it follows that if $\mathsf{D}(G)=\mathsf{D}^*(G)$, then

$$\left\lfloor\frac{1}{2}\mathsf{D}(G)+1\right\rfloor\leq 1+\sum_{i=1}^{r}\left\lfloor\frac{n_i}{2}\right\rfloor.$$

We give an example where $\daleth(H)<2+\max\Delta(H)$. For details of the example see [23, Proposition 4.1.2].

Example 5.11 Let G be an abelian group and $r, n\in\mathbb{N}_{\geq 3}$ with $n\neq r+1$. Let $e_1,\dots,e_r\in G$ be independent elements with $\operatorname{ord} e_i=n$ for all $i\in[1,r]$. We set $e_0=-(e_1+\dots+e_r)$ and $G_0=\{e_0,e_1\dots,e_r\}$. Then $\Delta(\mathcal{B}(G_0))=\{|n-r-1|\}$ yet $\daleth(\mathcal{B}(G_0))=0$.

To see this note that the only minimal zero-sum sequences are e_i^n for $i\in[0,r]$ and $W=\prod_{i=0}^{r}e_i$. To have a nontrivial relation, we at least need to have W^n, which factors also as $\prod_{i=0}^{r}e_i^n$.

We continue with a bound on $\daleth(H)$; this is a special case of [21, Theorem 5.1].

Theorem 5.12 *Let H be a Krull monoid with finite class group G such that each class contains a prime divisor. If $\exp(G)=n$ and $\mathsf{r}(G)=r$, then*

$$\daleth(H)\leq\max\left\{n,\ \frac{1}{3}\left(2\mathsf{D}(G)+\frac{1}{2}rn+2^r\right)\right\}.$$

In combination with the preceding result one obtains bounds for $\max\Delta(H)$ for various types of class groups. We formulate one explicitly.

Corollary 5.13 *Let H be a Krull monoid with finite class group $G\cong C_n^2$ with $n\geq 2$ such that each class contains a prime divisor. Then*

$$\max\Delta(H)\leq\frac{5n-4}{3}.$$

We recall that the lower bound for $\max\Delta(H)$ is $n-2$ for odd n and $n-1$ for even n whereas the simple upper bound given by $\mathsf{D}(C_n^2)-2$ is $2n-3$.

We point out that for this problem knowledge of the structure of minimal zero-sum sequences of maximal length seems insufficient. The extremal known examples are attained by minimal zero-sum sequences of length about $\mathsf{D}(C_n^2)/2$.

Up to now we only discussed $\Delta(H)$, that is the collection of all distance that can occur in some monoid. It is also an interesting question to study $\Delta(\mathsf{L}(a))$ for individual elements of $a\in H$. By definition it is clear that each $d\in\Delta(H)$ occurs in

$\Delta(\mathsf{L}(a))$ for some $a \in H$. Yet, passing to more than one distance, one gets interesting questions. For example, for distances $d_1, d_2 \in H$ one can ask if there exists some $a \in H$ such that $d_1, d_2 \in \Delta(\mathsf{L}(a))$. Or, for some fixed distance $d \in H$ one can ask what are all the other distances in the sets of lengths having d as a distance; formally, one can study similarly to $\mathscr{U}_k(H)$ the sets

$$\bigcup_{a \in H,\, d \in \Delta(\mathsf{L}(a))} \Delta(\mathsf{L}(a)).$$

Recently, Chapman, Gotti, and Pelayo [6] obtained the following result on this type of problem.

Theorem 5.14 *Let H be a Krull monoid with cyclic class group of order $n \geq 3$, and let $a \in H$. If $n - 2 \in \Delta(\mathsf{L}(a))$, then $\Delta(\mathsf{L}(a)) = \{n - 2\}$.*

We recall that $n - 2$ is the maximum of the set of distances for Krull monoid with cyclic class group n, assuming that each class contains a prime divisor. A similar result for elementary 2-groups is also known, see [27, Lemma 3.10].

6 Large Sets

Sets of lengths can be arbitrarily large. However, one can show that they are not arbitrarily complicated, in a sense to be made precise.

The construction we saw in Lemma 4.1, when we recalled that there cannot be a global bound on the size of sets of lengths in non-half-factorial monoids, suggests that there is some additive structure to large sets of lengths. Indeed, this is the case for various classes of monoids. We recall the result and related relevant notions.

Definition 6.1 A nonempty subset L of $\mathbb{Z}$ is called an almost arithmetic multiprogression (AAMP for short) with bound $M \in \mathbb{N}_0$, difference $d \in \mathbb{N}$ and period $\mathscr{D}$ (where $\{0, d\} \subset \mathscr{D} \subset [0, d]$) if

$$L = y + (L' \cup L^* \cup L'') \subset y + \mathscr{D} + d \cdot \mathbb{Z}$$

with $0 \in L^* = [0, \max L^*] \cap (\mathscr{D} + d \cdot \mathbb{Z})$ and $L' \subset [-M, -1]$ and $L'' \subset \max L^* + [1, M]$. One calls L^* the central part, and L' and L'' the beginning and the end part, respectively.

The notion of AAMP turns out, as we see below, to be natural for describing sets of lengths of Krull monoids with finite class group, and also other monoids. Informally, one can imagine an AAMP as a union of several slightly shifted copies of an arithmetic progressions where at the beginning and the end some elements might be removed. The definition of AAMP contains the following special cases.

Definition 6.2

- an AAMP with bound $M = 0$ is called an arithmetic multiprogression (AMP for short).
- an AAMP with period $\mathscr{D} = \{0, d\}$ is called an almost arithmetic progression (AAP for short).
- an AAMP with bound $M = 0$ and period $\mathscr{D} = \{0, d\}$ is called an arithmetic progression (AP for short).

The notion of AP just recalled of course coincides with the usual notion of a finite arithmetic progression. The notion of arithmetic multiprogression should not be confused with that of multidimensional arithmetic progressions, which is typically defined as a sumset of several arithmetic progressions.

Some care needs to be taken when saying that some set is or is not an AAMP. In fact, one has:

- every nonempty finite set $L \subset \mathbb{Z}$ is an AAP with bound $\max L - \min L$ (and period $\{0, 1\}$).
- every nonempty finite set $L \subset \mathbb{Z}$ is an AMP with period $-\min L + L$.

Thus, it is crucial to restrict bound and period in some way to make saying that a set is an AAMP meaningful.

The importance of the notion of AAMP in this context is mainly due to the following result, a Structure Theorem for Sets of Lengths (STSL). This result is due to Geroldinger [17], except that there a slightly different notion of AAMP was used; the current version was obtained in [13].

Theorem 6.3 *Let H be a Krull monoid with finite class group. There is some $M \in \mathbb{N}_0$ and a nonempty finite set $\Delta^* \subset \mathbb{N}$ such that for each $a \in H$ its set of lengths $\mathsf{L}(a)$ is a AAMP with bound M and difference d in Δ^*.*

A crucial point in this result is that the bound and the set of differences depend on the monoid, and not on the element. Indeed, by the transfer results recalled in Sect. 2.3 they depend on the class group or more precisely the subset of classes containing prime divisors, only.

This result was generalized in several ways and is known to hold for various other classes of monoids, too (see [23, Chap. 4]). Even sticking to Krull monoids it holds under the weaker condition that only finitely many classes contain prime divisors, or still weaker, that the Davenport constant of the set of classes containing prime divisors is finite (see Theorem 6.22).

The Structure Theorem for Sets of Lengths raises various follow-up questions. On the one hand, it is a natural question to ask if this description is a natural one or if there could be a simpler one. On the other hand, the result contains a bound M and a set of differences Δ^* and the question arises what are the actual values of these parameters. We discuss this in the remainder of this section.

6.1 The Relevance of AAMPs

Realizations results for sets of lengths prove that in a certain sense Theorem 6.3 is optimal. We recall such a realization result from [42]; for earlier result of this form see [23, Sect. 4.8].

Theorem 6.4 *Let $M \in \mathbb{N}_0$ and let $\emptyset \neq \Delta^* \subset \mathbb{N}$ be a finite set. Then, there exists a Krull monoid H with finite class group such that the following holds: for every set L that is an AAMP with difference $d \in \Delta^*$ and bound M there is some $y_{H,L}$ such that*

$$y + L \in \mathscr{L}(H) \text{ for all } y \geq y_{H,L}.$$

This result implies the existence of Krull monoids with finite class group whose system of sets of lengths contains all possible sets whose maximum and minimum are not too far apart. (Though, this was known already earlier.)

Corollary 6.5 *Let $M \in \mathbb{N}_0$. Then, there exists a Krull monoid H with finite class group such that $L \in \mathscr{L}(H)$ for every $L \subset \mathbb{N}_{\geq 2}$ with* $\max L - \min L \leq M$.

In [42] some explicit conditions on the class group were obtained that guarantee that the above results hold. For example, it is known that $\mathscr{B}(C_p^r)$ for p a prime greater than 5 and $r \geq 21(M^2 + \max \Delta^*)$ fulfills the conditions of Theorem 6.4 and thus of the corollary, too. This motivates the following problem.

Problem 6.6 Can one determine a function $f : \mathbb{N} \to \mathbb{N}$ such that for G a finite abelian group with $|G| \geq f(M)$ one has that $\mathscr{L}(G)$ contains each finite set $L \subset \mathbb{N}_{\geq 2}$ with $\max L - \min L \leq M$?

The author believes that such a function exists and a solution of this problem should be well within reach of current methods and results. The appeal of having such a result would be that it would give a precise way to express the informal idea that $\mathscr{L}(G)$ contains all possible sets that are 'small' relative to G.

The result that $\mathscr{L}(G)$ for infinite G contains every finite set $L \subset \mathbb{N}_{\geq 2}$ could be thought of as a limiting case of this result, for an infinite group every finite set $L \subset \mathbb{N}_{\geq 2}$ is 'small.' In fact, a positive answer to this problem would even yield a proof of the result for infinite torsion groups.

We do not recall a proof of Theorem 6.4 but still recall some simple constructions that show how AAMPs arise naturally in this context (cf. Lemmas 3.3 and 4.1).

Lemma 6.7 *Let $g \in G$ be an element of order $n \geq 3$. Then $\mathsf{L}((g(-g))^{kn})$ is an AP with difference $n - 2$ and length k, more specifically it is $2k + (n - 2) \cdot [0, k]$.*

Proof The only minimal zero-sum sequences over the set $\{-g, g\}$ are $(-g)g$, g^n, and $(-g)^n$. The only factorizations of $(g(-g))^{kn}$ are thus $(g^n(-g)^n)^{k-j}(g(-g))^{nj}$ for $j \in [0, k]$; their lengths are $2(k - j) + jn$.

Based on this lemma we give explicit examples of richer structures arising as sets of lengths; we choose to really fix some parameters to avoid confusion from having many parameters.

Example 6.8 Let $e_1, e_2, g, h \in G$ be independent elements of order 2, 2, 10 and 14 respectively, then

$$\mathsf{L}((g(-g))^{10k}(h(-h))^{14k}) = \{4k\} \cup (4k + 8 + 4 \cdot [0, 5k - 4]) \cup \{24k\}$$

is an AAP with difference 4 and bound 8, and

$$\begin{aligned}\mathsf{L}((e_1e_2(e_1 + e_2))^2(g(-g))^{10k}(h(-h))^{14k}) &= \{4k + 2, 4k + 3\} \\ &\cup (4k + 10 + \{0, 1\} + 4 \cdot [0, 5k - 4]) \cup \{24k + 2, 24k + 3\}\end{aligned}$$

is an AAMP with difference 4, period $\{0, 1, 4\}$ and bound 8.

6.2 Some Special Cases

As discussed for a general result the notion of AAMP seems inevitable. However, for special classes of groups simpler descriptions can be obtained. This is of course the case for class groups C_1 and C_2 where the system of sets of lengths consists of singletons only (see Theorem 3.4), but it is certainly also the case for C_2^2 and C_3 where by Theorem 5.6 one has that $\Delta(G) = \{1\}$, which implies that all sets are intervals.

In recent work of Geroldinger and the author [26] a characterization of all groups was obtained for which the more restrictive notions AP, AAP, or AMP suffice to describe the system of sets of lengths of $\mathcal{B}(G)$. We recall the result. (The definition and relevance of the set $\Delta^*(G)$, used in the result below, is recalled later in this section; the exact definition is not really crucial for the result below, and it could be replaced by $[1, |G|]$ for example.)

Theorem 6.9 *Let G be a finite abelian group.*

1. *The following statements are equivalent:*
    - *All sets of lengths in $\mathcal{L}(G)$ are arithmetical progressions.*
    - *G is cyclic of order $|G| \le 4$ or isomorphic to a subgroup of C_2^3 or isomorphic to a subgroup of C_3^2.*
2. *The following statements are equivalent:*
    - *There is a constant $M \in \mathbb{N}$ such that all sets of lengths in $\mathcal{L}(G)$ are* AAPs *with bound M.*
    - *G is isomorphic to a subgroup of C_3^3 or isomorphic to a subgroup of C_4^3.*

3. *The following statements are equivalent:*
   - *All sets of lengths in $\mathscr{L}(G)$ are AMPs with difference in $\Delta^*(G)$.*
   - *G is cyclic with $|G| \le 5$ or isomorphic to a subgroup of C_2^3 or isomorphic to a subgroup of C_3^2.*

In several of these cases it is even possible to give a complete description of $\mathscr{L}(G)$. We already discussed the first point several times; for the following ones see [23, Theorem 7.3.2], and for the last one [27, Proposition 3.12].

Proposition 6.10

1. $\mathscr{L}(C_1) = \mathscr{L}(C_2) = \big\{\{m\} \colon m \in \mathbb{N}_0\big\}$.
2. $\mathscr{L}(C_3) = \mathscr{L}(C_2 \oplus C_2) = \big\{y + 2k + [0, k] \colon y,\ k \in \mathbb{N}_0\big\}$.
3. $\mathscr{L}(C_4) = \big\{y + k + 1 + [0, k] \colon y,\ k \in \mathbb{N}_0\big\} \cup \big\{y + 2k + 2 \cdot [0, k] \colon y,\ k \in \mathbb{N}_0\big\}$.
4. $\mathscr{L}(C_2^3) = \big\{y + (k + 1) + [0, k] \colon y \in \mathbb{N}_0,\ k \in [0, 2]\big\}$
   $\cup \big\{y + k + [0, k] \colon y \in \mathbb{N}_0,\ k \ge 3\big\} \cup \big\{y + 2k + 2 \cdot [0, k] \colon y,\ k \in \mathbb{N}_0\big\}$.
5. $\mathscr{L}(C_3^2) = \{[2k, l] \colon k \in \mathbb{N}_0, l \in [2k, 5k]\}$
   $\cup \{[2k + 1, l] \colon k \in \mathbb{N}, l \in [2k + 1, 5k + 2]\} \cup \{\{1\}\}$.

However, to obtain results of this complete form becomes quite difficult. We recall a quite precise yet not complete description for the group of order 5 from [26].

Lemma 6.11 *Let G be a cyclic group of order $|G| = 5$. Then every $L \in \mathscr{L}(G)$ has one of the following forms:*

- *L is an arithmetical progression with difference* 1.
- *L is an arithmetical progression with difference* 3.
- *L is an AMP with period* $\{0, 2, 3\}$ *or with period* $\{0, 1, 3\}$.

6.3 The Set of Differences

The formulation of the Structure Theorem of Sets of Lengths contains a set Δ^*. We give an overview on the current knowledge about these sets. Of course, given the way the result is phrased this set cannot be determined uniquely; for one thing, if some set Δ^* is admissible for some bound M, then any superset of it would work, too.

Yet, there is a natural choice for the set Δ^* in the STSL for Krull monoids with finite class group, it is

$$\Delta^*(H) = \{\min \Delta(S) \colon S \subset H \text{ a divisor-closed submonoid with } \Delta(S) \ne \emptyset\}.$$

We recall that a submonoid $S \subset H$ is called divisor-closed if for each $s \in S$ every $a \in H$ with $a \mid s$ (in H) is in fact an element of S.

The result holds true for this set and it can be shown that $\mathcal{L}(H)$ contains AAMPs with difference d for each $d \in \Delta^*(H)$, so that it is not "too large." The details of the proof of the STSL provide further justification for considering this set as the natural choice.

It should be noted though that in general this is not a minimal choice. If L is an AAMP with difference d, period $\mathcal{D}$ and bound M, then L is also an AAMP with difference md, period $\mathcal{D} + d \cdot [0, m-1]$, and bound M. Thus, if the STSL holds for some set Δ^* that contains elements d, d' with $d \mid d'$, then one could omit d without effect on the result.

Thus, one could in principle "simplify" the set $\Delta^*(H)$ by omitting elements that are a divisor of an element already in the set. Yet doing so rather obscures the situation without yielding a true simplification.

Similarly, setting $D = \operatorname{lcm} \Delta^*(H)$ one can even replace the set of differences by a unique difference and get the following reformulation of the STSL.

Corollary 6.12 *Let H be a Krull monoid with finite class group. There is some $M \in \mathbb{N}_0$ and some $D \in \mathbb{N}$ such that for each $a \in H$ its set of lengths $\mathsf{L}(a)$ is a AAMP with bound M and difference D.*

While somewhat simpler to state, this formulation captures the reality of the situation not as well as the common one.

By transfer results as recalled in Sect. 2.3 one can get that

$$\Delta^*(H) = \{\min \Delta(G_0) \colon G_0 \subset G_P,\ \Delta(G_0) \neq \emptyset\}$$

where as usual $G_P \subset G$ denotes the subset of classes containing prime divisor and G the class group. (Some extra care is needed to check that divisor-closed submonoids actually are preserved in this way.)

For $|G| \geq 3$, one denotes by $\Delta^*(G) = \{\min \Delta(G_0) \colon G_0 \subset G,\ \Delta(G_0) \neq \emptyset\}$; this matches the usual convention that $\Delta^*(G) = \Delta^*(\mathcal{B}(G))$.

By Lemma 4.1 we know that $\min \Delta(G) = 1$ for $|G| \geq 3$. Thus $1 \in \Delta^*(G)$. Moreover the following constructions of elements of $\Delta^*(G)$ are classical.

Lemma 6.13 *Let G be a finite abelian group with $|G| \geq 3$.*

1. $[1, \mathsf{r}(G) - 1] \subset \Delta^*(G)$.
2. $d - 2 \in \Delta^*(G)$ *for each* $3 \leq d \mid \exp(G)$.
3. $|n - r - 1| \in \Delta^*(C_n^r)$ *for* $n \geq 2$, $r \geq 1$, *and* $n \neq r + 1$.

In particular, $\max \Delta^*(G) \geq \max\{\mathsf{r}(G) - 1, \exp(G) - 2\}$.

Proof We only give a sketch for details see [23]. For the first point, let $d \in [2, r]$ and let $e_1, \ldots, e_d \in G$ be independent elements of the same order, which we denote by n; note that by the definition of the rank such elements exist. Further, let $e_0 = \sum_{i=1}^n e_i$. It follows that $W_j = e_o^j \prod_{i=1}^d e_i^{n-j}$ for $j \in [1, n]$ and e_i^n for $i \in [1, d]$ are the only minimal zero-sum sequences. One has $W_j W_k = W_{j+k} \prod_{i=1}^d e_i^n$ for $j + k \leq n$, and

$W_j W_k = W_{j+k-n} W_n$ for $j + k > n$ are the only nontrivial relations. The former relations yield a distance of $(d + 1) - 2 = d - 1$.

For the second point, we consider the set $\{-g, g\}$ for an element of order g; cf. Lemma 6.7.

For the third point, we consider the example given in Example 5.11.

Recently, Geroldinger and Zhong [30] proved that in fact the inequality above is an equality; partial results and relevant techniques appeared in various papers, including [14, 41].

Theorem 6.14 *Let H be a Krull monoid with finite class group G.*

1. *If $|G| \le 2$, then $\Delta^*(H) = \emptyset$.*
2. *If $2 < |G| < \infty$, then $\max \Delta^*(H) \le \max\{\exp(G) - 2, \mathsf{r}(G) - 1\}$. If every class contains a prime divisor then equality holds.*

For the case of infinite class group it was proved by Chapman, Schmid, Smith [8] that if each class contains a prime divisor then $\Delta^*(H) = \mathbb{N}$.

For groups G where the rank is large relative to the exponent the set $\Delta^*(G)$ is completely determined by the preceding theorem.

Corollary 6.15 *Let G be a finite abelian group. If $\mathsf{r}(G) - 1 \ge \exp(G) - 2$, then $\Delta^*(G) = [1, \mathsf{r}(G) - 1]$.*

Moreover, directly from the above results, for $\exp(G) - 2 = \mathsf{r}(G)$ the set $\Delta^*(G)$ must still be an interval, namely $[1, \exp(G) - 2]$, yet for groups with $\mathsf{r}(G) < \exp(G) - 2$ the set $\Delta^*(G)$ could have gaps. Indeed, it frequently does have gaps, as the result below shows (it is a direct consequence of [41, Theorem 3.2] and [30]).

Theorem 6.16 *Let H be a Krull monoid with class group G such that each class contains a prime divisor. Suppose that $\exp(G) - 3 \ge \mathsf{r}(G)$ and that G does not have a subgroup isomorphic to $C_{\exp(G)}^2$. Then $\Delta^*(H)$ is not an interval, as $\exp(G) - 3 \notin \Delta^*(H)$ while $\{1, \exp(G) - 2\} \subset \Delta^*(H)$.*

The type of groups for which the problem of determining $\Delta^*(G)$ in more detail has received most attention are cyclic groups. In this case $\Delta^*(G)$ shows a rich structure that is not yet fully understood, despite various partial results.

For G a cyclic group of order n we have, by the results above, that $\max \Delta^*(G) = n - 2$, and it was proved by Geroldinger and Hamidoune [24] that the second largest element of $\Delta^*(G)$ is $\lfloor n/2 \rfloor - 1$ for $n \ge 4$.

Recently several further elements were determined by Plagne and the author [38]; we state a simplified version of the result (the actual result goes down to a tenth, rather than a fifth, of the order of the group).

Theorem 6.17 *Let G be a cyclic group of order at least n_0 (where $n_0 = 250$ is a possible choice). We have*

$$\Delta^*(G) \cap \mathbb{N}_{\geq |G|/5} = \mathbb{N} \cap \left\{ |G|-2, \frac{|G|-2}{2}, \frac{|G|-3}{2}, \frac{|G|-4}{2}, \frac{|G|-4}{3}, \right.$$
$$\left. \frac{|G|-6}{3}, \frac{|G|-4}{4}, \frac{|G|-5}{4}, \frac{|G|-6}{4}, \frac{|G|-8}{4} \right\}.$$

An important tool in obtaining this result is the determination of $\min \Delta(G_0)$ for G_0 a set with $|G_0| = 2$. The key case, to which all other cases can be reduced, is that $G_0 = \{e, ae\}$ where e is a generating element and $\gcd(a, \operatorname{ord} e) = 1$.

In this case, one can express $\min \Delta(G_0)$ in terms of the continued fraction expansion of $(\operatorname{ord} e)/a$. More specifically, one has the following results [5, Theorem 2.1].

Theorem 6.18 *Let* $G = \langle e \rangle$ *with* $\operatorname{ord} e = n > 3$. *Further, let* $a \in [2, n-1]$ *and let* $[a_0, a_1, \ldots, a_m]$ *be the continued fraction expansion of* n/a *of odd length (that is* m *is even). Then*

$$\min \Delta(\{e, ae\}) = \gcd(a_1, a_3, \ldots, a_{m-1}).$$

The continued fraction expansion mentioned in the result is the standard continued fraction expansion, except for the fact that one allows the last term to equal 1, which allows to always achieve that m is even.

As a consequence of this, one obtains the following elements that correspond precisely to those a for which the continued fraction expansion has length 3.

Remark 6.19 Let $G = \langle e \rangle$ with $\operatorname{ord} e = n > 3$. Further, let $b, c \in [1, n-1]$ such that $(n-b)/c$ and $(n-b-c)/(bc)$ are positive integers. Then

$$\min \Delta\left(\left\{e, \frac{n-b}{c} e\right\}\right) = \frac{n-b-c}{bc}.$$

Moreover, it can be shown that if $\min \Delta(\{e, ae\})$ is 'large' then it must be of that form (cf. [5, Corollary 3.2] and [38]).

Theorem 6.20 *Let* G *be a cyclic group,* e *be a generating element of* G *and* $a \in [1, |G|]$ *such that* $\gcd(a, |G|) = 1$.

Then $\min \Delta(\{e, ae\}) > \sqrt{|G|}$ *if and only if there exist some positive integers* c_1 *and* c_2 *such that*

$$a = \frac{|G| - c_1}{c_2}$$

and the quantity

$$d_a = \frac{|G| - (c_1 + c_2)}{c_1 c_2}$$

is integral and satisfies $d_a > \sqrt{|G|}$. *Indeed, in this case* $\min \Delta(\{e, ae\}) = d_a$.

These results already explain the presence of several of the elements we mentioned in Theorem 6.17. Specifically one gets the elements

$$\left\{|G|-2, \frac{|G|-3}{2}, \frac{|G|-4}{3}, \frac{|G|-5}{4}, \frac{|G|-4}{4}\right\}$$

for (c_1, c_2) equaling $(1, 1)$, $(1, 2)$, $(1, 3)$, $(1, 4)$, and $(2, 2)$ respectively.

Furthermore, for every subgroup G' of G, one gets that $\exp(G') - 2$ is an element of $\Delta^*(G')$ and thus of $\Delta^*(G)$. This yields the elements

$$\left\{\frac{|G|-4}{2}, \frac{|G|-6}{3}, \frac{|G|-8}{4}\right\},$$

considering subgroups of order $|G|/2, |G|/3, |G|/4$, respectively. In addition, $(|G|-6)/4$ is in $\Delta^*(G)$ as $(\exp(G')-3)/2$ is in $\Delta^*(G')$ for G' a subgroup of order $|G|/2$.

It remains to construct $\{(|G|-2)/2\}$. This element can be shown to equal $\min \Delta(\{e, -e, (|G|/2)e\})$. In this way we have given some arguments for the presence of all these elements. Of course it remains to show that there are no other elements. We do not discuss this here.

For other types of groups the set $\Delta^*(G)$ is less well understood. But, it is for example known for $n \geq 5$ that $\{n-3, n-2\} \subset \Delta^*(C_n^2)$ and $\max(\Delta^*(C_n^2) \setminus \{n-3, n-2\}) = \lfloor n/2 \rfloor - 1$ (see [41, Corollary 3.7]). Further results of this form can be obtained for more general groups under assumptions; see [41, Theorem 3.2] and [29]. We end with a specific problem and a general remark on further work.

Problem 6.21 Is there a finite abelian group G such that $\Delta^*(G)$ is an interval and $\exp(G) \geq 2\mathsf{r}(G) + 2$?

For $n \leq 2r + 1$, it follows that $\Delta^*(C_n^r) = [1, \max\{n-2, r-1, \}]$ as $[1, r-1]$ and $[\max\{1, n-r-1\}, n-2]$ are contained in it.

Having some information about the differences $\Delta^*(H)$ at hand a next natural question would be to determine which periods can appear in the STSL. Beyond the information contained in the complete results on $\mathcal{L}(H)$, for special cases which we recalled above, not too much is known on this problem. However, given the recent progress on the problem of determining $\Delta^*(G)$ and associated descriptions of sets yielding the relevant distances, it might now be a good time to approach this problem.

6.4 The Bound in the STSL

Having discussed the set of differences we turn to the other parameter in the STSL, the bound. A lot less is known about it. Geroldinger and Grynkiewicz [20, Theorem 4.4.2] showed the following refinement and generalization of Theorem 6.3.

Theorem 6.22 *Let H be a Krull monoid with subset of classes containing prime divisors G_P such that $\mathsf{D}(G_P)$ is finite (and at least 3). Let*

$$M = (2\mathsf{D}(G_P) - 5)\mathsf{D}(G_P)^2 + \frac{1}{2}\mathsf{D}(G_P)^4)^{\frac{\mathsf{D}(G_P)(\mathsf{D}(G_P)-1)}{2}}.$$

For each $a \in H$ its set of lengths $\mathsf{L}(a)$ is an AAMP with bound M and difference $d \in \Delta(H)$.

The condition that $\mathsf{D}(G_P) \geq 3$ is no actual restriction as otherwise the monoid is half-factorial. As mention in Sect. 2 finiteness of G_P implies finiteness of $\mathsf{D}(G_p)$. Thus, the result includes the case that only a finite number of classes contains prime divisors. We highlight that in this result the set of differences is $\Delta(H)$ not $\Delta^*(H)$. However, in case the class group is finite we can combine the results to get that every set of lengths is an AAMP with difference in $\Delta^*(H)$ and still have an explicit bound.

The bound above, being of the form $\exp(c \log(\mathsf{D}(G_P))\mathsf{D}(G_P)^2)$, grows quite fast in terms of the Davenport constant. It is not at all clear what the actual order of magnitude of the bound should be. Below we give a simple example showing that the dependence is at least of quadratic order.

Example 6.23 Let $n \geq 6$ be even, such that $n/2$ is odd. Let $C_{n/2} \oplus C_n = \langle e_1 \rangle \oplus \langle e_2 \rangle$. For sufficiently large k, one has that the set of lengths of $(e_1(-e_1))^{kn/2}(e_2(-e_2))^{kn}$ is an AAP with difference 1 and bound (at least) $(n-3)(n/2-3)$ while $\mathsf{D}(C_{n/2} \oplus C_n) = 3n/2 - 1$.

To see this let d_1, d_2 be co-prime positive integers. Then, for all sufficiently large k_1, k_2 one has that $L = (a + d_1 \cdot [0, k_1]) + (b + d_2 \cdot [0, k_2])$ is an AAP with difference 1 and bound (at least) $(d_1 - 1)(d_2 - 1)$; recall that the Frobenius number of d_1, d_2 is $(d_1 - 1)(d_2 - 1) - 1$. Thus $a + b \in L$ while $a + b + (d_1 - 1)(d_2 - 1) - 1 \notin L$ so that when writing $L = y + (L' \cup L^* \cup L'')$ in the usual way with L^* an AP with difference 1, that is an interval, then $y \geq a + b + (d_1 - 1)(d_2 - 1)$ and $a + b \geq y - M$ implies that $M \geq (d_1 - 1)(d_2 - 1)$. Now, by Lemma 6.7 the set of length of $(g(-g))^{k \operatorname{ord} g}$ is an AP with difference $\operatorname{ord} g - 2$ of length k. And $\mathsf{L}((e_1(-e_1))^{kn/2}(e_2(-e_2))^{kn})) = (2k + (n/2 - 2) \cdot [0, k]) + (2k + (n-2) \cdot [0, k])$. If $n/2$ is odd, $n - 2$ and $n/2 - 2$ are co-prime. By the argument above we thus have an AAP with bound at least $(n-3)(n/2-3)$ in $\mathcal{L}(C_{n/2} \oplus C_n)$, and $\mathsf{D}(C_{n/2} \oplus C_n) = 3n/2 - 1$.

This example shows that the bound is at least of quadratic order in terms of the Davenport constant.

Problem 6.24 What is the (rough) order of magnitude of the bound in the STSL for $\mathcal{L}(G)$ (in terms of $\mathsf{D}(G)$)?

Initially, it would also be interesting to have an answer to this problem just for some special (infinite) family of groups, or in other more restricted scenarios.

There is very little evidence on which one might base conjectures regarding the size of the bound M. However, an effect that might limit the size of the bound is that elements divisible by prime divisors from many different classes tend to have very simple sets of lengths. We recall a result in this direction due to Geroldinger and Halter-Koch [23, Theorem 7.6.9]; their actual result is more precise.

Theorem 6.25 *Let H be a Krull monoid with finite class group, and let $\varphi : H \to F$ be its divisor theory. If $a \in H$ such that $\varphi(a)$ is divisible by a prime divisor from each nonzero class, then $\mathsf{L}(a)$ is an interval.*

Acknowledgments The author is very grateful to the referee for many useful remarks and corrections. This work was supported by the ANR project Caesar, project number ANR-12-BS01-0011.

References

1. D.F. Anderson, Elasticity of factorizations in integral domains: a survey, *Factorization in Integral Domains (Iowa City, IA, 1996)*, vol. 189, Lecture Notes in Pure and Applied Mathematics (Dekker, New York, 1997), pp. 1–29
2. N.R. Baeth, A. Geroldinger, Monoids of modules and arithmetic of direct-sum decompositions. Pac. J. Math. **271**(2), 257–319 (2014)
3. P. Baginski, A. Geroldinger, D.J. Grynkiewicz, A. Philipp, Products of two atoms in Krull monoids and arithmetical characterizations of class groups. Eur. J. Comb. **34**(8), 1244–1268 (2013)
4. L. Carlitz, A characterization of algebraic number fields with class number two. Proc. Am. Math. Soc. **11**, 391–392 (1960)
5. S. Chang, S.T. Chapman, W.W. Smith, On minimum delta set values in block monoids over cyclic groups. Ramanujan J. **14**(1), 155–171 (2007)
6. S.T. Chapman, F. Gotti, R. Pelayo, On delta sets and their realizable subsets in Krull monoids with cyclic class groups. Colloq. Math. **137**(1), 137–146 (2014)
7. S.T. Chapman, U. Krause, E. Oeljeklaus, On Diophantine monoids and their class groups. Pac. J. Math. **207**(1), 125–147 (2002)
8. S.T. Chapman, W.A. Schmid, W.W. Smith, On minimal distances in Krull monoids with infinite class group. Bull. Lond. Math. Soc. **40**, 613–618 (2008)
9. S.T. Chapman, W.W. Smith, Factorization in Dedekind domains with finite class group. Isr. J. Math. **71**(1), 65–95 (1990)
10. K. Cziszter, M. Domokos, A. Geroldinger, The interplay of invariant theory with multiplicative ideal theory and with arithmetic combinatorics. **170**, 43–95 (2016)
11. Y. Fan, Q. Zhong, Products of k atoms in Krull monoids. Monatshefte für Mathematik, to appear
12. M. Freeze, A. Geroldinger, Unions of sets of lengths. Funct. Approx. Comment. Math. **39(part 1)**, 149–162 (2008)
13. G. Freiman, A. Geroldinger, An addition theorem and its arithmetical application. J. Number Theory **85**(1), 59–73 (2000)
14. W. Gao, A. Geroldinger, Systems of sets of lengths. II. Abh. Math. Sem. Univ. Hamburg **70**, 31–49 (2000)
15. W. Gao, A. Geroldinger, On products of k atoms. Monatsh. Math. **156**(2), 141–157 (2009)
16. W. Gao, A. Geroldinger, D.J. Grynkiewicz, Inverse zero-sum problems. III. Acta Arith. **141**(2), 103–152 (2010)
17. A. Geroldinger, Über nicht-eindeutige Zerlegungen in irreduzible Elemente. Math. Z. **197**(4), 505–529 (1988)
18. A. Geroldinger, Additive group theory and non-unique factorizations, *Combinatorial Number Theory and Additive Group Theory*, Advanced Courses in Mathematics (CRM Barcelona, Birkhäuser Verlag, Basel, 2009), pp. 1–86
19. A. Geroldinger, D. Grynkiewicz, P. Yuan, On products of k atoms II. Mosc. J. Comb. Number Theory **5**, 71–126 (2015)
20. A. Geroldinger, D.J. Grynkiewicz, On the arithmetic of Krull monoids with finite Davenport constant. J. Algebr. **321**(4), 1256–1284 (2009)
21. A. Geroldinger, D.J. Grynkiewicz, W.A. Schmid, The catenary degree of Krull monoids I. J. Théor. Nombres Bordeaux **23**(1), 137–169 (2011)
22. A. Geroldinger, F. Halter-Koch, Congruence monoids. Acta Arith. **112**(3), 263–296 (2004)

23. A. Geroldinger, F. Halter-Koch, *Non-Unique Factorizations. Algebraic, Combinatorial and Analytic Theory* (Chapman & Hall/CRC, Boca Raton, 2006)
24. A. Geroldinger, Y.O. Hamidoune, Zero-sumfree sequences in cyclic groups and some arithmetical application. J. Théor. Nombres Bordeaux **14**(1), 221–239 (2002)
25. A. Geroldinger, F. Kainrath, A. Reinhart, Arithmetic of seminormal weakly Krull monoids and domains. J. Algebr. **444**, 201–245 (2015)
26. A. Geroldinger, W.A. Schmid, A characterization of class groups via sets of lengths
27. A. Geroldinger, W.A. Schmid, The system of sets of lengths in Krull monoids under set addition. Rev. Mat. Iberoam
28. A. Geroldinger, P. Yuan, The set of distances in Krull monoids. Bull. Lond. Math. Soc. **44**(6), 1203–1208 (2012)
29. A. Geroldinger, Q. Zhong, A characterization of class groups via sets of lengths II. J. Théor. Nombres Bordx
30. A. Geroldinger, Q. Zhong, The set of minimal distances of Krull monoids. Acta Arith. **173**, 97–120 (2016)
31. A. Geroldinger, Q. Zhong, The catenary degree of Krull monoids II. J. Aust. Math. Soc. **98**, 324–354 (2015)
32. F. Halter-Koch, Finitely generated monoids, finitely primary monoids, and factorization properties of integral domains, *Factorization in Integral Domains (Iowa City, IA, 1996)*, vol. 189, Lecture Notes in Pure and Applied Mathematics (Dekker, New York, 1997), pp. 31–72
33. F. Halter-Koch, An introduction to multiplicative ideal theory, *Ideal Systems*, vol. 211, Monographs and Textbooks in Pure and Applied Mathematics (Marcel Dekker Inc., New York, 1998)
34. F. Kainrath, Factorization in Krull monoids with infinite class group. Colloq. Math. **80**(1), 23–30 (1999)
35. U. Krause, On monoids of finite real character. Proc. Am. Math. Soc. **105**(3), 546–554 (1989)
36. G.J. Leuschke, R. Wiegand, *Cohen-Macaulay Representations*, vol. 181, Mathematical Surveys and Monographs (American Mathematical Society, Providence, 2012)
37. W. Narkiewicz, Finite abelian groups and factorization problems. Colloq. Math. **42**, 319–330 (1979)
38. A. Plagne, W.A. Schmid, On congruence half-factorial Krull monoids with cyclic class group
39. C. Reiher, A proof of the theorem according to which every prime number possesses property B. Thesis, University of Rostock (2010)
40. S. Savchev, F. Chen, Long zero-free sequences in finite cyclic groups. Discret. Math. **307**(22), 2671–2679 (2007)
41. W.A. Schmid, Arithmetical characterization of class groups of the form $\mathbb{Z}/n\mathbb{Z} \oplus \mathbb{Z}/n\mathbb{Z}$ via the system of sets of lengths. Abh. Math. Sem. Hamburg **79**, 25–35 (2009)
42. W.A. Schmid, A realization theorem for sets of lengths. J. Number Theory **129**(5), 990–999 (2009)
43. W.A. Schmid, Inverse zero-sum problems II. Acta Arith. **143**(4), 333–343 (2010)
44. D. Smertnig, Sets of lengths in maximal orders in central simple algebras. J. Algebr. **390**, 1–43 (2013)
45. P. Yuan, On the index of minimal zero-sum sequences over finite cyclic groups. J. Comb. Theory Ser. A **114**(8), 1545–1551 (2007)

Factorizations of Elements in Noncommutative Rings: A Survey

Daniel Smertnig

Dedicated to Franz Halter-Koch on the occasion of his 70th birthday

Abstract We survey results on factorizations of non-zero-divisors into atoms (irreducible elements) in noncommutative rings. The point of view in this survey is motivated by the commutative theory of nonunique factorizations. Topics covered include unique factorization up to order and similarity, 2-firs, and modular LCM domains, as well as UFRs and UFDs in the sense of Chatters and Jordan and generalizations thereof. We recall arithmetical invariants for the study of nonunique factorizations, and give transfer results for arithmetical invariants in matrix rings, rings of triangular matrices, and classical maximal orders as well as classical hereditary orders in central simple algebras over global fields.

1 Introduction

Factorizations of elements in a ring into atoms (irreducible elements) are natural objects to study if one wants to understand the arithmetic of a ring. In this overview, we focus on the semigroup of non-zero-divisors in noncommutative (associative, unital) rings. The point of view in this article is motivated by analogy with the commutative theory of nonunique factorizations (as in [4, 28, 50, 53]).

We start by giving a rigorous notion of *rigid factorizations* and discussing sufficient conditions for the existence of factorizations of any non-zero-divisor, in Sect. 3. In Sect. 4, we look at several notions of *factoriality*, that is, notions of unique

D. Smertnig (✉)
Institute for Mathematics and Scientific Computing, University of Graz,
NAWI Graz, Heinrichstrae 36, 8010 Graz, Austria
e-mail: daniel.smertnig@uni-graz.at

S. Chapman et al. (eds.), *Multiplicative Ideal Theory and Factorization Theory*,
Springer Proceedings in Mathematics & Statistics 170,
DOI 10.1007/978-3-319-38855-7_15

factorization, that have been introduced in the noncommutative setting. Finally, in Sect. 5 we shift our attention to nonunique factorizations and the study of arithmetical invariants used to describe them.

The investigation of factorizations in noncommutative rings has its origins in the study of homogeneous linear differential equations. The first results on the uniqueness of factorizations of linear differential operators are due to Landau, in [71], and Loewy, in [81]. Ore, in [89], put this into an entirely algebraic context by studying skew polynomials (also called Ore extensions) over division rings. He showed that if D is a division ring, then the skew polynomial ring $D[x; \sigma, \delta]$, where σ is an injective endomorphism of D and δ is a σ-derivation, satisfies an Euclidean algorithm with respect to the degree function. Hence, factorizations of elements in $D[x; \sigma, \delta]$ are unique up to order and *similarity*. We say that $D[x; \sigma, \delta]$ is *similarity factorial* (see Definition 4.1).

Jacobson, in [65], already describes unique factorization properties for principal ideal domains. He showed that PIDs are similarity factorial. In a further generalization, principal ideal domains were replaced by *2-firs*, and the Euclidean algorithm was replaced by the *2-term weak algorithm*. This goes back to work primarily due to P.M. Cohn and Bergman. The main reference is [39].

Factorizations in 2-firs, the 2-term weak algorithm, and the notion of *similarity factoriality* are the focus of Sect. 4.1. A key result is that the free associative algebra $K\langle X\rangle$ over a field K in a family of indeterminates X is similarity factorial. Here, K cannot be replaced by an arbitrary factorial domain, as $\mathbb{Z}\langle x, y\rangle$ is not similarity factorial. Brungs, in [21], studied the slightly weaker notion of *subsimilarity factoriality*. Using a form of Nagata's theorem, it follows that free associative algebras over factorial commutative domains are subsimilarity factorial.

Modular right LCM domains were studied by Beauregard in a series of papers and are also discussed in Sect. 4.1. Many results on unique factorizations in Sect. 4.1 can be derived from the Jordan–Hölder theorem on (semi-)modular lattices by consideration of a suitable lattice. Previous surveys covering unique factorizations in noncommutative rings, as considered in Sect. 4.1, are [32, 34, 36, 37]. We also refer to the two books [38, 39].

A rather different notion of [Noetherian] UFRs (unique factorization rings) and UFDs (unique factorization domains), originally introduced by Chatters and Jordan in [26, 29], has seen widespread adoption in ring theory. We discuss this concept, and its generalizations, in Sect. 4.2. Examples of Noetherian UFDs include universal enveloping algebras of finite-dimensional solvable Lie algebras over $\mathbb{C}$, various (semi)group algebras, and quantum algebras. In a UFR R, the semigroup of nonzero normal elements, $N(R)^\bullet$, is a UF-monoid. Thus, nonzero normal elements of R factor uniquely as products of prime elements.

Section 5 is devoted to the study of nonunique factorizations in noncommutative rings. Here, the basic interest is in determining arithmetical invariants that suitably measure, characterize, or describe the extent of nonuniqueness of the factorizations. A recent result by Bell, Heinle, and Levandovskyy, from [17], establishes that many interesting classes of noncommutative domains are finite factorization domains (FF-domains).

We recall several arithmetical invariants, as well as the notion of *[weak] transfer homomorphisms*. Transfer homomorphisms have played a central role in the commutative theory of nonunique factorizations and promise to be useful in the noncommutative setting as well. By means of transfer results, it is sometimes possible to reduce the study of arithmetical invariants in a ring to the study of arithmetical invariants in a much simpler object.

Most useful are transfer results from the non-zero-divisors of a noncommutative ring to a commutative ring or semigroup for which the factorization theory is well understood. Such transfer results exist for rings of triangular matrices (see [5, 22]), rings of matrices (see [44, 45]), and classical hereditary (in particular, maximal) orders in central simple algebras over global fields (see [22, 46–48, 93]). These results are covered in Sect. 5.4.

Throughout the text, we gather known examples from the literature and point out their implications for factorization theory. In particular, these examples demonstrate limitations of certain concepts or methods in the noncommutative setting when compared to the commutative setting.

As a note on terminology, we call a domain *similarity [subsimilarity,projectivity] factorial* instead of a *similarity-[subsimilarity,projectivity]-UFD*. This matches the terminology presently preferred in the commutative setting. Using an adjective to describe the property sometimes makes it easier to use it in writing. Moreover, this allows us to visibly differentiate factorial domains from the [Noetherian] UFRs and UFDs in the sense of Chatters and Jordan that are discussed in Sect. 4.2.

While an attempt has been made to be comprehensive, it would be excessive to claim the results contained in this article are entirely exhaustive. Many interesting results on nonunique factorizations are scattered throughout the literature, with seemingly little previous effort to tie them together under a common umbrella of a theory of (nonunique) factorizations.

Naturally, there are certain restrictions on the scope of the present treatment. For the reader who came expecting something else under the heading *factorization theory*, some pointers to recent work, which is beyond the scope of this article, but may conceivably be considered to be factorization theory, are given in Sect. 6.

2 Preliminaries

All rings are assumed to be unital and associative, but not necessarily commutative. All semigroups have a neutral element. A ring R is a *domain* if 0 is the unique zero-divisor (in particular, $R \neq 0$). A *right principal ideal domain (right PID)* is a domain in which every right ideal is principal. A *left PID* is defined analogously, and a domain is a *principal ideal domain (PID)* if it is both, a left and a right PID. We make similar conventions for other notions for which a left and a right variant exist, e.g., Noetherian, Euclidean, etc.

2.1 Small Categories as Generalizations of Semigroups

We will be interested in factorizations of non-zero-divisors in a ring R. Even so, it will sometimes be useful to have the notions of factorizations available in the more general setting of semigroups, or even more generally, in the setting of small categories. Thus, we develop the basic terminology in the very general setting of a cancellative small category. This generality does not cause any significant additional problems over making the definitions in a more restrictive setting, such as cancellative semigroups, or even the semigroup of non-zero-divisors in a ring. It may however be useful to keep in mind that the most important case for us will be where the cancellative small category simply is the semigroup of non-zero-divisors of a ring.

Here, a small category is viewed as a generalization of a semigroup, in the sense that the category of semigroups is equivalent to the category of small categories with a single object. In practice, we will however be concerned mostly with semigroups. Therefore, we use a notation for small categories that is reminiscent of that for semigroups. We briefly review the notation. See also [93, Sect. 2.1] and [22, Sect. 2] for more details.

Let H be a small category. A morphism a of H has a source $s(a)$ and a target $t(a)$. If a and b are morphisms with $t(a) = s(b)$ we write the composition left to right as ab. The objects of the category will play no significant role (they can always be recovered from the morphisms via the source and target maps). We identify the objects with their identity morphisms and denote the set of all identity morphism by H_0. We identify H with its set of morphisms. Accordingly, we call a morphism a of H simply an element of H and write $a \in H$.

More formally, from this point of view, a small category $H = (H, H_0, s, t, \cdot)$ consists of the following data: A set H together with a distinguished subset $H_0 \subset H$, two functions $s, t\colon H \to H_0$, and a partial function: $H \times H \to H$ such that:

(1) $s(e) = t(e) = e$, for all $e \in H_0$,
(2) $a \cdot b \in H$ is defined, for all $a, b \in H$ with $t(a) = s(b)$,
(3) $a \cdot (b \cdot c) = (a \cdot b) \cdot c$, for all $a, b, c \in H$ with $t(a) = s(b)$ and $t(b) = s(c)$,
(4) $s(a) \cdot a = a \cdot t(a) = a$, for all $a \in H$.

For $e, f \in H_0$, we define $H(e, \cdot) = \{\, a \in H \mid s(a) = e \,\}$, $H(\cdot, f) = \{\, a \in H \mid t(a) = f \,\}$, $H(e, f) = H(e, \cdot) \cap H(\cdot, f)$, and $H(e) = H(e, e)$.

To see the equivalence of this definition with the usual definition of a small category, suppose first that H is as above. Take as set of objects of a category $\mathscr{C}$ the set H_0, and, for two objects $e, f \in H_0$, set $\mathrm{Hom}_{\mathscr{C}}(e, f) = H(f, e)$. Define the composition on $\mathscr{C}$ using the partial map $\cdot$. Then $\mathscr{C}$ is a small category in terms of the usual definition, with composition written right to left and with $e \in \mathrm{Hom}(e, e)$ the identity morphism of the object e. Conversely, if $\mathscr{C}$ is a small category in the usual sense, set $H = \bigcup_{e,f \in Ob\mathscr{C}} \mathrm{Hom}_{\mathscr{C}}(e, f)$ and $H_0 = \{\, \mathrm{id}_e \mid e \in Ob\mathscr{C} \,\}$. For $a \in H$ with domain e and codomain f, set $s(a) = f$ and $t(a) = e$. The partial function $\cdot$ on H is defined via the composition of $\mathscr{C}$. Then H satisfies the properties above.

Let H be a small category. If $a, b \in H$ and we write ab, we implicitly assume $t(a) = s(b)$. The subcategory of units (isomorphisms) of H is denoted by $H^\times$. The small category H is a *groupoid* if $H = H^\times$, and it is *reduced* if $H^\times = H_0$. An element $a \in H$ is *cancellative* if it is both a monomorphism and an epimorphism, that is, for all b, c in H, $ab = ac$ implies $b = c$ and $ba = ca$ implies $b = c$. The subcategory of cancellative elements of H is denoted by $H^\bullet$. A functor f from H to another small category H' is referred to as a *homomorphism*. Two elements $a, b \in H$ are (two-sided) *associated* if there exist $\varepsilon, \eta \in H^\times$ such that $a = \varepsilon b \eta$.

Let H be a small category. A subset $I \subset H$ is a right ideal of H if $IH = \{ xa \mid x \in I,\ a \in H : t(x) = s(a) \}$ is a subset of I. A right ideal of H is called a *right H-ideal* if there exists an $a \in H^\bullet$ such that $a \in I$. A right ideal $I \subset H$ is *principal* if there exists $a \in H$ such that $I = aH$. An ideal $I \subset H$ is *principal* if it is principal as a left and right ideal, that is, there exist $a, b \in H$ such that $I = aH = Hb$. Suppose that every left or right divisor of a cancellative element is again cancellative. If $I \subset H$ is an ideal and $I = Ha = bH$ with $a, b \in H^\bullet$, then it is easy to check that also $I = aH = Hb$.

Let H be a semigroup. An element $a \in H$ is *normal* (or *invariant*) if $aH = Ha$. We write $N(H)$ for the subsemigroup of all normal elements of H. The semigroup H is *normalizing* if $H = N(H)$.

In the commutative theory of nonunique factorizations, a *monoid* is usually defined to be a cancellative commutative semigroup. Since the meaning of *monoid* in articles dealing with a noncommutative setting is often different, we will avoid its use altogether. The exception are compound nouns such as *Krull monoid*, *free monoid*, *free abelian monoid*, *monoid of zero-sum sequences*, and *UF-monoid*, where the use of *monoid* is universal and it would be strange to introduce different terminology.

2.2 Classical Maximal Orders

Classical maximal orders in central simple algebras over a global field will appear throughout in examples. Moreover, they are one of the main objects for which we are interested in studying nonunique factorizations. Therefore, we recall the setting. We use [90] as a general reference, and [40, 95] for strong approximation. For the motivation for calling such orders *classical* orders, and the connection to different notions of orders, see [85, Sect. 5.3].

Let K be a global field, that is, either an algebraic number field or an algebraic function field (of transcendence degree 1) over a finite field. Let S_{fin} denote the set of all non-archimedean places of K. For each $v \in S_{\text{fin}}$, let $\mathcal{O}_v \subset K$ denote the corresponding discrete valuation domain. A subring $\mathcal{O} \subset K$ is a *holomorphy ring* if there exists a finite subset $S \subset S_{\text{fin}}$ (and $\emptyset \neq S$ in the function field case) such that

$$\mathcal{O} = \mathcal{O}_S = \bigcap_{v \in S_{\text{fin}} \setminus S} \mathcal{O}_v.$$

The holomorphy rings in K are Dedekind domains which are properly contained in K and have quotient field K. The most important examples are rings of algebraic integers and S-integers in the number field case, and coordinate rings of nonsingular irreducible affine algebraic curves over finite fields in the function field case.

Let A be a central simple K-algebra, that is, a finite-dimensional K-algebra with center K which is simple as a ring. A *classical $\mathcal{O}$-order* is a subring $\mathcal{O} \subset R \subset A$ such that R is a finitely generated $\mathcal{O}$-module and $KR = A$. A *classical maximal $\mathcal{O}$-order* is a classical $\mathcal{O}$-order which is maximal with respect to set inclusion within the set of all classical $\mathcal{O}$-orders contained in A. A *classical hereditary $\mathcal{O}$-order* is a classical $\mathcal{O}$-order which is hereditary as a ring. Every classical maximal $\mathcal{O}$-order is hereditary.

If v is a place of K, the completion A_v of A is a central simple algebra over the completion K_v of K. Hence, A_v is of the form $A_v \cong M_{n_v}(D_v)$ with a finite-dimensional division ring $D_v \supset K_v$. The algebra A is *ramified at* v if $D_v \neq K_v$.

Isomorphism classes of right ideals and class groups. Let $\mathcal{F}^\times(\mathcal{O})$ denote the group of nonzero fractional ideals of $\mathcal{O}$. Let $K_A^\times$ denote the subgroup of $K^\times$ consisting of all $a \in K^\times$ for which $a_v > 0$, for all archimedean places v of K at which A is ramified. To a classical maximal $\mathcal{O}$-order R (or more generally, a classical hereditary $\mathcal{O}$-order), we associate the ray class group

$$\mathcal{C}_A(\mathcal{O}) = \mathcal{F}^\times(\mathcal{O}) \,/\, \{\, a\mathcal{O} \mid a \in K_A^\times \,\}.$$

This is a finite abelian group, with operation induced by the multiplication of fractional ideals.

Let $\mathrm{LF}_1(R)$ denote the (finite) set of isomorphism classes of right R-ideals. In general, $\mathrm{LF}_1(R)$ does not have a natural group structure. Let $\mathcal{C}(R)$ denote the set of stable isomorphism classes of right R-ideals. The set $\mathcal{C}(R)$ naturally has the structure of an abelian group, with operation induced from the direct sum operation. There is a surjective map of sets $\mathrm{LF}_1(R) \to \mathcal{C}(R)$, and a group homomorphism $\mathcal{C}(R) \to \mathcal{C}_A(\mathcal{O})$, $[I] \mapsto [\mathrm{nr}(I)]$. The homomorphism $\mathcal{C}(R) \to \mathcal{C}_A(\mathcal{O})$ is in fact an isomorphism (see [95, Corollary 9.5]). However, the map $\mathrm{LF}_1(R) \to \mathcal{C}(R)$ need not be a bijection in general. It is a bijection if and only if stable isomorphism of right R-ideals implies isomorphism. This holds if A satisfies the Eichler condition relative to $\mathcal{O}$ (see below). We will at some point need to impose the weaker condition that every stably free right R-ideal is free, that is, that the preimage of the trivial class under $\mathrm{LF}_1(R) \to \mathcal{C}(R)$ consists only of the trivial class. This condition will be of paramount importance for the existence of a transfer homomorphism from $R^\bullet$ to a monoid of zero-sum sequences over the ray class group $\mathcal{C}_A(\mathcal{O})$.

A ring over which every finitely generated stably free right module is free is called a *(right) Hermite ring*. (Using the terminology of [70, Chap. I.4], some authors require in addition that R has the invariant basis number (IBN) property. For instance, this is the case in [39, Chap. 0.4].) For a classical maximal $\mathcal{O}$-order R, every finitely generated projective right R-module is of the form $R^n \oplus I$ for a right ideal I of R. It follows that R is a Hermite ring if and only if every stably free right R-ideal is free.

Strong approximation and Eichler condition. Let $S \subset S_{\mathrm{fin}}$ be the set of places defining the holomorphy ring $\mathcal{O} = \mathcal{O}_S$. Denote by S_∞ the set of archimedean places of K. ($S_\infty = \emptyset$ if K is a function field.) We consider the places in $S_{\mathrm{fin}} \setminus S$ to be places arising from $\mathcal{O}$, since they correspond to maximal ideals of $\mathcal{O}$. We consider the places of $S_\infty \cup S$ to be places not arising from $\mathcal{O}$. The algebra A satisfies the *Eichler condition (relative to $\mathcal{O}$)* if there exists a place v not arising from $\mathcal{O}$ such that A_v is not a noncommutative division ring.

If K is a number field, and A does not satisfy the Eichler condition, then A is necessary a totally definite quaternion algebra. That is, $\dim_K A = 4$ and, for all $v \in S_\infty$, we have $K_v \cong \mathbb{R}$ and A_v is a division ring, necessarily isomorphic to the Hamilton quaternion algebra.

The Eichler condition is a sufficient condition to guarantee the existence of a strong approximation theorem for the kernel of the reduced norm, considered as a homomorphism of the idele groups. As a consequence, if A satisfies the Eichler condition, then the map $\mathrm{LF}_1(R) \to \mathscr{C}(R)$ is a bijection. In particular, every stably free right R-ideal is free. (See [40, 90, 95].)

On the other hand, if K is a number field, $\mathcal{O}$ is its ring of algebraic integers, and A is a totally definite quaternion algebra, then, for all but finitely many isomorphism classes of A and R, there exist stably free right R-ideals which are not free. The classical maximal orders for which this happens have been classified. (See [61, 92, 97].)

The strong approximation theorem is also useful in the determination of the image of the reduced norm of an order. Suppose that A satisfies the Eichler condition with respect to $\mathcal{O}$. Let $\mathcal{O}_A^\bullet$ denote the subsemigroup of all nonzero elements of $\mathcal{O}$ which are positive at each $v \in S_\infty$ which ramifies in A. Then, if R is a classical hereditary $\mathcal{O}$-order in A, the strong approximation theorem together with an explicit characterization of local hereditary orders implies that $\mathrm{nr}(R^\bullet) = \mathcal{O}_A^\bullet$. (See [90, Theorem 39.14] for the classification of hereditary orders in a central simple algebra over a quotient field of a complete DVR, and [95, Theorem 8.2] or [40, Theorem 52.11] for the globalization argument via strong approximation.)

Hurwitz quaternions. Historically, the order of Hurwitz quaternions has received particular attention. It is Euclidean, hence a PID, and therefore enjoys unique factorization in a sense. An elementary discussion of the Hurwitz quaternions (without reference to the theory of maximal orders) and their factorization theory can be found in [42]. We give [86, 97] as references for the theory of quaternion algebras over number fields.

Example 2.1 Let K be a field of characteristic not equal to 2. Usually, we will consider $K = \mathbb{Q}$ or $K = \mathbb{R}$. Let $\mathbb{H}_K$ denote the four-dimensional K-algebra with basis $1, i, j, k$, where $i^2 = j^2 = -1$, $ij = -ji = k$, and 1 is the multiplicative identity. This is a quaternion algebra, that is, a four-dimensional central simple K-algebra. On $\mathbb{H}_K$ there exists an involution, called *conjugation*, defined by K-linear extension of $\overline{1} = 1$, $\overline{i} = -i$, $\overline{j} = -j$, and $\overline{k} = -k$. The *reduced norm* $\mathrm{nr}\colon H_K \to K$ is defined by $\mathrm{nr}(x) = x\overline{x}$, for all $x \in \mathbb{H}_K$. Thus $\mathrm{nr}(a + bi + cj + dk) = a^2 + b^2 + c^2 + d^2$ if $a, b, c, d \in K$. If $K = \mathbb{R}$, then $\mathbb{H}_K$ is the division algebra of *Hamilton quaternions*.

The algebra $\mathbb{H}_{\mathbb{Q}}$ is a totally definite quaternion algebra over $\mathbb{Q}$. Let $\mathscr{H}$ be the classical $\mathbb{Z}$-order with $\mathbb{Z}$-basis $1, i, j, \frac{-1+i+j+k}{2}$ in $\mathbb{H}_{\mathbb{Q}}$. That is, $\mathscr{H}$ consists of elements $a + bi + cj + dk$ with a, b, c, d, either all integers or all half-integers. Then $\mathscr{H}$ is a classical maximal $\mathbb{Z}$-order, the order of *Hurwitz quaternions*. The ring $\mathscr{H}$ is Euclidean with respect to the reduced norm, and hence a PID.

The unit group of $\mathscr{H}$ consists of the 24 elements

$$\mathscr{H}^{\times} = \left\{ \pm 1, \pm i, \pm j, \pm k, \frac{\pm 1 \pm i \pm j \pm k}{2} \right\}.$$

Up to conjugation by units of $\mathbb{H}_{\mathbb{Q}}$, the order of Hurwitz quaternions is the unique classical maximal $\mathbb{Z}$-order in $\mathbb{H}_{\mathbb{Q}}$. The algebra $\mathbb{H}_{\mathbb{Q}}$ is only ramified at 2 and ∞. Thus, for any odd prime number p, one has $\mathbb{H}_{\mathbb{Q}} \otimes_{\mathbb{Q}} \mathbb{Q}_p \cong M_2(\mathbb{Q}_p)$ and $\mathscr{H} \otimes_{\mathbb{Z}} \mathbb{Z}_p \cong M_2(\mathbb{Z}_p)$. Moreover, in this case, $\mathscr{H}/p\mathscr{H} \cong M_2(\mathbb{F}_p)$.

On the other hand, $\mathbb{H}_{\mathbb{Q}} \otimes_{\mathbb{Q}} \mathbb{R} \cong \mathbb{H}_{\mathbb{R}}$ is a division algebra. Similarly, for $p = 2$, the completion $\mathbb{H}_{\mathbb{Q}} \otimes_{\mathbb{Q}} \mathbb{Q}_2$ is isomorphic to the unique quaternion division algebra over $\mathbb{Q}_2$.

In the maximal order $\mathscr{H} \otimes_{\mathbb{Z}} \mathbb{Z}_2$, every right or left ideal is two-sided. The ideals of $\mathscr{H} \otimes_{\mathbb{Z}} \mathbb{Z}_2$ are linearly ordered, and each of them is a power of the unique maximal ideal, which is generated by $(1 + i)$. Note that this is not the case for p odd, since then $\mathscr{H} \otimes_{\mathbb{Z}} \mathbb{Z}_p \cong M_2(\mathbb{Z}_p)$.

3 Factorizations and Atomicity

We develop the basic notions of (rigid) factorizations in the very general setting of a cancellative small category. Moreover, we show how this notion is connected to chains of principal right ideals and recall sufficient conditions for a cancellative small category to be atomic.

We introduce the notions for a cancellative small category H. When we later apply them to a ring R, we implicitly assume that they are applied to the semigroup of non-zero-divisors $R^{\bullet}$. For instance, when we write "R is atomic," this means "$R^{\bullet}$ is atomic." and so on.

3.1 Rigid Factorizations

Let H be a cancellative small category.

Definition 3.1 An element $a \in H$ is an *atom* if $a = bc$ with $b, c \in H$ implies $b \in H^{\times}$ or $c \in H^{\times}$.

Viewing H as a quiver (a directed graph with multiple edges allowed), the atoms of H form a subquiver, denoted by $\mathscr{A}(H)$. We will often view $\mathscr{A}(H)$ simply as a set of atoms, forgetting about the additional quiver structure.

A *rigid factorization* of $a \in H$ is a representation of a as a product of atoms up to a possible insertion of units. We first give an informal description. We write the symbol $*$ between factors in a rigid factorizations, to distinguish the factorization as a formal product from its actual product in H. Thus, if $a \in H$ and $a = \varepsilon_1 u_1 \cdots u_k$ with atoms $u_1, \ldots, u_k$ of H and $\varepsilon_1 \in H^\times$, then $z = \varepsilon_1 u_1 * \ldots * u_k$ is a rigid factorization of a. If $\varepsilon_2, \ldots, \varepsilon_k \in H^\times$ are such that $t(\varepsilon_i) = s(u_i)$, then also $z = \varepsilon_1 u_1 \varepsilon_2^{-1} * \varepsilon_2 u_2 \varepsilon_3^{-1} * \ldots * \varepsilon_k u_k$ represents the same rigid factorization of a. The unit ε_1 can be absorbed into u_1, unless $k = 0$, that is, unless $a \in H^\times$.

If $a, b \in H$ and $t(a) = s(b)$, then two rigid factorization z of a and z' of b can be composed in the obvious way to obtain a rigid factorization of ab. We write $z * z'$ for this composition. In this way, the rigid factorizations themselves form a cancellative small category, denoted by $\mathsf{Z}^*(H)$.

More formally, we make the following definitions. See [93, Sect. 3] or [22, Sect. 3] for details. Let $\mathscr{F}^*(\mathscr{A}(H))$ denote the path category on the quiver $\mathscr{A}(H)$. Thus, $\mathscr{F}^*(\mathscr{A}(H))_0 = H_0$. Elements (*paths*) $x \in \mathscr{F}^*(\mathscr{A}(H))$ are denoted by

$$x = (e, u_1, \ldots, u_k, f)$$

where $e, f \in H_0$, and $u_i \in \mathscr{A}(H)$ with $s(u_1) = e$, $t(u_k) = f$, and $t(u_i) = s(u_{i+1})$ for $i \in [1, k-1]$. We set $s(x) = e$, $t(x) = f$, and the composition is given by the obvious concatenation of paths.

Denote by $H^\times \times_r \mathscr{F}^*(\mathscr{A}(H))$ the cancellative small category

$$H^\times \times_r \mathscr{F}^*(\mathscr{A}(H)) = \big\{ (\varepsilon, x) \in H^\times \times \mathscr{F}^*(\mathscr{A}(H)) \mid t(\varepsilon) = s(x) \big\},$$

where $\big(H^\times \times_r \mathscr{F}^*(\mathscr{A}(H))\big)_0 = \{ (e, e) \mid e \in H_0 \}$, which we identify with H_0, $s((\varepsilon, x)) = s(\varepsilon)$ and $t((\varepsilon, x)) = t(x)$. If $x = (e, u_1, \ldots, u_k, f)$, $y = (e', v_1, \ldots, v_l, f')$ in $\mathscr{F}^*(\mathscr{A}(H))$ and ε, $\varepsilon' \in H^\times$ are such that (ε, x), $(\varepsilon', y) \in H^\times \times_r \mathscr{F}^*(\mathscr{A}(H))$ with $t(x) = s(\varepsilon')$, we set

$$(\varepsilon, x)(\varepsilon', y) = (\varepsilon, (e, u_1, \ldots, u_k\varepsilon', v_1, \ldots, v_l f') \quad \text{if } k > 0,$$

and $(\varepsilon, x)(\varepsilon', y) = (\varepsilon\varepsilon', y)$ if $k = 0$.

On $H^\times \times_r \mathscr{F}^*(\mathscr{A}(H))$ we define a congruence relation $\sim$ by $(\varepsilon, x) \sim (\varepsilon', y)$ if and only if

(1) $k = l$,
(2) $\varepsilon u_1 \cdots u_k = \varepsilon' v_1 \cdots v_l \in H$, and
(3) there exist $\delta_2, \ldots, \delta_k \in H^\times$ and $\delta_{k+1} = t(u_k)$, such that

$$\varepsilon' v_1 = \varepsilon u_1 \delta_2^{-1} \quad \text{and} \quad v_i = \delta_i u_i \delta_{i+1}^{-1} \quad \text{for all } i \in [2, k].$$

Definition 3.2 The quotient category $\mathsf{Z}^*(H) = H^\times \times_r \mathscr{F}^*(\mathscr{A}(H))/\sim$ is called the *category of rigid factorizations* of H. The class of (ε, x) (as above) in $\mathsf{Z}^*(H)$ is denoted by $\varepsilon u_1 * \ldots * u_k$. There is a natural homomorphism

$$\pi = \pi_H\colon \mathsf{Z}^*(H) \to H, \quad \varepsilon u_1 * \ldots * u_k \mapsto \varepsilon u_1 \cdots u_k.$$

For $a \in H$, the set $\mathsf{Z}^*(a) = \mathsf{Z}^*_H(a) = \pi^{-1}(a)$ is the set of *rigid factorizations of* a. If $z = \varepsilon u_1 * \ldots * u_k \in \mathsf{Z}^*(H)$, then $|z| = k$ is the *length* of the (rigid) factorization z.

Remark (1) If H is a cancellative semigroup, then $H^\times \times_r \mathscr{F}^*(\mathscr{A}(H))$ is the product of $H^\times$ and the free monoid on $\mathscr{A}(H)$. If moreover H is reduced, then $\mathsf{Z}^*(H)$ is the free monoid on $\mathscr{A}(H)$. Hence, in this case, rigid factorizations are simply formal words on the atoms of H. In particular, if H is a reduced commutative cancellative semigroup, we see that rigid factorizations are ordered, whereas the usual notion of factorizations is unordered.

(2) While complicating the definitions a bit, the presence of units in the definition of $\mathsf{Z}^*(H)$ allows for a more uniform treatment of factorizations. It often makes it unnecessary to treat units as a (trivial) special case. In particular, with our definitions, every unit has a unique (trivial) rigid factorization of length 0.

3.2 *Factor Posets*

Let H be a small category.

Another useful way of viewing rigid factorizations is in terms of chains of principal left or right ideals. Suppose that, for $a, b \in H^\bullet$, we have $aH \subset bH$ if and only if there exists $c \in H^\bullet$ such that $a = bc$.[1] If $a \in H^\bullet$ and $b \in H^\bullet$, then $aH = bH$ if and only if there exists an $\varepsilon \in H^\times$ such that $a = b\varepsilon$, that is, a and b are right associated.

For $a \in H^\bullet$, let

$$[aH, H] = \left\{ bH \mid b \in H^\bullet \text{ such that } aH \subset bH \subset H \right\}$$

denote the set of all principal right ideals containing aH which are generated by a cancellative element. Note that $[aH, H]$ is naturally a partially ordered set via set

[1] We may always force this condition by replacing H by the subcategory of all cancellative elements. Note that then principal right ideals aH have to be replaced by $aH^\bullet$. Sometimes it can be more convenient work with H with $H^\bullet \neq H$, because typically we will have $H = R$ a ring and $H^\bullet = R^\bullet$ the semigroup of non-zero-divisors. In this setting, sufficient conditions for the stated condition to be satisfied are for $R^\bullet$ to be Ore, or R to be a domainWe may always force this condition by replacing H by the subcategory of all cancellative elements. Note that then principal right ideals aH have to be replaced by $aH^\bullet$. Sometimes it can be more convenient work with H with $H^\bullet \neq H$, because typically we will have $H = R$ a ring and $H^\bullet = R^\bullet$ the semigroup of non-zero-divisors. In this setting, sufficient conditions for the stated condition to be satisfied are for $R^\bullet$ to be Ore, or R to be a domain.

inclusion. This order reflects left divisibility in the following sense: Left divisibility gives a preorder on the cancellative left divisors of a. The corresponding poset, obtained by identifying right associated cancellative left divisors of a, is order anti-isomorphic to $[aH, H]$. We call $[aH, H]$ the *(right) factor poset* of a.

An element $a \in H^\bullet$ is an atom if and only if $[aH, H] = \{aH, H\}$. Rigid factorizations of a, that is, elements of $\mathsf{Z}^*(a)$, are naturally in bijection with finite maximal chains in $[aH, H]$. For instance, a rigid factorization $z = u_1 * \ldots * u_k$ of a corresponds to the chain

$$aH = u_1 \cdots u_k H \subsetneq u_1 \cdots u_{k-1} H \subsetneq \ldots \subsetneq u_1 u_2 H \subsetneq u_1 H \subsetneq H.$$

Thus, naturally, properties of the set of rigid factorizations of a correspond to properties of the poset $[aH, H]$.

In particular, we are interested in $[aH, H]$ being a lattice (or, stronger, a sublattice of the right ideals of H). If the factor poset $[aH, H]$ is a lattice, we are interesting in it being (semi-)modular or distributive. For a modular lattice the Schreier refinement theorem holds: Any two chains have equivalent refinements. For semimodular lattices of finite length one has a Jordan–Hölder theorem (and finite length of a semimodular lattice is already guaranteed by the existence of one maximal chain of finite length). Thus, if all factor posets are (semi-)modular lattices, we obtain unique factorization results for elements. This point of view will be quite useful in understanding and reconciling results on unique factorization in various classes of rings, such as 2-firs, modular LCM domains, and LCM domains having RAMP (see Sect. 4.1).

Remark Given an element $a \in H^\bullet$, we have defined $[aH, H]$ in terms of principal right ideals of H. We may similarly define $[Ha, H]$ using principal left ideals. If $b \in H^\bullet$ and $bH \in [aH, H]$, then there exists $b' \in H^\bullet$ such that $a = bb'$. This element b' is uniquely determined by bH up to left associativity, that is, Hb' is uniquely determined by bH. Hence, there is an anti-isomorphism of posets

$$[aH, H] \to [Ha, H], \quad bH \mapsto Hb'.$$

3.3 Atomicity, BF-Categories, and FF-Categories

Let H be a cancellative small category.

Definition 3.3

(1) H is *atomic* if the set of rigid factorizations, $\mathsf{Z}^*(a)$, is nonempty, for all $a \in H$. Explicitly, for every $a \in H$, there exist $k \in \mathbb{N}_0$, atoms $u_1, \ldots, u_k \in \mathcal{A}(H)$, and a unit $\varepsilon \in H^\times$ such that $a = \varepsilon u_1 \cdots u_k$.
(2) H is a *BF-category* (a category with *bounded factorizations*) if the set of lengths, $\mathsf{L}(a) = \{\, |z| \mid z \in \mathsf{Z}^*(a) \,\}$, is nonempty and finite, for all $a \in H$.
(3) H is *half-factorial* if $|\mathsf{L}(a)| = 1$, for all $a \in H$.

(4) H is an *FF-category* (a category with *finite factorizations*) if the set of rigid factorizations, $\mathsf{Z}^*(a)$, is nonempty and finite, for all $a \in H$.

Obviously, any FF-category is a BF-category. Analogous definitions are made for BF-semigroups, BF-domains, etc., and FF-semigroups, FF-domains, etc.

Remark The definition of an FF-category here is somewhat ad hoc in that it relates only to rigid factorizations, but this is in line with [17]. It is a bit restrictive in that a PID need not be an FF-domain (see Example 5.11). It may be more accurate to talk of a *finite rigid factorizations* category.

The following condition for atomicity is well known. A proof can be found in [93, Lemma 3.1].

Lemma 3.4 *If H satisfies both, the ACC on principal left ideals and the ACC on principal right ideals, then H is atomic.*

Remark Suppose for a moment that H is a small category which is not necessarily cancellative. If H satisfies the ACC on right ideals generated by cancellative elements, then $H^\bullet$ satisfies the ACC on principal right ideals. (If $a, b \in H^\bullet$ with $aH = bH$, then a and b are right associated, and hence also $aH^\bullet = bH^\bullet$.) Hence $H^\bullet$ is atomic. Phrasing the condition in this slightly more general way is often more practical. For instance, if R is a Noetherian ring, then $R^\bullet$ is atomic.

A more conceptual way of looking at the previous lemma is the following. By the duality of factor posets, the ACC on principal left ideals is equivalent to the restricted DCC on principal right ideals. That is, the ACC on principal left ideals translates into the DCC on $[aH, H]$ for $a \in H$. Thus, $[aH, H]$ has the ACC and DCC. Hence, there exist maximal chains in $[aH, H]$ and any such chain $[aH, H]$ has finite length. From this point of view, it is not surprising that the ACC on principal right ideals by itself is not sufficient for atomicity, as the following example shows:

Example 3.5 A domain R is a *right Bézout domain* if every finitely generated right ideal of R is principal. R is a *Bézout domain* if it is both, a left and right Bézout domain. Trivially, every PID is a Bézout domain.

Let R be a Bézout domain which is a right PID but not a left PID. (Such a domain, which is moreover simple, was constructed by P. M. Cohn and Schofield in [41].) Then R does not satisfy the ACC on principal left ideals. (For otherwise it would satisfy the ACC on finitely generated left ideals, and hence be left Noetherian. This would in turn imply that it is a left PID.) However, an atomic Bézout domain satisfies the ACC on principal left ideals and the ACC on principal right ideals. (This follows from the Schreier refinement theorem.) Hence R is not atomic.

A function $\ell\colon H \to \mathbb{N}_0$ is called a *(right) length function* if $\ell(a) > \ell(b)$ whenever $a = bc$ with $b, c \in H$ and $c \notin H^\times$. If H has a right length function, then it is easy to see that H satisfies the ACC on principal right ideals, as well as the restricted DCC on principal right ideals. In fact, if H has a right length function, then $[aH, H]$ has finite length for all $a \in H$. Thus, the length of a factorization of a is bounded by $\ell(a)$, and we have the following.

Lemma 3.6 *If H has a right length function, then H is a BF-category.*

4 Unique Factorization

It turns out to be nontrivial to obtain a satisfactory theory of factorial domains (also called unique factorization domains, short UFDs) in a noncommutative setting. Many different notions of factoriality have been studied. They cluster into two types.

First, there are definitions based on an elementwise notion of the existence and uniqueness of factorizations. For such a definition, typically, every non-zero-divisor has a factorization which is in some sense unique up to order and an equivalence relation on atoms. Usually, such classes of rings will contain PIDs but will not be closed under some natural ring-theoretic constructions, such as forming a polynomial ring or a ring of square matrices. This will be the focus of Sect. 4.1.

Second, definitions have been studied which start from more ring-theoretic characterizations of factorial commutative domains. Here, one does not necessarily obtain elementwise unique factorization results. Instead, one has unique factorization for normal elements into normal atoms. On the upside, this type of definition tends to behave better with respect to natural ring-theoretic constructions. This will be discussed in Sect. 4.2.

4.1 Similarity Factorial Domains and Related Notions

We first discuss the notions of similarity factoriality and n-firs. These have mainly been studied by P.M. Cohn and Bergman. (Although it seems that Bergman did not publish most of the results outside of his thesis [15].) We mention as general references for this section [15, 38, 39] as well as the two surveys [32, 33, 36, 37].

Brungs, in [21], introduced the weaker notion of subsimilarity factorial domains. This permits a form of Nagata's theorem to hold. Beauregard has investigated right LCM domains and the corresponding notion of projectivity factoriality. These works will also be discussed in this section.

Let R be a domain and $a, b \in R^{\bullet}$. We call a and b *similar* if $R/aR \cong R/bR$ as right R-modules. Fitting, in [49], observed that $R/aR \cong R/bR$ if and only if $R/Ra \cong R/Rb$, and hence the notion of similarity is independent of whether we consider left or right modules. (This duality has later been extended to the factorial duality by Bergman and P.M. Cohn, see [37] or [39, Theorem 3.2.2].)

If R is commutative, and $R/aR \cong R/bR$ for $a,\ b \in R$, then we have $aR = \operatorname{ann}(R/aR) = \operatorname{ann}(R/bR) = bR$, and thus a and b are similar if and only if they are associated. For noncommutative domains it is no longer true in general that $R/aR \cong R/bR$ implies that a and b are left-, right-, or two-sided associated.

Definition 4.1 A domain R is called *similarity factorial* (or, a *similarity-UFD*) if

(1) R is atomic, and
(2) if $u_1 \cdots u_m = v_1 \cdots v_n$ for atoms $u_1, \ldots, u_m, v_1, \ldots, v_n \in R$, then $m = n$ and there exists a permutation $\sigma \in \mathfrak{S}_m$ such that u_i is similar to $v_{\sigma(i)}$, for all $i \in [1, m]$.

Remark (1) A note on terminology. It is more common to refer to similarity factorial domains as *similarity-UFDs*. P.M. Cohn calls a similarity-UFD simply a UFD. We use the terminology *similarity factorial domains*, because using the adjective "factorial" over the noun "UFD" is more in line with the modern development of the terminology in the commutative setting.

In [22], a similarity factorial domain is called $\mathsf{d}_{\mathrm{sim}}$-*factorial*. This follows a general system: In [22], distances between rigid factorizations are introduced. Each distance d naturally gives rise to a corresponding notion of d-factoriality by identifying two rigid factorizations of an element if they have distance 0. The distance $\mathsf{d}_{\mathrm{sim}}$ is defined using the similarity relation. See Sect. 5.1 below for more on this point of view.

(2) Let R be a ring which is not necessarily a domain. We call R *(right) similarity factorial* if $R^\bullet$ is atomic, and factorizations of elements in $R^\bullet$ are unique up to order and similarity of the atoms. In general, it is no longer true that right and left similarity are the same.

Example 4.2 (1) Every PID is similarity factorial. This is immediate from the Jordan–Hölder theorem.

(2) Let K be a field. In the free associative K-algebra $R = K\langle x, y\rangle$, the elements x and y are similar but not associated. We will see below that $K\langle x, y\rangle$ is similarity factorial. However, factorizations are not unique up to order and associativity, as

$$x(yx + 1) = (xy + 1)x$$

shows.

(3) Let R be a classical maximal $\mathbb{Z}$-order in a definite quaternion algebra over $\mathbb{Q}$. Suppose that R is a PID. Then R is similarity factorial. For every prime number p which is unramified in R, there exist $p + 1$ atoms with reduced norm p. These $p + 1$ atoms are all similar, but, since $R^\times$ is finite, for sufficiently large p, they cannot all be right-, left-, or two-sided associated. For instance, this is the case for $R = \mathscr{H}$, the ring of Hurwitz quaternions.

One may be tempted to require factorizations to be unique up to order and, say, two-sided associativity of elements. This is referred to as *permutably factorial* in [22]. However, Examples (2) and (3) above show that such a notion is often too restrictive.

If R is a PID, then R is similarity factorial. However, when looking for natural examples of similarity factorial domains, one should consider a more general class of rings than PIDs, namely that of 2-firs. The motivation for this is the following: If K is a field and $R = K\langle x, y\rangle$ is the free associative K-algebra in two indeterminates, then $xR \cap yR = \mathbf{0}$. Hence $xR + yR \cong R^2$ is a nonprincipal right ideal of R. Thus R is not a PID. However, P.M. Cohn has shown that R is an atomic 2-fir and hence, in particular, similarity factorial (see below).

Definition 4.3 Let $n \in \mathbb{N}$. A ring R is an *n-fir* if every right ideal of R on at most n generators is free, of unique rank. A ring R is a *semifir* if R is an n-fir, for all $n \in \mathbb{N}$.

It can be shown that the notion of an n-fir for $n \in \mathbb{N}$ is symmetric (see [39, Theorem 2.3.1]). Thus R is an n-fir if and only if every left ideal of R on at most n generators is free, of unique rank. Any n-fir is of course an m-fir, for all $m < n$. A ring R is a *right fir* (free right ideal ring) if all right ideals of R are free, of unique rank. R is a *fir* if it is a left and right fir. Any fir is atomic (see [39, Theorem 2.2.3]).

The case which is particularly important for the factorization of elements is that of a 2-fir. (More generally, over a $2n$-fir one can consider factorizations of $n \times n$-matrices.) A ring R is a 1-fir if and only if it is a domain. Thus, in particular, any 2-fir is a domain.

Theorem 4.4 ([39, Theorem 2.3.7]) *For a domain R, the following conditions are equivalent*

(a) R is a 2-fir.
(b) For $a, b \in R^\bullet$ we have $aR \cap bR = mR$ for some $m \in R$, while $aR + bR$ is principal if and only if $m \neq 0$.
(c) If $a, b \in R$ are such that $aR \cap bR \neq \mathbf{0}$, then $aR + bR$ is a principal right ideal of R.
(d) For all $a \in R^\bullet$, $[aR, R]$ is a sublattice of the lattice of all right ideals of R.

It follows from (c) that a 2-fir is a right Ore domain if and only if it is a right Bézout domain. In particular, a commutative ring is a 2-fir if and only if it is a Bézout domain.

Note that (d) implies that $[aR, R]$ is a modular lattice, for all $a \in R^\bullet$. The Schreier refinement theorem for modular lattices then implies that finite maximal chains of $[aR, R]$ are unique up to perspectivity. In particular, if $[aR, R]$ contains any finite maximal chain, then $[aR, R]$ has finite length.

Since $[aR, R]$ is a sublattice of the lattice of right ideals of R, the uniqueness of maximal chains up to perspectivity translates into the factors of a maximal chain being isomorphic as modules (up to order). Translated into factorizations, this implies that the factorizations of nonzero elements in R are unique up to order and similarity. More generally, one obtains a similar result for factorizations of full matrices in $M_n(R)$ over a $2n$-fir R. A matrix $A \in M_n(R)$ is *full* if it cannot be written in the form $A = BC$ with B an $n \times r$-matrix and C and $r \times n$-matrix where $r < n$. Over an n-fir, any full matrix $A \in M_n(R)$ is cancellative (see [39, Lemma 3.1.1]). A *full atom* is a (square) full matrix which cannot be written as a product of two non-unit full matrices.

Theorem 4.5 ([39, Chap. 3.2]) *If R is a $2n$-fir, any two factorization of a full matrix in $M_n(R)^\bullet$ into full atoms are equivalent up to order and similarity of the atoms. In particular, if R is an atomic 2-fir, then R is similarity factorial.*

Remark (1) A commutative atomic 2-fir is an atomic Bézout domain, and hence a PID. However, noncommutative atomic 2-firs need not be PIDs. The free associative algebra $K\langle x, y\rangle$ over a field K provides a counterexample.
(2) If R is a semifir, then products of full matrices are full (see [39, Corollary 5.5.2]), so that the full matrices form a subsemigroup of $M_n(R)^\bullet$.
(3) Let R be a commutative Noetherian ring with no nonzero nilpotent elements. If $M_n(R)$ is similarity factorial, for all $n \geq 2$ (equivalently, $M_2(R)$ is similarity factorial), then R is a finite direct product of PIDs (see [45] or Theorem 5.19). This is a partial converse to the theorem above.
(4) Leroy and Ozturk, in [80], introduced F-algebraic and F-independent sets to study factorizations in 2-firs. In particular, they obtain lower bounds on the lengths of elements in terms of dimensions of certain vector spaces.

A sufficient condition for a domain to be an atomic right PID, respectively an atomic n-fir, is the existence of a right Euclidean algorithm, respectively an n-term weak algorithm.

A domain R is right Euclidean if there exists a function $\delta\colon R \to \mathbb{N}_0 \cup \{-\infty\}$ such that, for all $a, b \in R$, if $b \neq 0$, there exist $q, r \in R$ such that $a = bq + r$ and $\delta(r) < \delta(b)$. Equivalently, if $a, b \in R$ with $b \neq 0$, and $\delta(b) \leq \delta(a)$, then there exists $c \in R$ such that

$$\delta(a - bc) < \delta(a). \tag{1}$$

Any right Euclidean domain is a right PID and moreover atomic. Thus, right Euclidean domains are similarity factorial. The atomicity follows since the least function defining the Euclidean algorithm induces a right length function on $R^\bullet$ (see [39, Proposition 1.2.5]). By contrast, we recall that a right PID need not be atomic (see Example 3.5). See [18, Sect. 3.2.7] for a discussion of Euclidean domains. An extensive discussion of Euclidean rings can be found in [39, Chap. 1.2].

Example 4.6 (1) Let D be a division ring, σ an injective endomorphism of D and δ a (right) σ-derivation (that is, $\delta(ab) = \delta(a)\sigma(b) + a\delta(b)$, for all $a, b \in D$). The skew polynomial ring $D[x; \sigma, \delta]$ consists of elements of the form

$$\sum_{n\in\mathbb{N}_0} x^n a_n \qquad \text{with } a_n \in D, \text{ almost all zero.}$$

The multiplication is defined by $ax = x\sigma(a) + \delta(a)$. We set $D[x; \sigma] = D[x; \sigma, 0]$ and $D[x; \delta] = D[x; \mathrm{id}_D, \delta]$ if δ is a derivation.
Using polynomial division, it follows that $D[x; \sigma, \delta]$ is right Euclidean with respect to the degree function. If σ is an automorphism, then $D[x; \sigma, \delta]$ is also left Euclidean, by symmetry.
In particular, if K is a field and x is an indeterminate, then $B_1(K) = K(x)[y; -\frac{d}{dx}]$ is Euclidean. If the characteristic of K is 0, then $K(x)$ naturally has a faithful right $B_1(K)$-module structure, with y acting, from the right, as the formal derivative $\frac{d}{dx}$. In this way, $B_1(K)$ can be interpreted as the ring of linear differential operators (with rational functions as coefficients) on $K(x)$.

From the fact that $B_1(K)$ is similarity factorial, one obtains results on the uniqueness of factorizations of homogeneous linear differential equations, as in [71, 81].

(2) The ring of Hurwitz quaternions, $\mathscr{H}$, is Euclidean with respect to the reduced norm. This leads to an easy proof of Lagrange's Four-Square theorem, in the same way that the ring of Gaussian integers $\mathbb{Z}[i]$ can be used to obtain an easy proof of the Sum of Two Squares theorem (see [90, Theorem 26.6]).

Free associative algebras in more than one indeterminate over a field are not PIDs and hence not Euclidean. However, in the 1960s, P. M. Cohn and Bergman developed the more general notion of an *(n-term) weak algorithm* (see [39]), which can be used to prove that a ring is an atomic n-fir. We recall the definition, following [39, Chap. 2].

A *filtration* on a ring R is a function $v\colon R \to \mathbb{N}_0 \cup \{-\infty\}$ satisfying the following conditions:

(1) For $a \in R$, $v(a) = -\infty$ if and only if $a = 0$.
(2) $v(a - b) \le \max\{v(a), v(b)\}$, for all $a, b \in R$.
(3) $v(ab) \le v(a) + v(b)$, for all $a, b \in R$.
(4) $v(1) = 0$.

Equivalently, a filtration is defined by a family $\{0\} = R_{-\infty} \subset R_0 \subset R_1 \subset R_2 \subset \dots$ of additive subgroups of R such that $R = \bigcup_{i \in \mathbb{N}_0 \cup \{-\infty\}} R_i$, for all $i, j \in \mathbb{N}_0 \cup \{-\infty\}$ it holds that $R_iR_j \subset R_{i+j}$, and $1 \in R_0$. The equivalence of the two definitions is seen by setting $R_i = \{\, a \in R \mid v(a) \le i \,\}$, respectively, in the other direction, by setting $v(a) = \min\{\, i \in \mathbb{N}_0 \cup \{-\infty\} \mid a \in R_i \,\}$.

Let R be a ring with filtration v. A family $(a_i)_{i \in I}$ in R with index set I is *right v-dependent* if either $a_i = 0$ for some $i \in I$, or there exist $b_i \in R$, almost all zero, such that

$$v\left(\sum_{i \in I} a_i b_i\right) < \max_{i \in I} v(a_i) + v(b_i).$$

If $a \in R$ and $(a_i)_{i \in I}$ is an family in R, then a is *right v-dependent on* $(a_i)_{i \in I}$ if either $a = 0$ or there exist $b_i \in R$, almost all zero, such that

$$v\left(a - \sum_{i \in I} a_i b_i\right) < v(a) \quad \text{and} \quad v(a_i) + v(b_i) \le v(a) \text{ for all } i \in I.$$

Definition 4.7 For $n \in \mathbb{N}$, a filtered ring R satisfies the *n-term weak algorithm* if, for any right v-dependent family $(a_i)_{i \in [1,m]}$ of $m \le n$ elements with $v(a_1) \le v(a_2) \le \dots \le v(a_m)$, there exists a $j \in [1, m]$ such that a_j is right v-dependent on $(a_i)_{i \in [1,j-1]}$. R satisfies the *weak algorithm* if it satisfies the n-term weak algorithm, for all $n \in \mathbb{N}$.

The asymmetry in the definition is only an apparent one. A filtered ring R satisfies the n-term weak algorithm with respect to the notion of right v-dependence if and only if the same holds true with respect to left v-dependence (see [39, Proposition 2.4.1]).

If R satisfies the n-term weak algorithm, then it also satisfies the m-term weak algorithm for $m < n$. If R satisfies the 1-term weak algorithm, then R is a domain and $v(ab) = v(a) + v(b)$, for all $a, b \in R\setminus\{0\}$. If moreover $R_0 \subset R^\times \cup \{0\}$, that is R_0 is a division ring, then v induces a length function on $R^\bullet$. In this case, R is a BF-domain. If R satisfies the n-term weak algorithm for $n \geq 2$, then R is a domain with $R_0 \subset R^\times \cup \{0\}$ a division ring.

Of particular interest is the 2-term weak algorithm. Explicitly, it says that for two elements $a, b \in R$ which are right v-dependent, if $b \neq 0$ and $v(b) \leq v(a)$, then there exists $c \in R$ such that $v(a - bc) < v(a)$. Comparing with Eq. (1), we see that the existence of a 2-term weak algorithm implies that a Euclidean division algorithm holds for elements a and b which are right v-dependent.

Theorem 4.8 ([39, Proposition 2.4.8], [38, Proposition 2.2.7]) *Let R be a filtered ring with n-term weak algorithm, where $n \geq 2$. Then R is an n-fir and satisfies the ACC on n-generated left, respectively right, ideals. In particular, R is similarity factorial.*

We also note in passing that if R is a filtered ring with weak algorithm then R is not only a semifir but even a fir (see [39, Theorem 2.4.6]).

Example 4.9 A standard example shows that a right Euclidean domain need not be a left PID. Let K be a field, and let σ be the endomorphism of the rational function field $K(x)$ given by $\sigma(x) = x^2$ and $\sigma|_K = \mathrm{id}_K$. Then the skew polynomial ring $R = K(x)[y; \sigma]$ is right Euclidean, but does not even have finite uniform dimension as a left module over itself, as it contains an infinite direct sum of left ideals (see [85, Example 1.2.11(ii)]). However, since R is right Euclidean, it has a 2-term weak algorithm. Hence R is an atomic 2-fir and in particular similarity factorial.

The notions of n-fir, similarity factoriality, and [n-term] weak algorithm are symmetric, while being a right PID and being right Euclidean are nonsymmetric concepts.

Before we can state one of the main theorems on the existence of a weak algorithm, we have to recall A-rings (for a ring A), tensor A-rings, and coproducts of A-rings. Let A be a ring. An *A-ring* is a ring R together with a ring homomorphism $A \to R$. If V is an A-bimodule, we set $V^{\otimes 0} = A$ and inductively $V^{\otimes n} = V^{\otimes (n-1)} \otimes_A V$, for all $n \in \mathbb{N}$. The *tensor A-ring* $A[V]$ is defined as $A[V] = \bigoplus_{n \in \mathbb{N}_0} V^{\otimes n}$, with multiplication induced by the natural isomorphisms $V^{\otimes m} \otimes_A V^{\otimes n} \to V^{\otimes (m+n)}$. If V is a free right A-module with basis X, then the free monoid X^* generated by X is a basis of the right A-module $A[V]$. In this case, every $f \in A[V]$ has a unique representation of the form

$$f = \sum_{x \in X^*} x a_x \qquad \text{with } a_x \in A, \text{ almost all zero.} \tag{2}$$

Note however that elements of A need not commute with elements from X.

If V is a free right A-module with basis X, and a bimodule structure is defined on V by means of $\lambda x = x\lambda$ for all $\lambda \in A$ and $x \in X$, then $A\langle X \rangle = A[V]$ is the *free A-ring* on X. By the choice of bimodule structure, elements from A commute with elements from X in $A\langle X\rangle$. If R and S are A-rings, the coproduct $R *_A S$ in the category of A-rings is the pushout of the homomorphisms $A \to R$ and $A \to S$ in the category of rings.

If D is a division ring, V is a D-bimodule, and R and S are filtered D-rings with $R_0 \cong S_0 \cong D$, then $D[V]$ as well as $R *_D S$ are naturally filtered. If X is a set, one defines the *free R-ring* $R_D\langle X\rangle$ on the D-centralizing indeterminates X as $R_D\langle X\rangle = R *_D D\langle X\rangle$. In $R_D\langle X\rangle$, elements of D commute with elements of X.

Theorem 4.10 ([39, Chap. 2.5]) *Let D be a division ring.*

(1) Let V be a D-bimodule. Then the tensor D-ring $D[V]$ satisfies the weak algorithm relative to the natural filtration.
(2) Let R, S be D-rings with weak algorithm, where $R_0 \cong S_0 \cong D$. Then the coproduct $R *_D S$ in the category of D-rings satisfies the weak algorithm relative to the natural filtration.
(3) Let R be a ring with weak algorithm and $R_0 \cong D$. For any set X, the free R-ring $R_D\langle X\rangle = R *_D D\langle X\rangle$ on D-centralizing indeterminates X satisfies the weak algorithm relative to the natural filtration.

In particular, these rings are firs and hence similarity factorial.

Corollary 4.11 *If K is a field and X is a set of noncommuting indeterminates, then the free associative K-algebra $K\langle X\rangle$ satisfies the weak algorithm. In particular, $K\langle X\rangle$ is a fir and hence similarity factorial.*

In a similar fashion, the *inverse weak algorithm* can be used to show that power series rings in any number of noncommuting indeterminates are similarity factorial (see [31] or [39, Chap. 2.9]). A *transfinite weak algorithm* can be used to prove that certain semigroup algebras are right firs (see [39, Chap. 2.10]).

For classical maximal orders in central simple algebras over global fields, we have the following result on similarity factoriality.

Theorem 4.12 ([22, Corollary 7.14]) *Let R be a classical maximal $\mathcal{O}$-order over a holomorphy ring $\mathcal{O}$ in a global field. Suppose that every stably free right R-ideal is free. Then the following statements are equivalent:*

(a) R is similarity factorial.
(b) Every right R-ideal is principal.
(c) Every left R-ideal is principal.
(d) The ray class group $\mathcal{C}_A(\mathcal{O})$ is trivial.

4.1.1 Rigid Domains

A domain R is *rigid* if $[aR, R]$ is a chain, for all $a \in R^\bullet$. Rigid domains and rigid similarity factorial domains have been characterized by P.M. Cohn. Recall that a nonzero ring R is *local* if $R/J(R)$ is a division ring. Here, $J(R)$ is the Jacobson radical of R.

Theorem 4.13 ([39, Theorem 3.3.7]) *A domain is rigid if and only if it is a 2-fir and a local ring.*

Lemma 4.14 *For a domain R, the following statements are equivalent.*

(a) R is rigid and atomic.
(b) R is rigid and similarity factorial.
(c) R is rigidly factorial *in the sense of [22]. That is,* $|\mathsf{Z}^*(a)| = 1$, *for all* $a \in R^\bullet$.
(d) R is an atomic 2-fir and a local ring.

Proof (a) $\Leftrightarrow$ (b) $\Leftrightarrow$ (c) is trivial. The nontrivial equivalence (a) $\Leftrightarrow$ (d) follows from the previous theorem.

Note that a factorial commutative domain is rigid if and only if it is a discrete valuation ring. The extreme restrictiveness of rigid domains is what requires one to study notions of factoriality which are weaker than rigid factoriality, such as similarity factoriality, where some degree of refactoring is permitted. However, interesting rings which satisfy the equivalent conditions of Lemma 4.14 do exist: power series rings in any number of noncommuting indeterminates over a division ring (see [39, Theorems 2.9.8 and 3.3.2]).

4.1.2 Distributive Factor Lattices

Let R be a domain. Then R is a 2-fir if and only if the factor posets $[aR, R]$ for $a \in R^\bullet$ are sublattices of the lattice of principal right ideals. Hence, for all $a \in R^\bullet$, the factor lattice $[aR, R]$ is modular. For a commutative Bézout domain, in fact, the factor lattices are distributive, since the lattice of fractional principal ideals is a lattice-ordered group. In the noncommutative setting this is no longer true in general.

Example 4.15 Let $\mathscr{H}$ be the ring of Hurwitz quaternions. Then $\mathscr{H}$ is a PID and hence, in particular, a 2-fir. If $p \in \mathbb{P}\setminus\{2\}$ is an odd prime number, then $\mathscr{H}/p\mathscr{H} \cong M_2(\mathbb{F}_p)$. Thus, $[p\mathscr{H}, \mathscr{H}]$ is isomorphic to the lattice of right ideals of $M_2(\mathbb{F}_p)$. The lattice of right ideals of $M_2(\mathbb{F}_p)$ is in turn isomorphic to the lattice of $\mathbb{F}_p$-subspaces of $\mathbb{F}_p^2$. Hence, $[p\mathscr{H}, \mathscr{H}]$ is not distributive.

A domain R, which is a K-algebra over a field K, is an *absolute domain* if $R \otimes_K L$ is a domain, for all algebraic field extensions L of K. If R is moreover a 2-fir and $K(x)$ denotes the rational function field over K, the ring R is a *persistent 2-fir* if $R \otimes_K K(x)$ is again a 2-fir. For instance, the free associative K-algebra $K\langle X\rangle$ on a set of indeterminates X is an absolute domain and a persistent 2-fir.

Theorem 4.16 ([39, Theorem 4.3.3]) *Let K be a field and let R be a K-algebra that is an absolute domain and a persistent 2-fir. Then the factor lattice* $[aR, R]$ *is distributive, for all* $a \in R^\bullet$.

There is a duality between the category of finite distributive lattices and the category of finite partially ordered sets. It is given (in both directions), by mapping a distributive lattice X, respectively a partially ordered set X, to $\operatorname{Hom}(X, \{0, 1\})$ (see [39, Chap. 4.4]). Here $\{0, 1\}$ is to be considered as two-element distributive lattice, respectively partially ordered set, with $0 < 1$.

Under this duality, the distributive lattices that appear as factor lattices in a factorial commutative domain correspond to disjoint unions of finite chains. In contrast, in noncommutative similarity factorial domains, we have the following. (This seems to go back to Bergman and P. M. Cohn.)

Theorem 4.17 ([39, Theorem 4.5.2]) *Let K be a field and $R = K\langle x_1, \dots, x_n\rangle$ with $n \geq 2$ a free associative algebra. Let L be a finite distributive lattice. Then there exists $a \in R^\bullet$ with $[aR, R] \cong L$.*

On the other hand, if R is a PID, we have the following:

Theorem 4.18 ([39, Theorem 4.2.8]) *Let R be a PID. Then every factor lattice $[aR, R]$ for $a \in R^\bullet$ is distributive if and only if every element of $R^\bullet$ is normal.*

Thus, every left (or right) ideal I of R is already an ideal of R, and $I = aR = Ra$ for a normal element $a \in R$.

4.1.3 Comaximal Transposition/Metacommutation

In an atomic 2-fir R, it follows from the usual inductive proof of the Jordan–Hölder theorem that every rigid factorization of an element can be transformed into any other rigid factorization of the same element by successively replacing two consecutive atoms by two new ones. Using the arithmetical invariants that will be introduced in Sect. 5.1 for the study of nonunique factorizations, this means $\mathsf{c}^*(R^\bullet) \leq 2$.

To understand factorizations in such rings in more detail, the following question is of central importance: Given two atoms $u, v \in R^\bullet$, what can be said about atoms $u', v' \in R^\bullet$ such that $uv = v'u'$? Such a relation is referred to as *(comaximal) transposition* in the context of 2-firs when $uR \neq v'R$, that is $uR + v'R = R$ (see [39, Chaps. 3.2 and 3.5]). In [42], in the context of the ring of Hurwitz quaternions, this problem is referred to as *metacommutation* when $\operatorname{nr}(u)$ and $\operatorname{nr}(v)$ are coprime.

Example 4.19 Let R be a classical maximal $\mathcal{O}$-order in which every right [left] R-ideal is principal. Consider two atoms u and v of R. Suppose first $\operatorname{nr}(u) \not\simeq \operatorname{nr}(v)$. Then there exist atoms $u', v' \in R$ such that $uv = v'u'$, $\operatorname{nr}(u) \simeq \operatorname{nr}(u')$ and $\operatorname{nr}(v) \simeq \operatorname{nr}(v')$. Moreover, $v' * u'$ is uniquely determined. That is, if u and v have coprime reduced norms, then there is a unique (up to units) way of refactoring uv such that the order of reduced norm is exchanged.

If $\operatorname{nr}(u) \simeq \operatorname{nr}(v)$, then the situation is more complicated. The rigid factorization $u * v$ can be the unique factorization of uv, or there can be many different rigid factorizations. For instance, consider the ring $R = M_2(\mathbb{Z})$ and let $p \in \mathbb{P}$ be a prime number. Then $\begin{pmatrix} p^2 & 0 \\ 0 & 1 \end{pmatrix}$ has a unique rigid factorization, namely $\begin{pmatrix} p & 0 \\ 0 & 1 \end{pmatrix} * \begin{pmatrix} p & 0 \\ 0 & 1 \end{pmatrix}$. However, $\begin{pmatrix} p & 0 \\ 0 & p \end{pmatrix}$ has $p+1$ distinct rigid factorizations, given by $\begin{pmatrix} 1 & 0 \\ 0 & p \end{pmatrix} * \begin{pmatrix} p & 0 \\ 0 & 1 \end{pmatrix}$ and $\begin{pmatrix} p & x \\ 0 & 1 \end{pmatrix} * \begin{pmatrix} 1 & -x \\ 0 & p \end{pmatrix}$ with $x \in [0, p-1]$.

H. Cohn and Kumar have studied the comaximal transposition (metacommutation) of atoms with coprime norm in the Hurwitz quaternions in detail.

Theorem 4.20 ([30]) *Let $\mathscr{H}$ be the ring of Hurwitz quaternions, and let $p \neq q \in \mathbb{P}$ be prime numbers. Let $v \in \mathscr{A}(\mathscr{H})$ be an atom of reduced norm q, and let $\mathscr{A}_p$ denote the set of left associativity classes of atoms of reduced norm p. Metacommutation with v induces a permutation π of $\mathscr{A}_p$: If $\mathscr{H}^\times u \in \mathscr{A}_p$, there exist atoms u' and v' with $\operatorname{nr}(v') = q$, $\operatorname{nr}(u') = p$ and $uv = v'u'$, with the left associativity class $\mathscr{H}^\times u'$ of u' uniquely determined by $\mathscr{H}^\times u$. Then $\pi(\mathscr{H}^\times u) = \mathscr{H}^\times u'$.*

(1) The sign of π is the quadratic character $\left(\frac{q}{p}\right)$ of q modulo p.
(2) If $p = 2$ or $u \equiv n \mod p\mathscr{H}$ for some $n \in \mathbb{Z}$, then $\pi = \mathrm{id}_{\mathscr{A}_p}$. Otherwise, π has $1 + \left(\frac{\operatorname{tr}(v)^2 - q}{p}\right)$ fixed points.

4.1.4 Polynomial Rings

If D is a division ring, σ is an injective endomorphism of D, and δ is a σ-derivation, we have already noted that the skew polynomial ring $D[x; \sigma, \delta]$ is a right Euclidean domain, and hence similarity factorial. If R is a factorial commutative domain, then the polynomial ring $R[x]$ is factorial as well. This follows either from Gauss's lemma or from Nagata's theorem. The following two striking examples due to Beauregard show that a similar result cannot hold in the noncommutative setting in general.

Theorem 4.21 ([10]) *Let $\mathscr{H}$ denote the ring of Hurwitz quaternions. Then the polynomial ring $\mathscr{H}[x]$ is not half-factorial. Explicitly, with atoms $a = 1 - i + k$, $f = ax^2 + (2+2i)x + (-1+i-2k)$, and $h = \frac{1}{2}(1 - i + j + k)x^2 + (1+i)x + (-1+i)$, one has*

$$f\bar{f} = a\bar{a}h\bar{h}.$$

Theorem 4.22 ([11]) *Let $\mathbb{H}_{\mathbb{Q}}$ denote the Hamilton quaternion algebra with coefficients in $\mathbb{Q}$. Then $\mathbb{H}[x, y]$ is not half-factorial. Explicitly, with*

$$f = (x^2y^2 - 1) + (x^2 - y^2)i + 2xyj,$$

one has

$$f\bar{f} = (x^2 + i)(x^2 - i)(y^2 + i)(y^2 - i),$$

with all stated factors being atoms.

Note that this is quite independent of the precise definition of factoriality we are using. In particular, the second result implies that as long as we expect a factorial domain to be at least half-factorial and that division rings are (trivially) factorial domains, then it cannot be that polynomial rings over factorial domains are again always factorial domains.

4.1.5 Weaker Forms of Similarity and Nagata's Theorem

A basic form of Nagata's theorem in the commutative setting is the following: Let R be a commutative domain, and $S \subset R$ a multiplicative subset generated by prime elements. Then, if $S^{-1}R$ is factorial, so is R. In this way, one obtains that $\mathbb{Z}[x]$ is factorial from the fact that $\mathbb{Q}[x]$ is factorial.

A similar result cannot hold for similarity factoriality, as the following example from [35] shows. In $\mathbb{Z}\langle x, y\rangle$, we have

$$xyx + 2x = x(yx + 2) = (xy + 2)x.$$

However, $yx + 2$ is not similar to $xy + 2$ in $\mathbb{Z}\langle x, y\rangle$, as can be verified by a direct computation.

This provides a motivation to study weaker forms of equivalence relations on atoms than that of similarity. Two elements a, b in a domain R are called *(right) subsimilar*, if there exist injective module homomorphisms $R/aR \hookrightarrow R/bR$ and $R/bR \hookrightarrow R/aR$. Brungs, in [21], studied domains in which factorizations are unique up to permutation and subsimilarity of atoms.

Definition 4.23 A domain R is *subsimilarity factorial* (or a *subsimilarity-UFD*) if R is atomic, and factorizations of elements are unique up to order and subsimilarity of the atoms.

Brungs proved a form of Nagata's theorem using this notion and a, in general somewhat complicated, concept of prime elements (see [21, Satz 7]). In turn, he obtained the following.

Theorem 4.24 ([21, Satz 8]) *Let R be a commutative domain and X a set of non-commuting indeterminates. Then the free associative algebra $R\langle X\rangle$ is subsimilarity factorial if and only if R is factorial.*

In the same paper, Brungs showed that skew power series rings over right PIDs are right LCM domains. He used this to construct an atomic right LCM domain which is not half-factorial (see Example 4.25 below).

Motivated by Brungs' work, and with the goal of obtaining a variant of Nagata's theorem with a simpler notion of prime elements than the one Brungs was using, P. M. Cohn, in [35], introduced the notion of *(right) monosimilarity*. Let R be a ring, and call an element $a \in R$ *regular* if all divisors of a are non-zero-divisors. A right

R-module is *strictly cyclic* if it is isomorphic to R/aR for a regular element $a \in R$. The category $\mathscr{C}_R$ of strictly cyclic modules is the full subcategory of the category of right R-modules with objects the strictly cyclic right R-modules. If R is a 2-fir, then $\mathscr{C}_R$ is an abelian category.

Two regular elements $a, b \in R$ are called *(right) monosimilar* if there exist monomorphisms $R/aR \to R/bR$ and $R/bR \to R/aR$ in $\mathscr{C}_R$. In general, this is a weaker notion than subsimilarity. Indeed, if R is a domain, then a homomorphism f of strictly cyclic modules is a monomorphism in $\mathscr{C}_R$ if and only if its kernel (as homomorphism of R-modules) is torsion free. Within the class of 2-firs, the notions of subsimilarity and monosimilarity are equivalent to similarity.

In [37], P.M. Cohn gives a set of axioms for an equivalence relation on elements that is sufficient to obtain Nagata's theorem. These axioms are satisfied by the (right) monosimilarity relation, but in general not by the similarity relation. The main obstacle in the case of the similarity relation is that a and b being similar in $S^{-1}R$ does not imply that a and b are similar in R.

4.1.6 Stronger Forms of Similarity

A ring R is *permutably factorial* if $R^\bullet$ is atomic and factorizations in $R^\bullet$ are unique up to order and two-sided associativity of the atoms. This was studied in [22]. It is a rather strong requirement, but there are results for $R = T_n(D)$, the ring of $n \times n$ upper triangular matrices over an atomic commutative domain D, and $R = M_n(D)$ when D is commutative. See Sects. 5.4.1 and 5.4.2 below.

In [13], Beauregard studied right UFDs. A domain R is a *right unique factorization domain (right UFD)* if it is atomic, and factorizations are unique up to order and right associativity of the atoms. Note that Example 4.2(3) implies that there exist PIDs which are not right UFDs. Beauregard gives an example of a right UFD which is not a left UFD. In particular, while any right or left UFD is permutably factorial, the converse is not true. (This can also be seen by looking at $M_n(R)$ for R a commutative PID, $n \geq 2$, and using the Smith Normal Form.)

4.1.7 LCM Domains and Projectivity Factoriality

LCM domains and factorizations of elements therein were investigated by Beauregard in a series of papers (see [6–9, 12, 14]). A domain R is a *right LCM domain* if $aR \cap bR$ is principal, for all $a, b \in R$. A *left LCM domain* is defined analogously, and an *LCM domain* is a domain which is both, a right and a left LCM domain. By the characterization in Theorem 4.4, any 2-fir is an LCM domain.

If R is an LCM domain and $a \in R^\bullet$, then the poset $[aR, R]$ is a lattice with respect to the partial order induced by set inclusion (see [6, Lemma 1]). However, $[aR, R]$ need not be a sublattice of the lattice of all right ideals of R, that is, $bR + cR$ need not be principal for $bR, cR \in [aR, R]$.

A commutative domain is an atomic LCM domain if and only if it is factorial. Unfortunately, if R is an atomic right LCM domain, R need not even be half-factorial, as the following example shows:

Example 4.25 ([21] *or* [8, Remark 3.9]) Let $R = K[x]$ be the polynomial ring over a field K. Let $\sigma : R \to R$ be the monomorphism with $\sigma|_K = \mathrm{id}_K$ and $\sigma(x) = x^2$. The skew power series ring $S = R[\![y; \sigma]\!]$, consisting of elements of the form $\sum_{n=0}^{\infty} y^n a_n$ with $a_n \in R$, and with multiplication given by $ay = y\sigma(a)$ for $a \in R$, is a right LCM domain by [21, Satz9]. The equality $xy = yx^2$ shows that S is not half-factorial.

However, under an additional condition we do obtain unique factorization in a sense. For $a, b \in R^\bullet$, denote by $[a, b]_r$ a least common right multiple (LCRM), that is, a generator of $aR \cap bR$, and by $(a, b)_l$ a greatest common left divisor (GCLD), that is, a generator of the least principal ideal containing $aR + bR$. Note that $[a, b]_r$ and $(a, b)_l$ are only defined up to right associativity. A right LCM domain is called *modular* if, for all $a, b, c \in R^\bullet$,

$$[a, bc]_r = [a, b]_r \text{ and } (a, bc)_l = (a, b)_l \text{ implies } c \in R^\times.$$

If R is an LCM domain, the condition is equivalent to the lattice $[aR, R]$ being modular. Thus, any 2-fir is a modular LCM domain. However, the converse is not true. Any factorial commutative domain which is not a PID is a counterexample.

Let R be a domain. Beauregard calls two elements $a, a' \in R^\bullet$ *transposed*, and writes $a \text{ tr } a'$, if there exists $b \in R^\bullet$ such that

$$[a, b]_r = ba' \quad \text{and} \quad (a, b)_l = 1.$$

If this is the case, there exists $b' \in R^\bullet$ such that $ba' = ab'$ and $b \text{ tr } b'$. If R is an LCM domain, then $a \text{ tr } a'$ if and only if the interval $[aR, R]$ is down-perspective to $[ba'R, bR]$ in the lattice $[ba'R, R]$. If R is a 2-fir, then a and a' are transposed if and only if they are similar. The elements a and a' are *projective* if there exist $a = a_0$, $a_1, \ldots, a_n = a'$ such that, for each $i \in [1, n]$, either $a_{i-1} \text{ tr } a_i$ or $a_i \text{ tr } a_{i-1}$.

Definition 4.26 A domain R is *projectivity factorial* (or a *projectivity-UFD*) if R is atomic, and factorizations of elements are unique up to order and projectivity of the atoms.

Theorem 4.27 ([6, Theorem 2]) *If R is an atomic modular right LCM domain, then R is projectivity factorial.*

In [14], the condition of modularity has been weakened to the *right atomic multiple property (RAMP)*. A domain satisfies the RAMP if, for elements $a, b \in R$ with a an atom and $aR \cap bR \neq 0$, there exist $a', b' \in R$ with a' an atom such that $ab' = ba'$. One can check that, for an LCM domain, the RAMP is equivalent to the lattice $[aR, R]$ being lower semimodular, for all $a \in R^\bullet$. An atomic LCM domain is modular if and only if it satisfies both, the RAMP and LAMP, which is defined symmetrically (see [14, Theorem 3]).

Beauregard shows that, in a right LCM domain R, the RAMP is equivalent to the following condition: If $a, a' \in R^\bullet$ such that a is an atom, and a tr a', then a' is also an atom (see [14, Proposition 2]). He obtains the following generalization of the previous theorem.

Theorem 4.28 ([14, Theorem 1]) *If R is an atomic right LCM domain satisfying the RAMP, then R is projectivity factorial.*

If R is an atomic LCM domain, this theorem (as well as the previous one) can be deduced from the Jordan–Hölder theorem for semimodular lattices (see, for instance, [54]). To do so, note the following: If $a, a' \in R^\bullet$ and there exists $b \in R^\bullet$ such that interval $[aR, R]$ is projective to $[ba'R, bR]$ in the lattice $[ba'R, R]$, then the elements a and a' are projective.

Beauregard has also obtained a form of Nagata's theorem for modular right LCM domains (see [8]). He has moreover shown that an atomic LCM domain with conjugation is already modular (see [14, Theorem 4]). In [14, Example 3] he gives an example of an LCM domain which satisfies neither the RAMP nor the LAMP, and hence, in particular, does not have modular factor lattices.

Skew polynomial rings over total valuation rings provide another source of LCM domains. A subring V of a division ring D is called a *total valuation ring* if $x \in V$ or $x^{-1} \in V$ for each $x \in D^\bullet$.

Theorem 4.29 ([84]) *Let V be a total valuation ring, let σ be an automorphism of V and let δ be a σ-derivation on V such that $\delta(J(V)) \subset J(V)$, where $J(V)$ denotes the Jacobson radical of V. Then $V[x; \sigma, \delta]$ is an LCM domain.*

4.2 A Different Notion of UFRs and UFDs

A commutative domain is factorial if and only if every nonzero prime ideal contains a prime element. Based on this characterization, Chatters introduced Noetherian unique factorization domains (Noetherian UFDs) in [26]. Noetherian UFDs were generalized to Noetherian unique factorization rings (Noetherian UFRs) by Chatters and Jordan in [29].

Noetherian UFDs and UFRs, and generalizations thereof, have received quite a bit of attention and found many applications (e.g., [1, 20, 23, 24, 27, 55, 57, 58, 68, 69, 72, 79]). A large number of examples of Noetherian UFDs have been exhibited in the form of universal enveloping algebras of finite-dimensional solvable complex Lie algebras as well as various semigroup algebras. Moreover, Noetherian UFRs are preserved under the formation of polynomial rings in commuting indeterminates.

UFRs, respectively UFDs, which need not be Noetherian, were introduced by Chatters, Jordan, and Gilchrist in [24]. Many Noetherian Krull orders turned out not to be Noetherian UFRs in the sense of [24], despite having a factorization behavior similar to Noetherian UFRs. This was the motivation for Abbasi, Kobayashi, Marubayashi, and Ueda to introduce the notion of a (σ-)UFR in [1], which provides another generalization of Noetherian UFRs.

Let R be a prime ring. An element $n \in R$ is *normal* provided that $Rn = nR$. We denote the subsemigroup of all normal elements of R by $N(R)$. Since R is a prime ring, $N(R)^\bullet = N(R)\setminus\{0\}$ is a subset of the non-zero-divisors of R. An element $p \in R\setminus\{0\}$ is *prime* if p is normal and pR is a prime ideal. An element $p \in R\setminus\{0\}$ is *completely prime* if p is normal and pR is a completely prime ideal, that is, R/pR is a domain. If R is Noetherian and $p \in R$ is a prime element, the principal ideal theorem (see [85, Theorem 4.1.11]) implies that pR has height one.

Definition 4.30 ([24]) Let R be a ring.

(1) R is a *unique factorization ring*, short *UFR*, (in the sense of [24]) if it is a prime ring and every nonzero prime ideal of R contains a prime element.
(2) R is a *unique factorization domain*, short *UFD*, (in the sense of [24]) if it is a domain and every nonzero prime ideal of R contains a completely prime element.

Some remarks on this definition and its relation to the definitions of Noetherian UFRs and Noetherian UFDs in [29], respectively [26], are in order.

Remark (1) In [29], *Noetherian UFRs* were defined. A ring R is a *Noetherian UFR* in the sense of [29] if and only if it is a UFR in the sense of [24] and Noetherian.
(2) We will call a domain R a *Noetherian UFD* if it is a UFD and Noetherian. Except in Theorem 4.35, we will not use the original definition of a *Noetherian UFD* from [26]. A domain R is a *Noetherian UFD* in the sense of [26] if it contains at least one height one prime ideal and every height one prime ideal of R is generated by a completely prime element.
For a broad class of rings the two definitions of Noetherian UFDs agree. Suppose that R is a Noetherian domain which is not simple. If every nonzero prime ideal of R contains a height one prime ideal, then R is a Noetherian UFD in the sense of [26] if and only if it is a UFD in the sense of [24]. If R is a UFR or R satisfies the descending chain condition (DCC) on prime ideals, then every nonzero prime ideal contains a height one prime ideal. In general, it is open whether every Noetherian ring satisfies the DCC on prime ideals (see [56, Appendix, Sect. 3]).
(3) We warn the reader that a [Noetherian] UFR which is a domain need not be a [Noetherian] UFD: prime elements need not be completely prime. See Example 4.36 below.

From the point of view of factorization theory, UFRs and UFDs of this type are quite different from similarity factorial domains. UFRs have the property that the subsemigroup $N(R)^\bullet$ of nonzero normal elements is a UF-monoid (see Theorem 4.34). However, if R is a UFR, the prime elements of $N(R)^\bullet$ need not be atoms of $R^\bullet$. If R is a UFD, then prime elements of $N(R)^\bullet$ are indeed atoms in $R^\bullet$. However, since they also need to be normal, this is in some sense quite a restrictive condition. Nevertheless, many interesting examples of (Noetherian) UFRs and UFDs exist.

Example 4.31 (1) Universal enveloping algebras of finite-dimensional solvable Lie algebras over $\mathbb{C}$ are Noetherian UFDs (see [26]).
(2) Trace rings of generic matrix rings are Noetherian UFRs (see [72]).

(3) Let R be a commutative ring and G a polycyclic-by-finite group. It has been characterized when the group algebra $R[G]$ is a Noetherian UFR, respectively a Noetherian UFD. See [20, 23] and also [27]. There exist extensions of these results to semigroup algebras (see [68, 69]). Also see the book [66].
(4) Certain iterated skew polynomial rings are Noetherian UFDs. This has been used to show that many quantum algebras are Noetherian UFDs. See [79].
(5) Let R be a Noetherian UFR. Then also $M_n(R)$ for $n \in \mathbb{N}$ as well as $R[x]$ are Noetherian UFRs. It has been studied when $R[x; \sigma]$, with σ an automorphism, and $R[x; \delta]$ are UFRs (see [29]).

We refer to the survey [3] for more comprehensive results on the behavior of UFRs and UFDs under ring-theoretic constructions.

In [1], a generalization of Noetherian UFRs is introduced (even more generally, when R is a ring and σ is an automorphism of R, the notion of σ-UFR is defined). Let R be a prime Goldie ring and let Q be its simple Artinian quotient ring. For $X \subset R$, let $(R : X)_l = \{ q \in Q \mid qX \subset R \}$ and $(R : X)_r = \{ q \in Q \mid Xq \subset R \}$. For a right R-ideal I, that is, a right ideal I of R containing a non-zero-divisor, let $I_v = (R : (R : X)_l)_r$, and for a left R-ideal I, let ${}_vI = (R : (R : X)_r)_l$. A right [left] R-ideal I is called *divisorial* (or *reflexive*) if $I = I_v$ [$I = {}_vI$]. We refer to any of [3, 25, 87] for the definition of right [left] τ-R-ideals. (The terminology in [25] is slightly different in that such right [left] ideals are called *fermé*.) Recall that any right [left] τ-R-ideal is divisorial. In particular, if R satisfies the ACC on right [left] τ-R-ideals, it also satisfies the ACC on divisorial right [left] R-ideals.

Definition 4.32 ([1]) A prime Goldie ring R is a *UFR (in the sense of* [1]*)* if

(1) R is τ-Noetherian, that is, it satisfies the ACC on right τ-R-ideals as well as the ACC on left τ-R-ideals.
(2) Every prime ideal P of R such that $P = P_v$ or $P = {}_vP$ is principal.

Equivalent characterizations, including one in terms of the factorizations of normal elements, can be found in [1, Proposition 1.9].

Theorem 4.33 ([1, Proposition 1.9]) *Let R be a τ-Noetherian prime Goldie ring. Then the following statements are equivalent.*

(a) R is a UFR in the sense of [24].
(b) R is a UFR in the sense of [1] and the localization $(N(R)^\bullet)^{-1}R$ is a simple ring.

Following P.M. Cohn, a cancellative normalizing semigroup H is called a *UF-monoid* if $H/H^\times$ is a free abelian monoid. Equivalently, H is a normalizing Krull monoid in the sense of [51] with trivial divisor class group.

Theorem 4.34 ([1, 24]) *If R is a UFR in the sense of [1] or a UFR in the sense of [24], then $N(R)^\bullet = N(R)\setminus\{0\}$ is a UF-monoid. Explicitly, every nonzero normal element $a \in N(R)^\bullet$ can be written in the form*

$$a = \varepsilon p_1 \cdots p_n$$

with $n \in \mathbb{N}_0$, a unit $\varepsilon \in R^\times$, and prime elements $p_1, \ldots, p_n$ of R. This representation is unique up to order and associativity of the prime elements.

Remark The unique factorization property for normal elements has been taken as the definition of another class of rings, studied by Jordan in [67]. Jordan studied *Noetherian UFN-rings*, that is, Noetherian prime rings R such that every nonzero ideal of R contains a nonzero normal element and $N(R)^\bullet$ is a UF-monoid.

Noetherian UFDs in the sense of [26] can be characterized in terms of factorizations of elements. If P is a prime ideal of a ring R, denote by $C(P) \subset R$ the set of all elements of R whose images in R/P are non-zero-divisors.

Theorem 4.35 ([26]) *Let R be a prime Noetherian ring. Set $C = \bigcap C(P)$, where the intersection is over all height one primes P of R. The following statements are equivalent:*

(a) R is a Noetherian UFD in the sense of [26].
(b) Every nonzero element $a \in R^\bullet$ is of the form $a = cp_1 \cdots p_n$ for some $c \in C$ and completely prime elements $p_1, \ldots, p_n$ of R.

We note that property (b) of the previous theorem also holds for Noetherian UFDs in the sense of [24]. If $C \subset R^\times$, then $R = N(R)$, and hence $R^\bullet$ is a UF-monoid.

In a ring R which is a UFR, a prime element p of R is an atom of $N(R)^\bullet$ but need not be an atom in the (possibly larger) semigroup $R^\bullet$. On the other hand, if R is a UFD, the additional condition that R/pR be a domain forces p to be an atom.

Example 4.36 ([29]) Let $\mathbb{H}_\mathbb{Q}$ be the Hamilton quaternion algebra with coefficients in $\mathbb{Q}$. The ring $R = \mathbb{H}_\mathbb{Q}[x]$ is a Noetherian UFR and a domain, but R is no UFD. The element x^2+1 is central and generates a height one prime ideal, but $(x^2+1)R$ is not completely prime. Thus, R is not a UFD, even though it is Euclidean. The element x^2+1 is an atom in $N(R)^\bullet$. However, in $R^\bullet$, it factors as $x^2+1 = (x-i)(x+i)$.

Thus many interesting rings are UFRs but not UFDs. This is especially true in the case of classical maximal orders in central simple algebras over global fields. In this case, all but finitely many associativity classes of prime elements of $N(R)^\bullet$ are simply represented by the prime elements of the center of R. We elaborate on this in the following example:

Example 4.37 (1) Let $\mathcal{O}$ be a holomorphy ring in a global field K, and let A be a central simple K-algebra with $\dim_K A = n^2 > 1$. Let R be a classical maximal $\mathcal{O}$-order.
If $\mathfrak{p}$ is a prime ideal of $\mathcal{O}$ such that $\mathfrak{p}$ is unramified in R (i.e., $\mathfrak{p}R$ is a maximal ideal of R), then $\mathfrak{p}R$ is a height one prime ideal of R, and $R/\mathfrak{p}R \cong M_n(\mathcal{O}/\mathfrak{p})$. Thus, if $\mathfrak{p} = p\mathcal{O}$ is principal, then p is a prime element of R which is not completely prime. Recall that at most finitely many prime ideals of $\mathcal{O}$ are ramified in R. Thus,

R is not a UFD. However, R is a Noetherian UFR if and only if the normalizing class group of R, that is, the group of all fractional R-ideals modulo the principal fractional R-ideals (generated by normalizing elements), is trivial.

(2) Elaborating on (1) in a specific example, the ring of Hurwitz quaternions $\mathcal{H}$ is Euclidean and a Noetherian UFR, but not a UFD. The only completely prime element in $\mathcal{H}$ (up to right associativity) is $1 + i$. If p is an odd prime number, then p is a prime element of $\mathcal{H}$ which is not completely prime, since $M_2(\mathcal{H}/p\mathcal{H}) \cong M_2(\mathbb{F}_p)$. A complete set of representatives for associativity classes of prime elements of $\mathcal{H}$ is given by $\{1 + i\} \cup \mathbb{P}\backslash\{2\}$. If p is an odd prime number, the $p + 1$ maximal right $\mathcal{H}$-ideals containing the maximal $\mathcal{H}$-ideal $p\mathcal{H}$ are principal and correspond to right associativity classes of atoms of reduced norm p. Thus $|\mathsf{Z}^*_{\mathcal{H}}(p)| = p + 1$. However, all atoms of reduced norm p are similar. As already observed, $\mathcal{H}$ is similarity factorial.

(3) If R is a commutative Dedekind domain with class group G, and $\exp(G)$ divides r, then $M_r(R)$ is a Noetherian UFR, but not a UFD if $r > 1$.

We say that a prime ring R is *bounded* if every right R-ideal and every left R-ideal contains a nonzero ideal of R. Recall that every prime PI ring is bounded. In [55], Gilchrist and Smith showed that every bounded Noetherian UFD which is not commutative is a PID. This was later generalized to the following:

Theorem 4.38 ([24]) *Let R be a UFD in the sense of [24]. Let $C = \bigcap C(P)$, where the intersection is over all height one prime ideals P of R. If $C \subset R^\times$, then R is duo. That is, every left or right ideal of R is an ideal of R. Moreover, if R is not commutative, then R is a PID.*

Hence, "noncommutative UFDs are often PIDs," as the title of [55] proclaims.

5 Nonunique Factorizations

We now come to nonunique factorizations. We have already noted that a ring R satisfying the ascending chain condition on principal left ideals and on principal right ideals is atomic. In particular, this is true for any Noetherian ring. Thus, we can consider rigid factorizations of elements in $R^\bullet$. However, the conditions which are sufficient for various kinds of uniqueness of factorizations are much stricter. Hence, a great many natural examples of rings have some sort of nonunique factorization behavior.

5.1 Arithmetical Invariants

The study of nonunique factorizations proceeds by defining suitable arithmetical invariants intended to capture various aspects of the nonuniqueness of factorizations.

The following invariants are defined in terms of lengths of factorizations, and have been investigated in commutative settings before.

Definition 5.1 (*Arithmetical invariants based on lengths*) Let H be a cancellative small category.

(1) $\mathsf{L}(a) = \{\, |z| \mid z \in \mathsf{Z}^*(a) \,\}$ is the *set of lengths* of $a \in H$.
(2) $\mathscr{L}(H) = \{\, \mathsf{L}(a) \mid a \in H \,\}$ is the *system of sets of lengths* of H.

Let H be atomic.

(3) For $a \in H \setminus H^\times$,

$$\rho(a) = \frac{\sup \mathsf{L}(a)}{\min \mathsf{L}(a)}$$

is the *elasticity* of a, and $\rho(a) = 1$ for $a \in H^\times$.
(4) $\rho(H) = \sup\{\, \rho(a) \mid a \in H \,\}$ is the *elasticity* of H.
(5) The invariants

$$\rho_k(H) = \sup \big\{ \sup \mathsf{L}(a) \mid a \in H \text{ with } \min \mathsf{L}(a) \le k \big\},$$

for $k \in \mathbb{N}_{\ge 2}$, are the *refined elasticities* of H.
(6) For $k \in \mathbb{N}_{\ge 2}$,

$$\mathscr{U}_k(H) = \bigcup_{\substack{L \in \mathscr{L}(H) \\ k \in L}} L$$

is the union of all sets of lengths containing k.
(7) If $a \in H$ with $k, l \in \mathsf{L}(a)$ and $[k, l] \cap \mathsf{L}(a) = \{k, l\}$, then $l - k$ is a *distance* of a. We write $\Delta(a)$ for the *set of distances* of a.
(8) The *set of distances* of H is $\Delta(H) = \bigcup_{a \in H} \Delta(a)$.

Example 5.2 (1) Let $\mathbb{H}_\mathbb{Q}$ denote the Hamilton quaternion algebra with coefficients in $\mathbb{Q}$, and let $\mathscr{H}$ denote the ring of Hurwitz quaternions. Beauregard's results (Theorems 4.21 and 4.22) imply $\rho_2(\mathscr{H}[x]) \ge 4$ and $\rho_2(\mathbb{H}_\mathbb{Q}[x, y]) \ge 4$. Hence $\rho(\mathscr{H}[x]) \ge 2$ and $\rho(\mathbb{H}_\mathbb{Q}[x, y]) \ge 2$.
(2) If $A_1(K) = K\langle x, y \mid xy - yx = 1 \rangle = K[x][y; -\frac{d}{dx}]$ denotes the first Weyl algebra over a field K of characteristic 0, the example $x^2 y = (1 + xy)x$ of P. M. Cohn shows $\rho_2(A_1(K)) \ge 3$, and hence $\rho(A_1(K)) \ge \frac{3}{2}$.

Recall that H is *half-factorial* if $|\mathsf{L}(a)| = 1$, for all $a \in H$ (equivalently, H is atomic, and $\Delta(H) = \emptyset$ or $\rho(H) = 1$). Since all the invariants introduced so far are defined in terms of sets of lengths, they are trivial if H is half-factorial.

It is more difficult to make useful definitions for the more refined arithmetical invariants, such as catenary degrees, the ω-invariant, and the tame degree, in a noncommutative setting. In [22], a formal notion of distances between rigid factorizations was introduced. This allows the definition and study of catenary degrees and monotone catenary degrees.

Definition 5.3 (*Distances*) Let H be a cancellative small category. A *global distance on* H is a map $\mathsf{d}\colon \mathsf{Z}^*(H) \times \mathsf{Z}^*(H) \to \mathbb{N}_0$ satisfying the following properties.

(D1) $\mathsf{d}(z, z) = 0$, for all $z \in \mathsf{Z}^*(H)$.
(D2) $\mathsf{d}(z, z') = \mathsf{d}(z', z)$, for all $z, z' \in \mathsf{Z}^*(H)$.
(D3) $\mathsf{d}(z, z') \le \mathsf{d}(z, z'') + \mathsf{d}(z'', z')$, for all $z, z', z'' \in \mathsf{Z}^*(H)$.
(D4) For all $z, z' \in \mathsf{Z}^*(H)$ with $s(z) = s(z')$ and $x \in \mathsf{Z}^*(H)$ with $t(x) = s(z)$ it holds that $\mathsf{d}(x * z, x * z') = \mathsf{d}(z, z')$, and for all $z, z' \in \mathsf{Z}^*(H)$ with $t(z) = t(z')$ and $y \in \mathsf{Z}^*(H)$ with $s(y) = t(z)$ it holds that $\mathsf{d}(z * y, z' * y) = \mathsf{d}(z, z')$.
(D5) $\big||z| - |z'|\big| \le \mathsf{d}(z, z') \le \max\{|z|, |z'|, 1\}$, for all $z, z' \in \mathsf{Z}^*(H)$.

Let $L = \{ (z, z') \in \mathsf{Z}^*(H) \times \mathsf{Z}^*(H) : \pi(z) = \pi(z') \}$. A *distance on* H is a map $\mathsf{d}\colon L \to \mathbb{N}_0$ satisfying properties (D1), (D2), (D3), (D4), and (D5) under the additional restrictions on z, z', and z'' that $\pi(z) = \pi(z') = \pi(z'')$.

Let us revisit the notion of factoriality using distances as a tool. We follow [22, Sect. 3]. If d is a distance on H, we can define a congruence relation $\sim_{\mathsf{d}}$ on $\mathsf{Z}^*(H)$ by $z \sim_{\mathsf{d}} z'$ if and only if $\pi(z) = \pi(z')$ and $\mathsf{d}(z, z') = 0$. That is, two factorizations are identified if they are factorizations of the same element and are at distance zero from each other.

Definition 5.4 Let H be a cancellative small category, and let d be a distance on H. The quotient category $\mathsf{Z}_{\mathsf{d}}(H) = \mathsf{Z}^*(H)/\sim_{\mathsf{d}}$ is called the *category of* d*-factorizations*. The canonical homomorphism $\pi\colon \mathsf{Z}^*(H) \to H$ induces a homomorphism $\pi_{\mathsf{d}}\colon \mathsf{Z}_{\mathsf{d}}(H) \to H$. For $a \in H$, we call $\mathsf{Z}_{\mathsf{d}}(a) = \pi_{\mathsf{d}}^{-1}(a)$ the *set of* d*-factorizations of* a. We say that H is d*-factorial* if $|\mathsf{Z}_d(a)| = 1$, for all $a \in H$.

Example 5.5 (1) We may define a so-called *rigid distance* d^*. Informally speaking, $\mathsf{d}^*(z, z')$ is the minimal number of replacements, deletions, and insertions of atoms necessary to pass from z to z'. (The actual definition is more complicated to take into account the presence of units and the necessity to replace, delete, or insert longer factorizations than just atoms.) If $\mathsf{d}^*(z, z') = 0$, then $z = z'$, and hence $\mathsf{Z}_{\mathsf{d}^*}(H) = \mathsf{Z}^*(H)$. We say that H is *rigidly factorial* if it is d^*-factorial.

(2) Let $\sim$ be an equivalence relation on the set of atoms $\mathscr{A}(H)$ such that $v = \varepsilon u \eta$ with $\varepsilon, \eta \in H^\times$ implies $u \sim v$. Then, comparing atoms up to order and equivalence with respect to $\sim$ induces a global distance $\mathsf{d}_\sim$ on $\mathsf{Z}^*(H)$. Let R be a domain, $H = R^\bullet$, and consider for the equivalence relation $\sim$ one of similarity, subsimilarity, monosimilarity, or projectivity. Then R is $\mathsf{d}_\sim$-factorial if and only if it is similarity [subsimilarity, monosimilarity, projectivity] factorial.

(3) If, in (2), we use two-sided associativity as the equivalence relation on atoms, we obtain the *permutable distance* d_p. We say that H is *permutably factorial* if it is d_p-factorial. For a commutative cancellative semigroup H, the permutable distance is just the usual distance.

Having a rigorous notion of factorizations and distances between them at our disposal, it is now straightforward to introduce catenary degrees.

Definition 5.6 (*Catenary degree*) Let H be an atomic cancellative small category, d a distance on H, and $a \in H$.

(1) Let $z, z' \in \mathsf{Z}^*(a)$ and $N \in \mathbb{N}_0$. A finite sequence of rigid factorizations $z_0, \dots, z_n \in \mathsf{Z}^*(a)$, where $n \in \mathbb{N}_0$, is called an *N-chain (in distance* d*)* between z and z' if

$$z = z_0, \; z' = z_n, \; \text{ and } \; \mathsf{d}(z_{i-1}, z_i) \le N \; \text{ for all} i \in [1, n].$$

(2) The *catenary degree (in distance* d*) of* a, denoted by $\mathsf{c}_{\mathsf{d}}(a)$, is the minimal $N \in \mathbb{N}_0 \cup \{\infty\}$ such that for any two factorizations $z, z' \in \mathsf{Z}^*(a)$ there exists an N-chain between z and z'.
(3) The *catenary degree (in distance* d*) of* H is

$$\mathsf{c}_{\mathsf{d}}(H) = \sup\{\, \mathsf{c}_{\mathsf{d}}(a) \mid a \in H \,\} \in \mathbb{N}_0 \cup \{\infty\}.$$

To abbreviate the notation, we write c^* instead of $\mathsf{c}_{\mathsf{d}^*}$, c_p instead of $\mathsf{c}_{\mathsf{d}_p}$, and so on.

Note that H is d-factorial if and only if it is atomic and $\mathsf{c}_{\mathsf{d}}(H) = 0$. Hence, the catenary degree provides a more fine grained arithmetical invariant than those derived from sets of lengths.

Example 5.7 If R is an atomic 2-fir, it follows from the usual inductive proof of the Jordan–Hölder theorem that $\mathsf{c}^*(R^\bullet) \le 2$. Since R is similarity factorial, $\mathsf{c}_{\mathrm{sim}}(R^\bullet) = 0$, where $\mathsf{c}_{\mathrm{sim}}$ denotes the catenary degree with respect to the similarity distance. However, $\mathsf{c}^*(R^\bullet) = 0$ if and only if R is rigid. More generally, if R is an atomic modular LCM domain, then $\mathsf{c}^*(R^\bullet) \le 2$, and $\mathsf{c}_{\mathrm{proj}}(R^\bullet) = 0$, where the latter stands for the catenary degree in the projectivity distance.

The definitions of the monotone and the equal catenary degree can similarly be extended to the noncommutative setting. For the permutable distance, it is also possible to introduce an ω_p-invariant $\omega_p(H)$ and a tame degree $\mathsf{t}_p(H)$ (see [22, Sect. 5]). Unfortunately, these notions are not as strong as in the commutative setting.

5.2 *FF-Domains*

Faced with an atomic domain with nonunique factorizations, a first question one can ask is when R is a BF-domain, that is, $|\mathsf{L}(a)| < \infty$, for all $a \in R^\bullet$, respectively an FF-domain, that is, $|\mathsf{Z}^*(a)| < \infty$, for all $a \in R^\bullet$. A useful sufficient condition for R to be a BF-domain is the existence of a length function (see Lemma 3.6).

In [17], Bell, Heinle, and Levandovskyy give a sufficient condition for many important noncommutative domains to be FF-domains. Let K be a field and R a K-algebra. A *finite-dimensional filtration* of R is a filtration of R by finite-dimensional K-subspaces.

Theorem 5.8 ([17, Corollary 1.2]) *Let K be a field, $\overline{K}$ an algebraic closure of K, and let R be a K-algebra. If there exists a finite-dimensional filtration on R such that the associated graded ring* $\mathrm{gr}\ R$ *has the the property that* $\mathrm{gr}\ R \otimes_K \overline{K}$ *is a domain, then R is an FF-domain.*

The proof of the theorem proceeds by (classical) algebraic geometry.

Definition 5.9 Let K be a field and $n \in \mathbb{N}$. For $i, j \in [1, n]$ with $i < j$, let $c_{i,j} \in K^\times$ and $d_{i,j} \in K\langle x_1, \ldots, x_n \rangle$. The K-algebra

$$R = K\langle x_1, \ldots, x_n \mid x_j x_i = c_{i,j} x_i x_j + d_{i,j},\ i < j \in [1, n] \rangle$$

is called a *G-algebra* (or *PBW algebra*, or *algebra of solvable type*) if

(1) the family of monomials $\mathscr{M} = (x_1^{k_1} \cdots x_n^{k_n})_{(k_1,\ldots,k_n) \in \mathbb{N}_0^n}$ is a K-basis of R, and
(2) there exists a monomial well-ordering $\prec$ on $\mathscr{M}$ such that, for all $i < j \in [1, n]$ either $d_{i,j} = 0$, or the leading monomial of $d_{i,j}$ is smaller than $x_i x_j$ with respect to $\prec$.

Remark The family of monomials $\mathscr{M}$ is naturally in bijection with $\mathbb{N}_0^n$. A monomial well-ordering on $\mathscr{M}$ is a total order on $\mathscr{M}$ such that, with respect to the corresponding order on $\mathbb{N}_0^n$, the semigroup $\mathbb{N}_0^n$ is a totally ordered semigroup, and such that $\mathbf{0}$ is the least element of $\mathbb{N}_0^n$. By Dickson's lemma, this implies that the order is a well-ordering.

Corollary 5.10 ([17, Theorem 1.3]) *Let K be a field. Then G-algebras over K as well as their subalgebras are FF-domains. In particular, the following algebras are FF-domains:*

(1) Weyl algebras and shift algebras,
(2) universal enveloping algebras of finite-dimensional Lie algebras,
(3) coordinate rings of quantum affine spaces,
(4) q-shift algebras and q-Weyl algebras,

as well as polynomial rings over any of these algebras.

In addition, explicit upper bounds on the number of factorizations are given in [17, Theorem 1.4].

The following example shows that even for very nice domains (e.g., PIDs) one cannot in general expect there to be only finitely many rigid factorizations for each element.

Example 5.11 Let Q be a quaternion division algebra over a (necessarily infinite) field K with $\mathrm{char}(K) \neq 2$. Let $a \in Q^\times \setminus K^\times$. We denote by $\overline{a}$ the conjugate of a. Then $\mathrm{nr}(a) = a\overline{a} \in K^\times$ and $\mathrm{tr}\,(a) = a + \overline{a} \in K$. For all $c \in Q^\times$,

$$f = x^2 - \mathrm{tr}\,(a)x + \mathrm{nr}(a) = (x - cac^{-1})(x - c\overline{a}c^{-1}) \in Q[x],$$

and thus $|\mathsf{Z}^*(f)| = |Q^\times|$ is infinite. Hence $Q[x]$ is not an FF-domain. However, being Euclidean, $Q[x]$ is similarity factorial, that is, $|\mathsf{Z}_{\mathrm{sim}}(f)| = 1$, for all $f \in Q[x]^\bullet$.

Remark Another sufficient condition for a domain or semigroup to have finite rigid factorizations is given in [93, Theorem 5.23.1].

5.3 Transfer Homomorphisms

Transfer homomorphisms play an important role in the theory of nonunique factorizations in the commutative setting. A transfer homomorphism allows us to express arithmetical invariants of a ring, semigroup, or small category in terms of arithmetical invariants of a possibly simpler object.

In the commutative setting, a particularly important transfer homomorphism is that from a commutative Krull monoid H to the monoid of zero-sum sequences $\mathcal{B}(G_0)$ over a subset G_0 of the class group G of H. In particular, if $H = \mathcal{O}^\bullet$ with $\mathcal{O}$ a holomorphy ring in a global field, then $G_0 = G$ is a finite abelian group. This allows one to study the arithmetic of H through combinatorial and additive number theory. (See [50].)

In a noncommutative setting, transfer homomorphisms were first explicitly used by Baeth, Ponomarenko, Adams, Ardila, Hannasch, Kosh, McCarthy, and Rosenbaum in the article [19]. They studied nonunique factorizations in certain subsemigroups of $M_n(\mathbb{Z})^\bullet$ and $T_n(\mathbb{Z})^\bullet$. Transfer homomorphisms for cancellative small categories have been introduced in [93], where the main application was to classical maximal orders in central simple algebras over global fields. This has been developed further in [22], where arithmetical invariants going beyond sets of lengths were studied.

Implicitly, the concept of a transfer homomorphism was already present in earlier work due to Estes and Matijevic (in [44, 45]), who essentially studied when $\det\colon M_n(R)^\bullet \to R^\bullet$ is a transfer homomorphism, and Estes and Nipp (in [46–48]), who essentially studied when the reduced norm in a classical hereditary $\mathcal{O}$-order over a holomorphy ring $\mathcal{O}$ is a transfer homomorphism. Unfortunately, their results seem to have been largely overlooked so far.

We recall the necessary definitions. See [22, Sect. 2] for more details.

Definition 5.12 (*Transfer homomorphism*) Let H and T be cancellative small categories. A homomorphism $\phi\colon H \to T$ is called a *transfer homomorphism* if it has the following properties:

(T1) $T = T^\times \phi(H) T^\times$ and $\phi^{-1}(T^\times) = H^\times$.
(T2) If $a \in H$, $b_1, b_2 \in T$ and $\phi(a) = b_1 b_2$, then there exist $a_1, a_2 \in H$ and $\varepsilon \in T^\times$ such that $a = a_1 a_2$, $\phi(a_1) = b_1 \varepsilon^{-1}$, and $\phi(a_2) = \varepsilon b_2$.

We denote by $T_n(D)$ the ring of $n \times n$ upper triangular matrices over a commutative domain D. To study $T_n(D)^\bullet$, weak transfer homomorphisms were introduced by Bachman, Baeth, and Gossell in [5].

Definition 5.13 (*Weak transfer homomorphism*) Let H and T be cancellative small categories, and suppose that T is atomic. A homomorphism $\phi\colon H \to T$ is called a *weak transfer homomorphism* if it has the following properties:

(T1) $T = T^\times \phi(H) T^\times$ and $\phi^{-1}(T^\times) = H^\times$.
(WT2) If $a \in H$, $n \in \mathbb{N}$, $v_1, \ldots, v_n \in \mathscr{A}(T)$ and $\phi(a) = v_1 \cdots v_n$, then there exist $u_1, \ldots, u_n \in \mathscr{A}(H)$ and a permutation $\sigma \in \mathfrak{S}_n$ such that $a = u_1 \cdots u_n$ and $\phi(u_i) \simeq v_{\sigma(i)}$ for each $i \in [1, n]$.

(Weak) transfer homomorphisms map atoms to atoms. If $a \in H$, property (T2) of a transfer homomorphism allows one to lift rigid factorizations of $\phi(a)$ in T to rigid factorizations of a in H. For a weak transfer homomorphism, (WT2) allows the lifting of rigid factorizations of $\phi(a)$ up to permutation and associativity. These properties are sufficient to obtain an equality of the system of sets of lengths of H and T (see Theorem 5.15 below).

To obtain results about the catenary degree, in the case where ϕ is a transfer homomorphism, we need additional information about the fibers of the induced homomorphism $\phi^*\colon \mathsf{Z}^*(H) \to \mathsf{Z}^*(T)$.

Definition 5.14 (*Catenary degree in the permutable fibers*) Let H and T be atomic cancellative small categories, and let d be a distance on H. Suppose that there exists a transfer homomorphism $\phi\colon H \to T$. Denote by $\phi^*\colon \mathsf{Z}^*(H) \to \mathsf{Z}^*(T)$ its natural extension to the categories of rigid factorizations.

(1) Let $a \in H$, and let $z, z' \in \mathsf{Z}^*(a)$ with $\mathsf{d}_p(\phi^*(z), \phi^*(z')) = 0$. We say that an N-chain $z = z_0, z_1, \ldots, z_{n-1}, z_n = z' \in \mathsf{Z}^*(a)$ of rigid factorizations of *a lies in the permutable fiber of* z if $\mathsf{d}_p(\phi^*(z_i), \phi^*(z)) = 0$, for all $i \in [0, n]$.
(2) We define $\mathsf{c}_\mathsf{d}(a, \phi)$ to be the smallest $N \in \mathbb{N}_0 \cup \{\infty\}$ such that, for any two $z, z' \in \mathsf{Z}^*(a)$ with $\mathsf{d}_p(\phi^*(z), \phi^*(z')) = 0$, there exists an N-chain (in distance d) between z and z', lying in the permutable fiber of z. Moreover, we define the *catenary degree in the permutable fibers*

$$\mathsf{c}_\mathsf{d}(H, \phi) = \sup\left\{ \mathsf{c}_\mathsf{d}(a, \phi) \mid a \in H \right\} \in \mathbb{N}_0 \cup \{\infty\}.$$

For the following basic result on [weak] transfer homomorphisms, see [22] and also [5, 93].

Theorem 5.15 *Let H and T be cancellative small categories. Let $\phi\colon H \to T$ be a transfer homomorphism, or let T be atomic and $\phi\colon H \to T$ a weak transfer homomorphism.*

(1) H is atomic if and only if T is atomic.
(2) For all $a \in H$, $\mathsf{L}_H(a) = \mathsf{L}_T(\phi(a))$. In particular $\mathscr{L}(H) = \mathscr{L}(T)$, and all arithmetical invariants from Definition 5.1 coincide for H and T.
(3) If ϕ is a transfer homomorphism and H is atomic, then

$$\mathsf{c}_\mathsf{d}(H) \le \max\left\{\mathsf{c}_p(T), \mathsf{c}_\mathsf{d}(H, \phi)\right\}.$$

(4) *If ϕ is an isoatomic weak transfer homomorphism (that is, $\phi(a) \simeq \phi(b)$ implies $a \simeq b$) and T is atomic, then $\mathsf{c}_p(H) = \mathsf{c}_p(T)$. If, moreover, T is an atomic commutative semigroup, then $\omega_p(H) = \omega_p(T)$ and $\mathsf{t}_p(H) = \mathsf{t}_p(T)$.*

The strength of a transfer result comes from being able to find transfer homomorphism to a codomain T which is significantly easier to study than the original category H. Monoids of zero-sum sequences have played a central role in the commutative theory, and also turn out to be useful in studying classical maximal orders in central simple algebras over global fields. We recall their definition and some of the basic structural results about their arithmetic.

Let $(G, +)$ be an additively written abelian group, and let $G_0 \subset G$ be a subset. In the tradition of combinatorial number theory, elements of the multiplicatively written free abelian monoid $\mathscr{F}(G_0)$ are called *sequences over* G_0. The inclusion $G_0 \subset G$ extends to a homomorphism $\sigma\colon \mathscr{F}(G_0) \to G$. Explicitly, if $S = g_1 \cdot \ldots \cdot g_l \in \mathscr{F}(G_0)$ is a sequence, written as a formal product of elements of G_0, then $\sigma(S) = g_1 + \cdots + g_l \in G$ is its sum in G. We call S a *zero-sum sequence* if $\sigma(S) = 0$. The subsemigroup

$$\mathscr{B}(G_0) = \big\{ S \in \mathscr{F}(G_0) \mid \sigma(S) = 0 \big\}$$

of the free abelian monoid $\mathscr{F}(G_0)$ is called the *monoid of zero-sum sequences over* G_0. (See [50] or [53, Chap. 2.5].)

The semigroup $\mathscr{B}(G_0)$ is a Krull monoid. It is of particular importance in the theory of nonunique factorizations since every commutative Krull monoid [domain] H possesses a transfer homomorphism to a monoid of zero-sum sequences over a subset of the class group of H. Thus, problems about nonunique factorizations in H can often be reduced to questions about $\mathscr{B}(G_0)$.

Factorization problems in $\mathscr{B}(G_0)$ are studied with methods from combinatorial and additive number theory. Motivated by the study of rings of algebraic integers, the case where $G_0 = G$ is a finite abelian group has received particular attention. We recall some of the most important structural results in this case. See [50, Definition 3.2.2] for the definition of an almost arithmetical multiprogression (AAMP).

Theorem 5.16 *Let G be a finite abelian group, and let $H = \mathscr{B}(G)$ be the monoid of zero-sum sequences over G.*

(1) *H is half-factorial if and only if $|G| \le 2$.*
(2) *The set of distances, $\Delta(H)$, is a finite interval, and if it is nonempty, then $\min \Delta(H) = 1$.*
(3) *For every $k \in \mathbb{N}$, the union of sets of lengths containing k, $\mathscr{U}_k(H)$, is a finite interval.*
(4) *There exists an $M \in \mathbb{N}_0$ such that for every $a \in H$ the set of lengths $\mathsf{L}(a)$ is an AAMP with difference $d \in \Delta(H)$ and bound M.*

The last result, (4), is called the *Structure Theorem for Sets of Lengths*, and is a highly nontrivial result on the general structure of sets of lengths. We give a short motivation for it. Suppose that H is a cancellative semigroup and an element a

has two factorizations of distinct length, say $a = u_1 \cdots u_k$ and $a = v_1 \dots v_l$ with $k < l$ and atoms $u_i, v_j \in \mathscr{A}(H)$. That is, $\{k, l = k + (l-k)\} \subset \mathsf{L}(a)$. Then $a^n = (u_1 \cdots u_k)^i (v_1 \cdots v_l)^{n-i}$, for all $i \in [0, n]$. Hence the arithmetical progression $\{k + i(l-k) \mid i \in [0, n]\}$ with difference $l - k$ and length $n + 1$ is contained in $\mathsf{L}(a^n)$. Additional pairs of lengths of a give additional arithmetical progressions in $\mathsf{L}(a^n)$.

If everything is "nice," we might hope that this is essentially the only way that large sets of lengths appear. Consequently, we would expect large sets of lengths to look roughly like unions of long arithmetical progressions. The Structure Theorem for Sets of Lengths implies that this is indeed so in the setting above: If $a \in H$, then $\mathsf{L}(a)$ is contained in a union of arithmetical progressions with some difference $d \in \Delta(H)$, and with possible gaps at the beginning and at the end. The size of these gaps is uniformly bounded by the parameter M which only depends on H and not the particular element a.

5.4 *Transfer Results*

In this section, we gather transfer results for matrix rings, triangular matrix rings, and classical hereditary and maximal orders in central simple algebras over global fields.

5.4.1 Matrix Rings

For R a $2n$-fir, factorizations in $M_n(R)$ have been studied by P. M. Cohn. In the special case where R is a commutative PID, the existence of the Smith normal form implies that $\det\colon M_n(R)^\bullet \to R^\bullet$ is a transfer homomorphism. This was noted in [19].

Let R be a commutative ring. In [44, 45], Estes and Matijevic studied when $M_n(R)$ has *[weak] norm-induced factorization*, respectively *determinant-induced factorization*. Here, $M_n(R)$ has determinant-induced factorization if for each $A \in M_n(R)$ and each $r \in R^\bullet$ which divides $\det(A)$, there exists a right divisor of A having determinant r. We do not give the definition of [weak] norm-induced factorization, but recall the following:

Proposition 5.17 *Let R be a commutative ring and $n \in \mathbb{N}$. Consider the following statements:*

(a) $M_n(R)$ has norm-induced factorization.
(b) $M_n(R)$ has determinant-induced factorization.
(c) $\det\colon M_n(R)^\bullet \to R^\bullet$ is a transfer homomorphism.

Then (a) $\Rightarrow$ (b) $\Rightarrow$ (c). If R is a finite direct product of Krull domains, then also the converse implications hold.

Proof The implications (a) $\Rightarrow$ (b) $\Rightarrow$ (c) follow immediately from the definitions and the fact that a matrix $A \in M_n(R)$ is a zero-divisor if and only if $\det(A) \in R$ is a zero-divisor. Suppose that R is a finite direct product of Krull domains. Then (b) $\Rightarrow$ (a) holds by [44, Proposition 5], and (c) $\Rightarrow$ (b) can be deduced from [44, Lemma 2].

In the characterization of rings R for which $M_n(R)$ has norm-induced factorization, the notion of a *Towber ring* (see [74, 96]) appears. We do not recall the exact definition, but give a sufficient as well as a necessary condition for R to be Towber when R is a commutative Noetherian domain. There is a small gap between the sufficient and the necessary condition.

Let R be a commutative Noetherian domain. If $\mathrm{gldim}(R) \le 2$ and every finitely generated projective R-module is isomorphic to a direct sum of a free module and an ideal of R, then R is a Towber ring. Conversely, if R is a Towber ring, then $\mathrm{gldim}(R) \le 2$ and every finitely generated projective R-module of rank at least 3 is isomorphic to a direct sum of a free module and an ideal.

Theorem 5.18 ([44]) *Let R be a commutative Noetherian ring with no nonzero nilpotent elements. Then the following statements are equivalent:*

(a) $M_n(R)$ has norm-induced factorization, for all $n \in \mathbb{N}$.
(b) $M_2(R)$ has norm-induced factorization.
(c) R is a Towber ring, that is, R is a finite direct product of Towber domains.
(d) $\mathrm{gldim}(R) \le 2$, each projective module P of constant rank $r(P)$ is stably equivalent to $\bigwedge^{r(P)} P$, and stably free finitely generated projective R-modules are free.

Moreover, the statements above imply the following statements (e)–(h). If R is a finite direct product of Noetherian integrally closed domains, then the converse holds, and any of the above statements (a)–(d) is equivalent to any of the statements (e)–(h).

(e) $M_n(R)$ has determinant-induced factorization, for all $n \in \mathbb{N}$.
(f) $M_2(R)$ has determinant-induced factorization.
(g) $\det\colon M_n(R)^\bullet \to R^\bullet$ is a transfer homomorphism, for all $n \in \mathbb{N}$.
(h) $\det\colon M_2(R)^\bullet \to R^\bullet$ is a transfer homomorphism.

Proof The equivalences (a) $\Leftrightarrow$ (b) $\Leftrightarrow$ (c) $\Leftrightarrow$ (d), and more, follow from [44, Theorem 1]. The remaining implications follow from Proposition 5.17.

Theorem 5.19 ([45]) *Let R be a commutative Noetherian ring with no nonzero nilpotent elements. Then the following statements are equivalent:*

(a) $M_n(R)$ is permutably factorial, for all $n \in \mathbb{N}$.
(b) $M_2(R)$ is permutably factorial.
(c) $M_n(R)$ is similarity factorial, for all $n \in \mathbb{N}$.
(d) $M_2(R)$ is similarity factorial.
(e) R is a finite direct product of PIDs.

Proof Here, (a) ⇒ (b) and (c) ⇒ (d) are trivial. Since associated elements are similar, (a) ⇒ (c) and (b) ⇒ (d) are also clear. The key implication (d) ⇒ (e) follows from [45, Theorem 2]. Finally, (e) ⇒ (a) follows using the Smith Normal Form. (The implication (e) ⇒ (c) can also be deduced from Theorem 4.5.)

In [45], the ring $M_n(R)$ is called *determinant factorial* if factorizations of elements in $M_n(R)^\bullet$ are unique up to order and associativity of the determinants of the atoms. If $M_n(R)$ is similarity factorial, then it is determinant factorial (by [45, Proposition 5].

Theorem 5.20 ([45]) *Let R be a commutative Noetherian ring with no nonzero nilpotent elements. Then the following statements are equivalent:*

(a) $M_n(R)$ is determinant factorial, for all $n \in \mathbb{N}$.
(b) $M_2(R)$ is determinant factorial.
(c) $R^\bullet$ is factorial and for all $n \in \mathbb{N}$ and $U \in \mathcal{A}(M_n(R)^\bullet)$ we have $\det(U) \in \mathcal{A}(R^\bullet)$.
(d) $R^\bullet$ is factorial and for all $U \in \mathcal{A}(M_2(R)^\bullet)$ we have $\det(U) \in \mathcal{A}(R^\bullet)$.
(e) R is a finite direct product of factorial Towber domains.

Proof The implications (a) ⇒ (b) and (c) ⇒ (d) are clear. The equivalences (a) ⇔ (c) and (b) ⇔ (d) follow from [45, Proposition 1]. Finally, (b) ⇒ (e) is the key implication and follows from [45, Theorem 1], and (e) ⇒ (a) follows from [45, Corollary to Proposition 1] or Theorem 5.18.

The following example from [44] forms the basis of a key step in [45]. We recall it here, as it demonstrates explicitly that a matrix ring over a factorial commutative domain need not even be half-factorial.

Example 5.21 (1) Let R be a commutative ring containing elements x, y, z which form a regular sequence in any order. (E.g., if R is a regular local ring of dimension at least 3, three elements from a minimal generating set of the maximal ideal of R will do. Also $R = K[x, y, z]$ with K a field works.)

Consider the ring $M_2(R)$. In [44] it is shown that the matrix

$$A = \begin{pmatrix} x^2 & xy - z \\ xy + z & y^2 \end{pmatrix},$$

which has $\det(A) = z^2$, has no right factor of determinant z. Let $\operatorname{adj}(A)$ denote the adjugate of A. Then

$$A \operatorname{adj}(A) = z^2 1_{M_2(R)} = \begin{pmatrix} z & 0 \\ 0 & 1 \end{pmatrix}^2 \begin{pmatrix} 1 & 0 \\ 0 & z \end{pmatrix}^2.$$

Hence $\rho_2(M_2(R)^\bullet) \geq 4$. In particular, for the elasticity we have $\rho(M_2(R)^\bullet) \geq 2$.

(2) Let K be a field. Then $M_2(K[x])$ is permutably, similarity, and determinant factorial. The ring $M_2(K[x, y])$ is determinant factorial but neither similarity nor permutably factorial. For $n \geq 3$, the ring $M_2(K[x_1, \ldots, x_n])$ is not even half-factorial.

5.4.2 Rings of Triangular Matrices

For a commutative domain R and $n \in \mathbb{N}$, let $T_n(R)$ denote the ring of $n \times n$ upper triangular matrices. The study of factorizations in $T_n(R)^\bullet$ turns out to be considerably simpler than in $M_n(R)^\bullet$.

Theorem 5.22 *Let R be an atomic commutative domain and let $n \in \mathbb{N}$.*

(1) Suppose R is a BF-domain and $n \geq 2$. Then $\det\colon T_n(R)^\bullet \to R^\bullet$ *is a transfer homomorphism if and only if R is a PID.*
(2) The map $T_n(R)^\bullet \to (R^\bullet)^n$ sending a matrix $(a_{i,j})_{i,j\in[1,n]} \in T_n(R)^\bullet$ to the vector of its diagonal entries $(a_{i,i})_{i\in[1,n]}$ is an isoatomic weak transfer homomorphism. Moreover, for atoms of $T_n(R)^\bullet$, associativity, similarity, and subsimilarity coincide, $\mathsf{c}_p(T_n(R)^\bullet) = \mathsf{c}_p(R^\bullet)$, $\mathsf{t}_p(T_n(R)^\bullet) = \mathsf{t}(R^\bullet)$, and $\omega_p(T_n(R)^\bullet) = \omega(R^\bullet)$. In particular, $T_n(R)$ is permutably [similarity, subsimilarity, determinant] factorial if and only if R is factorial.

Remark (1) The existence of the transfer homomorphism, in case R is a PID, was shown in [19]. The characterization of when the determinant is a transfer homomorphism, in case R is a BF-domain, as well as the existence of a weak transfer homomorphism, is due to [5, Theorems 2.8 and 4.2]. The isoatomicity and transfer of catenary degree, tame degree, and ω_p-invariant can be found in [22, Proposition 6.14].
(2) That $T_n(R)$ is determinant factorial if and only if $R^\bullet$ is factorial was not stated before, but is easy to observe. If R is factorial, then $T_n(R)^\bullet$ is permutably factorial and hence determinant factorial. For the converse, suppose that $T_n(R)^\bullet$ is determinant factorial, and consider the embedding $R^\bullet \to T_n(R)^\bullet$ that maps $a \in R^\bullet$ to the matrix with a in the upper left corner, ones on the remaining diagonal, and zeroes everywhere else.
(3) In general, there does not exist a transfer homomorphism from $T_2(R)^\bullet$ to any cancellative commutative semigroup (see [5, Example 4.5]). This was the motivation for the introduction of weak transfer homomorphisms.

5.4.3 Classical Hereditary and Maximal Orders.

Earlier results of Estes and Nipp in [46–48] on *factorizations induced by norm factorization (FNF)* can be interpreted as a transfer homomorphism. The following is proved for central separable algebras in [47]. We state the special case for central simple algebras.

Theorem 5.23 *Let $\mathcal{O}$ be a holomorphy ring in a global field K, and let A be a central simple K-algebra. Assume that A satisfies the Eichler condition with respect to $\mathcal{O}$. If R is a classical hereditary $\mathcal{O}$-order in A, $x \in R$, and $a \in \mathcal{O}$ is such that $a \mid \mathrm{nr}(x)$, then there exists a left divisor y of x in R, and $\varepsilon \in \mathcal{O}^\times$ such that $\mathrm{nr}(y) = a\varepsilon$. Moreover, ε can be taken arbitrarily subject to the restriction that $a\varepsilon$ is positive at each archimedean place of K which ramifies in A.*

The proof in [47] proceeds by localization and an explicit characterization of classical hereditary orders over complete DVRs. For quaternion algebras, more refined results, not requiring the Eichler condition but instead requiring that every stably free right R-ideal is free, can be found in [46, 48].

Let $\mathcal{O}_A^\bullet$ denote the subsemigroup of all nonzero elements of $\mathcal{O}$ which are positive at each archimedean place of K which ramifies in A. Recall that $\mathrm{nr}(R^\bullet) = \mathcal{O}_A^\bullet$ if R is a classical hereditary $\mathcal{O}$-order.

Corollary 5.24 *With the conditions as in the previous theorem, $\mathrm{nr}\colon R^\bullet \to \mathcal{O}_A^\bullet$ is a transfer homomorphism. The semigroup $\mathcal{O}_A^\bullet$ is a Krull monoid with class group $\mathcal{C}_A(\mathcal{O})$. Each class in $\mathcal{C}_A(\mathcal{O})$ contains infinitely many prime ideals. Hence, there exists a transfer homomorphism $R^\bullet \to \mathcal{B}(\mathcal{C}_A(\mathcal{O}))$. In particular, the conclusions of Theorem 5.16 hold for $R^\bullet$ in place of H.*

Proof By the previous theorem, $\mathrm{nr}\colon R^\bullet \to \mathcal{O}_A^\bullet$ is a transfer homomorphism. The semigroup $\mathcal{O}_A^\bullet$ is a regular congruence submonoid of $\mathcal{O}_A^\bullet$ (see [53, Chap. 2.11]). As such it is a commutative Krull monoid, with class group $\mathcal{C}_A(\mathcal{O})$. Each class contains infinitely many prime ideals by a standard result from analytic number theory. (See [88, Corollary 7 to Proposition 7.9] or [53, Corollary 2.11.16] for the case where $\mathcal{O}$ is the ring of algebraic integers in a number field. The general number field case follows by a localization argument. For the function field case, use [53, Proposition 8.9.7].) Hence there exists a transfer homomorphism $\mathcal{O}_A^\bullet \to \mathcal{B}(\mathcal{C}_A(\mathcal{O}))$. Since the composition of two transfer homomorphisms is a transfer homomorphism, it follows that there exists a transfer homomorphism $R^\bullet \to \mathcal{B}(\mathcal{C}_A(\mathcal{O}))$.

A different way of obtaining the result in Corollary 5.24 in the case that R is a classical maximal order is given in Theorem 5.27 (1) below. It relies on the global ideal theory of R. In this way, we also obtain information about the catenary degree in the permutable fibers.

We first extend the result about the transfer homomorphism for commutative Krull monoids into a setting of noncommutative semigroups, respectively cancellative small categories. This general result then includes, as a special case, the transfer homomorphism for normalizing Krull monoids obtained in [51] as well as the desired theorem. We follow [22, 93].

A *quotient semigroup* is a semigroup Q in which every cancellative element is invertible, that is, $Q^\bullet = Q^\times$. Let Q be a quotient semigroup and $H \subset Q$ a subsemigroup. Then H is an *order* in Q if $Q = H(H \cap Q^\bullet)^{-1} = (H \cap Q^\bullet)^{-1}H$. Two orders H and H' in Q are equivalent if there exist $x, y, z, w \in Q^\bullet$ such that $xHy \subset H'$ and $zH'w \subset H$. A *maximal order* is an order which is maximal with respect to set inclusion in its equivalence class. Let H be a maximal order. A subset $I \subset Q$ is a *fractional right H-ideal* if $IH \subset I$, and there exist $x, y \in Q^\bullet$ such that $x \in I$ and $yI \subset Q$. If moreover $I \subset H$, then I is a *right H-ideal*.

For a fractional right H-ideal $I \subset Q$, we define $I^{-1} = \{ x \in Q \mid IxI \subset I \}$, and $I_v = (I^{-1})^{-1}$. The fractional right H-ideal I is called *divisorial* if $I = I_v$. A divisorial right H-ideal I is *maximal integral* if it is maximal within the set of proper divisorial right H-ideals. Analogous definitions are made for (fractional) left H-ideals. If H and H' are equivalent maximal orders, we call a subset $I \subset Q$ a *[fractional] (H, H')-ideal* if it is both, a [fractional] left H-ideal and a [fractional] right H'-ideal. A *[fractional] H-ideal* is a [fractional] (H, H)-ideal. We say that H is *bounded* if every fractional left H-ideal and every fractional right H-ideal contains a fractional H-ideal.

The additional restrictions imposed in the following definition ensure that the set of maximal orders equivalent to H has a "good" theory of divisorial left and right ideals.

Definition 5.25 ([93, Definition 5.18]) Let H be a maximal order in a quotient semigroup Q. We say that H is an *arithmetical maximal order* if it has the following properties:

- **(A1)** H satisfies both the ACC on divisorial left H-ideals and the ACC on divisorial right H-ideals.
- **(A2)** H is bounded.
- **(A3)** The lattice of divisorial fractional left H-ideals is modular, and the lattice of divisorial fractional right H-ideals is modular.

Let H be an arithmetical maximal order in a quotient semigroup Q, and let α denote the set of maximal orders in its equivalence class. We define a category $\mathcal{F}_v = \mathcal{F}_v(\alpha)$ as follows: the set of objects is α, and for $H', H'' \in \alpha$, the set of morphisms from H' to H'', denoted by $\mathcal{F}_v(H', H'')$, consists of all divisorial fractional (H', H'')-ideals. If $I \in \mathcal{F}_v(H', H'')$ and $J \in \mathcal{F}_v(H'', H''')$, the composition $I \cdot_v J \in \mathcal{F}_v(H', H''')$ is defined by $I \cdot_v J = (IJ)_v$. In terms of our point of view from the preliminaries, $\mathcal{F}_v(\alpha)_0 = \alpha$, and for a divisorial fractional (H', H'')-ideal I we have that $s(I) = H'$ is the left order of I, and $t(I) = H''$ is the right order of I.

With these definitions, $\mathcal{F}_v$ is an *arithmetical groupoid*, the precise definition of which we omit here. By $\mathcal{I}_v = \mathcal{I}_v(\alpha)$, we denote the subcategory of $\mathcal{F}_v(\alpha)$ with the same set of objects, but where the morphisms are given by divisorial (H', H'')-ideals. Set $\mathcal{H}_H = \{ q^{-1}(aH)q \mid a \in H^\bullet,\ q \in Q^\bullet \}$ (as a category).

The subcategory $\mathcal{F}_v(H)$ of all divisorial fractional H-ideals is a free abelian group. If $H' \in \alpha$, then there is a canonical isomorphism $\mathcal{F}_v(H) \to \mathcal{F}_v(H')$. We identify, and call this group $\mathbb{G}$. One can define a homomorphism, the abstract norm, $\eta\colon G \to \mathbb{G}$. Set $P_{H^\bullet}$ to be the quotient group of $\eta(\mathcal{H}_H)$ as a subgroup of $\mathbb{G}$.

Theorem 5.26 ([93, Theorem 5.23] and [22, Corollary 7.11]) *Let H be an arithmetical maximal order in a quotient semigroup Q and let α denote the set of maximal orders of Q equivalent to H. Let $\eta\colon \mathcal{F}_v(\alpha) \to \mathbb{G}$ be the abstract norm of $\mathcal{F}_v(\alpha)$, let $C = \mathbb{G}/P_{H^\bullet}$, and set $C_M = \{ [\eta(I)] \in C \mid I \in \mathcal{I}_v(\alpha)$ maximal integral$\}$. Assume that*

(**N**) *a divisorial fractional right H-ideal I is principal if and only if $\eta(I) \in P_{H^\bullet}$.*

Then there exists a transfer homomorphism $\theta\colon H^\bullet \to \mathcal{B}(C_M)$. Let d be a distance on $H^\bullet$ that is invariant under conjugation by normalizing elements. Then $\mathsf{c}_{\mathsf{d}}(H^\bullet, \theta) \le 2$.

Remark (1) The result can be proven in the more general setting of saturated subcategories of arithmetical groupoids (see [93, Theorem 4.15] or [22, Theorem 7.8]). The strong condition (**N**) cannot be omitted. We discuss the condition in our application to classical maximal orders in central simple algebras over global fields below.

(2) In a saturated subcategory of an arithmetical groupoid (here, $\mathcal{I}_v$ in $\mathcal{F}_v$), elements (i.e., divisorial one-sided ideals) enjoy a kind of unique factorization property. The boundedness guarantees the existence of the abstract norm, which provides a useful invariant in describing these factorizations. This was originally proven by Asano and Murata in [2]. It is a generalization of a similar result for (bounded) Dedekind prime rings, where the one-sided ideals of the equivalence class of a Dedekind prime ring form the so-called *Brandt groupoid*. This unique factorization of divisorial one-sided ideals is the key ingredient in the proof of the previous result.

(3) We note in passing that every arithmetical maximal order is a BF-semigroup (see [93, Theorem 5.23.1]). For a commutative cancellative semigroup H the following is true: If H is v-Noetherian (satisfies the ACC on divisorial ideals), then H is a BF-monoid. It seems to be unknown whether every order H which satisfies (A1) is a BF-semigroup, even in the special case where H is a maximal order.

(4) If G is a lattice-ordered group, then G is distributive as a lattice. From this, one concludes that a commutative cancellative semigroup that is a maximal order (i.e., completely integrally closed) and satisfies (A1) is already an arithmetical maximal order (that is, a commutative Krull monoid).
If $H = R$ with R a Dedekind prime ring, or more generally, a Krull ring in the sense of Chamarie (see [25]), then (A3) holds. It is open whether there exist maximal orders which satisfy (A1) and (A2) but not (A3). It would be interesting to know such examples or sufficient and/or necessary conditions on H for (A1) and (A2) to imply (A3).

Applied to classical maximal orders in central simple algebras over global fields, we have the following. (See also Corollary 5.24.)

Theorem 5.27 ([22, 93]) *Let $\mathcal{O}$ be a holomorphy ring in a global field K, A a central simple algebra over K, and R a classical maximal $\mathcal{O}$-order of A.*

(1) Suppose that every stably free right R-ideal is free. Then there exists a transfer homomorphism $\theta\colon R^\bullet \to \mathcal{B}(\mathcal{C}_A(\mathcal{O}))$. Moreover, $\mathsf{c}_{\mathsf{d}}(R^\bullet, \theta) \le 2$ for any distance d on $R^\bullet$ which is invariant under conjugation by normalizing elements.
In particular, the conclusions of Theorem 5.16 hold for $R^\bullet$ in place of H. If R is not half-factorial, then $\mathsf{c}_{\mathrm{sim}}(R^\bullet) = \mathsf{c}_p(R^\bullet) = \mathsf{c}^*(R^\bullet) = \mathsf{c}_p\big(\mathcal{B}(\mathcal{C}_A(\mathcal{O}))\big)$.

(2) *Let K be a number field and $\mathcal{O}$ its ring of algebraic integers. If there exist stably free right R-ideals that are not free, then there exists no transfer homomorphism $\theta\colon R^\bullet \to \mathcal{B}(G_0)$, where G_0 is any subset of an abelian group. Moreover,*

  (i) $\Delta(R^\bullet) = \mathbb{N}$.
  (ii) *For every $k \geq 3$, we have $\mathbb{N}_{\geq 3} \subset \mathcal{U}_k(R^\bullet) \subset \mathbb{N}_{\geq 2}$.*
  (iii) $\mathsf{c}_{\mathsf{d}}(R^\bullet) = \infty$ *for every distance* d *on* $R^\bullet$.

Remark (1) The importance of the condition for every stably free right R-ideal to be free was noted already by Estes and Nipp (see [46, 48]). That the absence of this condition not only implies that nr, respectively θ, is not a transfer homomorphism, but that the much stronger result in (2) holds, first appeared in [93]. In the setting of (2), arithmetical invariants are infinite and hence the factorization theory is radically different from the case (1), where all arithmetical invariants are finite.

(2) Throughout this section we have required that $\mathcal{O} = \mathcal{O}_S$ is a holomorphy ring defined by a finite set of places $S \subset S_{\text{fin}}$. This is the most important case. However, most results go through, with possibly minor modifications, for $\mathcal{O} = \mathcal{O}_S$ with an infinite set $S \subsetneq S_{\text{fin}}$.
Theorem 5.23 remains true without changes. In Corollary 5.24 and Theorem 5.27(1) it is not necessarily true anymore that every class of $\mathcal{C}_A(\mathcal{O})$ contains infinitely many prime ideals. However, by a localization argument, every class, except possibly the trivial one, contains at least one nonzero prime ideal. Accordingly, there exists a transfer homomorphism $R^\bullet \to \mathcal{B}(C_M)$ with C_M either equal to $\mathcal{C}_A(\mathcal{O})$ or to $\mathcal{C}_A(\mathcal{O})\setminus\{0\}$.

It was noted in [47], that Theorem 5.23 can be extended to a more general setting of classical hereditary orders over Dedekind domains whose quotient fields are not global fields. In fact, using a description of finitely generated projective modules over hereditary Noetherian prime (HNP) rings, one can extend the construction of the transfer homomorphism to bounded HNP rings. We refer to [82] for background on hereditary Noetherian prime (HNP) rings.

If R is a HNP ring, one can define a class group $G(R)$. If R is a Dedekind prime ring, then simply $G(R) = \ker(\text{udim}\colon K_0(R) \to \mathbb{Z})$. Let $G_0 \subset G(R)$ denote the subset of classes $[I] - [R]$, where I is a right R-ideal such that the composition series of R/I consists precisely of one tower of R.

Theorem 5.28 ([94]) *Let R be a bounded hereditary Noetherian prime ring. Suppose that every stably free right R-ideal is free. Then there exists a transfer homomorphism $\theta\colon R^\bullet \to \mathcal{B}(G_0)$.*

6 Other Results

Finally, we note some recent work which is beyond the scope of this article, but may conceivably be considered to be factorization theory.

In a noncommutative setting, even in the (similarity) factorial case, many interesting questions in describing factorizations in more detail remain. Factorizations of (skew) polynomials over division rings have received particular attention. This is especially true for Wedderburn polynomials (also called W-polynomials). Some recent work in this direction due to Haile, Lam, Leroy, Ozturk, and Rowen is [62, 75–78]. In [73], Leroy shows that factorizations of elements in $\mathbb{F}_q[x;\theta]$, where θ is the Frobenius automorphism, can be computed in terms of factorizations in $\mathbb{F}_q[x]$. We also note [16, 59].

I. Gelfand and Retakh, using their theory of quasideterminants and noncommutative symmetric functions, have obtained noncommutative generalizations of Vieta's theorem. This allows one to express coefficients of polynomials in terms of pseudo-roots. We mention the surveys [52, 91] as starting points into the literature in this direction. In [43], a connection is made between the theory of quasideterminants, noncommutative symmetric functions, and W-polynomials.

Motion polynomials are certain polynomials over the ring of dual quaternions. They have applications in the the study of rational motions and in particular the construction of linkages in kinematics. This approach was introduced by Hegedüs, Schicho, and Schröcker in [63, 64] and has since been very successful. See the survey [83] or also the expository article [60].

We mainly discussed the semigroup of non-zero-divisors of a noncommutative ring, and, in Sect. 4.2, the semigroup of nonzero normal elements. The factorization theory of some other noncommutative semigroups, which do not necessarily arise in such a way from rings, has been studied. We mention polynomial decompositions (see [98]) and other subsemigroups of rings of matrices (see [19]) over the integers.

Acknowledgments I thank the anonymous referee for his careful reading. The author was supported by the Austrian Science Fund (FWF) project P26036-N26.

References

1. G.Q. Abbasi, S. Kobayashi, H. Marubayashi, A. Ueda, Noncommutative unique factorization rings. Commun. Algebra **19**(1), 167–198 (1991)
2. K. Asano, K. Murata, Arithmetical ideal theory in semigroups. J. Inst. Polytech. Osaka City Univ. Ser. A. Math. **4**, 9–33 (1953)
3. E. Akalan, H. Marubayashi, Multiplicative ideal theory in non-commutative rings, in *Multiplicative Ideal Theory and Factorization Theory* (Springer, Berlin, 2016) preprint
4. D.D. Anderson (ed.), *Factorization in Integral Domains*, vol. 189. Lecture Notes in Pure and Applied Mathematics (Marcel Dekker Inc, New York, 1997)
5. D. Bachman, N.R. Baeth, J. Gossell, Factorizations of upper triangular matrices. Linear Algebra Appl. **450**, 138–157 (2014)
6. R.A. Beauregard, Right LCM domains. Proc. Am. Math. Soc. **30**, 1–7 (1971)

7. R.A. Beauregard, Right-bounded factors in an LCM domain. Trans. Am. Math. Soc. **200**, 251–266 (1974)
8. R.A. Beauregard, An analog of Nagata's theorem for modular LCM domains. Can. J. Math. **29**(2), 307–314 (1977)
9. R.A. Beauregard, Left versus right LCM domains. Proc. Am. Math. Soc. **78**(4), 464–466 (1980)
10. R.A. Beauregard, Unique factorization in the ring $R[x]$. J. Aust. Math. Soc. Ser. A **53**(3), 287–293 (1992)
11. R.A. Beauregard, When is $F[x, y]$ a unique factorization domain? Proc. Am. Math. Soc. **117**(1), 67–70 (1993)
12. R.A. Beauregard, Factorization in LCM domains with conjugation. Can. Math. Bull. **37**(3), 289–293 (1994)
13. R.A. Beauregard, Right unique factorization domains. Rocky Mountain J. Math. **24**(2), 483–489 (1994)
14. R.A. Beauregard, Rings with the right atomic multiple property. Commun. Algebra **23**(3), 1017–1026 (1995)
15. G.M. Bergman, Commuting elements in free algebras and related topics in ring theory. Ph.D. thesis, Harvard University, 1968
16. J. Bergen, M. Giesbrecht, P.N. Shivakumar, Y. Zhang, Factorizations for difference operators. Adv. Differ. Equ. **57**, 6 (2015)
17. J.P. Bell, A. Heinle, V. Levandovskyy, On noncommutative finite factorization domains. Trans. Am. Math. Soc. (2015) to appear
18. A.J. Berrick, M.E. Keating, *An Introduction to Rings and Modules with K-Theory in View*, vol. 65. Cambridge Studies in Advanced Mathematics (Cambridge University Press, Cambridge, 2000)
19. N.R. Baeth, V. Ponomarenko, D. Adams, R. Ardila, D. Hannasch, A. Kosh, H. McCarthy, R. Rosenbaum, Number theory of matrix semigroups. Linear Algebra Appl. **434**(3), 694–711 (2011)
20. K.A. Brown, Height one primes of polycyclic group rings. J. Lond. Math. Soc. (2) **32**(3), 426–438 (1985)
21. H.-H. Brungs, Ringe mit eindeutiger Faktorzerlegung. J. Reine Angew. Math. **236**, 43–66 (1969)
22. N.R. Baeth, D. Smertnig, Factorization theory: from commutative to noncommutative settings. J. Algebra **441**, 475–551 (2015)
23. A.W. Chatters, J. Clark, Group rings which are unique factorisation rings. Commun. Algebra **19**(2), 585–598 (1991)
24. A.W. Chatters, M.P. Gilchrist, D. Wilson, Unique factorisation rings. Proc. Edinb. Math. Soc. (2), **35**(2), 255–269 (1992)
25. M. Chamarie, Anneaux de Krull non commutatifs. J. Algebra **72**(1), 210–222 (1981)
26. A.W. Chatters, Noncommutative unique factorization domains. Math. Proc. Camb. Philos. Soc. **95**(1), 49–54 (1984)
27. A.W. Chatters, Unique factorisation in P.I. group-rings. J. Aust. Math. Soc. Ser. A **59**(2), 232–243 (1995)
28. S.T. Chapman (ed.), *Arithmetical Properties of Commutative Rings and Monoids*, vol. 241. Lecture Notes in Pure and Applied Mathematics (Chapman & Hall/CRC, Boca Raton, 2005)
29. A.W. Chatters, D.A. Jordan, Noncommutative unique factorisation rings. J. Lond. Math. Soc. (2), **33**(1), 22–32 (1986)
30. H. Cohn, A. Kumar, Metacommutation of Hurwitz primes. Proc. Am. Math. Soc. **143**(4), 1459–1469 (2015)
31. P.M. Cohn, Factorization in non-commutative power series rings. Proc. Camb. Philos. Soc. **58**, 452–464 (1962)
32. P.M. Cohn, Noncommutative unique factorization domains. Trans. Am. Math. Soc. **109**, 313–331 (1963)
33. P.M. Cohn, Rings with a weak algorithm. Trans. Am. Math. Soc. **109**, 332–356 (1963)

34. P.M. Cohn, Errata to: noncommutative unique factorization domains. Trans. Am. Math. Soc. **119**, 552 (1965)
35. P.M. Cohn, Factorization in general rings and strictly cyclic modules. J. Reine Angew. Math. **239**(240), 185–200 (1969)
36. P.M. Cohn, Correction to: unique factorization domains (Am. Math. Mon. 80, 1–18). Am. Math. Mon. **80**(1115), 1973 (1973)
37. P.M. Cohn, Unique factorization domains. Am. Math. Mon. **80**, 1–18 (1973)
38. P.M. Cohn, *Free Rings and their Relations*, vol. 19, 2nd edn. London Mathematical Society Monographs (Academic Press Inc., Harcourt Brace Jovanovich Publishers, London, 1985)
39. P.M. Cohn, *Free Ideal Rings and Localization in General Rings*, vol. 3, New Mathematical Monographs (Cambridge University Press, Cambridge, 2006)
40. C.W. Curtis, I. Reiner, *Methods of Representation Theory, vol. II*. Pure and Applied Mathematics (Wiley, New York, 1987) (With applications to finite groups and orders, A Wiley-Interscience Publication)
41. P.M. Cohn, A.H. Schofield, Two examples of principal ideal domains. Bull. Lond. Math. Soc. **17**(1), 25–28 (1985)
42. J.H. Conway, D.A. Smith, *On Quaternions and Octonions: Their Geometry, Arithmetic, and Symmetry* (A K Peters Ltd., Natick, 2003)
43. J. Delenclos, A. Leroy, Noncommutative symmetric functions and W-polynomials. J. Algebra Appl. **6**(5), 815–837 (2007)
44. D.R. Estes, J.R. Matijevic, Matrix factorizations, exterior powers, and complete intersections. J. Algebra **58**(1), 117–135 (1979)
45. D.R. Estes, J.R. Matijevic, Unique factorization of matrices and Towber rings. J. Algebra **59**(2), 387–394 (1979)
46. D.R. Estes, G. Nipp, Factorization in quaternion orders. J. Number Theory **33**(2), 224–236 (1989)
47. D.R. Estes, Factorization in hereditary orders. Linear Algebra Appl. **157**, 161–164 (1991)
48. D.R. Estes, Factorization in quaternion orders over number fields, in *The Mathematical Heritage of C. F. Gauss* (World Scientific Publishing, River Edge, 1991), pp. 195–203
49. H. Fitting, Über den Zusammenhang zwischen dem Begriff der Gleichartigkeit zweier Ideale und dem Äquivalenzbegriff der Elementarteilertheorie. Math. Ann. **112**(1), 572–582 (1936)
50. A. Geroldinger, Additive group theory and non-unique factorizations, in *Combinatorial Number Theory and Additive Group Theory*. Advanced Courses in Mathematics. CRM Barcelona (Birkhäuser Verlag, Basel, 2009), pp. 1–86
51. A. Geroldinger, Non-commutative Krull monoids: a divisor theoretic approach and their arithmetic. Osaka J. Math. **50**(2), 503–539 (2013)
52. I. Gelfand, S. Gelfand, V. Retakh, R.L. Wilson, Quasideterminants. Adv. Math. **193**(1), 56–141 (2005)
53. A. Geroldinger, F. Halter-Koch, *Non-unique factorizations. Algebraic, combinatorial and analytic theory*, vol. 278. Pure and Applied Mathematics (Boca Raton) (Chapman & Hall/CRC, Boca Raton, 2006)
54. G. Grätzer, J.B. Nation, A new look at the Jordan-Hölder theorem for semimodular lattices. Algebra Universalis **64**(3–4), 309–311 (2010)
55. M.P. Gilchrist, M.K. Smith, Noncommutative UFDs are often PIDs. Math. Proc. Camb. Philos. Soc. **95**(3), 417–419 (1984)
56. K.R. Goodearl, R.B. Warfield Jr., *An Introduction to Noncommutative Noetherian Rings*, vol. 61, 2nd edn. London Mathematical Society Student Texts (Cambridge University Press, Cambridge, 2004)
57. K.R. Goodearl, M.T. Yakimov, From quantum Ore extensions to quantum tori via noncommutative UFDs (2012) Preprint
58. K.R. Goodearl, M.T. Yakimov, Quantum cluster algebras and quantum nilpotent algebras. Proc. Natl. Acad. Sci. USA **111**(27), 9696–9703 (2014)
59. A. Heinle, V. Levandovskyy, Factorization of z-homogeneous polynomials in the first (q)-Weyl Algebra (2014) preprint

60. G. Hegedüs, Z. Li, J. Schicho, H.-P. Schröcker, From the fundamental theorem of algebra to Kempe's universality theorem. Int. Math. Nachr. **229**, 13–26 (2015)
61. E. Hallouin, C. Maire, Cancellation in totally definite quaternion algebras. J. Reine Angew. Math. **595**, 189–213 (2006)
62. D. Haile, L.H. Rowen, Factorizations of polynomials over division algebras. Algebra Colloquium **2**(2), 145–156 (1995)
63. G. Hegedüs, J. Schicho, H.-P. Schröcker, Construction of overconstrained linkages by factorization of rational motions, in *Latest Advances in Robot Kinematics*, ed. by J. Lenarcic, M. Husty (Springer, Netherlands, 2012), pp. 213–220
64. G. Hegedüs, J. Schicho, H.-P. Schröcker, Factorization of rational curves in the study quadric. Mech. Mach. Theory **69**(1), 142–152 (2013)
65. N. Jacobson, *The Theory of Rings*. American Mathematical Society Mathematical Surveys, vol. I (American Mathematical Society, New York, 1943)
66. E. Jespers, J. Okniński, *Noetherian Semigroup Algebras*, vol. 7. Algebras and Applications (Springer, Dordrecht, 2007)
67. D.A. Jordan, Unique factorisation of normal elements in noncommutative rings. Glasgow Math. J. **31**(1), 103–113 (1989)
68. E. Jespers, Q. Wang, Noetherian unique factorization semigroup algebras. Commun. Algebra **29**(12), 5701–5715 (2001)
69. E. Jespers, Q. Wang, Height-one prime ideals in semigroup algebras satisfying a polynomial identity. J. Algebra **248**(1), 118–131 (2002)
70. T.Y. Lam, *Serre's Problem on Projective Modules*. Springer Monographs in Mathematics (Springer, Berlin, 2006)
71. E. Landau, Ein Satz über die Zerlegung homogener linearer Differentialausdrücke in irreductible Factoren. J. Reine Angew. Math. **124**, 115–120 (1902)
72. L. Le Bruyn, Trace rings of generic matrices are unique factorization domains. Glasgow Math. J. **28**(1), 11–13 (1986)
73. A. Leroy, Noncommutative polynomial maps. J. Algebra Appl. **11**(4), 1250076, 16 (2012)
74. D. Lissner, A. Geramita, Towber rings. J. Algebra **15**, 13–40 (1970)
75. T.Y. Lam, A. Leroy, Algebraic conjugacy classes and skew polynomial rings, in *Perspectives in Ring Theory (Antwerp, 1987)*, vol. 233. NATO Advanced Science Institutes Series C: Mathematical and Physical Sciences (Kluwer Academic Publishers, Dordrecht, 1988), pp. 153–203
76. T.Y. Lam, A. Leroy, Principal one-sided ideals in Ore polynomial rings, in *Algebra and Its Applications (Athens, OH, 1999)*, vol. 259. Contemporary Mathematics (American Mathematical Society, Providence, 2000), pp. 333–352
77. T.Y. Lam, A. Leroy, Wedderburn polynomials over division rings. I. J. Pure Appl. Algebra **186**(1), 43–76 (2004)
78. T.Y. Lam, A. Leroy, A. Ozturk, Wedderburn polynomials over division rings. II, in *Noncommutative Rings, Group Rings, Diagram Algebras and their Applications*, vol. 456. Contemporary Mathematics (American Mathematical Society, Providence, RI, 2008), pp. 73–98
79. S. Launois, T.H. Lenagan, L. Rigal, Quantum unique factorisation domains. J. Lond. Math. Soc. (2) **74**(2), 321–340 (2006)
80. A. Leroy, A. Ozturk, Algebraic and F-independent sets in 2-firs. Commun. Algebra **32**(5), 1763–1792 (2004)
81. A. Loewy, Über reduzible lineare homogene Differentialgleichungen. Math. Ann. **56**(4), 549–584 (1903)
82. L.S. Levy, J.C. Robson, *Hereditary Noetherian Prime Rings and Idealizers*, vol. 174. Mathematical Surveys and Monographs (American Mathematical Society, Providence, 2011)
83. Z. Li, T.-D. Rad, J. Schicho, H.-P. Schröcker, Factorization of rational motions: a survey with examples and applications, in *Proceedings of the 14th IFToMM World Congress (Taipei, 2015)* (2015) preprint
84. H. Marubayashi, Ore extensions over total valuation rings. Algebras Represent. Theory **13**(5), 607–622 (2010)

85. J.C. McConnell, J.C. Robson, *Noncommutative Noetherian Rings*, vol. 30. With the cooperation of L. W. Small, revised edition. Graduate Studies in Mathematics (American Mathematical Society, Providence, 2001)
86. C. Maclachlan, A.W. Reid, *The Arithmetic of Hyperbolic 3-Manifolds*, vol. 219. Graduate Texts in Mathematics (Springer, New York, 2003)
87. H. Marubayashi, F. Van Oystaeyen, *Prime Divisors and Noncommutative Valuation Theory*, vol. 2059. Lecture Notes in Mathematics (Springer, Heidelberg, 2012)
88. W. Narkiewicz, *Elementary and Analytic Theory of Algebraic Numbers*, 3rd edn. Springer Monographs in Mathematics (Springer, Berlin, 2004)
89. O. Ore, Theory of non-commutative polynomials. *Ann. Math. (2)*, **34**(3), 480–508 (1933)
90. I. Reiner, *Maximal Orders*. London Mathematical Society Monographs, No. 5 (Academic Press (A subsidiary of Harcourt Brace Jovanovich, Publishers), London-New York, 1975)
91. V. Retakh, From factorizations of noncommutative polynomials to combinatorial topology. Central Eur. J. Math. **8**(2), 235–243 (2010)
92. D. Smertnig, A note on cancellation in totally definite quaternion algebras. J. Reine Angew. Math. **707**, 209–216 (2015)
93. D. Smertnig, Sets of lengths in maximal orders in central simple algebras. J. Algebra **390**, 1–43 (2013)
94. D. Smertnig. Factorizations in bounded hereditary Noetherian prime rings (2016) preprint http://arxiv.org/abs/1605.09274
95. R.G. Swan, Strong approximation and locally free modules, in *Ring Theory and Algebra, III (Proceedings of Third Conference, University Oklahoma, Norman, Okla., 1979)*, vol. 55. Lecture Notes in Pure and Applied Mathematics (Dekker, New York, 1980), pp. 153–223
96. J. Towber, Complete reducibility in exterior algebras over free modules. J. Algebra **10**, 299–309 (1968)
97. M.-F. Vignéras, Simplification pour les ordres des corps de quaternions totalement définis. J. Reine Angew. Math. **286**(287), 257–277 (1976)
98. M. Zieve, P. Müller, On Ritt's polynomial decomposition theorems (2008) preprint

Index

S. Chapman et al. (eds.), *Multiplicative Ideal Theory and Factorization Theory*,
Springer Proceedings in Mathematics & Statistics 170,
DOI 10.1007/978-3-319-38855-7

MIX
Papier aus verantwortungsvollen Quellen
Paper from responsible sources
FSC® C105338

If you have any concerns about our products,
you can contact us on
ProductSafety@springernature.com

In case Publisher is established outside the EU,
the EU authorized representative is:
Springer Nature Customer Service Center GmbH
Europaplatz 3, 69115 Heidelberg, Germany

Printed by Libri Plureos GmbH
in Hamburg, Germany